P9-DVI-956

Parent Birds and Their Young

From a "parent" to her
"young" / with love
Mom
5-14-81

NUMBER TWO

The Corrie Herring Hooks Series

Parent

Birds and Their Young

Alexander F. Skutch

University of Texas Press · Austin and London

Library of Congress Cataloging in Publication Data

Skutch, Alexander Frank, 1904–
    Parent birds and their young.

    (The Corrie Herring Hooks series; no. 2)
    Bibliography: p.
    Includes index.
    SUMMARY: "A detailed survey . . . of the family life
and reproductive behavior of birds, from the formation of
pairs to the young birds' attainment of independence."
    1. Birds—Behavior. 2. Parental behavior in animals.
[1. Birds—Habits and behavior. 2. Parental behavior in
animals] I. Title.
QL698.3.S56   598.2'5   75-2195
ISBN 0-292-76424-3

Copyright © 1976 by Alexander F. Skutch
All rights reserved
Printed in the United States of America

Second Printing, 1979

TO
WALTER AND ELENA ARP
with gratitude and affection

CONTENTS

PLATES

TABLES

FIGURES

PREFACE

Of all the spectacles that the natural world presents to us, no other is so open to observation yet so intimate, so exciting and endearing, as the rearing of a family of young birds. Equipped with only binoculars, patience, and perhaps a blind for hiding ourselves, we can watch the whole drama unfold before us, from the building of the nest and laying of the eggs to the hatching of the young, their nurture, development, and first hesitant flight. The birth and rearing of mammals are carried on in much greater secrecy, in dim burrows or under cover of darkness, and, in the case of some of the larger kinds, close watching is not exempt from danger. The life histories of many invertebrates, especially social insects, are fascinating; but to learn about them is usually more difficult than to follow the nesting of birds, which seem almost to invite us to watch them rear their young.

As is evident from the writings of Aristotle, Plutarch, and other classical authors, the ancients were already familiar with some of the salient features in the reproduction of birds, as, no doubt, were the illiterate savages who long antedated them. Nevertheless, by far the greater part of our detailed, accurate knowledge of the breeding of birds of many kinds has been gathered in the present century by an increasing number of dedicated students, amateur and professional, who have carefully followed, recorded, and photographed the activities of nesting birds, not only in their dooryards but also in distant places. Not until quite recently did men and women spend long periods, sometimes extending to years, amid the discomforts and isolation, and sometimes perils, of tropical forests, remote islands, Arctic tundra, and Antarctic ice, not to shoot and skin birds for museum collections, but to follow patiently, day by day, the establishment of their territories and the progress of their nesting.

Modern observations on the family life of birds are to be found chiefly in the journals issued quarterly by the ornithological societies: the *Ibis* in Britain, the *Auk*, *Condor*, and *Wilson Bulletin* in North America, the *Emu* in Australia, the *Ostrich* in South Africa, the *Journal für Ornithologie* in Germany, and a number of others. Published as they are received, these reports on nesting habits are scattered at random amid articles dealing with many other aspects of ornithology, from anatomy and physiology to migration and vocalizations. We also have a growing number of popular books and scientific monographs dealing largely with the breeding biology of single species or groups of related species. Ornithological textbooks contain chapters on nests, eggs, incubation, and other phases of the reproductive cycle; but these accounts are necessarily rather abbreviated, for they are included in volumes that treat many other subjects.

The present book was written to give a more

comprehensive and detailed account of the parental activities of birds and the responses of their young. The method of treatment is by subject rather than by taxonomic division. Examples are selected from the whole world of birds and include aspects of avian family life that are omitted from works attempting to survey every branch of ornithology or to give a conspectus of the birds of the world.

Although for more than four decades I have studied this subject in the field, chiefly in tropical America, only a fraction of the observations in this book are my own. I have drawn freely from the writings of many other naturalists, and it seems but fair to acknowledge my sources, not only to give credit where it is due, but also to help the interested reader to follow any subject in more detail by referring to the original publications. Full citations will be found at the back of the book. For readability, I have used English names for the birds mentioned in the text and have given corresponding scientific names in the index, thereby avoiding their repetition from chapter to chapter. I have tried to make this book readable by the amateur naturalist as well as useful as a reference work for the investigator.

El Quizarrá
Costa Rica

For illustrations in this book I thank the following: Edward A. Armstrong, George A. Bartholomew, Walter J. Breckenridge, Nicolas E. and Elsie C. Collias, William G. Conway and the New York Zoological Society, H. J. Frith, Albert E. Gilbert, Samuel A. Grimes, Joseph A. Hagar, Donald A. Hammer, John B. D. Hopcraft, Thomas R. Howell, Gordon L. Maclean, Ronald K. Murton, S. Dillon Ripley II, Helmut Sick, Barbara K. Snow, Leslie M. Tuck, John Warham, and Milton W. Weller. The sources of these photographs, drawings, and diagrams are given in the captions; those not otherwise credited are by the author. Figures 1, 2, 3, 10, 11, 12, 14, 16, and 17 were redrawn by Nancy McGowan.

I am grateful to George M. Sutton for reading the manuscript and making many helpful suggestions.

Parent Birds and Their Young

CHAPTER 1

The Formation of Pairs

Before an animal can become a parent, it must make certain preparations, both internal and external. Internal preparations, which go forward without the conscious control and largely without the knowledge of the organism in which they occur, include the formation and maturation of the reproductive organs and accessory structures, such as mammary glands, necessary for procreation. The study of the organs and processes involved in this part of the preparation for parenthood is the task of anatomy and physiology and is beyond the scope of this book.

External preparations for parenthood, which fall within the province of animal behavior, vary greatly from group to group within the animal kingdom. In some creatures they are simple and short; in others, elaborate and prolonged. Among men, for example, they are made complex by custom, law, sentiment, and ideals, although by nature they are slight and brief. Except in a few species of brood parasites, such as cowbirds and cuckoos, birds cannot become parents in the offhand manner that irresponsible humans do. With birds, preparation for parenthood includes winning a nuptial partner, acquiring a breeding territory or at least a nest site, and patiently incubating from ten to as many as eighty days. Not only must birds exert themselves to hatch

out their chicks; but also, as a rule, the eggs do not even mature in the female until the earlier stages of the breeding cycle, such as pair formation and nest building, have provided adequate stimuli. Thus, the safeguards against casual and irresponsible parenthood, which tribal custom, religion, and civil law have striven to impose upon mankind, are in birds provided by nature.

The order in which the several steps in the preparation for parenthood are carried out varies from species to species. In some birds the first step is a migratory journey, often thousands of miles and at times crossing great oceans, vast deserts, or lofty mountain ranges. But many kinds of birds, especially those that nest in the milder parts of the earth, never migrate; even among those that do, a number pair before they reach their breeding ground. In migratory birds, the male usually secures a territory and then a mate before nest building begins, but in a few species he builds, or at least starts, a nest before he wins a partner. Thus, in a broad survey of feathered animals, the order in which one considers these matters will be more or less arbitrary. Were we to deal largely or exclusively with the birds of high northern latitudes, it would be most natural to begin with migration. But, since we are not limiting ourselves to northern birds, it seems best to start with pair formation. At what age do birds acquire mates and begin to breed? How long do their unions last?

Age of Breeding

The age at which birds become capable of reproduction varies greatly. Probably the majority of small songbirds nest in the year following their birth; thus, if they hatched late in one nesting season and begin to breed early in the following season, they will be considerably less than one year of age. In Ohio, some Song Sparrows began to lay their eggs when only ten and a half months old (Nice 1937). In equatorial regions, where some species nest through much of the year, small birds may become sexually mature far earlier than this. In Colombia, male Rufous-collared Sparrows were ready to breed when five to eight months old, and some females nested at the age of five months (A. H. Miller 1959a). Captive Bengalese Finches

sometimes lay at the surprisingly early age of two and a half months (Eisner 1960). At the other extreme is the Royal Albatross, which does not nest until nine or ten years old (Richdale 1952). Most seafowl seem to allow several years to pass before they breed. The North Atlantic Gannet takes from four to six years to reach reproductive maturity; the Fulmar apparently even longer; even the little Common Puffin, three or four years (table 1). In many birds, as in certain mammals, the female breeds at an earlier age than the male.

Land birds seem on the whole to mature more quickly than sea birds, although we need more information on this matter. In the Satin Bower-Bird, which is about a foot in length, the male acquires full breeding plumage when from four to seven years of age, although he is sexually active before this (A. J. Marshall 1954). Some small songbirds, including the Purple Finch and Yellow-crowned Euphonia, may win mates and nest before they

TABLE 1
Age at Which Birds Begin to Breed

Species	Age[a]	Locality	Reference
Ostrich	3–5	—	Schneider *in* Lack 1968
Yellow-eyed Penguin	(2) 3	New Zealand	Richdale 1951
Royal Albatross	8–11	New Zealand	Richdale 1952
Laysan Albatross	7	Midway Atoll	Rice & Kenyon 1962
Short-tailed Shearwater	6	Australia	Serventy 1956
Cape Pigeon	5	Southern seas	Beck 1969
Gannet	(4) 5–6	Scotland	Nelson 1966
White Stork	3–5	Europe	Lack 1968
Bewick's Swan	4–5	Northern Europe	Scott 1970
Shelduck	22 mo.	England	Hori 1964
Oystercatcher	4 (3–5)	Wales	Harris 1967
Arctic Tern	3	Great Britain	Horobin 1969
Black-headed Gull	♂2–4, ♀1–3	Latvia	Mihelsons 1968
Common Puffin	3–4	Great Britain	Lockley 1953
Common Swift	2	England	Lack 1956a
Red-throated Bee-eater	1 (2)	Nigeria	Fry 1972
Bicolored Antbird	♂1+, ♀6.5 mo.	Panama	Willis 1967
Rook	(1) 2	England	Coombs 1960
Pied Flycatcher	1–2 (3)	Germany	Creutz 1955
Common Starling	♂(1)2, ♀1–2	Canada	Collins & de Vos 1966
Red-winged Blackbird	♂2, ♀1 (2)	North America	Orians 1961a
Song Sparrow	10.5–12.5 mo.	Ohio	Nice 1937
Rufous-collared Sparrow	5–10 mo.	Colombia	A. H. Miller 1959a
Zebra Finch	3 mo.	Australia	Immelman 1963

[a] Approximate ages are in years, unless months (mo.) are indicated. Age when a minority breed is given in parentheses.

have put on the bright adult plumage. Birds of most kinds require much longer to reach reproductive maturity than mammals of corresponding size. As a rule, they do not breed until long after they are full grown, whereas many mammals reproduce as soon as they attain adult size, or even before they cease to grow. Thus, while a small songbird may not breed until it is nine or ten months old, voles, whose bodies are of about the same size, may do so at the age of seven weeks in the male and less than a month in the female (Bourlière 1954). Gannets, fulmars, and albatrosses take far longer to reach reproductive maturity than horses and cows, whose bodies are many times heavier. The significance of birds' long delay in starting to breed will be considered in chapter 34.

Many birds exhibit certain forms of parental behavior long before they are able to become parents. When only a few months old, and sometimes much less, they may feed nestlings, brood, or participate in nest building. Young passerines seem most often to engage in such activities when stimulated by the example of older birds or when closely associated with helpless nestlings or fledglings, with the result that at times they help their parents attend their younger siblings—a matter that will receive our attention in chapter 29.

Birds that do not nest until several years old, especially those that breed in colonies, may gradually work up to full reproductive activity. When only a year old, they may stay away from the nesting areas —far from land in the case of seafowl. In the following breeding season they may visit established colonies, prospect for new colony sites, perhaps even toy with nest materials or build imperfect nests. Yellow-eyed Penguins, which normally reproduce at three years of age, may lay in the preceding year, although they are often careless about incubation, especially the males, and few of their eggs hatch (Richdale 1951).

The Engagement Period

Among the activities of immature birds are pair formation, courtship display, and sometimes helping at nests of older birds. Many birds choose their partners long before they are sexually mature, with

the conspicuous exception of small migratory species that reach their breeding area only a short while before nesting begins. In at least three species of wood warblers in the Central American highlands—the Collared Redstart, Slate-throated Redstart (especially the race *aurantiacus*), and Pink-headed Warbler—young birds hatched in April and May have acquired adult coloration and go in pairs by August or September, although no nests will be built until the following March or April (Skutch 1954). In the North Temperate Zone, nonmigratory passerines often find partners at a similarly early age. In California, young of the Wren-Tit (Erickson 1938) and Plain Titmouse (Dixon 1949) pair in the fall, soon after separating from their parents. Common Bluebirds hatched in May sometimes have partners by September (Thomas 1946). In Europe, Bearded Tits join in pairs when they are only two and a half months old and still wear juvenal plumage, although they will not breed until nine months later. Jackdaws, which develop more slowly, first choose mates at the age of about one year, a full twelve months before they are ready to nest (Lorenz 1952). Shelducks pair in spring when about 10 months old, although they will not breed until the following year (Hori 1964).

The interval between taking a mate and the beginning of sexual activity is often called the "engagement period." Such an interval of "betrothal" occurs in practically all birds that form pairs, as opposed to those (considered in chapter 3) in which the female merely visits an assembly of displaying males to have her eggs fertilized. In small migrants that come from afar to nest in high latitudes, this interval may be only a few weeks or even days; however, in many birds it is longer, up to eighteen months in the Royal Albatross (Richdale 1951), and one or even two years in the Laysan Albatross (Fisher 1972). The constancy of the engaged couple during this period varies with the species. Lorenz believed that young Jackdaws are nearly always true to their first love; among his tame daws at Altenberg, he noticed only one instance of change of partners during the year of betrothal. In certain other birds, there appears to be more changing about until a satisfactory partner is found and the serious business of rearing a family begins. Thus in Wren-Tits the pairs formed by juveniles in the fall do not appear to be permanent. In Common Bluebirds the bonds between immature individuals are less binding than those between mature birds that have already nested together; in Blackbirds in England the pairs formed in winter often break apart as the breeding season approaches (D. W. Snow 1956). Possibly some of these early attachments fail to persist because as the young birds mature they discover that they have chosen companions of their own sex, a matter discussed later in this chapter.

In Yellow-eyed Penguins, individuals too young to breed and older unemployed birds often change their partners at intervals of a few weeks. Richdale (1951), the great authority on these birds, refers to such associations as "keeping company" rather than engagements, a term that suggests their looser, less serious character. Among animals of whatever kind, it is certainly not undesirable that youngsters become familiar with a number of individuals of the opposite sex before selecting a permanent partner.

Pair Formation without Initial Hostility

When pair formation long precedes nesting, it is often accomplished in so obscure a fashion that it escapes the bird watcher, as commonly happens with the tropical American birds among which I have dwelt for many years. Pair formation is most readily followed in migratory or wandering birds that mate only a short while before they breed. As our first example, let us take the Yellow Bunting carefully studied by Diesselhorst (1950). In the spring when they are seeking partners, these European birds can distinguish the sex of other individuals of their kind solely by appearance, by voice, or by behavior, and of course by all three together. Males are usually brighter than females, but the yellowest breeding females grade into the dullest adult males. When, in early spring, a female in search of a partner flies into a bush near an unmated male who is singing on his territory, he continues a while longer, apparently unconcerned. Soon his song becomes lower or stops, and he may reveal by movements or calls that he has noticed the new-

comer. Presently he flies to the ground, followed by the female, who alights beside him. Both run about in a stooping posture, pecking the ground and picking up small particles, such as pebbles and short pieces of straw, and dropping them again. Although occasionally one of the buntings may pick up and swallow something edible, their actions seem to symbolize the gathering of nest materials. Sometimes the ceremony ends when the female suddenly rushes at the male and both fly away.

This ceremony does not invariably lead to pairing, for the female may sooner or later depart to enact it with another male; and a male may be visited by a succession of females before one stays with him. Evidently the personal element, what we might call compatibility or affection, influences the choice of a mate. The female Yellow Bunting appears also to pay attention to the adequacy of the nest sites that a male's territory offers, and she shows strong attachment to the spot where she has nested in past years. Yet one female, who found an unmated yearling male in possession of the territory where in the preceding year she had nested with a male who died, left the yearling to go elsewhere, despite her evident preference for her former home. We are not sure whether the factors that influence the female's final choice of a mate are physiological or more subtly psychological, but it is evident that the problem of a bird seeking a partner involves more than the recognition of the species and sex of other individuals.

A single ceremony of the sort that we have described seems not to suffice for welding the pair bond between a male and a female Yellow Bunting. It must be repeated again and again, over an interval of days or weeks, before the two have forged a firm nuptial bond and are ready to proceed with nesting. Diesselhorst noticed the symbolic picking up of material by both sexes only early in the season. A male who has remained long unmated, or one who has lost his mate, proceeds differently to win a partner. Now, if a female alights on the ground in his territory, he runs back and forth before her with his body held upright, wings slightly drooping, tail fanned, and the feathers of his crown, throat, and rump puffed out. After a few seconds of this, he suddenly flies off. At other times, he picks up one or several long straws, runs about before the female with them in his bill, sometimes singing softly, then drops them. The female signifies her acceptance of the male by following him in flight.

In contrast to the Yellow Bunting, which can distinguish sex by appearance, voice, and behavior, the Cedar Waxwing is poorly provided with means for sexual discrimination. Since male and female wear similar plumage and neither has much of a song, they apparently rely upon behavior to reveal sex. Unlike many other birds, the sociable, wandering waxwings pair within the flock before the establishment of breeding territories. A male who feels the urge to win a partner hops sideward toward a flock mate who rests on the same slender, horizontal perch. If the waxwing so addressed happens to be another male or a female disinclined to pair, it gives a threat display with lowered body, partly out-fluffed feathers, raised crest, and open bill, which may be vigorously snapped at the would-be suitor. The unresponsive waxwing may even attack the bird who makes overtures to it or it may fly away. If, however, the courting waxwing happens to hop toward a responsive female, she hops away and then back to him. Often the male provides himself with a berry or an insect before he begins to court, and he offers this to the receptive female. Taking the gift, she holds it in her bill while hopping away and back, then returns it to him. Now he hops away from and then back to her, passing the food to her once more. Between hops, the male frequently bows. This highly stereotyped courtship dance, which gives the impression of mechanical toys in action, may last from one to five minutes. Usually the female terminates the display by swallowing the male's offering (Putnam 1949, Crouch 1936).

In the gregarious Swallow-Tanager, sole member of its widespread tropical South American family, the gorgeous black-faced blue-green male is readily distinguished from the female, who is less intense green and lacks black on head and throat. As in the Cedar Waxwing, pairs are usually formed before territories are established, but in this case the female may take the initiative. Schaefer (1953) described how a female flew toward two males, one fully adult and the other in transitional plumage,

who were sitting close together on a high building in northern Venezuela.. The "female arrives and alights 1.5 meters away. Immediately she begins curtseying . . . , takes a blade of grass in her beak, and approaches the older male with bobbing movements. While the young male remains passive, the old one first draws back, then also begins curtseying until the female, followed by the gallant, flies to the flat roof immediately below. Now the female takes up some small particles, possibly pebbles. The old male approaches her from behind and, trembling with excitement, feeds her in a symbolized action but without uttering a sound; during this ceremony, the female stoops a little, spreads the tail, and lets her wings hang down." Thenceforth, the chosen male does not allow his mate out of his sight; he follows her wherever she goes and rarely perches more than a yard from her.

Pair Formation Complicated by Antagonism

From the accounts of the courtship of the Yellow Bunting, Cedar Waxwing, and Swallow-Tanager, it appears that even at the first meeting of eligible individuals there is little antagonism or mistrust on either side. The waxwing receives a mildly hostile reception only if he makes advances to another male or possibly a female unready to pair. In these birds, pair formation involves recognition of species and of sex, and doubtless also the more subtle matter of finding a compatible individual. In many birds, however, a further complication arises. The bird who is approached by an individual seeking a mate is antagonistic to the stranger who suddenly intrudes upon his zealously defended territory or ventures within the "individual distance" beyond which many social birds try to hold their companions. Often there appears to be mistrust, perhaps even fear, in both the interested individuals. At the same time, there may be uncertainty of the sex of the approaching bird, so that the other does not know whether it will confront a rival or a potential mate. This complication of factors leads to a situation that is often difficult to interpret.

As a first example of pair formation in birds of this kind, we may take the Song Sparrow, a territory-holding species in which the sexes are quite alike in plumage but only the male normally sings.

When another Song Sparrow of either sex alights near a male who has been singing freely to advertise his possession of an area suitable for breeding and his need of a mate, he flies threateningly toward the intruder, of whose sex he may be ignorant. The newcomer must reveal its sex by its behavior. If it is a male in search of a territory, he may stand his ground and give a hostile display, whereupon the first male attacks him. If the intruder is a more timid male or a female not ready to breed, it flees when the territory holder threatens. If, however, the newcomer is a female Song Sparrow seeking a mate, she holds her ground in the face of the attack and reveals her sex by trilling. The male then gives his courtship display, which consists in flying down and colliding with her, then immediately flying away with a loud song. Sometimes this "pouncing" is a severe collision, but at other times the male merely brushes past the female, who early in the season usually trills when subjected to this boisterous treatment, although later she may utter a threat note (Nice 1943).

In the European Robin, the sexes are alike in plumage, but an experienced observer can distinguish some females by their somewhat slimmer aspect and more nervous demeanor. Possibly the robins themselves can discriminate the sex of others of their species with greater ease. Females sing, although in the breeding season they do so less than the males. When a female in search of a mate suddenly appears in the territory of a singing, unpaired male, he flies up to her and postures aggressively. Instead of retreating before his onslaught, as another male might do, she stands firm. If he retires with loud song, she follows, often singing more or less herself. If the male holds his ground when she advances to him, she may posture as he does. With greater or less excitement, this performance may last from one to four hours, with intermissions while the birds appear to pick food from the ground. The excited singing and posturing by the male and female rarely continue as much as two days. When, soon or late, the excitement dies away, the male accepts the intruding female as his mate. By her very persistence in the face of rebuffs she wins the right to share his territory (Lack 1953). Apparently the male robin's initial antagonism to the female is due, not to his failure to recognize her sex, but to his reluctance to admit any other robin into his vigilantly guarded territory. If he at first confuses her with a male, why does he not fight her as he would fight another male?

In the Snow Bunting, the situation is different from that in the Song Sparrow and European Robin, for the sexes differ strikingly in their breeding plumage. Nevertheless, an unmated male reacts similarly to either sex that enters his territory on the melting snow of the Arctic tundra. In either case, he assumes a threatening attitude and with head drawn in between his shoulders calls *p E E E*. He may hold this posture for a considerable time, until the approaching bird is only a few yards away. If the newcomer happens to be a female in search of a mate, she alights near him, undeterred by his threat. Thereupon, the male abruptly changes his tactics. Assuming an erect, strangely stretched attitude, with widely spread tail and partly expanded wings, he runs quickly away from the female, thus exposing to her view the contrasting black and white areas of his upper plumage, wings, and tail. After running several yards, he returns unostentatiously to the female and repeats the display. She may preen or peck at the vegetation with the hasty, more or less incomplete movements typical of displacement activities. By her refusal either to retreat in the face of his territorial defense or to fight, as another male would have done, she has demonstrated her sex by behavior no less than by plumage and has won acceptance as his mate (Tinbergen 1939, 1958).

Pairing in the Red-necked or Northern Phalarope strikingly resembles the course of events observed in its Arctic neighbor, the Snow Bunting, despite the fact that the roles of the sexes are reversed. In this small shore bird, the female, who is much brighter than the male, leaves him to incubate the eggs and care for the young. She arrives first on the breeding ground, claims a territory, and calls persistently to advertise her presence. She vigorously attacks other females that visit her pond, and sometimes two of them fight together. When a male alights near an unmated female, she reacts to his arrival as though he were a competitor of her own sex, flying toward him, often with song, and, after alighting, threatening him with the same flat, menacing posture that she adopts when attacking another female. But when a foot or so from the threatened male, she suddenly stretches her neck and swims away. She may alternate repeatedly between approach in the forward threat posture and withdrawal with raised head. When finally the male is accepted, the two forage together. The female, sometimes with the male's help, scrapes a number of depressions in the ground; and in one of these she finally lays eggs for him to attend (Tinbergen 1958).

In various colonial-nesting herons and egrets, pair formation takes much the same course as in the three passerine birds that we have just considered, except that the male often starts a nest before he has acquired a partner, and the territory that he defends is restricted to the incipient nest and the space closely surrounding it in a crowded colony. Standing on or beside his structure, he alternately works with the nest sticks and gives displays, in which stretching and contorting his slender neck, clicking his long mandibles together, and hissing are, in several members of the family, conspicuous features. This display, so striking in a bird resplendent in nuptial plumes, warns other males to remain aloof but strongly attracts unmated females. For some time, however, the building male repulses every intruder of either sex with threats and, if necessary, direct attacks. If, after an absence to forage or to fetch another stick, he finds that a female has settled on his nest, he may drive her away fiercely. But gradually his hostility subsides, until finally he accepts a female who approaches him with an appropriate appeasement display. In the Cattle Egret, however, a female may force the issue by landing on an unfinished nest beside a male who has been displaying for only two or three days, or she may even quell his aggression by alighting squarely on his back and repeatedly striking his head (Meyerriecks 1960, Lancaster 1970).

In the Anhinga, pair formation follows much the same course as in herons, with the still unmated male starting a platform of sticks on which he displays by bowing, waving his wings, and stretching and contorting his snakelike neck. But, unlike the herons and egrets, he was not seen to threaten an

approaching female, probably because in the Anhinga striking differences in plumage facilitate recognition of sex (T. T. Allen 1961).

Courtship of the Hawfinch has points in common with that of the Cedar Waxwing: it begins in the winter flock and the male takes the initiative. Since in this European finch the female is somewhat duller than the male, he may have no problem of identification, yet he must proceed discreetly. Hawfinches are wary of each other's powerful beaks and each insists on keeping flock mates beyond its "individual distance." With great caution and elaborate posturing, the male approaches a female foraging on the ground. If she snaps at him, he jumps back and hops rapidly away. Despite his timidity, he is persistent and approaches again, hopping directly toward her, whereupon she takes flight. He pursues, and, after circling through the garden at top speed, they alight close together. Finally, he maneuvers into a position in front of the female and gives the appeasement display, which consists of a bow so deep that it brings his inverted head beneath his body. With more ceremonies, and a display of the magnificent black, white, and purple of one wing, he at last comes so close to the female that by stretching forth his neck to the utmost he can touch the tip of her great bill with his own. This is the first step toward feeding her, which he will do when the "engaged" birds have become sufficiently accustomed to each other to permit a close approach (Mountfort 1957). Similarly, in the Black-headed Gull, a male and female are at first distrustful of each other. Gradually they reach an understanding and mutual confidence by means of repeated threat and appeasement ceremonies as they stand together on the ground before they have chosen their breeding territory (Tinbergen 1958).

Recognition of Species and Sex

Since, among free birds, pairs almost invariably consist of individuals of the same species, the formation of even temporary pairs involves the bird's ability to distinguish individuals of its own kind from those of other kinds. In chapter 25 we shall consider how birds learn to recognize their own species. Suffice it to say here that at a very early age birds develop attachments to the creatures with

Hosted with love at yourwebsite.

which they are then closely associated, be it their parents, birds of another kind, or even a human. Thenceforth they choose as their social companions, and later as their mates, beings of the kind to which they first became attached.

The ability of birds to distinguish specific differences is remarkable. In many parts of the world, one finds in the same area species so similar in appearance that their field identification puzzles the experienced ornithologist, yet these confusingly similar species scarcely ever interbreed. Nearly always species of similar appearance that inhabit the same area have quite different utterances, both calls and songs. The bird watcher learns to depend on voice more than on plumage to distinguish these allied species and, apparently, so do the birds (Skutch 1951). In the Galápagos Islands, however, different species of the famous Darwin's finches (Geospizinae) that breed in the same locality show no constant differences in coloration or song and perhaps little in size. The various closely related species do, however, differ in the size and shape of their bills, and the birds seem to recognize other members of their own species largely, or solely, by their bills (Lack 1947).

Not only must a bird seeking a mate find an individual of its own species, but it must also choose one of the appropriate sex. Wherever, as in the majority of the birds that never pair and likewise in many that form lasting pairs, the males and females differ conspicuously in plumage, size, or other visible features, each individual should have no difficulty in recognizing the sex of other mature individuals of its kind. In a number of these birds, the capacity to discriminate the sexes by visual clues has been amply demonstrated by experiment no less than by field observations (Chapman 1935, Noble and Vogt 1935). Among woodpeckers, the sexes are generally alike in plumage except for certain small, sharply defined areas on the head. In the Yellow-shafted Flicker, sexual recognition marks take the form of black malar stripes or "mustaches" that the male has and the female lacks. When Noble (1936) fastened black mustaches on a mated female flicker, her partner treated her as a rival of his own sex and tried to drive her away; when the artificial mustaches were removed, he accepted her again.

In most of the birds whose pairing we have already followed, the male and female differ in appearance or voice, or in both. Even when these visual and auditory features do not suffice for sex discrimination (at least by the human observer), there is a notable difference in the reaction that an individual ready to pair makes to the advances of a bird of the other sex, as in the Cedar Waxwing and egrets. In all these species, the problem of winning a mate seems to consist less in species and sex recognition than in finding a well-disposed individual and in overcoming the timidity of the prospective partner.

In another group of birds, exemplified by certain penguins, terns, and grebes, ornithologists have as yet discovered no differences in appearance, voice, or courtship behavior that suffice to distinguish the sexes. Especially when birds quite alike in appearance form twosomes while sexually immature, it must be difficult for them to ascertain each other's sex. In these circumstances, do not two juveniles of the same sex sometimes associate as a pair until, with greater familiarity, they discover their error and separate? Indeed, before sexual responses awaken, two juveniles obviously of the same sex may become close companions, as in the case of two cockerels that we once had. In the Spotted Antbird, two males may form a pair that persists for years, "so the birds themselves clearly do not distinguish the sexes well" (Willis 1972).

This trial-and-error theory of pair formation has been assailed by Richdale (1951) as a result of his long experience with sea birds, especially Yellow-eyed Penguins. After five years of familiarity with these birds, he learned to distinguish their sexes by means of certain subtle differences in their heads, that of the female being more graceful. He believed that penguins can recognize on sight the sex of other penguins, including that of young birds who have just acquired adult plumage. They also recognize their associates as individuals. In this species, the formation of pairs that will breed is often a protracted process, which may occur at any season, although the majority of pairs are formed in the winter when these penguins do not nest. Evidently they do not require so much time merely to discover each other's sex. Their long preliminary association before they begin to breed and their occasional changes of partners in the nonbreeding season must be largely concerned with making sure that the male and female are temperamentally adjusted to each other. Richdale believed such "affinity" to be an important factor in pair formation in penguins, just as Diesselhorst considered it to be of some influence in Yellow Buntings. Evidently, then, pair formation involves a good deal more than the problem of specific and sexual recognition. It remains true, as Richdale wrote two decades ago, that the study of pair formation is only in its infancy. We require many more painstaking studies of a wide variety of birds before we can reach firm conclusions.

Mating in Birds with Enclosed Nests

In birds that nest in holes but cannot make them, such as the Pied Flycatcher and Redstart of Europe, and those of which the male builds or at least begins an enclosed nest before he has won a mate, including a number of weaverbirds and wrens, the formation of pairs follows a course quite different from that in the majority of birds. When a female appears in the territory of such a male, his first concern is to show her his nest cavity or the nest that he has started or perhaps almost finished. He may direct her attention to the cavity or nest by special displays, such as flying repeatedly to it or clinging at the doorway with spread wings, or he may lead her to his still unlined nest with profuse singing, as in the Northern House Wren and Winter Wren. Then, when she is established therein and has started to incubate, he may find another hole, or begin another nest, and try to attract another female to it. After seeing a female established in one hole, a male Pied Flycatcher sometimes displays so persistently at a neighboring one that the female deserts the eggs she has just laid and follows him to the new cavity (Creutz 1955).

Of the forty-three species of European passerines in which polygyny has been reported as frequent or occasional, twenty-three breed in holes, niches, or covered nests (Haartman 1969a). Why should such a large proportion of polygynous species have enclosed nests? Various suggestions have been offered. One is that because the female enjoys a safer nest site she has less need of a male's help in defending

and rearing her brood than has a female breeding in a more vulnerable open nest.

I believe polygyny to be so strongly associated with enclosed nests because of the way that pairs are formed. The male bird who has found a hole or built a nest in which he desires to settle a female frequently reminds one of a real-estate agent trying to sell a house, rather than of a lover trying to win a bride. Without waiting to become intimate with the newly arrived female, he hustles her to the house of which he is eager to dispose. She, on her part, seems more interested in the home site than in a partner, and whether she stays or leaves appears to depend more on the nest that is offered to her than on the quality of the male who offers it. If she approves the site, she may promptly line the nest and lay her eggs. A female Northern House Wren deposited her first egg only four days after accepting a male and his nest (Kendeigh 1941). There was no long engagement period during which, as in many other birds, the male and female become attached to each other, forming personal as well as sexual bonds. Indeed, Kendeigh reported an observation that suggested the male house wren fails to recognize his mate when he suddenly meets her on a foraging expedition, although individual recognition of nuptial partners is frequent among birds of many other kinds. It is not surprising that, while the female he knows so slightly is incubating her eggs and no longer attracts him sexually, the male loses interest in her and turns his attention to finding or building another nest and winning another mate. This seems the most probable explanation of the high incidence of polygyny in birds with enclosed nests.

However, males who mate hastily after preparing a nest are by no means invariably unfaithful to their partners. In the Blackfaced Weaver of West Africa, males weave globular, roofed nests in immense colonies; only as his structure nears completion does each one accept a mate, who promptly lays in it while the male completes the nest. Yet in this species polygyny seems not to occur (Morel et al. 1957), possibly because the multitude of birds nesting in a compact group must fly afar to gather food for their young, and, without the full-time assistance of a male, the female could not nourish her brood.

The association of polygyny with breeding in holes or in covered nests appears to be more frequent in the Old World than in the New. In their survey of the mating systems of passerine birds of the United States and Canada, Verner and Willson (1969) list only fourteen species as normally polygynous, with at least 5 percent of the breeding males having two or more mates simultaneously. Only two of these, the Northern House Wren and Winter Wren, breed in crannies, bird boxes, and similar situations, while two more, the Long-billed and Short-billed marsh wrens, build globular nests with a side entrance. The last two, together with nearly all the rest of the fourteen polygynous species, breed in grasslands or marshes, habitats that encourage polygyny by the concentration of the sun's radiant energy. In such places, the food that the sun directly or indirectly produces is in a thin horizontal layer rather than diffused through the greater depth of woodland. This concentration of the seeds or small animals with which the female nourishes her brood enables her to gather food so efficiently that she can dispense with the help of a male, who is therefore freed to win additional consorts. Although polygynous males occasionally feed their offspring, they obviously cannot give full attention to the broods of all their mates.

Of the many tropical American birds, both passerine and nonpasserine, that breed in holes or covered nests, I know of none that is polygynous. All the wrens that have been studied in the tropics appear to be monogamous. The only exceptions to monogamy that have come to my attention among tropical American hole-nesters are certain woodcreepers, including the Tawny-winged Dendrocincla and Plain-brown Dendrocincla, and an ovenbird, the Buffy Tuftedcheek. The former appear to be promiscuous, or to form no regular attachments between the sexes; for the latter we have observations on only a single nest, which may have been abnormal. Doubtless the rarity of polygyny among neotropical birds that build covered nests or breed in holes is to be attributed to the fact that they mate long before the nesting season begins, if not permanently. Although polygyny is rare among neotropical birds with enclosed nests, some of them, espe-

cially among the flycatchers with pensile nests, do not pair at all and will be considered in chapter 3.

Patterns of Pair Formation

When we analyze the various examples of pair formation, we recognize three distinct patterns:

1. When two mate-seeking birds of opposite sexes first meet, each behaves differently toward the other than it would toward an individual of its own sex. Usually the female seeks the male, who has proclaimed his sex, possession of territory, and need of a partner by tireless singing, as in the Yellow Bunting, Pied Flycatcher, Northern House Wren, and Yellow Wagtail (S. Smith 1950). In such cases, the sexes need not be distinguishable (by the human eye) by appearance, although often they are. Sometimes, however, the first steps toward pair formation are taken within the flock, as in the Swallow-Tanager (in which the female takes the initiative) and the Hawfinch (in which the male makes the first advances). In these cases, the sexes often differ in appearance.

2. When the female approaches a male who has revealed his sex by singing or by starting a nest, he at first responds to her as to another male, usually by attacking or threatening her. However, she wins acceptance by behaving differently from a male, often by passively resisting his threats instead of either fighting back or fleeing, as another male might do. Sometimes she voices a distinctive note. Examples are the Song Sparrow, Snow Bunting, and European Robin, in which the female is attracted to a territory-holding male by his song. In herons and egrets, she is attracted by his displays, often given on a partly built nest. The Northern Phalarope follows this pattern, except that the roles of the sexes are reversed, as they are also in incubation and care of the young.

3. When the two sexes are alike both in appearance and behavior it may require days of subtle interaction before two birds trying to pair determine that each belongs to the opposite sex. This seems to be the way pairs are formed in certain penguins, grebes, and terns.

It will be noticed that I have characterized these three patterns of pair formation by the birds' behavior rather than by their ability to recognize sex. Doubtless when the male and female differ in plumage or voice they promptly become aware of each other's sex. But the fact that a male at first treats an intruding female much as he would treat another male is certainly not proof that he is not already aware that she is a female. His initial hostility to her may be simply an expression of the strength of his determination to defend his territory. His first need is to acquire control of a breeding area by driving out all others of his kind. His aloofness must be gradually overcome, at least to the extent of admitting a nuptial partner. In human terms, he must become reconciled to the idea of sharing his territory with a mate. The female's behavior, her passive resistance, may be primarily a method of mitigating his truculence rather than revealing her sex to him. Richdale's experience with Yellow-eyed Penguins, Lack's with European Robins, Tinbergen's with Herring Gulls, and mine with several kinds of American flycatchers should make us wary of asserting that where we can detect no difference in the appearance of the sexes, the birds themselves detect none. Subtle differences that we recognize after long practice may be more promptly evident to the birds themselves. The situation among fishes is instructive. The Jewel Fish studied by Noble and Curtis (1939) evidently distinguished each other's sex by movements so subtle that they escaped these experienced observers. Male Guppies learn by experience to recognize females. A male reared alone will attempt to mate with another male as readily as with a female; but, after the two sexes have lived together for a long while in the same tank, the males pursue only individuals of the opposite sex (Noble 1938).

Decreased Singing after Pairing

After a songful bird has won a mate, he sings far less or hardly at all, as has been noticed in the Song Sparrow (Nice 1943), Snow Bunting (Tinbergen 1939), White-crowned Sparrow (Blanchard 1941), European Redstart (Buxton 1950), California Quail (Genelly 1955), and other birds (Andrew 1961). I have repeatedly noticed that, while a pair of Rufous-breasted Wrens is working together at a nest, the male is almost silent, but, if his mate goes off temporarily and he continues to build alone, he sings freely. In many species, after the female begins to incubate and the male is alone much of the time, the frequency of his songs increases.

To account for the sudden diminution of song by a male who has just won a mate, two explanations occur to me, one psychological and the other evolutionary. Singing may be in part an expression of loneliness, stress, or an unfilled need, as it often is in ourselves; when the male bird's need is filled and his loneliness overcome, he sings less. His increased songfulness, after his partner's domestic occupations leave him alone much of the time, lends strength to this interpretation. On the other hand, his comparative silence after winning a mate may be an adaptation. By dropping out of the singing contest, he ceases to attract females to himself and may make it easier for the still unmated, freely singing males of his kind to win partners. Moreover, the silence of the newly mated bird, who no longer advertises his availability, should favor monogamy. On this view, the decrease in songfulness after pairing is an adaptation brought about by natural selection: the races in which the newly mated male ceases to compete with his neighbors rear more progeny than those in which he continues to attract the attention of unattached females by his song. Since the psychological explanation is not incompatible with the view that decreased song is adaptive, both explanations may be correct.

Coincident with the decrease in songfulness of a male who has won a mate, he frequently becomes more aggressive toward other males, or even toward intruding females, as has been observed in the Yellow Wagtail (S. Smith 1950).

CHAPTER 2

The Stability of Pairs

In the preceding chapter, we considered how pairs arise. Now we wish to know how long they endure. With reference to the permanence of the bond between the nuptial partners, birds may be divided into those that remain paired throughout the year, those that pair only for the breeding season, and those that never pair but associate only fleetingly with a sexual partner. In the present chapter we shall survey the first two classes, leaving the third for the following chapter.

Constantly Paired Birds

Edmund Selous remarked long ago that birds live in pairs whenever they can, and I believe that we may regard the constantly paired state as that most congenial to the majority of them, that toward which they naturally incline. However, the tendency toward enduring monogamous bonds is counteracted by several powerful factors, chief among which are the necessity to migrate in order to avoid a season of rigor and scarcity, and the numerical disparity of the sexes in the breeding population, which under a regime of strict monogamy might condemn many members of the more numerous sex to lifelong celibacy.

Birds that live in pairs throughout the year are most numerous in mild climates where many kinds remain on, or at least closely associated with, their nesting territories at all seasons. Of the birds about me as I write, many finches, tanagers, wood warblers, gnatcatchers, wrens, American flycatchers, antbirds, woodpeckers, parrots, and others are found two by two in every month. At high latitudes, constantly paired birds are much less numerous but by no means absent. Although most constantly mated birds are not gregarious, some of them associate in flocks. When a party of Scarlet Macaws or Red-lored Parrots flies overhead, it is evident at a glance that the birds are grouped in twos, although there are usually a few lone individuals and likewise some agitated trios, which evidently represent matrimonial tangles. In other parrots, however, twosomes are not evident in the flock, although it may well be that the individuals who compose it are mostly paired. When two birds remain closely associated during the long season when they do not nest and their reproductive processes are dormant, they are held together by personal rather than sexual attraction. They seem to be the natural archetypes of Platonic love.

It is often said of these constantly paired birds that they mate "for life." While this may well be true of many or even most of them, the evidence on which such statements are based is scanty and for many species wholly lacking. Obviously, the condition of being almost always paired is not incompatible with frequent shifts of partners, and only prolonged observation of marked or individually known birds can settle this point. Perhaps what we should find, if we had enough statistics, is a continuous gradation from species in which change of mates never or scarcely ever occurs to those in which divorce is rather frequent.

In a long-continued study of Yellow-eyed Penguins, Richdale (1951) found one male and female who remained faithful to each other for eleven years; yet in this New Zealand bird he found a divorce rate of 18 percent. Thus, in nearly one-fifth of the instances when two birds who nested together in one year returned to breed in the same colony in the following year, and so might have resumed their association, they actually chose different partners. Although Richdale considered these penguins to be constantly paired, in winter when they spend much time at sea the bond between male and female is evidently much looser than in many other birds of this category, as one deduces from the fact that on winter evenings individuals nearly always come ashore at different times. Were the two more closely associated in the winter months, changes of nuptial partners might be less frequent. Adelie Penguins normally retain the same nest sites and keep the same mates from year to year, but Chinstrap Penguins are not quite so faithful (Sladen 1955).

Royal Albatrosses apparently mate for life. Richdale (1947) knew one partnership to remain unbroken for ten years and he discovered no instance of divorce, but he followed the histories of far fewer of the albatrosses than of the penguins. Likewise, Lockley (1942) wrote that Manx Shearwaters normally retain the same mates as long as they live. More recent studies revealed that, among 240 pairs of which both members were caught in successive years, only 2 pairs had separated (Perrins et al. 1973). In these shearwaters and albatrosses, it would be surprising if a male and female who have nested together could manage to keep close company as they forage over many miles of open sea, often in tempestuous weather. Possibly we are deal-

ing here not with constant partnership but with a high incidence of remating, based on persisting attachment to the nest site that these marine birds have used in past years. The same considerations apply to the Herring Gull, which according to Tinbergen (1953) seems to pair for life. In another gull, the Kittiwake, nearly half of the birds changed mates from one year to the next (Coulson and White 1958).

Chiefly from familiarity with semidomesticated individuals, Lorenz (1952) concluded that Jackdaws retain the same mates as long as they live. The same is true of the Thick-billed Nutcracker, of which mated birds live together on their territory throughout the year, supporting themselves in winter and early spring on their stores of buried nuts, which they retrieve with great accuracy even through a thick layer of snow (Swanberg 1956). Geese, shelducks, and whistling ducks are also reputed to form nuptial unions that are dissolved only by death (Delacour and Mayr 1945). It is said that, if a Greylag Goose loses its mate, she or he will thenceforth remain a widow or widower. Among gallinaceous birds, Mallee-Fowl establish enduring monogamous bonds. One pair was proved to remain together for six years, and half a dozen other banded pairs continued intact for four or five years (Frith 1962).

Among smaller birds, House Sparrows in England show lifelong attachment to the mate with whom they have successfully bred, as likewise to their nest site (Summers-Smith 1958). Marsh Tits, Coal Tits, and Willow Tits in England, and Plain Titmice in California remain paired and on their territories throughout the year, and probably they keep the same mate until one dies (Morley 1950, Ruttledge 1946, Gibb 1956, Dixon 1949). Lifelong partnerships are also formed by Wren-Tits (not members of the titmouse family) carefully watched in California by Erickson (1938), and by resident White-crowned Sparrows studied in the same state by Blanchard (1941). In the arid southwestern United States, Cactus Wrens keep the same mates and territories as long as they live (Anderson and Anderson 1963). In the central and eastern United States, Cardinals, Common Bluebirds, and Carolina Wrens often remain with their mates through the winter months. In certain other species, including the Common Mockingbird, European Robin, and Hairy Woodpecker, the members of a pair often occupy separate, mutually respected territories during the winter months, thereby avoiding competition for food when it is scarce.

It is almost certain that, when we have as detailed studies for many tropical birds as are now available for birds of northern lands, numerous additional examples of lifelong matrimonial constancy will be revealed. It is among the smaller and more rapidly maturing of these constantly mated birds that we find the most precocious pairing, sometimes at the age of only a few months. Natural selection should favor the retention of the same mate from year to year because, as we shall see, individuals who have previously nested together do so more successfully than newly formed pairs, even when composed of birds of the same age (p. 278).

Seasonally Paired Birds

Turning now to seasonally paired birds, we find that they are mostly migrants, or at least they abandon their territories and flock for part of the year, although neither migration nor flocking is invariably incompatible with constant matedness. In these birds, a male and female are closely associated from the beginning of the engagement period, which may precede nesting by some weeks or only a few days, until the close of the breeding season, in which several broods may be reared. Less frequently, the male and female change partners between the first and second broods. Or else, as in certain surface-feeding ducks, pairs may be formed in autumn or winter and persist until nesting begins in the following spring, perhaps far from the region where the ducks mated; in such species, the male deserts his partner after she begins to incubate, leaving her to attend her ducklings alone. In other birds, such as the Pied Flycatcher, a male may associate closely with a single female only until her eggs are laid, after which he turns his attention to another (Haartman 1956).

Since in all these seasonally mated birds the bond between the sexes is definitely dissevered after the close of the breeding season, if the same male and female are together in successive years, this must be regarded as remarriage rather than a continuous wedded state. Accordingly, if they take different partners in successive years, it seems more correct to say that they have failed to remarry rather than that they have become divorced, because in such birds separation at the end of the breeding season is the normal procedure. The remating of the same individuals in later years is promoted by the strong tendency of migratory birds to return each spring to the same locality where they have already nested, often to their last year's territory. This brings a male and female together again, even if they have wintered in different countries or perhaps on different continents. The personal recognition of a former mate and an acquired preference for him or her may also favor remating, but it is difficult for a human observer to know which is stronger: a bird's fondness for a former mate or its fondness for a known territory.

On the other hand, females that commonly arrive on the breeding ground after the males have established themselves sometimes find that their last year's partner has already contracted an alliance with an earlier arrival. Considering the great hazards of migration, and the large number of birds that fail to return, males that waited long for former mates would often be disappointed. Such a practice would be so detrimental to reproduction that natural selection would operate against it.

An unusually high proportion of remating occurred in the Brewer's Blackbirds studied for six years in California by L. Williams (1952). In forty-five cases in which both members of a pair returned in a succeeding year, forty-two unions were between the same individuals and only three changes of partners were noticed. This gives a constancy in remating of 93 percent. The high fidelity of Brewer's Blackbirds is the more surprising because, in consequence of the preponderance of females, many of the males had several partners, sometimes as many as four, whose nests they usually attended. Yet in the following year the great majority of these males were paired with their first, or primary, partner of the preceding season. In the intervening months, these blackbirds lived in a flock that foraged within a few miles of their nesting area, and, although at midwinter they did not seem to be paired, perhaps former mates maintained a certain liaison with each other.

Among the Great Tits studied for many years in Holland by Kluijver (1951), a male and female remated in successive years in fifty-five of the seventy-one cases in which both survived the winter, thus showing a constancy of 77 percent. Like the Brewer's Blackbirds, these tits stayed in the same neighborhood throughout the year. Some of the pairs remained intact through the winter, but more often they were formed anew in the spring. Of thirty-six pairs of English Skylarks of which both members survived from one year to the next, seventeen, or 47 percent, had the same mates in the second year as in the first. Faithfulness to nuptial partners was associated with a high degree of site tenacity in this partly migratory bird (Delius 1965). In the migratory Red-winged Blackbirds studied for six years in Wisconsin by Nero (1956), nine of sixteen marked females mated with the same male two years in succession, and two of these nine were paired with the same male in three consecutive seasons. In another migratory species, the Northern House Wren, which Kendeigh (1941) and his associates studied for many years in Ohio, remating in the following year occurred in 42 percent of the cases in which both members of the pair survived and returned to the same locality. In the Song Sparrows studied in the same state by Nice (1943), the incidence of remating was decidedly lower and occurred in only 27 percent of the cases in which both members of a pair returned. At the opposite extreme from the Brewer's Blackbirds, in about six hundred pairs of Pied Flycatchers studied over a period of six years by Creutz (1955) in Germany, the same male and female were never found together in successive years.

Less numerous and usually less closely associated with man, terrestrial nonpasserines have been less frequently banded and followed for long periods. In a long-continued study in Canada of individually recognizable woodpeckers of four species—Hairy, Downy, Yellow-bellied Sapsucker, and Yellow-shafted Flicker—Lawrence (1966) discovered that they retained the same mates as long as they lived. This constancy, however, seemed a consequence of attachment to the nest site rather than to the nuptial partner, for all, including the sedentary Hairy and Downy, needed to renew the pair bond at the outset of each breeding season. Tropical woodpeckers remain more closely associated with their mates throughout the year, often sleeping in the same hole at all seasons, as in the Golden-naped Woodpecker and Olivaceous Piculet, species which both seem to pair for life (Skutch 1969b). Of 1,173 nesting adult Piping Plovers banded over a period of twenty years by Wilcox (1959), 288 were caught again, but only 39 pairs consisted of the same individuals for more than one year. Only 2 pairs persisted into the third summer.

Although it is readily understandable why birds that perform long migratory journeys so often take different partners in successive years, we should expect them to remain constant to their mates at least throughout a single breeding season, even if they raise several broods, for they stay continuously in the same locality. Studies of banded birds have shown that the majority of multibrooded songbirds retain the same partners throughout the nesting season. Nevertheless, change of mates between broods has been recorded in the Song Sparrow, European Robin, American Robin, Gray Catbird, Common Bluebird, and other species, in most of which it appears to be exceptional. In the Northern House Wren, however, change of mates between the first and second broods is frequent, for in only 40 percent of the possible cases did a male and female who had raised the first brood together remain together for the next brood. Three pairs in every five broke up after rearing the first brood, the male and female choosing different partners for the second. In Kendeigh's (1941) study, a wren was slightly more likely to have the same mate in two successive years than to have it for both broods of the same year. One reason for these frequent divorces—the word seems appropriate in this context—was that in half the recorded instances the male wren failed to help his mate to feed the young of the first brood after they had left the nest box, so that he was ready to breed again before she was free to do so and he took a different partner. Even when both parents attended the fledglings, they often divided the brood between them, each going a different way, and sometimes these wanderings led to new matrimonial alliances.

While in the wedded state, birds, like other animals, may or may not remain strictly faithful to their nuptial partner. Although the bond between drake and duck is not as enduring as in many other birds, while it lasts the duck of certain species is strongly attached to her mate and resents the attentions of other drakes. When a pair of Common Eiders is approached by a strange drake, whether or not accompanied by his mate, the female is annoyed. Stretching her neck toward the intruding drake, she voices querulous notes and continues to point to him with her bill until her mate, who may be somnolent, is aroused and drives away the offender, thereby preventing improper intimacies. Likewise, Herring Gulls, for all the crudeness of certain aspects of their lives, are strictly monogamous, the female resisting every advance of a strange male, often driving him away (Tinbergen 1958, 1953). Similarly, the female Laysan Albatross will accept only her own mate and flees from other males that attempt to rape her (Fisher 1971). Among passerines, both sexes of the Red-winged Blackbird appear to preserve strict matrimonial fidelity (Nero 1956). Although mated male Song Sparrows often pounce rudely upon the female of a neighboring pair, they seem not to have sexual relations with her (Nice 1943).

Certain other birds, more wanton, are unable to resist the attractions of neighbors of the opposite sex. Such "stolen matings," as ornithologists call the resulting lapses from matrimonial fidelity, have been noticed in the European Cormorant, Gray Heron, Rook, Ovenbird, Yellow Bunting, and Pied Flycatcher (Nice 1943, Creutz 1955). In the absence of their own mates, female Bicolored Antbirds in the tropical forest often solicit food from unmated males (Willis 1967). In all these birds, such sexual escapades seem not, as a rule, to bring discord into the relations of the mated pair.

Bonds between Mates

For many birds, especially those that are seasonally paired, attachment to the same territory and service at the same nest suffice to hold the pair together until their young no longer need them. In numerous other birds, above all those that live in pairs throughout the year, vocal reassurances, delicate attentions, repeated caresses, and shared displays help to bind the two more firmly together.

It is above all in the constantly paired birds of the tropics that vocalizations have been elaborated to serve as bonds between mates. The numerous wrens of tropical America, especially those of the large genus *Thryothorus*, are outstanding examples of this. Foraging amid densest vegetation where visibility is narrowly circumscribed, mates keep in touch by sound rather than by sight, but, instead of using simple, unmelodious call notes, they depend upon song. Often this is antiphonal, the partners singing alternately to each other. Pairs differ much in their skill in articulating their separate verses into a single harmonious composition, but those most adept synchronize their contributions so perfectly that the listener must stand between them, and hear the notes come now from this side and now from that, in order to be convinced that the continuous flow of musical notes issues from more than one throat. Often the phrases of one wren, doubtless the male, are somewhat longer than those of his mate.

Outstanding masters of antiphonal singing are the Rufous-breasted, Striped-breasted, and Buff-breasted wrens. A briefer, less melodious, but exceedingly skillful performance is given by the Plain Wren, also known as the Chinchirigüí, which owes the latter name to its antiphonal singing. One member of the pair repeats *chincheery*, with short intervals that its mate fills with the monosyllable *gwee*. Unless the birds happen to be on opposite sides of the hearer, they sound like a single wren rapidly proclaiming its name amid the dense weeds of a neglected field. Because constantly mated wrens employ song to maintain contact with each other, they sing at all seasons and in the gloomiest weather when most other birds are silent. The situation in wrens and some other constantly mated birds lends weight to the theory that song evolved from contact notes (Andrew 1961).

Among other birds that duet antiphonally with great skill are quails. Often I have heard the Marbled Wood-Quail's mellow undulating song, sounding like *burst the bubble*, issuing over and over from the tropical forest in the evening twilight. The observations of Chapman (1929) make it probable that this stirring performance is given by two quails singing alternately, fitting their notes with consummate art into an unbroken flow of sound. Bobwhites and California Quails also call antiphonally (Stokes and Williams 1968).

Other birds sing responsively without articulating their verses into a single song. The songs of the two partners may be similar or so different that they seem to come from different species. Mated Buff-throated Saltators answer each other with dulcet notes, the male seeming to sing *cheery cheery*, his partner answering, in a weaker voice, *cheer to you*. These responsive verses resemble in tone quality, but are much shorter than, the male's long-continued, flowing advertising song. Females of the Blue-black Grosbeak, Yellow-tailed Oriole, and Melodious Blackbird also sing sweetly in response to the song of their mates, sometimes while they sit in the nest.

Foraging amid tangled thickets, paired Tyrannine Antbirds sing to each other with soft, cozy, little trills, that of the female distinctly higher in pitch than that of her mate. As they flit amid trees and shrubbery, Black-fronted Tody-Flycatchers communicate by means of clear, resonant trills, sometimes given simultaneously and sometimes in sequence. Even while sitting in her pensile nest, the female often answers her mate with a trill quite similar to his. Paired Streaked-headed Woodcreepers also exchange trills, which are long drawn, melodious, and delightfully clear, with no perceptible difference in their voices. Similar in form, the answering trills of the related Spotted-crowned Woodcreeper of the highlands have a quite different tone quality, being squeaky and plaintive, as though these birds suffered from the cold of the heights.

Among the birds that answer each other with contrasting utterances are the Buff-rumped War-

blers, of which the male delivers a jubilant, ringing crescendo, his mate a beautiful, soft warble. The greatest contrast in the notes of the two sexes is found in the Chisel-billed Cacique, a yellow-billed, black icterid that lurks in the most impenetrable thickets of tropical America. The male's full, mellow double whistle elicits a prolonged churring response from his unseen mate. Although I had long surmised that these so different notes belonged to the same species, I lacked proof of this until I watched an incubating female churr dryly in response to her mate's liquid inquiry.

Still other paired birds sing simultaneously. Although devoid of melody, the harshly garbled notes of a pair of Yellow-bellied Elaenias calling together may be designated a duet. As they flit restlessly through weedy fields and low thickets, mated Yellow Flycatchers reassure each other with an almost continuous chatter of soft, pleasant notes. Barred Antshrikes duet with dry rattles; Rufous-fronted Thornbirds with clear, ringing notes all very much alike. Duetting Banded-backed Wrens pour forth their harsh notes with vehement gusto. The related Rufous-naped Wrens sing in unison with more melodious notes, the partners keeping perfect time in their most intricate musical figures. After they fly from their burrows in the chill highland dawn, mated Blue-throated Green Motmots, who sleep together throughout the year, raise their voices together with delightfully clear, mellow notes. Very prolonged is the duetting of Gray-necked Wood-Rails, which has earned for this bird the onomatopoeic name "chirincoco." Heard at a distance, the rails' incessantly reiterated *chirincoco-co-co, chirincoco-co* is stirringly beautiful, but when the duettists are too near the listener the voice of one or both of them seems strained and cracked. Far from trying to achieve harmonious unison, a pair of Donacobiuses of South American marshes, perching close together and rhythmically swinging their fanned tails from side to side, pour forth contrasting notes, those of the male liquid and ringing, those of his mate with a sizzling or grating quality—as though a cicada duetted with a wren.

The foregoing are a few examples of dual singing or calling that I have heard in tropical America. Many others, among megapodes, moorhens, cuck-

PLATE 1. A male Scrub Jay feeds his incubating mate, an attention that strengthens the pair bond (photo by S. A. Grimes).

oos, trogons, barbets, shrikes, cuckoo-shrikes, honey-eaters, and other birds have been described, chiefly from the tropics of both hemispheres (Power 1966, Diamond and Terborgh 1968, Diamond 1972).

Certain constantly paired birds have special utterances that they use only or chiefly as they come together after a temporary separation. When a Vermilion-crowned Flycatcher flies up beside its mate, they salute each other with a soft, chiming sequence of somewhat trilled notes, at the same time fluttering their wings. Similarly, Tropical Kingbirds greet their mates with high-pitched trills and twitters, also delivered with partly spread, quivering wings. When one of the Black-striped Sparrows that frequent our garden alights beside its mate, the two simultaneously utter a rather whining note, rapidly repeated with falling inflection. Differing greatly from the sparrow's song and its call or alarm notes, this vocalization appears to be reserved for this special occasion. Yellow-thighed Finches greet each other with a long-continued, rapid flow of tinkling notes, all very much alike and quite different from the male's song. Among northern finches, the Brown Towhee, which in certain parts of its wide range remains paired throughout the year, also has a special call that is used chiefly, although not exclusively, as a greeting when the partners rejoin each other. Described as "a succession of eight or nine rather distressed-sounding squeaking notes, somewhat as one might squeak with one's lips," it was designated the "mate-call" by Quaintance (1941).

The feeding of one partner by the other also strengthens the bond between them (plate 1). This practice has been variously called "courtship feeding" and "nuptial feeding." The latter term is preferable because it is more inclusive. I believe that most birds who feed a member of the opposite sex whom they are trying to win continue to do so after the nuptial bond has been forged; indeed, the acceptance of food is often equivalent to saying "yes" to the suitor. In some birds, nuptial feeding seems not to begin until the eggs have been laid.

Feeding the female may serve various ends: it may help her to obtain substances that she needs to form her eggs; it may enable her to incubate much more constantly than she could if obliged to find all her own food; it may lead to the male's prompt discovery and feeding of the nestlings—matters to which we shall presently return. But in many birds nuptial feeding starts long before the eggs begin to enlarge rapidly, and, if continued into the incubation period, it is not sufficiently generous to reduce significantly the time that the female devotes to foraging. In a few birds, like crossbills and siskins that nest in freezing weather and hornbills that immure the female in the nest, nuptial feeding is indispensable for successful breeding. However, among birds in general its chief importance seems to be as a bond between the mates and, in species of which the male does not incubate, as a preparation for feeding the young.

Although by no means universal, nuptial feeding is very widespread among birds, both those that pair permanently and those that pair seasonally. It is frequent in birds of prey, rails, gulls and terns, pigeons, parrots, owls, kingfishers, bee-eaters, hornbills, jacamars, toucans, and many passerine families, including antbirds, swallows, crows and jays, titmice, tree creepers, nuthatches, dippers, certain members of the thrush family, Old World flycatchers, waxwings, silky-flycatchers, shrikes, wood warblers, tanagers, finches, and buntings. In numerous other families, it has been recorded only rarely: I have seen it in the Turquoise-browed Motmot, White-fronted Nunbird, Pale-legged Ovenbird, Tropical Pewee, Rufous-browed Wren, Turquoise Dacnis, and Melodious Blackbird.

Nearly always the male feeds the female, but rarely the roles are reversed. I witnessed a single instance of this in the White-flanked Antwren and Tawny-bellied Euphonia; Lack (1940) saw it once in the European Robin and Vegetarian Tree-Finch of the Galápagos Islands. M. K. Rowan (1955) occasionally saw a female Red-winged Starling feed her mate while the pair were renovating their nest after the loss of newly hatched young. As we have already noticed, a male and female Cedar Waxwing in courtship pass food back and forth between them, but the item is first brought by the male and presented to the female. Among button-quails the female normally feeds the male, who incubates the eggs and attends the young. Strangely enough, the only instance of nuptial feeding that I have found for hummingbirds is of the female feeding the male, which occurs in the Andean Hillstar and serves to overcome the antagonism that prevails among individuals of this species (Dorst 1962). Sometimes, as in the Squirrel Cuckoo, the male presents an insect to the female and both hold on to it while he mounts her.

Paired birds also intensify their intimacy by preening each other (plate 2). This practice is now generally known as "allopreening"; between mates it is usually, if not always, reciprocal, in which case it may be called "mutual preening." Since birds billing each other's feathers usually sit or stand close together for considerable intervals, mutual preening may establish more intimate relations than nuptial feeding, which is more quickly accomplished. Birds that insist on maintaining "individual distance," or a space between themselves and their companions—usually determined by the reach of their bills—do not often engage in allopreening, which is more compatible with the temperament of those that habitually rest in contact with each other. Most often a bird allopreens feathers that its mate cannot reach with its own bill, especially those of its head and neck, although other regions of the body may be so treated. Birds performing this service may or may not remove loose feathers, vermin, or other foreign matter from the plumage, but the value of the practice in establishing mutual confidence can hardly be doubted.

PLATE 2. Female Wood Pigeon caressing her mate on the nest (photo by R. K. Murton).

PLATE 3. Some birds preen their young as well as their mates. Before brooding, a Manx Shearwater nibbles its chick's plumage. The parent closes its eyes with the nictitating membrane as a protection against loosened down (photo by John Warham).

Among birds that preen each other are penguins, albatrosses, petrels, cormorants, boobies and gannets, ibises and spoonbills, herons and egrets, Hammerheads, screamers, Magpie-Geese, whistling ducks, New World and Old World vultures, kites, eagles, caracaras, wood-quails and bobwhites, rails and coots, terns and kittiwake gulls, murres and Razorbills, pigeons and doves, parrots and parakeets, anis among cuckoos, swifts, and toucans. Among passerines, allopreening appears to be less widespread, or less frequently noticed, than among the mostly larger nonpasserines. It occurs in the Slaty Castlebuilder (an ovenbird), Chestnut-backed Antbird, Jackdaw, Rook, magpies, Bearded Tit, Wren-Tit, babblers, Banded-backed Wrens and certain other species of *Campylorhynchus*, Plumed Helmet-Shrike, Rothschild's Grackle among starlings, white-eyes, Bullfinch, Cuban Grassquit, waxbills and mannikins, and Scaly-crowned Weaver.

Birds that allopreen do not always confine this service to their mates. Sometimes they also preen their young, as has been recorded of the Little Penguin, Rockhopper Penguin, Manx Shearwater (plate 3), several kinds of boobies, Shag, European Coot, blue waxbills, and others. They may likewise preen grown members of their flock or social group other than their mates, as in Hammerheads, whistling ducks, Marbled Wood-Quails, and anis; the last mentioned often perch in a compact row, each individual nibbling at the feathers of its companions closely pressed on either side. Moreover, chicks or nestlings may preen their parents and siblings, as in the Rockhopper Penguin, White Booby, and blue waxwings. Fledgling white-eyes often preen each other. A ten-day-old American Coot preened a slightly younger coot chick.

Birds that allopreen sometimes nibble the feathers of other species, but such interspecific preening has been noticed chiefly among captive individuals. A European Spoonbill preened a Crested Screamer in a zoo. Blue waxbills preen not only their mates and dependent young but also adults and fledglings of such other species as the Avadavat, Gold-breasted Waxbill, and Black-capped Waxbill, although the latter reciprocate, if at all, only after they have been receiving this attention for a long while. Between the ages of two and three weeks, Virginia Rails preened a Killdeer chick and a downy young Ruddy Duck as well as other chicks of their own kind. A month-old Sora Rail preened the head of a young Eared Grebe (Goodwin 1965, Nice 1962).

Most curious of all, cowbirds—including the Brown-headed, Shiny, Red-eyed, Giant, and Bay-winged—solicit preening from birds of other species by presenting to them a deeply bowed head with out-fluffed feathers. Neither cowbirds nor most of the birds that they so approach are known to allopreen, but, surprisingly, Brown-headed Cowbirds

PLATE 4. Greeting ceremony of the Australasian Gannet, a mutual display that binds mates together (photo by John Warham).

are sometimes preened, in the open as well as in the aviary, by House Sparrows, Red-winged Blackbirds, and meadowlarks, although birds of these species seem never to preen each other. The cowbirds' solicitation of preening is evidently an appeasement display. By inducing its victims to engage with it in an activity that serves as a bond between mates, the cowbird may mitigate their hostility and make it easier to slip its eggs into their nests, but the matter is far from clear (Selander and La Rue 1961, Selander, 1964*b*, Payne 1969*c*).

Some birds feed but do not preen their mates; others preen but do not feed them; while yet others perform both services. Perhaps, in addition, pairs may sing or call responsively to each other. Among those that both preen and feed their partners are Bobwhites, rails, pigeons, parrots, and crows. Once I watched a Chestnut-mandibled Toucan give its mate a fruit that it brought in the tip of its enormous bill, then bring up from its throat or crop four or five other items that it passed to its partner. After swallowing these offerings, the recipient preened the donor's head and neck. The sexes of these birds could not be distinguished. Since its bill is almost as long as its body, a toucan perches at some distance from the companion that it preens.

The mutual displays of certain birds strengthen the bond between mates (plate 4), especially in the interval between pairing and the beginning of incubation. We have already mentioned the Donacobius's tail-wagging display, accompanied by contrasting notes from the two partners. The extraordinarily elaborate displays of grebes are performed largely on the water. The male and female, alike in plumage and behaving almost identically, posture, dive, make porpoiselike leaps, rush side by side over the pond, and perform a variety of other antics. Not only do such shared exercises help to hold the pair together until the eggs and then the young provide a common interest; but they also evidently help to synchronize the sexual responses of the partners and thus ensure that the eggs will be fertile (J. S. Huxley 1914, Storer 1969).

If no accidents befall them, birds live much longer than mammals of similar size. Small songbirds have survived in captivity for over twenty years, pigeons upward of thirty years, parrots for over half a century. Accordingly, those that pair "for life" may, with good fortune, keep the same partner for many years—perhaps long enough to celebrate a "golden" wedding anniversary!

CHAPTER 3

Neglectful Fathers

In the great majority of birds, the male is interested in his offspring. Usually he helps to feed and protect them, and in many species he takes a share in incubating the eggs. Indeed, he may assume the major part of this task, and in a few groups of birds he alone is responsible for hatching out and attending the young. In contrast to these attentive fathers, there are some, scattered widely among the families of birds, that take not the slightest interest in nest, eggs, or chicks. These neglectful fathers include some of the most splendid and famous of birds, such as pheasants, hummingbirds, cotingas, manakins, birds of paradise, and bower-birds. A conservative estimate, based on our present far from complete knowledge of the breeding patterns of birds, places the number of species in which the male remains aloof from nest and young at no less than five hundred. At the end of this chapter, we shall consider why so many of these male birds have become so brilliant and lavishly adorned.

It would be wrong, however, to infer that all male birds who neglect their offspring have developed bright colors and elaborate courtship rites. In the tropical forest that I face as I write are a number of species whose nests I have watched carefully without ever seeing a second parent in attendance, although the sexes are nearly or quite alike, nor have

I discovered special displays. Among them are several small, plainly attired American flycatchers, a bright brown cotinga, a brown woodcreeper, and a manakin. One of the more brightly clad of the birds in this group is a flycatcher, the Sulphur-rumped Myiobius, which as it flits through the woodland often drops its olive wings and fans its black tail, exposing its bright yellow rump. Both sexes have this habit. It frequently utters a sharp *psit*, but I have heard its simple song only rarely.

More obvious is the method of courtship of another flycatcher, the Oleaginous Pipromorpha, a small, olive green bird of undistinguished appearance. Through about half the year, groups of males are found daily in certain parts of the forest, where each tirelessly repeats a simple refrain devoid of melody: *whip wit whip wit wit chip chip chip chip chip chip*. While singing, they flip up their wings one at a time and often fly from one perch in the underwood to another. After the female pipromorpha has built her beautiful, pear-shaped nest of green moss, attached to a slender, dangling vine or the cordlike root of an epiphyte, she seeks a male to fertilize her developing eggs. Then, all alone, she incubates the two or three white eggs and rears the young in their softly lined nursery with a round doorway in the side.

All these birds of which the male remains aloof from the nest, from the plainest and dullest to the most ornate and brilliant, have one thing in common: without a male's help, the female is quite capable of raising enough progeny to perpetuate the species. This is the point of departure for every deviation from a monogamous family in which the parents cooperate in attending the young. In polygynous species, one or more of the male's mates may receive more or less help in caring for her brood, although this does not always occur. In the birds we are now about to consider, the male quite ignores his offspring; he may not even know where the nests of his temporary mates are situated.

Two factors, singly or together, enable the unaided female to produce enough offspring to perpetuate her race: a low annual mortality and abundant, easily gathered food. A large proportion of these solitary females live throughout the year in evergreen tropical woodland, avoiding the hazards of migration and untroubled by winter's cold and dearth. Many of these females, including manakins, cotingas, and birds of paradise, subsist largely on easily gathered fruits, although most include insects and other small creatures in their diets and feed them to their nestlings. Others, notably hummingbird females, nourish themselves and their young with nectar from flowers, minute insects, and spiders. Yet others, especially unaided females of certain small American flycatchers, are almost wholly insectivorous. And, of course, when the chicks are precocial, able to pick up their own food under parental guidance, a father's attendance becomes less necessary, although he might provide valuable protection from enemies. Accordingly, we find neglectful fathers rather common among gallinaceous and other precocial species.

The birds that we are now about to consider are often called "promiscuous." The term is unfortunate, for among ourselves it is applied to disorderly sexual behavior. Actually, many of these birds exhibit admirable order in their mating habits, although it is not the order that prevails among the majority of birds and, ideally, in monogamous human societies. "Nonpairing" seems a better designation for birds of this kind.

Hummingbirds

The largest of the families in which the males are, with few known exceptions, neglectful of their offspring is the hummingbirds, with about 320 species. These feathered gems are confined to the Western Hemisphere. Although they occur from Alaska to Tierra del Fuego, they are most numerous in the tropics, where they abound not only in warm lowland forests but also, in amazing variety, on high mountains where nights are always cold. I have known some hummingbirds for years without discovering any special means employed by the males to attract the females. Many kinds, however, have obvious methods of courtship, consisting of songs, flight displays, or a combination of the two.

Hummingbirds do not belong to the great group of songbirds (Oscines) and we rarely think of them as songsters. Many of them, however, have utterances that, in view of their special function, must be classified as songs. Some of these "songs" are no more than monotonously reiterated squeaks, but a few are so sweetly varied that the appreciative listener spontaneously applies this name to them. If their voices were a little stronger, these hummingbirds might become famous as musicians.

I still vividly recall my delight many years ago as I listened to a Wine-throated Hummingbird singing on a bushy slope high on a Guatemalan mountain. By its intensity, variety of phrasing, and rising and falling cadences, the song reminded me of the higher passages of a small finch's song, especially that of the White-collared Seedeater of the lowlands. This surprising performance often continued without interruption for thirty or forty seconds. As he poured forth his sweet, impassioned little lay, the hummingbird spread the stiff feathers of his gorget to form a scaly shield with projecting lower corners and turned his head from side to side. When the bird faced me, his gorget glowed with an intense magenta light, but, as he slowly turned his head away, the color was extinguished; the shield, viewed from the side, was velvety black. At certain angles, it sent metallic green reflections to my eyes. At times the diminutive songster vibrated his wings in ecstasy and either floated slowly to another perch or hung motionless in the air on invisible wings, all without interrupting his verses. Sometimes he made a long, looping flight, returning to the twig from which he started, continuing to sing during the whole journey.

Another remarkable hummingbird songster is the Band-tailed Barbthroat, a rather plainly attired, brownish species that performs, not high on a flowery mountainside, but in the dark undergrowth of wet lowland forest. When I first heard this hummingbird, I surmised that the notes came from some songbird new to me. I was convinced of my error only when, after much searching, I glimpsed the songster perching a yard above the ground on a fallen dead branch, rhythmically wagging his white-tipped tail up and down. The thin, plaintive, almost trilling song continued for four or five seconds. Another brownish hummingbird that performs in the undergrowth of lowland forest, the Long-tailed Hermit, utters a tirelessly repeated squeak, one of the least engaging of all the hummingbird songs that I have heard. In musical quality, most hummingbird songs fall between these extremes. Although only exceptionally melodious, they often have simple rhythms that are pleasing.

Some hummingbirds, including the Little Hermit, Green Violet-ear, and Blue-throated Golden-tail, sing through most or all of the day, but others have a more limited song period. The Rufous-tailed Hummingbirds about our house repeat their slight phrases chiefly in the dawn and only rarely after sunrise. The Blue-chested Hummingbirds in our garden send forth their more animated utterances early in the morning and again late in the afternoon; through much of the day they are silent, except in periods of exceptional ardor, when they proclaim themselves through most of the sunny hours. Once I heard the notes of a Rufous-tailed Hummingbird persistently repeated in the late afternoon. This unexpected performance demanded investigation, which revealed that the supposed Rufous-tail was in fact a Blue-chested Hummingbird who had adopted the phrasing of his close neighbors, although he followed the singing schedule of his own kind. This observation, together with the great diversity of the songs of different males of the same species in the same locality, has convinced me that hummingbirds' songs are largely learned rather than innate. Each male adopts the song dialect of his associates in a singing assembly (Skutch 1972).

Hummingbirds rarely sing in solitude. Usually a number of individuals of the same kind gather in a courtship assembly, where each performs on his favorite perch or at least within a limited space, in hearing of several of his neighbors. From time to time, one songster invades another's territory, giving rise to a spirited chase in which additional individuals may join. Such pursuits have given hummingbirds a reputation for pugnacity; but, if one watches a while, he will find every member of the assembly back at his usual post, without a ruffled feather, singing as persistently as ever. Day after day, throughout a long breeding season, one finds the same perches occupied, presumably by the same individuals. These assemblies persist in the same place for many successive years, perhaps until the changing environment causes their abandonment, as I have noticed in a number of species. What more convincing evidence could one have that the seemingly belligerent hummingbirds rarely injure each other? Perhaps they are exceptionally playful rather than exceptionally pugnacious birds.

Since hummingbirds are supreme masters of flight, able to hover motionless and to fly backward as hardly any other bird can, one expects flight displays to enter largely into their courtship—indeed, the observer might expect to witness elaborate aerial performances more frequently than he does. These displays may supplement persistent singing or be the chief method of attracting the females. The Little Hermit of tropical lowlands is one of the most tireless of songsters. He also displays by floating a few inches above a resting individual of his species, evidently a female, with his head and tail turned upward, thus resembling a tiny boat with high-peaked bow and stern. In this attitude, he oscillates back and forth, shoots rapidly from side to side over a longer course, rotates in the air, and undulates up and down, while his invisibly vibrating wings emit a humming sound.

More spectacular is the flight display of the

Broad-tailed Hummingbird, who seems to depend chiefly on his aerial performances for winning a female. Even the ordinary flight of the male Broad-tail is accompanied by a buzzing shriller and more insistent than most members of the family make. His display consists in tracing in the air a U with arms that may be twenty or thirty yards high. The downward dive is made at astonishing speed, and sometimes the nadir, or turning point, of his course is just in front of a female who perches obscurely near the ground on a flowery mountainside. He may trace three gigantic U's in succession. At times a male and a female fly the U-shaped course together. In other species of hummingbirds, a male and female accompany each other through intricately looping, generally horizontal courses, pausing here and there to face each other in the air. The climax of these breathtaking courtship displays has scarcely ever been witnessed by ornithologists (Wagner 1954).

In hummingbirds whose courtship is largely "static," consisting in tireless singing from a perch, the male and female are often alike or differ little in plumage. When the "dynamic" mode of courtship is followed, the male seems usually to be much more brilliant than the female, wearing an intensely colored, expandable gorget—magenta, red, purple, or coppery—that he flashes in front of her as he rapidly traces a long U-shaped or more open course. This is true of the eight species that nest north of the extreme southern United States. As he darts over a female in an aerial power dive, the male Anna's Hummingbird faces into the sun, so that the rays strike upon his iridescent gorget and make it sparkle more brilliantly (Hamilton 1965). Exceptional in my experience is the Wine-throated Hummingbird, which sings persistently yet has a well-developed metallic gorget and is much more highly colored than the female.

In nearly all hummingbird species that have been carefully studied, the female, after fecundation of her eggs, hatches them and rears the young without a mate's help. In the Sparkling Violet-ear of the Andes, however, incubation and feeding of young by the male was reported by Schäfer (1954a); but such participation in nesting may be exceptional even in this species. Year after year, I have watched

a male Bronzy Hermit show great and persistent interest in his mate's nest fastened by cobweb beneath a strip of banana leaf, but his only service is to drive away other hermits (Skutch 1972). Aside from such rare cases, female hummingbirds rear their families in solitude.

Manakins

We have given much attention to hummingbirds because they are, at least in the Americas, the most widespread and familiar of the birds that form courtship assemblies. Although they do not shun observation, it is often difficult to interpret what one sees, because in many species the male and female are similar in plumage, and, even when the sexes are quite different, it may be impossible to distinguish such small birds that move so swiftly. Usually one cannot be sure whether a male is chasing a rival or pursuing a female.

Another family in which the male remains aloof from the nest is the manakins. Like hummingbirds, manakins are found only in the New World, but they are confined to the tropical parts of the American continents and they avoid the cold highlands. They abound in humid lowland forests. Typical manakins are very small birds, hardly larger than typical hummingbirds, but they have short bills and tails and relatively stout bodies, and they lay much larger eggs. The males are nearly always more richly colored than the greenish or olive females; many are so strikingly attired that the Germans have given the name Schmuckvögel—jewel birds—to the family. Of all birds that form courtship assemblies, manakins are perhaps the most satisfactory to study, because their stations are often low, many are far from shy and can be watched without concealment or from a simple blind, and in most species adult males are readily distinguished from females. Beginning with Chapman's (1935) classic account of the courtship of the Golden-collared or Gould's Manakin in Panama, we have had a growing number of careful studies of the nuptial habits of this family.

Although most of the sixty species of manakins still await study by enterprising naturalists, we already know enough to trace a series from those with quite simple to those with amazingly complex modes of courtship. Among the former is the brown Thrush-like Manakin, a large, atypical member of the family that lives in the dark undergrowth of humid forests from southern Mexico to southern Brazil. Slow and deliberate in its movements, it attracts attention chiefly by the beautiful whistle of three ascending notes that the male utters at a leisurely rate as he flits from bush to bush near the ground. This is his chief, and apparently his only, method of attracting the female. Without his help, she builds, often in a low spiny palm, a bulky open cup of dead leaves lined with fine rootlets or other filaments. Here, still unaided, she incubates her two black-spotted buffy white eggs and rears her nestlings (Skutch 1969*b*).

The male Blue-crowned Manakin is a very small, velvety black bird with a bright cobalt cap. His courtship station in the forests of southern Central America is an area in the undergrowth, about twenty to thirty feet in diameter, where he is to be found day after day resting on a slender, horizontal branch of a sapling, from about six to thirty feet above the ground. He is not alone but has a number of neighbors scattered through the underwood close enough to be heard, even if foliage often screens them from his view. At intervals, he proclaims his presence with various combinations of a clear little trill and a rather loud, harsh *k'wek*. Although the female often repeats the trill, the harsh note is peculiar to the male. He flaps his wings, makes looping flights from branch to branch, and from time to time descends near the ground to fly back and forth erratically through the undergrowth. If one watches carefully, he will see the male alight occasionally on a certain slender twig, often part of a fallen dead branch, only a foot or so above the ground. This is his nuptial perch, on which the green female settles when she visits him. Tracing a graceful sigmoid course through the undergrowth, the male alights on her back, where, beating his wings and uttering a low, harsh note, he consummates their mating.

Although the Blue-crowned Manakin employs a number of sounds and movements to attract a female, his courtship is primitive and simple compared with that of a related species that inhabits the same forests. This is the Yellow-thighed Manakin, a

PLATE 5. Female Yellow-thighed Manakin incubating two eggs. After visiting a displaying male to have her eggs fertilized, the female manakin attends her nest with no help from him.

short, stout black bird with a flaming scarlet head, bright yellow eyes, and lemon-colored thighs. While the Blue-crown performs his courtship activities within a small area, the Yellow-thighed Manakin centers his on a single slender, bare, horizontal branch, well above the ground and sometimes high in a great tree. This branch is both the stage where he displays and the nuptial perch where he receives the female. He draws attention by uttering not only a variety of sharp whistles but also surprisingly loud, snapping sounds, like the detonations of small firecrackers, which he seems to make by striking together the thickened shafts of his wing feathers. Single *snap*'s are given as he darts back and forth between neighboring twigs, while a whole salvo of *snap*'s is produced by rapidly beating his wings while he perches. Prominent among his displays are

FIGURE 1. Some courtship postures of the Golden-headed Manakin. *Above, left*: Male displaying in crouched posture to a female visiting his display perch. *Below*: Male slides backward toward female, displaying his scarlet thighs. *Above, right*: Two males displaying to each other (redrawn from Sick 1960).

a series of swift about-faces and a sort of dance that consists in sliding backward along his perch with his hindparts raised and his legs stretched high to show off his yellow pantaloons. After performing these spectacular antics for the benefit of a female who has come to his special perch, he makes a long circling flight and with a sharp *eeee* alights on her back.

Often four or five Yellow-thighed Manakins associate to form a courtship assembly. Each member has his own horizontal display perch, from twenty to one hundred or more feet from that of his nearest neighbor. During the long hours when no female visits the assembly, two males often rest close together on some convenient branch between their display perches. Here they go through their antics in a mild, subdued manner, as though courting each other, but without the energy they show in the presence of a female. When a female arrives, each male goes to his own station and tries to draw her to himself by vigorous acrobatics. Unchosen males do not interfere with the fortunate one whom the silent olive green visitor selects. Figure 1 shows some courtship postures of a closely related species, the Golden-headed Manakin.

Very different are the courtship habits of *Manacus*, a genus with five species widespread in tropical America. As an example, I shall take the species that I have studied most, the Orange-collared Manakin of southern Costa Rica. As the dry season begins, each brilliant black, orange, and olive green male prepares a stage for his nuptial activities by picking up in his bill and carrying away all the fallen leaves, twigs, and other loose debris on a spot of ground beneath light second-growth woods or a short way within the heavy forest. When his task is completed, he has a bare area, or "court," roughly circular or elliptical and a foot or two in diameter. If you drop a small leaf or bright flower on this patch of ground that seems to have been swept, you may see the manakin come and remove it. Usually a number of similar courts, at times a dozen or more, each belonging to a different male, are scattered through the same part of the woods, rarely as close as eight feet apart. These courts are found in the same area year after year.

Each court is situated among slender, upright stems of saplings and shrubs, two or more of which, finger thick, stand close beside it. The male manakin's most conspicuous activity is jumping back and forth between these saplings, over or around the bare area and a foot or two above it. Each brisk leap is accompanied by an explosive *snap* that carries far through the woods and seems to be made by striking together the thickened shafts of his wing feathers. A single performance consists of from two to about fourteen jumps and usually includes a descent to the bare ground, from which the bird springs up abruptly to the accompaniment of a growling *grrrt*. These bouts of jumping are too strenuous to be constantly repeated. While waiting near his court, the manakin often calls *cheeu* or rapidly beats his wings to produce a snapping roll. Even in ordinary flight, his wings make a loud rustle or *whirr*. Often, when no female is in sight, two males from neighboring courts rest amicably close together at some intermediate point.

The approach of a greenish female sends each male to his own court to shoot back and forth with resounding detonations, trying to entice her to it. Choosing one of the competing males, she joins in his dance, springing from side to side as he does. The two pass and repass each other above the court to the accompaniment of *snap*'s that seem to be made only by the male. Usually, after a few jumps, the visitor suddenly departs, but sometimes she clings to one of the upright stems beside the court and he alights on her back for sexual union. Then she goes off alone. Her slight, shallow nest is some-times in or near an assembly ground, but often it is far beyond hearing of the explosive *snap*'s. There in solitude she incubates her two mottled eggs and raises her tiny nestlings.

Male Yellow-thighed Manakins, as we have learned, sometimes direct subdued versions of their courtship stunts to each other when no female is in sight. Perhaps it was from some such beginning that the amazing group displays of the lovely manakins of the genus *Chiroxiphia* evolved. The Blue-backed Manakin of South America engages in at least two types of display. The first is a "cartwheel" dance, performed by two males who simultaneously fly backward through a vertical orbit, successively occupying the same positions in space (fig. 2). The second is a "bouncing" dance in which from one to four males bounce up and down and back and forth, often passing over each other, on a nearly horizontal limb. In the latter case, dancers may drop out, leaving only one, who mounts the waiting female. A noteworthy feature of the performances of this and other species of *Chiroxiphia* is the frequent participation, along with adults in full nuptial attire, of young males in transitional plumage, which in some other kinds of manakins is rarely seen. Blue-backed Manakins bite away some of the foliage surrounding their display perches, thereby making themselves more visible while they dance and making it easier for them to detect approaching enemies (Lamm 1948, Gilliard 1959c, D. W. Snow 1963).

A related species, the Swallow-tailed Manakin or *Dansador* of Brazil, has a charming performance in which three adult males line up on an inclined twig, all pressed close together and facing in the same direction with bodies horizontal and heads stretched forward. Beside the uppermost of the three a female or a young male stands upright. The adult male farthest from this greenish spectator rises a few inches in the air, hovers facing the group on the twig, then settles next to the motionless spectator in the space that the other two adults have made for him by sliding down the twig. After alighting, he turns to face in the same direction as his companions, the lower and outermost of whom now rises to hover and then take his place beside the spectator (fig. 3). To the accompaniment of lively

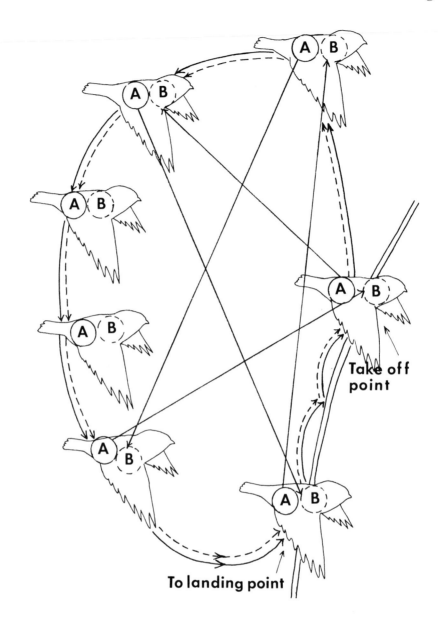

FIGURE 2. Cartwheel dance of Blue-backed Manakins. The participants are one adult and one subadult male, whose reciprocating positions are indicated by the straight lines (redrawn from Gilliard 1959).

sounds, the three performers continue this circular movement with mechanical regularity (Sick 1959*a*, 1967).

The White-ruffed Manakin is another species in which three or four adult males share the same stage—in this case a prostrate, moss-covered log in the damp mountain forests of southern Central America. With tail turned up against his back and

plumage puffed out—looking like a tiny blue-black balloon with a gleaming white patch in front—each manakin approaches the log with an undulatory flight, so slow that he seems barely to avoid stalling. Alighting on the mossy carpet, where one or more of his companions may already be standing, he bends forward with his white ruff widely spread and seems to examine the log attentively with one

FIGURE 3. Courtship display of Swallow-tailed Manakins. *Left*: Three adult males perch, bowing forward, before a watching female or young male. *Right*: The lowermost of the three males has left his perch and hovers in front of the female before alighting beside her, in the space that the two other males have made for him by sliding down the perch (redrawn from Sick 1960).

eye. Despite much watching, I did not learn what he does when a female arrives (Skutch 1967*a*).

Our survey of the courtship habits of a few of the hummingbirds and manakins permits us to draw certain conclusions that have been found to apply to other birds that gather in assemblies, or leks, to attract females:

1. The males are often much brighter or more lavishly adorned than the females of their species. In other families of birds, they may also be much larger.

2. The display movements or postures are such as to show to best advantage any striking ornament or color that the bird may possess, such as the thighs of the Yellow-thighed Manakin. This applies to all forms of display, not merely to courtship display in assemblies.

3. The members of an assembly either dwell in amity, or, if they clash, their bouts are formal and wounds are rarely inflicted. This is a condition essential to the existence of the assembly, which could not persist if its members often injured each other. Although they are rivals for the privilege of mating with the female, they are also cooperators in the endeavor to attract her to the locality where her eggs may be fertilized. The "rules" of the assembly must do justice to both of these aspects.

4. The males perform in the same place year after year. The persistence of the assembly in a traditional spot makes it easier for the females to find it.

5. The female freely chooses the male with whom she will mate, and her choice is respected by his neighbors, who rarely interfere with the nuptial act. When we recall the tension under which the unchosen males must be at this time, we admire their restraint, which is innate and reveals a long hereditary discipline, an old and well-established culture.

6. After coition, the sexes separate, and the male takes no part in nesting.

Bower-Birds

At this point, I am tempted to describe the spectacular courtship performances of Prairie Chickens, Sage Grouse, certain pheasants and cotingas, birds

PLATE 6. Male Tooth-billed Bower-bird or Stagemaker at his court. It is late in the season, and he has not removed all fallen leaves from the "stage" that he has cleared on the forest floor (photo by John Warham).

of paradise, and other neglectful fathers. But the reflection that they are less exciting to read about than to watch makes it easier to pass them by and proceed to consider a group of birds that construct elaborate stages for their nuptial displays—the bower-birds. These birds, found only in New Guinea and Australia, are not the only ones that prepare the setting for their courtship exercises, for, as we have seen, Orange-collared Manakins and related species clear a space on the ground and Blue-backed Manakins remove some of the foliage around their display perches. The Magnificent Bird of Paradise does both, clearing fallen leaves and other debris from its display area and pulling living foliage from the surrounding bushes and vines to

permit more light to fall upon his gorgeous plumage (Rand 1940). Among the cotingas, Calfbirds defoliate the tree in which they display singly or in couples (B. K. Snow 1961, 1972). But no other bird makes such elaborate stages as the bower-birds, which are related to the birds of paradise but less ornate. Some are very plainly attired, but they compensate for their lack of personal adornment by lavishly ornamenting their bowers. Gilliard (1969) has pointed out that the less colorful the bower-bird himself, the more elaborate and profusely ornamented his bower.

According to the form of their display grounds, bower-birds may be divided into stage makers, mat makers, avenue builders, and maypole builders. The

Tooth-billed Bower-bird or Stagemaker, a bird about ten inches long, inhabits tropical rain forest in a limited area of northeastern Queensland, Australia. It differs from most birds with similar courtship habits in that the sexes are alike, both being clad in brown and fawn color. As the breeding season approaches in August, each male prepares his stage by clearing all fallen leaves and other removable litter from a circle or oval of ground from three to eight feet in diameter, amid the undergrowth of the forest. Then with his strong toothed bill he laboriously cuts through the petioles of large living leaves and brings them to the bare area, where he lays them with the pale underside upward to gain maximum contrast with the dark soil where they rest. Usually they are well distributed over the court, rarely with any overlap. As the leaves wither, they are removed and replaced by others that are gathered each morning (plate 6). Discarded leaves along with litter originally removed from the display ground may form a rim one foot high around it. The Stagemaker appears to indulge in no striking antics such as manakins use, but he sings persistently from a perch above his arena, expertly imitating the notes of a large number of his avian neighbors as well as the sounds of frogs and insects. His loud, far-carrying voice guides females to his court adorned with inverted leaves (A. J. Marshall 1954).

A mat maker is Archbold's Bower-bird, which lives in mossy forests in the mountains of New Guinea, 7,000 to 12,000 feet (2,150 to 3,650 meters) above sea level. According to Gilliard (1959b), who discovered the dance stage of this rare bird, it is a mat of ferns three to five feet in diameter, decorated around the edge with piles of black beetle wings, snail shells, and blocks of resin. Above and around this stage the builder drapes a curtain of thin yellow bamboo strands and wilted ferns, hung from nearly horizontal vines on which are scatterings of black charcoal and black bark, both the size of golf balls. Here he also places some marble-sized green berries and sometimes a big whitish snail shell. The gold-crested black male who owns this bower spends much time bringing new ornaments and rearranging his treasures so that they show to best advantage. While resting in the trees above it, he delivers a wide variety of ventriloquial notes. When

these calls attract a crestless black female, he drops to his ferny mat and prostrates himself upon it. Then, holding a piece of vine in his bill and churring constantly, he crawls toward the female with his half-opened wings and partly spread tail pressed against the ferns. When he comes near her, she flies over his prostrate form, where she hovers and whips her wings violently, making a noise like tearing cardboard, then continues to a perch on the other side of the bower, toward which the male now creeps. One such display continued for twenty-two minutes.

Of the avenue builders, the best known is the Satin Bower-bird of eastern Australia. About twelve inches long, the male is clad in velvety black feathers, the exposed edges of which reflect a lovely lilac blue in the sunlight. The female is largely olive green. Although Satin Bower-birds flock through much of the year, when ready to build his bower a male in full breeding plumage separates from the group and claims a territory from which he drives other males. He gathers twigs and coarse grass and lays them on the ground to form a platform into which he sticks other twigs to make two upright walls, which are usually from ten to fourteen inches high, slightly less than this in length, and three or four inches thick. These sturdy walls stand four or five inches apart, and the space between them is the avenue, which is often more or less arched over by the incurved tips of the sticks.

Adjoining the avenue, usually at one end, is an extensive platform of sticks and grass on which the male Satin Bower-bird deposits the colorful things that he collects from the surrounding area. He prefers objects that are blue, the color of his and the female's eyes, yet he includes in his collection objects that are greenish yellow, yellow, brown, or gray. He disdains green or red things. Almost any item of convenient size and preferred color may be accepted by the Satin Bower-bird, whose collection contains a motley array of flowers, parrot feathers, berries, fragments of glass and pottery, scraps of rag and paper, land shells, pieces of snakeskin, and so forth. As a finishing touch to this amazing display ground, the Satin Bower-bird may paint the inside of the avenue's walls with charcoal, fruit pulp, stolen laundry blue, or some other colored material,

PLATE 7. The Great Gray Bower-bird is an avenue builder. The male is displaying before a female (*upper left*). He stands beside his treasure-trove with a display object in his bill, presenting his rose-lilac nuchal crest to the female, who watches him through the avenue of interlaced twigs (photo by John Warham).

which he mashes in his bill and mixes with saliva. While he applies the "paint," he holds between his separated mandibles a small wad of fibrous bark, which seems to act as a sponge and regulates the outflow of the pigment. This is one of the few well-authenticated instances of the use of a tool by a bird or indeed by any nonhuman animal. A few other kinds of avenue builders, including the Spotted Bower-bird and Regent Bower-bird, paint their bowers, but in none does it appear to be an invariable custom.

When a greenish female arrives at his bower, the male Satin Bower-bird picks up one of his colorful treasures and holds it in his bill while he displays on the platform with his hoard lying around him. The passive spectator rests at the other end of the bower or within it, looking down the avenue at the performer. Plumage glittering in the sunshine, he spreads his tail fanwise, stiffens his wings, and stretches forth his lowered head, all the while producing a rhythmic, whirring sound like that of a mechanical toy. The usual orientation of the avenue in a north-south line permits each bird to view the other without looking into the rising sun. The consummation of this superb display, which has rarely been seen, occurs between the walls of the bower. The female then lays two or rarely three eggs in the shallow, leaf-lined bowl of sticks that she has built in a leisurely fashion, and without a mate's help she hatches them and raises the nestlings.

Remarkable as are the establishments of the avenue builders, those of certain maypole builders excite our wonder even more. One who, without knowledge of bower-birds, stumbled upon one of these elegant arrangements in the dim undergrowth of a remote Papuan mountain forest might, if imaginative, attribute it to fairies or to a race of men with a more delicate and fastidious taste than most primitive humans possess. Imagine coming unexpectedly, as the Italian naturalist Odoardo Beccari did one day in 1872, in view of a miniature wigwam or conical hut, some two feet high, built around a slender sapling that stands in a level space, beneath trees so dense that scarcely a ray of sunshine penetrates their crowns. The hut is made wholly of the thin, shiny, living stems of a single kind of or-

chid (*Dendrobium amblyornidis*) that grows profusely on the surrounding trees. In one side of the wigwam is a low, wide doorway, through which is visible a cone of moss that the bird has piled around the base of the supporting sapling. Between this central core and the outer wall is a wide, circular passageway. Before the doorway the bird has transplanted living moss to make a lawn or meadow on which he has arranged vivid objects—flowers, fruits, fungi, insects—in neat piles, each containing items of the same color—red, pink, yellow, blue, or black. The bird's flowers and fruits are fresh, for, as Beccari saw, the feathered gardener promptly removes those that have withered or faded. The maker and custodian of this elfin garden is a reddish brown bird the size of an American Robin. Wholly lacking in adornment, he cannot be distinguished from the female of his kind. He has been variously called the Brown Gardener and the Vogelkop Gardener Bower-bird, the latter for his home in the Arfak Mountains in the "bird's head" peninsula of western New Guinea.

Nearly a century after Beccari drew and glowingly described the Brown Gardener's enchanting establishment, the seasoned explorer Thomas Gilliard (1969) watched from a blind the male bird's excited behavior when a female visited him. Rushing in and out of his bower, sometimes squatting or crouching in it, he emitted the most extraordinary flood of sounds, "almost seeming to be communicating in a complex avian language with the female." An accomplished mimic, the bower builder made the forest resound with the calls of a variety of birds, along with "rapping, ticking sounds and windy creaking noises too numerous and varied to describe." The female seemed more interested in her human observer than in the frantically calling bower-bird; she peered into the porthole of the blind in which Gilliard was hidden—a display of intelligent curiosity such as very few birds of any kind make.

The nest and eggs of this extraordinary bird seem never to have been described. The only member of the bower-bird family in which the male is known to attend the young is the Green Catbird, which builds no bower.

Sexual Selection

Although a single chapter can contain only a few examples of the many beautiful or fantastic courtship displays of nonpairing male birds, perhaps we have sufficient background to try to answer the question raised at the beginning: Why are so many neglectful fathers so lavishly adorned? Because, you may suggest, they remain aloof from the nest, and their bright colors are not likely to betray it to nest robbers.

Unless taken by surprise, swiftly mobile birds are well able to escape most of their enemies, so that concealing coloration is less important for their own safety than for that of the immobile eggs and nestlings that they attend. When only the female incubates in an open nest, she is often much duller than her mate, as in many finches and wood warblers; but if the nest is covered or in a hole, a brilliant male may incubate and brood, as is true of woodpeckers and trogons, including the gorgeous Quetzal. Many colorful male birds who never sit in the nest do help to feed the nestlings; but protective coloration seems less necessary for the food-bringing parent than for the sitting one. Even the dullest object attracts attention when in motion, and it behooves any bird, bright or dull, to make sure that no enemy is watching as it approaches its nest. Yet some neglectful fathers are so brilliant that even with the utmost caution their visits to the nest would unduly jeopardize it. Moreover, certain cotingas, birds of paradise, and pheasants are burdened with such long wattles or such extravagant plumage that they seem physically incapable of performing parental offices with efficiency, even if they were so inclined. Indeed, they seem handicapped in their daily activities, including their quest of food. It is certain that some of these male birds could never have become so ornate if their attendance at the nest were necessary for the survival of the species.

We have, then, discovered the condition that permits birds to acquire lavish ornamentation if any factors are operating to produce this effect, but we have not yet learned what these factors are. The most convincing explanation that has been offered is Darwin's (1871) theory of sexual selection. We have seen that in courtship assemblies the female freely chooses the male with whom she will mate. And if the females rather consistently select the most brilliant males, or those that give the most elaborate or vigorous displays, the qualities of these males will probably be transmitted to the offspring, and the species will evolve in the direction of greater beauty or more striking displays, or the two together. If these attributes are sex linked, as in many birds, they will be inherited only by the male offspring.

Outstanding beauty is by no means confined to those male birds that remain aloof from the nest; we find it in many monogamous birds who faithfully attend their offspring, and the theory of sexual selection may likewise be applicable to them. But there are two reasons why we should expect the effects of female choice, or whatever other factors produce striking coloration and ornamentation in birds, to be greater in polygamous or nonpairing species than in monogamous species. In the latter, a male fathers no more offspring than he and his single partner can nourish. In the former, a male who proves to be outstandingly attractive to the females may fertilize the eggs of a large number of them, with the result that he engenders more progeny than the monogamous male, and his peculiar qualities spread more rapidly through the population, accelerating evolution.

The second reason for the more striking coloration of nonpairing birds is somewhat harder to explain. The argument is based upon the greater likelihood of hybridization when enduring pairs are not formed, as likewise upon the capacity to survive of pure stocks as opposed to hybrids between them. Monogamous birds, as was shown in chapter 1, usually have a longer or shorter engagement period before they begin to rear a family. Presumably, if members of different species happen to become engaged, they will during this interval of intimate association discover their mistake and separate before nesting starts. Nonpairing females, on the contrary, may visit a displaying male, watch him for a short while, then invite coition, and, in such circumstances, hybridization is more likely to occur.

Accordingly, if hybridization is to be avoided, the males of closely related species of nonpairing birds must be distinguished by far more striking, easily recognized features than in the case of monogamous birds. Despite the often great differences in the color patterns and courtship displays of related species, interspecific and even intergeneric hybrids (although far from common) occur more frequently in hummingbirds, manakins, grouse, birds of paradise, and other nonpairing birds than in predominantly monogamous families (Mayr 1942). I suspect that it is chiefly the young, inexperienced individuals who are responsible for these mixtures. Once, while I watched a male Orange-collared Manakin, a female Blue-crowned Manakin approached his court. A novice in ornithology might have mistaken this greenish female for a greenish female Orange-collared Manakin. But the male that I was watching obviously did not make this mistake; he remained indifferent to her.

If hybrids between two related stocks are more successful in living and reproducing than either of the original stocks, they should gradually supplant the pure stocks in regions where they arise with some frequency. But, if they are less well adapted than the pure stocks, then the individuals with a tendency to hybridize will, in the long run, leave fewer progeny, and this tendency will be reduced in the population (Sibley 1957). Pure stocks often have special adaptations to their peculiar life conditions that are lost by hybridization, so that the hybrids are penalized in the struggle for existence. Since the less strikingly colored males are more likely to deceive females of a related species than are the more strikingly colored males, the elimination of strains with a tendency to hybridize will involve the removal of the plainer males, or those with the less distinctive displays. Thus natural selection, in the broad sense, should reinforce sexual selection in producing in the males of nonpairing species distinctive and often beautiful nuptial adornments to which the females are innately attuned.

In birds that court in assemblies, we find the conditions in which sexual selection should produce the most pronounced effects: no limit is placed on male adornment by the necessity to attend a nest; competition between males is keen and an outstandingly successful bird can beget a large number of offspring; unusual color patterns and displays are favored by natural selection because they diminish hybridization. Likewise, courtship assemblies offer to the naturalist unusually favorable opportunities to look for evidence of selection by the females, for often he can keep several males in view while a female makes her choice. Yet this evidence is not to be gathered without great patience, for one may watch the courtship stations of hummingbirds, manakins, bower-birds, and others for many hours without witnessing coition. And, especially in small birds, the courting males are often so much alike to the human eye that no ground for preference can be detected. While watching manakins, I have several times seen a female mate with the most persistent performer within my field of view, although he looked just like his rivals. Probably, in these manakins the color pattern has long been stabilized, and sexual selection now acts chiefly to improve the courtship displays and ensure frequent, vigorous performances by the males.

One of the best known of the birds that form courtship assemblies is the Ruff, a member of the sandpiper family whose display grounds, or leks, are situated in low meadows in northern Eurasia. Each lek is dotted with a number of bare circles, a foot or two in diameter, from which the flowery meadow herbage has been worn by the constant trampling of the Ruff whose display area it is. Here he runs about, darts, springs, kicks, and whirrs his wings in periodic bursts of excitement. At times two neighbors rush at each other, leaping high into the air as they strike together, but these clashes are usually momentary and rarely, if ever, disastrous. Ruffs display a degree of individual variation in coloration that is most unusual in wild birds, so that the Reeve (as the female is called) who visits a lek has a diversity of patterns from which to choose. In a gathering of more than twenty Ruffs that Selous (1927) carefully watched, the visiting Reeves showed a distinct preference for two that he regarded as outstandingly handsome. He believed that before sunrise on these frosty April mornings, while learned ornithologists lay sleeping in their warm beds, he had gathered weighty evidence in support of Darwin's theory of sexual selection.

Some biologists have objected to this theory because, they claim, it ascribes to birds and other animals a capacity to choose and an aesthetic sense that we have no ground for attributing to them. We need not suppose that the female bird who visits a courtship assembly deliberately assesses the qualifications of each claimant for her attention, weighs the good points of each against his shortcomings, and selects the one with the highest score. We must conceive of her choice as more spontaneous, just as ours often is when we are offered several attractive articles and take the one that most pleases us without first analyzing the ground of our preference. If an assembly contains only one male capable of stimulating the female to the point of acceptance, and she flies past others to reach this single adequate individual, she has no more made a choice than a lock does when it yields to the first proper key that is tried in it. But if, as seems frequently to happen, a female selects a certain male after watching others who, in this male's absence, she would have accepted, she has made a true choice.

Naturalists have commonly failed to realize all the great implications of sexual selection. They have regarded it as a form of natural selection, but it is this only in the sense that everything in nature is natural. Natural selection (an unfortunate term) acts by ruthlessly eliminating the unfit, picking them out for destruction without offering special advantages to the fittest. Sexual selection, on the contrary, picks out the most attractive individuals for reproduction and racial survival, and often also for constant companionship, without harming the less attractive. Sexual selection is positive selection, whereas natural selection is negative selection. Sexual selection seems to be the only example in the whole animal kingdom, apart from man, of the selection by individuals of individuals for their personal qualities. Otherwise, animals choose food, or habitat, or materials for nests, but seldom, if ever, favor individuals as individuals rather than simply as members of their species.

When considering sexual selection and the aesthetic sense of birds, we must not forget that other factors have been at work to give them bright colors and arresting patterns. Many of the visual features of animals that most delight us serve them for mutual recognition, are used in threat display against others of their kind, or warn predators that they are dangerous or unpalatable, as is particularly true in insects. Brilliant attire may even fall in the category of concealing coloration, as is true of the beautiful green plumage of many birds. Only a careful study of the uses that birds make of their color patterns can tell us whether these are effective in courting, in threat displays, as recognition marks, as camouflage, or perhaps for several of these purposes alternately as occasion may demand. As J. S. Huxley (1938*a*, *b*) pointed out, visual characters that are used exclusively in nuptial display, and not also in threat display, are beautiful and delicate rather than conspicuous at a distance. Or if the visual pattern serves two functions, such as recognition and nuptial stimulation, it may be boldly conspicuous as a whole and delicately beautiful in its details, as in the male Amherst Pheasant.

By an aesthetic sense, we mean the capacity to be pleasurably excited by beautiful sights, melodious sounds, and perhaps also fragrant scents. The sights and sounds that please us aesthetically do so by virtue of their intrinsic qualities; they are valued for their own sakes, not merely as signs that certain other advantages, such as food or physical comfort, are available there. When the gardens of the bower-birds first became known to the learned world, they were regarded as playgrounds and accepted as proof of the exquisite aesthetic sensibility of their makers. Later, when it was proved by observation and experiment that gardens and bowers are most zealously, although not exclusively, attended by males with active sex glands, there was a tendency to doubt that they reveal aesthetic sensibility. Yet it is but fair to remember that, in ourselves, not only does beauty excite love, but when in love we often have a heightened appreciation of music, painting, and the other arts. If we deny that birds possess the capacity for aesthetic enjoyment, then the exquisite details of plumage of many, the beautiful songs that others pour forth not only in the breeding season but at other times, and the tastefully decorated gardens of the bower-birds all remain incomprehensible to us.

CHAPTER 4

Territory and Its Significance

To have a fixed abode, a definite place for each activity and each possession, stabilizes life, promotes efficiency, and brings a sense of security. This is as true of birds and other animals as of men. And in a world crowded with creatures pressed for living space and hungry for food, the abode and its contents must be protected, especially against other animals of the same kind. Among men, protection is afforded by custom, law, and police force; only rarely in a well-governed country must a man defend his home or his land with muscle or weapons—many, indeed, seem to have lost the capacity to do so. Among other animals, the tenant must often fight for its land or at least intimidate the rival claimant with an impressive display. However, as we shall see, prior possession, especially among birds, confers a psychological or quasi-moral advantage not readily overcome by superior physical force. This need for a place of one's own, where others of one's kind may not intrude, and this readiness to defend it against trespassers, are the essence of the territoriality so widespread in the animal kingdom.

Any area, no matter how large or small, that is defended by an individual bird, a pair, or a closely knit group, or in which the resident is dominant over an intruder, is called a "territory" by ornithologists. A plot of ground over which an individual or a family of birds habitually forages but does not defend against others of its kind, or defends only in part, is distinguished as the "home range" of these birds. The home range often surrounds or includes the territory. Thus, woodpeckers vigorously defend the immediate vicinity of their nest hole, but they forage over a much wider area from which they make little attempt to exclude other woodpeckers.

Kinds of Territories

Of territories, we may recognize four principal kinds, according to their use: (a) rearing a family, (b) courtship only, (c) self-maintenance (chiefly foraging) only, and (d) sleeping only. The most complete territory serves all the needs of a pair or a family of birds during the breeding season or, in many constantly mated birds, throughout the year. In it the birds forage, court, build their nest, and rear their young, nourishing them with food found within the territory. Such a territory might be compared to a pioneer farm, supplying all the needs of a human family. From this complete territory every gradation can be found to territories that are little more than nest sites and perhaps also mating stations. The defended space may supply part of the food of the resident pair and their young, while the parents fly abroad for the remainder, often to an area where they forage in common with others of their species; it is like a home with a garden plot that furnishes only part of the needs of its inmates, who go to market or to shops for the remainder. Or the territory may be quite devoid of suitable food, as in the case of all marine birds that perforce breed on some spot of solid land but derive their sustenance from the high sea, which lacks fixed points to delimit territory. In their crowded multitudes and dependence upon distant sources of food, these sea-fowl remind us of urban people. In Palm-Chats, Sociable Weaverbirds, and Monk Parakeets, it is probable that each nesting compartment in the bulky communal structure is defended as a territory by the resident pair, so that these birds, which remain to be carefully studied, may be regarded as the apartment dwellers of the avian world.

The second type of territory is the courtship station, which may be a tree or a branch, as in certain manakins, birds of paradise, and hummingbirds; or it may be a bare patch of ground, as in some other manakins and birds of paradise, Cock-of-the-Rock, and Ruff. The male owner of the courtship post repulses intruders of his own sex but tries to attract females, which he may fertilize either on his station, as in manakins, or at a distance, as in hummingbirds. The holders of these courtship stations find their food at a distance, mostly in undefended areas, and the females whom they fecundate nearly always build at a distance. Their nests are usually well separated from each other and little is known of their territorial behavior. Our preceding chapter was devoted to birds of this class.

The third type of territory is used by a single bird for self-maintenance only, chiefly foraging, although on it he may also preen, bathe, sleep, and the like. If the bird is permanently resident, his territory may, as the breeding season approaches, be converted into a territory of the first type by the admission of a mate and perhaps some alteration of the boundaries. As was told in chapter 2, in winter the members of a pair of Common Mockingbirds, European Robins, and some other birds may occupy adjoining self-maintenance territories, which they may merge as spring arrives. But many self-maintenance territories are at a great distance from the breeding territory, held by a single male or female migratory bird whose mate may be far away. In the autumn, when the migrants come down from the north, I often hear singing by Summer Tanagers, Black-and-white Warblers, or Yellow Warblers, and excited calling by Yellow-bellied Flycatchers. These newcomers to Central America are claiming their winter territories and defying rivals. After they have settled for the winter months, each solitary individual on his own defended plot, they rarely sing. Seldom will a wintering male permit a female to share his foraging territory, as White Wagtails do in Israel, where these migrants are partly territorial and partly gregarious, depending upon the distribution of their food (Zahavi 1971).

In the communal roosts of many birds, it appears that each individual tries to occupy his preferred spot, repulsing or displacing others that desire it. This would account for some of the squabbling and shifting around that one often notices at such roosts as night approaches. Other birds sleep singly in holes or nests that they make for this purpose, repelling other individuals, even their mates, who may try to share it with, or perhaps take it from, them. This is true of the Bananaquit, which builds cozy globular nests in trees and shrubs for sleeping as well as for breeding. If one partner attempts to enter the other's dormitory, a tussle ensues, and only one remains for the night. Since the individual's dormitory is often situated in the pair's breeding territory, it might be regarded as a territory within a territory, belonging to one bird instead of two. Similarly, certain woodpeckers, such as the Red-crowned, always insist upon sleeping singly, repelling from their holes the mate and fledglings who share the breeding territory with them.

The Nature and Defense of Nesting Territories

Our present interest is in breeding territories, occupied by a mated pair, or, less commonly, by a polygamous family. Some writers, including Mayr (1935), Nice (1941), and Hinde (1956), have recognized three classes of breeding territories, distinguished by their size, whether or not they are used for mating, and whether or not they supply food. However, as we have seen, the territory may provide all or some or none of the food for its occupants; the several classes merge into each other and it is difficult to delimit them sharply. To be able to nourish the young from sources of food close to the nest is advantageous to all birds, and doubtless most try to breed as near their foraging areas as they can find adequate nest sites; but certain marine birds are obliged to nest far from their food supply.

The classical breeding territory, recognized long ago by Aristotle and made familiar to modern ornithologists chiefly by the researches and writings of the English bird watcher H. E. Howard (1920), contains a nest site situated in the midst of, or at least beside, an area that will provide most or all of the food needed by the parents and their prospective family. Arriving in the spring, the male bird sings from points in and around the margin of his chosen plot, thereby defining its boundaries, warning other males of his kind that he will resist their intrusion, and inviting a female to come and join him.

At first, the newly arrived male often claims an area larger than he needs and can defend. As more males arrive and contest his possession, he may relinquish outlying parts to them. But the more his plot of ground contracts, the more obstinately and effectively he guards it. In many species of birds, the struggle for possession of territory is carried on largely by means of competitive singing, posturing, and threats, without coming to blows. When sparring occurs, it rarely leads to the injury of either contestant, for the one who feels himself worsted takes flight, and the victor has no interest in pursuing his rival beyond his own domain. Fatal encounters between birds of the same species occur chiefly in cages, from which the defeated individual cannot escape, and are rare in natural conditions.

When two territory-holding birds fight along the frontier that separates them, it has frequently been noticed that each is dominant on his own side of the boundary. A can overcome B in A's territory, but on B's side of the often invisible line that separates the two domains B is victorious. Thus they oscillate back and forth across their frontier. The rule that birds fight most successfully on their home ground applies even to nestlings. The single nestling Red-footed Booby that occupies each arboreal nest fiercely resists the intrusion of other young, and in its own nest it can subdue bigger young from other nests (Verner 1961).

Facts of this sort suggest that in contests between birds victory does not depend solely upon brute force, which can hardly be increased or diminished by leaping or flitting across an invisible line, but that some psychological factor, something akin to moral scruples in ourselves, powerfully influences their fighting ability.

These reversals of dominance led Willis (1967) to define a territory as "a space in which one animal or group dominates others which become dominant elsewhere." In his long-continued study of Bicolored Antbirds in the Panamanian forest, he noticed that these birds, which depend largely upon foraging army ants to stir up the insects on which they subsist, freely follow the ant swarms as they pass from the breeding territory of one pair to that of another pair, and that in their own territory the members of each pair are dominant over all the visiting antbirds, including those that will dominate them when they accompany the ant horde into a neighboring territory. Submissiveness no less than aggressiveness enters into territorial behavior, and both are present in the same individual bird, to become manifest at appropriate moments. Other ornithologists have observed that the farther a bird wanders from its own territory, the more timid it becomes and the lower in the social hierarchy it descends. Thus Steller's Jays fall progressively lower in the peck order as they visit feeding stations more and more distant from their own home ground, and similar behavior has been noticed in several species of titmice and other birds (J. L. Brown 1963).

It is sometimes said that on its own territory a bird is unconquerable by another of his kind. While this is in general true, exceptions have been noticed —indeed, what generalization can one make about living nature that may not be contradicted by his next observation? On two occasions Nero (1956) witnessed pugnacious intruders evict established male Red-winged Blackbirds from their own territories; one was an old resident of six years' standing. H. Young (1956) saw male American Robins repeatedly defeat another male on the latter's own domain yet fail to wrest any part of the area from the loser. The robin's ability to hold a territory depends upon his extreme persistence in staying there as much as on his pugnacity.

In some species, the male without an adequate territory fails to win a mate. Song, however, may be of less importance in attracting a partner than in warning neighboring males that the territory is claimed and will be defended. Although thrushes are among the best and most persistent of songsters, studies of both the American Robin and European Blackbird failed to show that their songs are of primary importance in gaining a mate (H. Young 1956, D. W. Snow 1956).

After a male has secured a partner and comes to an understanding with the owners of adjoining territories, he usually sings less exuberantly, although his separation from his mate while she incubates the eggs may bring about an increase of songfulness, for he carols most freely when alone. The female may help her mate to defend their territory, and then it often happens that she confines her attack to others of her own sex, while the male fights only other males. In other species, however, each partner chases intruders of either sex.

In some kinds of birds, the female does not learn, or at least does not respect, the boundaries of her mate's territory and may build her nest in an attractive site beyond them, in a neighbor's domain. When this happens, the male will extend his holding to include the nest if he can, as sometimes happens in meadowlarks, European Blackbirds, Chaffinches, Yellow Buntings, and other birds. Some females fail to build in the mate's territory because it contains no good nest site. A female Great Reed Warbler, whose mate's territory contained only cattails, which provide inadequate support for the heavy nest, built among the reeds (*Phragmites*) in the domain of a neighboring unpaired male; whereupon her own mate extended his holding to include her nest (Kluyver 1955). When the male is unable to enlarge his territory to include the nest of a mate who has built beyond it, matrimonial complications may arise. In species that are regularly or occasionally polygynous, such as Red-winged Blackbirds and White-crowned Sparrows, the females of a single male may divide his territory between them, each repelling his other mates from the subterritory surrounding her nest, sometimes singing and fighting as though they were males (Nero 1956, Blanchard 1941).

The Size and Use of Nesting Territories

The size of the territory in which a pair of birds nests varies immensely from species to species, being smaller when it contains a nest site only than when it contains both a nest site and a foraging area and, if of the latter sort, larger for big birds than for small ones. In the more crowded parts of the huge colony of Lesser Flamingos on Lake Natron in Africa, a square yard contained four or five nests, each of which, with the immediately surrounding mud flat, was defended as a territory (L. H. Brown and A. Root 1971). In the enormous colonies of another African bird, the Black-faced Weaver, the small globular nests, each the center of a pair's breeding territory, hang almost as thickly as oranges on a tree. Both of these colonial birds find their food at a distance from their nests, in areas that are the common feeding ground of their flocks. At the other extreme, the territory of one of the larger raptors, supplying food as well as a nest site, may extend over several square miles. Even in a single species in the same locality, the size of the territory may vary greatly with the character of the vegetation and the quantity of food that it supplies. The areas used by pairs of Ovenbirds in Ontario ranged from 0.8 to 4.3 acres (0.3 to 1.7 hectares). The smallest of these territories were in stands of aspen, those of intermediate size in mixed stands of conifers and broad-leafed trees, and the largest in maple woods (Stenger and Falls 1959).

In a large area of rather uniform vegetation, all but the peripheral territories are surrounded by others. They form a mosaic or patchwork, often composed of more or less regular polygons. In fluviatile birds, such as kingfishers, dippers, Torrent Flycatchers, and Torrent Ducks, the long, narrow territories form a linear series along the watercourse and hardly extend beyond the banks.

We have evidence that birds often claim territories larger than they need to supply food for themselves and their young in a year of average abundance. On the small islands of various sizes that dotted a lake in northern Minnesota, several species of songbirds occupied territories and successfully reproduced on areas that were much smaller than the territories that they commonly defend on the mainland. On an islet only 0.04 acre (0.016 hectare) in extent, a pair of Song Sparrows raised a brood year after year, and in at least one year they produced two broods. This island was only one-tenth the size of the smallest area known to be used by a breeding pair of Song Sparrows on the mainland, yet it was so well isolated and so well supplied with food that the resident sparrows did not leave it to forage elsewhere. Yellow Warblers nested on islands 0.1 acre (0.04 hectare) in area, whereas on the mainland their territories average about 0.4 acre (0.16 hectare). Red-eyed Vireos nested on islets 0.33 acre (0.13 hectare) in extent, which is about one-half the minimum area of their territories on the mainland.

In conformity with this trend, the larger the island and the greater the number of pairs of any one of these species that bred on it, the larger the area that each pair occupied. Although a single pair of Song Sparrows might be content with an islet of 0.04 acre (0.016 hectare), when three pairs inhabited a larger island their ranges averaged 0.3 acre (0.12 hectare); when five pairs were present, each had 0.52 acre (0.21 hectare). Similar but less extensive data were obtained for the vireo and the warbler. When four or five pairs were settled on an island, the area that each utilized differed little from that reported from the mainland. Evidently these islands were no richer in food than mainland areas occupied by the same species (Beer et al. 1956).

The area actually used by a nesting pair of territorial birds and the time spent by the pair, chiefly by the male, in defending it fluctuate greatly as the breeding season progresses. In many species, these are greatest while the territory is being established, the nest site chosen, and the nest built; they diminish during egg laying, incubation, and the nestling period. After the young leave the nest, territorial boundaries may be disregarded and territorial defense neglected while the parents busily attend their wandering brood. Preparations for a second brood may be accompanied by a resurgence of territorial zeal. Thus it happens that a pair engaged in building a late nest may wrest territory from a neighboring pair already feeding nestlings. If the latter start another nest after the first pair have hatched their eggs, they may regain the lost ground.

While Black-capped Chickadees in Utah were building their nests, they used areas that averaged about 5.8 acres (2.3 hectares). From this maximum, the area actually used dropped to about 0.9 acre (0.36 hectare) during the incubation and nestling periods, then rose to about 2 acres (0.8 hectare) after their young became mobile. Similarly, the adults spent about 42 percent of their time defending their territory at the beginning, 31 percent while building their nests, 6 percent during egg laying and incubation, and 4 percent or less while attending nestlings and fledglings (Stefanski 1967). In New York, five pairs of American Redstarts foraged over and apparently defended areas that averaged about 1 acre (0.4 hectare) during the stage of self-maintenance before nesting began, but while they fed nestlings the average size of the utilized area fell to 0.18 acre (0.07 hectare). After the young left the nest, the utilized area expanded again, greatly in one case. The parents seemed to follow the fledglings wherever they chose to go (Yarrow 1970). Eastern Kingbirds, Chipping Sparrows, Blue-gray Gnatcatchers, and certain other passerines also stayed closer to the nest while feeding nestlings, but European Blackbirds foraging for late broods ranged more widely (Odum and Kuenzler 1955, Root 1969, D. W. Snow 1956). Birds that bring food to their nestlings many times each day save much effort when they can find this food close around their nest.

Thus, from unrelated investigations, we have much evidence that at the outset of the breeding season birds of a number of kinds claim territories larger than they need and will later use. They may thereby keep other individuals of their species from breeding in the most favorable habitats. How are we to interpret this? Must we recognize in birds the rudiments of that thirst for aggrandizement that drove Cyrus, Alexander, and other ruthlessly ambitious rulers to extend their sway over vast empires? If the pairs with the largest territories laid most eggs and raised most young, we could easily ascribe the drive for territorial expansion to natural selection, but we lack evidence that they are more prolific than pairs with more modest claims. On the other hand, in view of the variations in the abundance of food from area to area within the same region and from year to year in the same area, it would indeed be a refined instinct that would prompt a bird to defend a plot of exactly the size it needed to support its family. The task of delimiting such an area would baffle the most learned ecologist. Yet, if the size of the territory is not somehow related to the birds' needs, we might expect it to vary capriciously, being too small to support the brood as often as unnecessarily large. However, when they can, birds seem to exceed rather than to fall short of their actual need of land.

The large territory may provide a margin of safety for the nesting pair. Oversized in an average year and far too ample in a season of exceptional abundance, it may be none too big in a year when food is scarce. If it contains a variety of resources, such as fruiting trees or shrubs of several kinds or plants of different species that support different kinds of insects, the birds may turn to another kind of food when the supply of one kind fails. For the survival of the species in a lean year, the successful breeding of a few species with large territories is better than the failure to reproduce of numerous pairs that try to rear young with inadequate sources of food.

The Significance of Territory

Tedious as is the work of marking birds and plotting their territories, this labor is easier than discovering what the territories contribute to the welfare of birds. Ornithologists agree widely that birds hold territories but warmly dispute their significance. The debate centers around the larger breeding territories that provide some or all of the occupants' food as well as a nest site. No one, I believe, doubts that it is advantageous for a gannet, flamingo, or penguin to hold full sovereignty over the tiny plot of ground where its nest rests, or for a bird to know during the day where it will go to rest at nightfall. Let us, then, begin our discussion by listing the possible advantages of controlling a territory that is more than a nest site and supplies at least a substantial fraction of the food needed by the resident family. It may accomplish one or more of the following:

1. Promote the safety of its occupants
2. Help the resident male to win and retain a mate

3. Shield the family from interference by intruding conspecific individuals
4. Ensure a supply of food for the young and facilitate its harvest
5. Reduce predation on nests
6. Reduce the incidence of diseases and parasites
7. Promote the fullest use of available habitats
8. Help to stabilize the population and prevent excessive multiplication of the species

We shall consider these supposed advantages of territoriality in this order.

1. To become perfectly familiar with a fixed abode, its recurrent perils and every tree and shrub that may serve as a refuge when danger threatens, undoubtedly increases the security of the resident birds. Likewise, to know where food is most likely to be found increases efficiency in foraging. By associating closely with resident birds who know their home range well, some migratory birds appear to improve their own chances of survival (Odum 1942). Territory certainly promotes the holder's personal welfare. This is the only reason why single birds establish winter territories instead of continuing to wander, and it is at least an accessory advantage of the breeding territory.

2. Although it is true that possession of territory helps territorial birds to win a mate and may even be indispensable for this end, this must be regarded as a derived rather than a primary factor in the territorial habit. If this habit conferred no other solid advantage, the holding of territory would probably not have become necessary for attracting a partner. As we have learned, Cedar Waxwings, Hawfinches, and many other birds pair within the flock, then together seek their nest site. Pairing may be accomplished at a great distance from the breeding area. After pairing in winter in the southern United States, the drake Wood Duck follows his mate wherever she goes, and, when in spring she sets forth for her birthplace or her last year's nesting area, he may accompany her to a region a thousand miles from his own birthplace and equally far from the place where they met (Stewart 1959).

Although mates may be won without a territory, the possession of a definite area—to which partners become strongly attached and to which, in migratory or wandering species, they try to return year after year—definitely helps to hold the pair to-gether. In some species, but by no means all, it appears to be the strongest bond between mates, more binding than personal attachment. This seems to be true of certain migratory woodpeckers, wrens, larks, and other birds. To work with a tried partner, with whom the bird has already successfully nested, promotes breeding efficiency (pp. 278–279).

3. Although a fairly large territory affords a measure of privacy for the mated pair, "stolen matings" sometimes occur among birds that stubbornly protect their domains and their partners. Intruders may interfere with coition by the mated pair, even in a bird as aggressively territorial as the African Red-winged Starling (M. K. Rowan 1955). On the other hand, birds that nest in crowded colonies seem to preserve a satisfactory degree of monogamous fidelity. The avoidance of adultery among birds, as among humans, appears to depend more upon what we might call their morals than upon external arrangements.

Strangely enough, colonial birds often savagely mistreat, sometimes kill, neighbors' chicks that wander from, or even remain upon, their nests, as has been recorded of albatrosses, boobies, terns, and gulls. Although territorial passerine birds that trespass on each other's domains may at times cause matrimonial complications, they are more likely to cherish than to harm the helpless young. When two females of the Buff-throated Saltator, a strongly territorial bird, somehow built nests only eight feet apart in neighboring coffee shrubs, the dominant female did her best to drive away the other. Yet, from time to time she fed her neighbor's nestling and once she even absentmindedly sat brooding it for five minutes (Skutch 1954).

The privacy that territory affords appears not to be a sufficient reason for the origin of the territorial habit but at most a minor derived advantage.

4. As Lack (1968) has shown, certain methods of food gathering—such as the aerial flycatching of swallows and swifts, the seed eating of certain weaverbirds, and the fruit eating of oropéndolas and swallow-tanagers—are not incompatible with colonial nesting. Marine birds nest colonially because the safe islets and continental cliffs on which they breed represent a minute fraction of the vast oceans over which they forage. But birds that glean insects from foliage, seize volitant insects by repeated sallies from a perch rather than in sustained flight, or gather a mixed diet of insects and small fruits must make many trips to the nest each day and, consequently, find it highly advantageous to have abundant food around their nest sites. If they nested too near each other, they might deplete the supplies close at hand, and the resulting necessity to forage at greater distances would seriously impair their ability to nourish their young. The dispersion of nests by territorial behavior prevents this difficulty.

Two objections have been raised against the value of territory as a source of food. The first is that territory holders attempt to exclude only members of their own species, or in some instances also individuals of closely related species, but as a rule they make no effort to drive away other species whose diets may broadly overlap their own. The fact that numerous species inhabiting the same district take the same foods sharpens rather than diminishes the need for territorial behavior, without which attractive areas might become so densely overcrowded with nesting birds of various kinds that food would be seriously depleted. Probably the competing species will also be territorial, thereby helping to achieve the uniform dispersal of the total avian population. Moreover, although the foods taken by diverse species in the same habitat may overlap broadly, they are unlikely to be identical in kind and size, so that sharing the same foraging area with other species does not pose so great a threat to a breeding pair's food supply as does sharing it with other members of the same species.

The second argument against the value of the territory as a source of food is that after the young leave the nest, at which time their need is at or near its maximum, territorial boundaries are likely to be disregarded. This argument is clearly without weight. If territorial behavior limits to ten the number of pairs breeding in an area of woodland or meadow where twenty might otherwise have settled, there will be twice as much food for each family, whether they remain on their territories or wander widely over the whole area.

5. The scattering of nests through concealing veg-

etation most probably reduces predation upon them. Nest pillagers, like people, appear to hunt where their efforts are most richly rewarded; if they found many nests of the same kind close together, they would doubtless continue to plunder them until most had been emptied. This is the reason why birds that nest in colonies nearly always select some situation difficult of access: a forbidding cliff, a seagirt island devoid of terrestrial predators, or, as in oropéndolas, a tree with a long, clean trunk that holds its spreading crown aloof from other trees.

To avoid disastrous predation, the opposite extremes of dispersion are most effective: the organisms or their nests should be either thinly scattered and well hidden, or they should be massed in enormous numbers. In the latter case, the predators may satiate themselves with easily taken victims yet destroy only a small, biologically negligible fraction of them. This is the strategy employed by the periodical cicadas of North America, which appear above ground in myriads at intervals of thirteen or seventeen years. Similarly, the Black-faced Weaver of Africa breeds in immense, crowded colonies inadequately protected in low thornbrush, sometimes containing more than a million nests, yet it achieves a remarkably high nesting success. It is significant that both the cicadas and the nests of the weaver are available to predators for only relatively brief periods at long intervals. Were either continuously present, the predatory population might increase so greatly that the strategy of predator satiation would fail.

6. We have scarcely any evidence that the scattering of breeding birds by territorial behavior reduces the incidence of diseases and parasites among them. Birds that breed in crowded colonies appear healthy enough.

7. A bird that encounters strong resistance when it tries to establish itself in the midst of others of its kind is likely to nest elsewhere, if it can find a favorable habitat. Territoriality evidently promotes the uniform distribution of a species over suitable terrain; but perhaps, even without territorial exclusiveness, the quest for food and nest sites would accomplish much the same result. Certain tropical birds, such as the elegant little tanagers of the genus *Tangara*, avoid clumping their nests, which are rather thinly scattered through the trees; yet they exhibit little of the behavior associated with territoriality in the north, such as singing from conspicuous posts or engaging in border conflicts. Tropical birds often regulate their affairs by means much more subtle and harder to detect than those employed by birds of the temperate zones. We have no evidence that territoriality promotes the geographical extension of birds, although it may well do so.

8. The stabilization of populations of birds is probably territoriality's most important contribution to their welfare, and it is certainly that most persistently questioned, especially by those who believe that animal numbers are regulated largely or wholly by density-dependent mortality. As J. L. Brown (1969) pointed out, territoriality will begin to be effective in reducing recruitment of the population only if potential breeders remain unsettled after all the best habitat of a certain species is occupied by territory holders who will admit no newcomers. When this happens, the excess individuals will be forced to nest in marginal situations where they will, on average, raise fewer young, and perhaps themselves succumb to predation, hunger, or disease. Potential breeders of both sexes still remaining after all second-class habitats have been filled will be unable to reproduce and, pushed into unsuitable areas, will probably themselves suffer a high mortality. Thus, by territorial exclusiveness, the species will avoid building up a disastrously high population, which may seriously damage its sources of food before it crashes.

That territorial behavior does help to regulate population density by excluding excess individuals from breeding after the habitat is fully occupied has been demonstrated by a number of careful studies, among which are those of the Song Sparrow in British Columbia by Tompa (1962, 1971), of the Skylark in England by Delius (1965), of the Black-capped Chickadee in Utah by Stefanski (1967), of the Red-winged Blackbird in California by Orians (1961), and of the Red Grouse in Scotland by Jenkins and his associates (Lack 1966). Delius found that most of the Skylarks of both sexes that failed to obtain territories did not move away but tended to remain in or near his study area as a nonterritorial, nonbreeding population, which he estimated to constitute roughly 10 percent of the total population. He concluded that "the Skylark's territorial behaviour in conjunction with the other factors mentioned seems to be a powerful density-regulating mechanism, which can, when the population reaches or approaches a maximum density, exclude a proportion of birds from breeding and therefore can adjust the reproductive rate of the population as a whole." The nonbreeding birds may form a reserve, able to breed in a later year. Such a reserve may assume great importance in northern lands, where severe winters or disasters on migration periodically decimate the population. But the Red Grouse driven from the breeding territories amid the heather mostly succumbed to predation, starvation, and disease.

Even if territorial behavior never excluded any bird from nesting—that is, if late comers stubbornly persisted in demanding territory until the earlier tenants contracted their holdings and made room for them—territoriality would cause at least a slight reduction of the reproductive rate. The resulting fighting would so delay nesting that fewer broods could be raised in a limited season, or fewer renesting attempts could be made if prior efforts failed. Continued territorial fighting may cause some birds to so neglect their nests and young that the size of their fledged broods is reduced. Occasionally parent birds permit their nestlings to starve while they persistently repel invaders, as happened at a nest of Red-winged Starlings (M. K. Rowan 1955). The prevalence of such "aggressive neglect" among birds remains to be explored (Ripley 1961).

Years ago, I was given a very convincing demonstration of how territoriality can prevent reproduction. Two pairs of Masked Tityras wanted to nest in a tall dead tree that seemed to contain enough old woodpecker holes to accommodate both of them, but their territorial exclusiveness prohibited such an arrangement. Day after day, the little bare-cheeked white birds perched among the leafless branches and argued in their queer grunty voices. At intervals, one would fly at another, who always avoided contact, so that never a feather was lost. Tiring of the inconsequential dispute, the four

would fly off together to forage at a distance, but soon they would return to resume their altercation. This nonviolent contest continued for over three weeks, until the breeding season drew to a close. Where two late broods of tityras might have been reared, not a single fledgling was produced. Since tityras were already abundant in the region, this result of their quarrel was not deplorable; further increase of the population would have created more serious difficulties in the following year.

In conclusion, a territory promotes the personal safety of the birds who occupy one long enough to become thoroughly familiar with every bush and tree where they may take refuge when danger threatens. It provides an enduring bond between the male and female who become strongly attached to their territory and reside there constantly or try to return to it year after year—this is true even if the territory is hardly more than a nest site, as in colonial birds. When the territory provides a supply of food close to the nest, it is much easier for foliage gleaners, and other birds that bring meals to their nests many times each day, to nourish their young. It evidently reduces predation by permitting birds to disperse their nests through concealing vegetation, where they are hard to find. It helps to distribute the breeding population rather uniformly through suitable habitats. It tends to stabilize the population and prevent excessive multiplication of the species by excluding from reproduction those individuals left without territories after the best, and then the marginal, areas have been claimed.

Although it is true that in many species males fail to win mates unless they first obtain an attractive territory, this seems to be a derived result of the territorial habit rather than the reason for it, for in many other species pairs are formed before territory is occupied. The first necessity of any animal is to adapt its breeding dispersion to its ecology. Details like monogamy or polygamy, prompt or delayed sexual maturity, or pairing before or after choosing a territory or a nest site, will in due course be adjusted to the prevailing system.

Of all the effects of territoriality, that of helping to adjust the population to its resources is among the most important and beneficial. Ultimately, the density of most populations of animals is determined by the availability of food. In the absence of other checks to its increase, the population will be adjusted to the carrying capacity of its habitat by density-dependent mortality, or deaths in proportion to the excess of individuals. But this is a very harsh and wasteful process, condemning many animals to slow starvation or to death from the diseases that so often afflict the undernourished. It often reduces a population to a level below that to which it would fall if the individuals had reproduced less profusely, since, before they die, the starving animals consume much food that might have kept others alive through the season of scarcity. If fewer individuals had been present at the outset of the lean months, more would have survived. Moreover, as a selective agent favoring over-all fitness, density-dependent mortality is of little value, because the individuals that it removes are mostly young animals that never have a chance to mature and prove the worth of their genetic endowment.

Any system that prevents excessive reproduction makes a more stable population, with longer average longevity and less extreme contrasts between the numbers present at the end of one breeding season and the beginning of the next. Because territoriality helps to adjust the rate of reproduction to the carrying capacity of the habitat, it must be regarded as beneficent. It seems to act harshly by excluding some birds from the best habitats and perhaps forcing them into inferior areas where food is scarcer and they are more vulnerable to predation; but it mitigates the extent and severity of density-dependent mortality and prevents many individuals from being raised, by an effort that taxes their parents' strength, only to live miserably until they die prematurely.

Long-lived marine birds prevent excessive multiplication by delaying reproductive maturity for several years and laying small sets of eggs, often only one or two. Land birds of the humid tropics reproduce slowly because their sets are small and a high percentage of their nests are destroyed by predators. Among the small land birds of higher latitudes, which lay larger sets of eggs and enjoy better reproductive success, territoriality helps greatly to stabilize populations. This combination of high reproductive potential with participation in nesting

limited by territorial behavior is admirably adapted to their ecology. It permits rapid recovery when the population is greatly reduced by a hard winter or a disaster on migration, yet places a check upon recruitment when the population rises so high that all the suitable areas are occupied by breeding pairs.

Interspecific Territoriality

Territorial defense is primarily an intraspecific affair, a contest for living space between birds of the same kind. Even if a bird of a different species takes much of the same kind of food as the territory holder, it is not as a rule chased away. This tolerance of competitors has caused some ornithologists to question the value of territory as a source of food. But a little reflection should convince us that the bird who attempted to exclude every possible competitor from its territory would face an overwhelming task. A favorable habitat contains dozens of avian species whose diets overlap more or less broadly, who may eat something that any of the others might eat. The bird who tried to drive away all these others would work itself into an exhausting frenzy without accomplishing its objective. The value of the food preserved by territorial behavior must be commensurate with the energy expended in its protection. The bird that spends more energy defending its food resources than these resources yield is destined to disappear. When, by territorial or other behavior, each species takes care of its own dispersion in the nesting season, the whole avian population of the district benefits.

Although nearly always, as writers on territory have emphasized, territorial exclusiveness is limited to members of the same species, occasionally a bird consistently repels individuals, especially males, of some other species from his territory, giving rise to the phenomenon of interspecific territoriality. This involves more than trying to chase birds of many other kinds from near the nest or young, as parents of the most diverse species commonly do. It is also different from fighting for a nest hole when cavities are in short supply. Interspecific territoriality arises when, in its spatial relations if in no others, a bird of one species regularly treats those of a different species much the same as it treats other individuals of its own species. When the two species occupy the same area, their territories may form a mosaic, with little overlapping of their boundaries.

Interspecific territoriality has been demonstrated in only a few species of birds. Most of the pairs of species that exhibit such behavior are closely related, often so similar that they are designated sibling species. Some differ so little that they have been classified as different races of the same species; they look so much alike that all but the expert may have difficulty distinguishing them, in the field or even in the hand. Among these pairs of closely similar species are the Eastern or Common and Western meadowlarks, the Great-tailed and Boat-tailed grackles, the Dusky and Gray flycatchers (Lanyon 1956, Selander and Giller 1961, N. K. Johnson 1963). The mutually intolerant Red-bellied and Golden-fronted woodpeckers closely resemble each other but can readily be separated by their head markings; the Plain Titmouse and Chestnut-backed Chickadee, which also exhibit interspecific territoriality, have colors so different that they can be distinguished at a glance (Selander and Giller 1959, Dixon 1954).

What causes birds of different species to treat each other as territorial rivals? Why does one species single out another, among the many that associate with it, for exclusion from its territory? It cannot be merely visual resemblance, for many species that look very much alike occur together without interspecific territoriality. If similarity of diet were a frequent source of interspecific territoriality, we would expect this behavior to be far more widespread and to arise among species quite unrelated, as it rarely does. Apparently, as N. K. Johnson (1963) pointed out, similarity in the vocalizations and (or) postures that two species use to advertise possession of territory, and to defend it, causes these two species to treat each other as rivals and exclude each other from their domains. The song or display that arouses the defensive reactions of the territory holder when given by another male of the same species will elicit the same response when given by a male of a related species. Despite the very different appearance of the Plain Titmouse and Chestnut-backed Chickadee, their vocalizations are sufficiently alike to stir up rivalry.

Although among the grackles, meadowlarks, and

probably others of these species-pairs, the males of one species will court and mate with the females of the other, the females themselves are more discriminating. They evidently distinguish the males by signs other than those upon which the latter depend for species recognition. Since the choice of a partner rests ultimately with the females, and they almost invariably select a mate of their own species, miscegenation rarely occurs. If hybridization were frequent, specific distinctness could not be preserved where closely similar species intermingle, and interspecific territoriality would become intraspecific territoriality.

Interspecific territoriality is not restricted to the breeding season. Some of the earliest observations of this behavior were made on transient and wintering birds in Egypt. After a short migration, each Mourning Chat establishes, on the fringe of the desert, a self-maintenance territory, from which it excludes not only other individuals of its own species but also Hooded Chats and White-tailed Chats (Hartley 1949). On the Bay of Suez, Simmons (1951) noticed territorial exclusiveness among five species of chats (*Oenanthe*), between one of them and a female Blue Rock-Thrush, and likewise among migratory shrikes (*Lanius*). Because in some cases the territorial rivals differed greatly in coloration, Simmons believed that similarity in mannerisms and outline are more important than plumage in arousing interspecific antagonisms. Since these individual territories are not used for pairing or nesting, their value to their holders is evidently as sources of food.

Interspecific territoriality appears to arise most often among pairs of species derived from a common ancestor that, after a long separation, have recently come together by range extensions following climatic or ecological changes. The eastern and western populations of meadowlarks, for example, were evidently separated by the last glaciation and since its retreat have met again in central North America. The Great-tailed Grackle, widespread in Mexico and Central America, appears to have invaded the range of the Boat-tailed Grackle by moving northeastward along the Gulf coast of Texas and Louisiana in the present century. The Chestnut-backed Chickadee has still more recently extended its range into that of the Plain Titmouse in central

California. The Red-bellied and Golden-fronted woodpeckers meet only in a narrow zone in Texas and Oklahoma. During the period of isolation, the two populations of the ancestral meadowlark, or the ancestral grackle, diverged enough in a number of characters, especially those that determine mate selection by the females, to behave as distinct species when they intermingled, but they retained their territorial habits sufficiently unaltered to incite mutual rivalry.

When species of the same genus have long been associated over a wide area, they generally manage to adjust themselves to each other in a way that minimizes conflicts, including those for territory. This is aided by character divergence: those features of appearance, voice, posture, and the like that serve for recognition drift apart until there is little danger that a bird will mistake an individual of the related species for one of its own, whether in mating or in matters territorial. At the same time, the related species may diverge more or less in foraging habits, as appears to be happening in the populations of the Plain Titmouse and Chestnut-backed Chickadee in the area east of San Francisco Bay (R. A. Rowlett MS). Frequently the voices of the related species become strikingly dissimilar while their appearance remains much the same; every experienced bird watcher can name pairs of species that he commonly distinguishes by ear rather than by eye.

Interspecific territoriality is so rare in populations that have long been stable that I have never noticed a fully developed example of it among either the permanently resident or the wintering birds in Central America. The congeneric Vermilion-crowned and Gray-capped flycatchers, quite similar in general aspect but with different head markings and contrasting voices, often build their roofed nests in the same small tree, although two active nests of either of these species have not been found close together. The closely similar White-fronted and Gray-chested doves have incubated simultaneously in nests only a few yards apart. Four species of tanagers of the genus *Tangara* frequent and often nest in our garden without ever displaying mutual intolerance, and numerous similar cases have come to my attention (Skutch 1951). The rarity of interspecific territoriality indicates that it is not advantageous to birds to extend their territorial exclusiveness too widely. By behavior that disperses breeding pairs of their own kind, many species improve their adjustment to their environment; the attempt to regulate the dispersion of associated species would wastefully dissipate their energy.

CHAPTER 5

Nesting Seasons

When do birds nest? Why do they select a particular season? In Europe and North America, where most students of bird life have lived, the answers are easy. Most birds nest in spring and summer. They do so because all conditions are then most favorable: warmth, full foliage for hiding nests, abundance of food, long days for gathering it for their young. These answers, which any intelligent observer of nature could give, are irreproachable. However, the analytical investigator might wish to go into more detail to learn what degree of temperature is necessary to start the nesting of each species and in just which months its nests and particular food are most abundant. But when we turn to other parts of the earth, especially the tropics, the answers to our questions are not so obvious. At a season when balmy air, abundant flowers, and profuse verdure seem to invite every bird to sing and nest, we may find the majority of them songless and nestless. Conversely, when drenching downpours, searing drought, or, at high altitudes, chilling nights would seem to discourage breeding, we find some or even many birds incubating their eggs or feeding their young. How shall we explain these facts?

A General Survey

Most birds the world over have a single annual

nesting season, which may be short or long, and in which one or several broods may be reared. A few birds may, as species, be found nesting in every month. In Great Britain this is true of the Wood Pigeon, Rock Dove, Stock Dove, Robin, and House Sparrow. In North America, the Barn Owl, Mourning Dove, and Red Crossbill breed almost throughout the year. In none of these species, except the last, is winter nesting common, and in the coldest weather it is not likely to be successful. In the tropics, year-long breeding is more common, although far from the rule. On the island of Trinidad, it has been reported of the Ruddy Ground-Dove, Barred Antshrike, Southern House Wren, Palm Tanager, and Blue Tanager, although each of these species nests more freely at certain seasons than at others (Snow and Snow 1964). In Central America, Rufous-tailed Hummingbirds nest throughout the year, as do Rusty-margined Flycatchers in Surinam (Haverschmidt 1971). In Java, the White-breasted Mannikin and the warbler *Prinia familiaris* breed in every month (Voous 1950).

Most of the species known to reproduce throughout the year are closely associated with human homes or cultivation; this is not only because their nests are more likely to be discovered, but also because man's activities provide more or less favorable conditions for them at all seasons. Although these and a number of other species are occupied with nests in every month, no individual is known to do so; it is doubtful whether any bird could resist the strain.

A few birds are known to have two breeding seasons in a year. Perhaps the most carefully investigated example among land birds is the Rufous-collared Sparrow, which from southern Mexico to Tierra del Fuego is one of the most familiar of birds. In Colombia, on the crest of the western Andes at an altitude of 6,500 feet (2,000 meters), it nests throughout the year, with pronounced peaks in mid-January and mid-June. Each adult undergoes a complete molt, lasting about two months, twice in the course of a year, and in the intervening four-month periods it may nest (A. H. Miller 1962). Farther north, in Central America, this sparrow also has two nesting seasons, in spring and autumn, but it has not been proved that the same individuals participate in both, as they do in Colombia (Skutch 1967a).

In the Valley of El General in Costa Rica, the little Variable Seedeater has two distinct breeding seasons, the principal one lasting from May to September, in the midst of the rainy season, and a minor one in December and January, as the wet season passes into the dry. The same individual may breed in both seasons in the same nest site. In the same region, three doves—the White-fronted, Gray-chested, and Blue Ground-Dove—have two breeding seasons, the first during the drier weather from January or February to April, the second, of minor importance, in the rainy months of July, August, and September (Skutch 1954, 1964a). In Kenya, close to the equator, van Someren (1947) gathered evidence that a number of sunbirds, flycatchers, warblers, thrushes, and other birds have two breeding seasons in favorable years, one in the long rains and the other during the short rains. Each of the nesting seasons of these land birds occurs at a fairly definite time of year and is ultimately linked to the annual course of the sun; but certain marine birds, to which we shall presently return, breed at less than yearly intervals, with no evident relationship to any astronomical cycle.

A few birds, as individuals, nest only at intervals greater than a year. Most unusual is the breeding season of the King Penguin, which can nest successfully only twice in three years. Since it incubates its single egg for nearly eight weeks, feeds its nestling for ten to thirteen months, and needs time to fatten at sea, molt, and court, it can do this by laying early the first year, later the second year, too late for success in the third year, and in the fourth year beginning a new cycle with early laying (Stonehouse 1960). With an almost equally long period of nestling dependence and an incubation period longer than that of the penguin, the larger albatrosses—the Royal and Wandering—lay only in alternate years, each female depositing her single egg before the young of the preceding season have left the colony. Likewise frigatebirds attend their young for so many months that they can nest successfully only in alternate years. Among raptors, the African Crowned Eagle continues to feed its single young so long that it is not able to breed in the following year, and the same appears to be true of the Harpy Eagle of tropical America (L. H. Brown 1966, Fowler and Cope 1964).

Whether long or short, the bird's nesting season must fit into a year (or rarely a longer interval) in which other indispensable, life-sustaining activities cannot be omitted, and only by considering all these activities together can we hope to attain a true understanding of breeding seasons. Some of these activities can be carried on simultaneously, others are mutually exclusive. With few exceptions, birds can eat, bathe, and preen at all stages of the nesting cycle. Breeding and molting, however, are for most birds mutually exclusive; if they undertake to breed in the midst of a molt, it is suspended, or at least retarded, until they have finished nesting, the reason being that to form new feathers requires food and energy, which would have to be diverted from the care of the young. Yet, if a female is liberally fed by her mate while she sits continuously or almost continuously on her eggs, she may molt while she incubates, as happens in certain hornbills and raptors. Even without being fed, Snow Petrels of both sexes molt both body and wing feathers while nesting in Antarctica—a surprising situation (Maher 1962). Migratory birds must store fat to fuel a long journey, and they cannot well do this at the same time that they feed their offspring. And, obviously, they cannot nest while they are migrating.

The birds that most successfully fit all these obligatory activities into their year, performing each at the most appropriate time, will survive and multiply more than those that do so poorly; so that natural selection will tend, for each species, to make the best possible fit. Although in evergreen forests good sites for nests, and materials for building them, seem to be adequate throughout the year, in deciduous woodlands, very arid regions, and marshes, lack of sites, concealment, or materials may make breeding impracticable at certain seasons. Eggs are least likely to freeze or spoil in warm, dry weather, so that we should expect incubation to be performed chiefly in such periods. The heaviest demand is made upon a bird's ability to find food while it is feeding dependent young, so that they should hatch at, or shortly before, the season of greatest abundance. After breeding, most birds renew their feathers, a process that requires adequate food; thus, an interval when food is not too difficult to procure must be reserved for the molt. This is especially necessary for grebes, flamingos, geese, ducks, and other birds that become flightless while molting.

Pitelka (1958) believed that a northern race of Steller's Jay used the most favorable season of the year—midsummer when food seemed most plentiful —for molting after early nesting. The molt, he pointed out, could not be omitted from the birds' annual cycle, whereas pairs of long-lived birds can skip breeding for a year without jeopardizing the existence of their kind. Although this may be true of this particular jay, we would expect most birds to use the period of greatest abundance for rearing their young, rather than for molting. When feeding young, each parent must supply enough food to form one or more bodies about equal to its own weight; whereas when molting, it builds feathers weighing only 5 to 10 percent of its body weight, rarely more, in an interval that is often longer than that of nestling and fledgling dependence. Nevertheless, it is imperative that a bird replace its worn feathers—which serve as house and raincoat—in a period of adequate food. While molting, a Brown-headed Cowbird's expenditure of energy, as measured by its oxygen consumption, may increase by

as much as 24 percent. About half of this increase is used for feather formation and the other half for maintaining body temperature while its thinner plumage gives poorer thermal insulation (Lustick 1970).

Since the seasons were not made for the birds, the birds must adjust their lives to the seasons. It will sometimes happen that, to make the best fit, some compromises between the several stages of the life cycle are necessary. Thus, to form eggs, no less than to nourish young, requires extra supplies of food, to procure which the female may have to delay laying. Consequently her nestlings will not hatch at the time most favorable for rearing them, resulting in less successful reproduction (Perrins 1970). Sometimes, to exploit the richest supplies for its young, a bird must lay and incubate in unfavorable weather, even amid snow or frost, as happens not only at high latitudes but also on lofty tropical mountains, as we shall presently see. Or it may have to refrain from breeding at a period quite appropriate for this activity, because it could not raise its young to independence and allow time for them and itself to molt, or to lay on fat for migration, before the advent of unfavorable weather.

Those aspects of the total environment that make it profitable for a bird to breed when it does are often known as the "ultimate factors" determining the nesting season. With some knowledge of a bird's habits and food, on the one hand, and of the resources of its habitat, on the other, it is often possible to point out what these factors are. We now have abundant evidence that whenever possible the breeding seasons of birds have been adjusted by natural selection to bring the young into the world at the time most favorable for rearing them, which as a rule is the time when their food is most readily procured. But to accomplish this, the birds must often regenerate their reproductive organs, build nests, and even incubate eggs at a season that is far from favorable. The Emperor Penguin, for example, incubates in the worst weather in the world, the height of the Antarctic winter, in order to get its slowly developing young to sea when the pack ice breaks up. Accordingly, we must look for some timing device—some stimulus or releasing mechanism —to prepare the bird internally and thus set it to courting, nest building, and laying far enough in advance of the favorable season to ensure that its young will hatch at the appropriate time. In large birds, especially, this may be very far in advance. The events or conditions that start the reproductive process at the appropriate time are known as the "proximate factors." Often these are more puzzling than the ultimate factors, especially in rather uniform tropical climates.

Nesting Seasons in the Temperate Zones

To obtain uniformity in the study of breeding seasons, we must adopt a consistent procedure for dating nests. The best practice is to attribute each nest to the month or week when the first egg was laid in it. In the North Temperate Zone, the peak of laying follows the northward march of settled mild weather. Accordingly, it is earlier in the south than in the north, and in maritime rather than in continental climates. In the United States, many species nest earlier in Florida and Texas than in New England or the Great Lakes states; in Europe, most species start breeding in Britain several weeks earlier than they do in the center of the continent at the same latitude. In years when spring comes early, they start sooner than when it is belated. Taking the North Temperate Zone as a whole, April, May, and June are the chief months for laying, and many birds are still feeding nestlings in July and August. In the South Temperate Zone, for which most of our data comes from southern Africa, the main breeding season extends from September to December or January (Moreau 1950).

As the sun swings northward toward the Tropic of Cancer, the North Temperate Zone, in contrast to certain other parts of our planet, shows the simultaneous improvement of all factors that might promote breeding. In these circumstances, it is difficult to decide which of the favoring factors is most influential. We know that increasing day length stimulates the maturation of the reproductive organs of birds. Might it not be that, as a factor regulating breeding, day length overrides all others, compelling all birds to nest as it increases, irrespective of the special needs of each? Or is perhaps rising temperature the compelling factor? The consid-

eration of certain species whose breeding seasons differ substantially from that of the majority, the exceptionally early and exceptionally late nesters, should throw light on these questions.

The very early nesters in the north include certain birds that depend heavily upon tree seeds, especially those of conifers, to nourish their young. Among these are crossbills, whose bills are highly specialized for extracting seeds from cones. In the North American and Eurasian forests where they breed, crossbills often lay in February, or even January, when prevailing temperatures are far below freezing, and their little open nests are surrounded by snow; they continue to breed through the spring, summer, and autumn if enough seeds remain (A. M. Bailey et al. 1953, Haartman 1969b). Pine Siskins also eat seeds, including those of conifers, alder, birch, and likewise herbaceous plants, and they start to nest in early April or sometimes March, even in Canada where snow is still frequent (Bent et al. 1968).

A number of corvids breed long before advancing spring brings mild weather. In the western United States, Clark's Nutcrackers, which nourish their young with seeds of the ponderosa pine, begin in early March, often amid snow, and continue until May (Mewaldt 1956). The Piñon Jay, also of western North America, includes in the diet of its young seeds of the piñon pine that it has cached the preceding autumn, and it nests chiefly from February to June (Balda and Bateman 1971). The Thick-billed Nutcracker, which feeds its young hazelnuts that it has buried in the fall, lays its eggs in snowy March in Scandinavia (Swanberg 1956). Stored seeds, whether of pine or hazel, germinate or decay as temperature rises, making it imperative for birds that depend on them to use them while they are still in "cold storage." Warm days cause the cones of certain conifers to expand and scatter their winged seeds, and birds able to extract them directly from beneath the scales can forage more efficiently before this occurs. Disregarding cold, and not waiting until days become longer than nights, these eaters of seeds begin nesting when their food supply is assured.

Other corvids that nest very early, often in freezing weather, are the Common Raven, Rook, Yellow-billed Magpie, Gray or Canada Jay, and Siberian Jay. These birds of almost omnivorous habit manage to find enough food for self and offspring before the vernal awakening of vegetable and insect life. In the northern United States, the Great Horned Owl may lay its eggs in January, the Barred Owl in late February. Probably for these nocturnal raptors hunting is somewhat easier before spring vegetation provides better concealment for the small mammals on which they prey. Among diurnal raptors, Bald Eagles and Red-tailed Hawks likewise breed early, in February or March in the northern United States.

More surprising is the early nesting of the Horned Lark, which even in the northern United States lays its eggs from March, or rarely February, to June or July. Its earliest nests are often destroyed by snow. This widespread species, which breeds from Arctic coasts to tropical highlands and even lowlands, prefers to nest on barren ground. In early spring, its food consists largely of fallen seeds. Both of these requirements are met early in the year, before rising temperatures have caused the last year's seeds to germinate and cover the ground with vegetation. An even earlier nester is Anna's Hummingbird, which in California begins around the winter solstice in December and continues until August. In the coastal belt, temperatures are sufficiently mild to permit many plants to bloom in winter and very early spring; even in the interior, where this tiny hummingbird has been found incubating its eggs during a light snowfall, flowers are not lacking at this season. In addition to sipping nectar, this hummingbird picks minute insects from bark and foliage, catches those attracted to sapsucker pits in the bark of trees, sips the sap that oozes from them, and plunders spider webs, so that evidently it does not lack food even during the shortest winter days (Bent 1940, Pitelka 1951).

In contrast to these hardy birds that appear to disregard cold if other conditions are favorable for breeding, for many other species, perhaps the majority of temperate-zone nesters, the time of laying is influenced by temperature. In southern England, an outburst of laying by Blackbirds and Song Thrushes has been noticed five days after the mean

temperature in mid-March rises above 40° F (4.5° C) (D. W. Snow 1958). In Ohio, Song Sparrows started to lay in numbers five days after three days with an average temperature of 73.2° F (22.9° C) (Nice 1937). Turning to a quite different bird in a very different environment, Rockhopper Penguins nesting on islands in southern oceans have been found to begin laying ten days earlier for each increase of 1.8° F (1° C) in the mean temperature of the surrounding sea (Warham 1972).

In North America, Cedar Waxwings and American Goldfinches start to nest later than other birds. The waxwing, which breeds chiefly from the second week of June to August, nourishes its young, after they are a few days old, chiefly with berries, which become more abundant as summer advances. Putnam (1949) found that nests begun after June 20 were more successful than those started earlier. In the northern United States, the eggs of goldfinches rarely appear before July and the peak of laying is late July and August. Goldfinches raise their young chiefly on seeds of composites, including thistles, which ripen most profusely in late summer and autumn. The European Goldfinch, also an eater of small seeds, likewise breeds late. In Great Britain, the Spotted Flycatcher, which gathers flying insects for its nestlings, has its peak of laying seven weeks later than the Robin, which nourishes its young largely with caterpillars. The flying insects, of course, follow the larvae as summer advances (Lack 1950).

Most North Temperate species have a single breeding season, which may come early or late. Two separate breeding seasons in the course of a year, such as regularly occur in certain tropical birds, have been noticed in rather exceptional circumstances in a few northern species. The Piñon Jay, as we have seen, usually nests early. But in 1969, when food was scarce in the spring months and few pairs nested, many did so in August and September, when piñon seeds and insects became plentiful after late summer rain (Ligon 1971). Although Tricolored Blackbirds normally lay from April to mid-June, they sometimes do so in autumn; but, from thousands of nests built at this season, scarcely any young were reared (Payne 1969a). The Gambel's

Quail has also nested after violent summer rain in a year when spring rains failed in the Mohave and Colorado deserts, and fall breeding has been recorded of a few other Californian birds (Orians 1960). Autumnal nesting is also reported of a number of grebes, herons, ducks, and other aquatic birds in a district of the province of Santiago del Estero, northwestern Argentina, where receding floodwaters leave multitudes of easily accessible fishes and frogs in May and June, when these birds are feeding their young. In other parts of Argentina, these same species breed in the spring and summer of the Southern Hemisphere (Olrog 1965).

From the foregoing examples, it is evident that, although most temperate-zone birds begin to breed in the mild, lengthening days of spring and early summer, they may do so in the brief, bleak days of winter or when autumn days are becoming shorter, if their particular food is abundant in these seasons.

Nesting Seasons in the Arctic and in Antarctica

In the harsh climate of the Arctic, the breeding season begins late and is of short duration. The latitude at which the last stunted forests give way to tundra varies with the region, being substantially higher, for example, in Eurasia and northwestern North America than in northeastern North America, but it corresponds roughly with the isotherm of 50° F (10° C) for the warmest month. Above this limit, the temperature not infrequently falls below the freezing point even in July, and days when the thermometer rises above 60° F (15.5° C) are rare. The short, chilly spring and summer last only from about late May to September.

Few migrant birds arrive in the Arctic before the second week of May, when much snow still whitens the landscape, and few lay their eggs before the last days of May or the first days of June. Much of the food that helps to sustain them when they first arrive, and helps to form their eggs, consists of herbage, buds, seeds, and even berries from the preceding year, preserved by refrigeration through the eight or nine long months of winter. Since the favorable season is so short, it is of utmost importance for the small migratory birds to start nesting soon after they arrive. Thinly scattered over vast expanses of tundra, many of these birds avoid the

protracted territorial disputes in which species that breed in middle latitudes frequently engage. Some of them, including wheatears, redpolls, and perhaps Lapland Longspurs, save precious days by reusing old nests preserved from earlier years by winter's cold, a practice impossible in warm, damp climates, where empty nests soon deteriorate.

The numerous species of shore birds lay three or four eggs, as they do at lower latitudes, but the few small passerines that come so far north to breed often produce sets of four or five to seven. To prevent freezing, the eggs must be covered as soon as they are laid, especially at night. This early start of incubation results in asynchronous hatching, with nestlings of varying ages in the same brood. Probably because of crowding as they grow bigger, young passerines often leave the nest before they can fly. One wonders how many of their scattered brood the parents succeed in rearing to independence. The Arctic summer is so brief that second broods, frequent at lower latitudes, seem to be attempted rarely and are probably even more rarely successful. To be ready for the long southward journey when the weather worsens in August or September, adults must begin the postnuptial molt as soon as their single brood can do without their constant care, and it seems that adults sometimes abandon late nests to join molting flocks. By October, only the hardiest birds remain amid snowy blasts (Soper 1946, Wynne-Edwards 1952, Irving 1960, J. Brown 1970).

In some years, widespread failure to breed has been noticed among Arctic birds, both passerine and nonpasserine. The causes for this appear to be various: delayed arrival on the Arctic nesting grounds because of unfavorable weather along the way; belated spring in the Arctic; a winter with scant precipitation followed by a dry spring, resulting in the early drying of the ponds and marshes that supply much food for Arctic birds, especially insectivorous species; and, for raptors, a low point in the cycle of abundance of lemmings, which are a principal food. The testes of male nonbreeders often mature normally, but the ovaries of the females tend to remain underdeveloped (A. J. Marshall 1952, Hobson 1972).

One may ask why so many birds undertake such

long, hazardous journeys to reach a land where the brief, cool, fickle summer does not always permit successful reproduction. Do they find here freedom from the predators that destroy so many nests in warmer lands? Snakes, those most persistent pillagers of nests, do not harass them here, but other predators abound. Arctic Foxes, weasels, Gyrfalcons, Peregrine Falcons, Snowy Owls, jaegers, and gulls prey insatiably upon birds and their nests. Since, in the absence of trees and tall shrubs on the tundra, many nests are built on the ground, they are in danger of being trampled by the larger quadrupeds and even by Arctic Hares. On the favorable side, insects are abundant in good years, and the continuous daylight above the Arctic Circle at the height of the nesting season permits parent birds to spend long hours gathering food for large broods. Despite the risks, many kinds of birds do succeed in maintaining large populations by breeding in the Arctic. Such is the pressure of life upon the earth's resources that, wherever any kind of organism can propagate in adequate numbers, there it will persist in living, however inhospitable the habitat may appear to us.

Except for the great cold, conditions in Antarctica are quite different from those in the Arctic. The center of the latter is occupied by an icebound, landlocked sea; the center of Antarctica is a high, glacier-covered continent, separated from other continents by stormy oceans. The Arctic tundra yields much food for birds during the more clement months; the Antarctic continent, which supports only two species of seed plants, supplies scarcely any, except in so far as birds prey upon other birds that derive all their nourishment from the sea. The fourteen species known to nest on the continent and its bordering ice shelves are all marine: four penguins, five petrels, a cormorant, a gull, a tern, a skua, and a sheathbill, the last being the only bird with unwebbed feet that reaches the shores of the Antarctic continent. Passerines are wholly absent from the continent and, except for the Antarctic Pipit, from neighboring islands. The Emperor Penguin breeds in the frigid darkness of the Antarctic winter. All the others, as far as is known, lay their eggs and rear their single broods in the warmer (but still chilly!) part of the year, when the zone of pack ice surrounding the continent is at its narrowest and open water, in which parents can find food, is closest to shore.

Nesting Seasons in the Tropics

At middle and high latitudes, the year is divided into a cold winter when days are short and a warmer summer when days are long, with intervening periods of transition known as spring and autumn. At low latitudes where seasonal differences in day length are absent (as at the equator) or slight, and monthly fluctuations of temperature of small magnitude, the chief contrasts that the year presents are the rainy season and the dry season. The dry season may be so short and relieved by such frequent showers that forests retain their foliage and fields their verdure; or it may be so long and severe that many trees become leafless, parched fields turn brown, and even semidesert conditions prevail. In mountainous tropical lands the two extremes may lie not far apart, separated by a range that intercepts the moisture-bearing winds. Most parts of the tropics have a single rainy season and a single dry season, which on opposite sides of the equator occur in complementary months. The rains follow the sun, sometimes tardily, with the wet season of each hemisphere coming chiefly when the sun is on that side of the equator. The intervening equatorial belt, a few degrees wide, has two rainy and two dry seasons, thereby effecting the transition between the wet seasons that occur at opposite times of the year. It is in relation to these alternating seasons of abundant and sparse or absent rainfall, and their effects upon vegetation, insects, and other sources of food, that we must consider the breeding seasons of tropical birds.

Although nearly everywhere in the tropics, except possibly in the most arid regions, some plants flower in every month, the dry season is preeminently the time of bloom. The advent of sunny skies while the earth still contains much moisture from long rainy months elicits a prompt response from herbaceous plants, which then adorn fields and roadsides with a profusion of color. As the drought increases, bright flowers become much rarer; drying herbs shed seeds that nourish pigeons and other ground gleaners. Deep-rooted trees continue to blossom later, and some that make the most spectacular displays flower in the driest weather. When the dry season is prolonged, the number of arboreal species in flower drops toward its end.

The first light showers of the returning rainy season bring a swift response from other arboreal or arborescent species, so that a second maximum of flowering occurs at this time, while those that bloomed earlier are ripening their fruits. Some of these woody plants that flower with returning showers, especially the abundant melastomes, yield ripe berries rather promptly, so that the early months of the rainy season are a period of abundance for fruit-eating birds. This period also brings a great increase of insects, as is evident to anyone who leaves a light exposed during the night. The person who feeds birds in his garden receives a convincing, if indirect, proof of the greater abundance of native foods early in the rainy season: although birds are nesting and feeding their young all around, the bananas they so eagerly sought at other times now lie largely neglected on the board.

In northern South America, and even more in Central America, the northward departure of migrants about the time the rains begin, in late March, April, and early May, leaves more of this increasing food supply for the resident nesting birds. A. H. Miller (1963) believed that in the western Andes of Colombia this reduced competition for food was a principal cause of the greatly increased breeding of the local birds at this season. My impression is that this is only a minor factor in parts of Central America where food appears to become superabundant in these months. At Amani in Tanzania, the peak of nesting comes in the months of the northern winter, when migrants from Europe and Asia are present in great numbers (Moreau 1936b, 1937).

The rainy season is above all a period of luxuriant vegetative growth rather than of flowering. As it continues, the fruits so plentiful at its onset diminish in abundance and insects also appear to become fewer. Although there is never an absolute lack of ripening fruits and active insects, the latter part of a prolonged wet season is a time of relative scarcity for birds, who now eagerly seek the food provided by their human friends. This lean period continues until the months of diminishing rain when flowering

and fruiting increase, as though to anticipate the approaching dry weather (D. W. Snow 1962a, Fournier and Salas 1966, Leck 1972).

In spring at extratropical latitudes, where daylight lengthens rapidly and with astronomical precision all around the earth, breeding seasons show a certain uniformity over great areas, although their exact delimitation and underlying causation present problems enough to the serious investigator. In the tropics, where nesting is more closely linked with rainfall and its effects than with day length and temperature, and where both the date and amount of precipitation may vary widely within short distances, especially in mountainous regions, the problem of breeding seasons is more complex and generalizations are precarious. Nevertheless, I shall hazard this one, to which exceptions will be noticed: In the rainier parts of the tropics, the principal breeding season starts as the dry season passes into the wet and reaches its peak early in the rainy season, when the greatest number of species have the greatest number of nests. Some birds start a few weeks before the rains, some a few weeks after they begin. The majority avoid nesting in the driest or the wettest months. Where two rainy seasons occur, the curve of nesting may or may not show a double peak. In extremely arid regions, the birds wait for the rains.

A tropical region for which a large number of nesting records have been gathered over many years is the Valley of El General on the Pacific slope of southern Costa Rica. In this mountain-rimmed valley at 9° north latitude and 2,000 to 3,000 feet (600 to 900 meters) above sea level, the annual rainfall ranges from 90 to more than 157 inches (2,300 to 4,000 mm) with an average of about 118 inches (3,000 mm). The dry season, relieved in some years by occasional showers, extends from December or January to March. April is a time of increasing rain; May a month of usually sunny mornings but frequent afternoon deluges that may make it the rainiest month of the year; the wettest and gloomiest months are usually September, October, and November, after which the rains taper off in December. Here the number of nests in which eggs are laid rises gradually from January to March, reaches a maximum in April and May, declines to a much

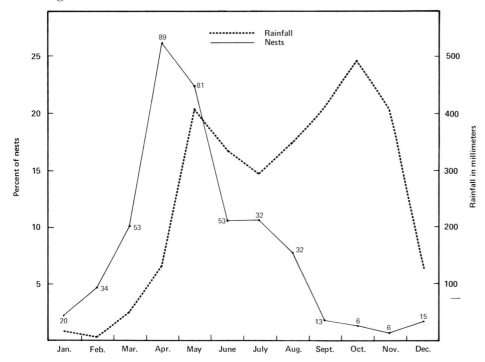

FIGURE 4. Rainfall and nesting in the Valley of El General, southern Costa Rica, 9° 15′ to 9° 30′ N. lat., 2,000 to 3,000 feet (600 to 900 meters) above sea level, a region with a single long rainy season. The broken line shows the average monthly rainfall in millimeters. The continuous line shows the percent of 1,357 nests of 140 species in which eggs were laid in each month. The numerals along the graph indicate the number of species represented in each monthly total (redrawn from Skutch 1950).

lower level in June, July, and August, and reaches a minimum in the last four months of the year (fig. 4).

The height of nesting activity in April, May, and June is comprised largely of birds whose feeding habits are not highly specialized, among which are tanagers, finches, honeycreepers, manakins, and flycatchers that include both fruits and insects in their diet, as well as more exclusively insectivorous birds, such as wood warblers, swallows, antbirds, woodcreepers, and certain other flycatchers. Birds with special diets or special modes of nesting often breed in different seasons. Those that nest earlier than the majority include some of the hawks, vultures, and larger woodpeckers, whose food is not so dependent on the rains. Kingfishers and other birds that nest in river banks do so in the dry early months, when their burrows are in less danger of being inundated or washed out by floodwaters. Doubtless, too, kingfishers find it easier to catch fish when rivers are

low and clear than when they are swollen and turbid. Nightjars that lay their eggs directly on the ground also do so early, while the earth is fairly dry. These groups of birds—the raptors, vultures, large woodpeckers, kingfishers, and nightjars—tend to breed early in other parts of the tropics, as well as in El General.

Probably to take advantage of seeds falling from drying vegetation, before these germinate with the returning rains, certain doves of clearings and second-growth thickets nest chiefly toward the end of the dry season, before the majority of other birds begin. Then, after an interval of quiescence while most other birds are breeding, they resume nesting on a reduced scale after the other species have finished. Thus they have two breeding seasons in a year, as already told. This is not true, however, of the Ruddy Quail-Dove, which forages on the ground in the forest and nests when most other birds do.

Small birds that feed their nestlings largely on

grass seeds, especially finches of the genera *Sporophila* and *Tiaris*, delay breeding until seeds are set on the grasses that spring up when the rains return. In the Valley of El General they nest chiefly from May to August, a month or two later than most of their neighbors. Except for the Yellow-faced Grassquit, whose roofed nests shelter eggs and nestlings from October's downpours, these seed eaters tend to avoid the latter part of the wet season. However, a few individuals lay again in December or January, before the herbage dries in the increasing sunshine.

It is above all the nectar sippers who demonstrate that the breeding seasons of tropical birds are regulated by the food supply. In El General, the tiny Bananaquit visits flowers far more regularly than do other honeycreepers of the region; and it nests practically throughout the year, but seldom in March, when drought has reduced flowering to a minimum, or in April and May, when vegetation runs to verdure and fruiting more than to blooming. The covered, rain-resisting nests of the Bananaquit hold eggs chiefly from June to February, thus including the months of least breeding by other passerines. At low and middle altitudes in Central America, hummingbirds lay their eggs throughout the year. The modestly attired hermit hummingbirds, which fasten their nests with cobweb beneath palm or banana leaves for shelter, prefer the rainier months for breeding; whereas those hummingbirds with open, exposed nests choose drier weather. Yet nests of both kinds are active during October's hardest deluges. Taking the hummingbirds of El General as a whole, the peak of nesting comes in December and January, when scarcely any passerines, except the Bananaquit, are breeding.

At higher altitudes in Central America, the divergence between the nesting season of the nectar drinkers and the rest of the birds is even more pronounced. In the highlands of western Guatemala, the rainy season lasts from about mid-May to mid-October; a dry season of increasing severity occupies the remainder of the year. In 1933, on the Sierra de Tecpán above 8,000 feet (2,500 meters), frost whitened open fields after most clear, windless nights from early November to early April. The great majority of the birds laid their eggs and reared their nestlings in the six or seven weeks between the last frosts at the beginning of April and the onset of cold, hard rains in mid-May. In contrast to the birds of lower altitudes in tropical America, they nested chiefly in the driest part of the year, but they thereby avoided both frosty nights and the chilling rains (fig. 5). Many of the trees, especially the abundant oaks and alders, had renewed their foliage during the preceding dry months and their fresh leaves evidently supported abundant insects, while deep-rooted shrubs still provided berries. The birds that delayed to lay until late May and June were chiefly ground foragers, who waited until renewed moisture had activated the small invertebrates in and beneath the ground litter, with which they nourished their nestlings.

Although one does not expect tropical birds to nest in the driest time of the year, a rather similar situation has been reported of another somewhat elevated region in Zambia, formerly Northern Rhodesia. Here the majority of the small birds begin to nest in September or even late August, at least two months before the rains start, and they have finished by the time the wet season is established. However, many trees renew their foliage, flower, and fruit in the driest months; herbage springs up on ground blackened by fires; and insects multiply on all this fresh verdure. Thus the birds have no lack of food. In this region, as in neighboring Malawi, formerly Nyasaland, vegetation has become adapted to growth under arid conditions, as it has not in certain other tropical areas, and the birds have taken advantage of this situation to nest in warm, dry weather (Moreau 1950).

To return to the Sierra de Tecpán, after July, I found no freshly laid eggs of any kind until October, when the hummingbirds started to breed as increasing sunshine called forth a great profusion of nectar-yielding flowers. Their peak of laying came in November, and they continued to nest until February when heavy nocturnal frosts and increasing drought made flowers scarce. Simultaneously with the hummingbirds nested a single passerine, the Cinnamon-bellied Flower-piercer, which, like them, varies a diet of nectar with minute insects caught in the air. From March to July, when all the other passerines were breeding, I found no evidence of nesting by the flower-piercers, or by any of the hum-

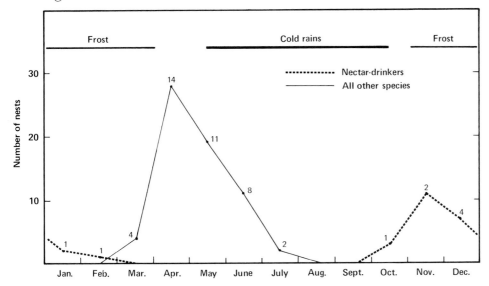

FIGURE 5. Climate and nesting on the Sierra de Tecpán, west central Guatemala, 14° 45′ N. lat., 7,000 to 10,000 feet (2,130 to 3,050 meters) above sea level, in 1933. The heavy horizontal bar denotes the season of frequent, cold rains. The light horizontal bars indicate the drier period when frosts formed on open fields on clear nights. The continuous line shows the months when eggs were laid in 64 nests of 24 species, excluding nectar drinkers. The broken line shows the monthly distribution of 24 nests of 4 species of nectar drinkers—3 hummingbirds and the Cinnamon-bellied Flower-piercer. The number of species represented in the monthly totals is indicated by the numerals along the lines (data from Skutch 1950).

mingbirds. Similar feeding habits caused simultaneous nesting at a peculiar season by unrelated birds. The thin, penetratingly cold air of frosty nights nearly 10,000 feet (3,000 meters) above sea level did not prevent breeding when food was abundant (Skutch 1950).

One of the most thorough studies of the nesting seasons of tropical birds was made in Trinidad by the Snows (1964). Here, at a latitude about one degree farther north than the Valley of El General, the march of the seasons is quite similar. In the Arima Valley in the north of the island, the annual rainfall of about 100 inches (2,540 mm) comes chiefly in the single long wet season that lasts from May until the year's end. The dry season, January to April, is mitigated by sporadic showers that are sometimes heavy. Gradually increasing from the first of the year, nesting by all species of land birds reaches its maximum early in the rainy season, in June, about two months later than in El General. The breeding of a number of species is substantially more prolonged than in El General, beginning earlier and lasting longer, so that the nesting of the is-

land's avifauna is spread somewhat more uniformly over the year, with a less pronounced peak. This may be because the Snows' records came from lower altitudes, from about 1,800 feet (550 meters) down to sea level, rather than above 2,000 feet (600 meters). Breeding seasons in the tropics tend to become more contracted as one rises from sea level into the highlands, as when one travels from the equator toward the poles.

In a much drier area of northeastern Venezuela, lying between 9° and 10° north latitude, some breeding occurs throughout the year, but the number of nests increases rapidly in April, just before the rainy season begins in May. The nesting season is at its height in May, June, and July, but afterward declines rapidly, although substantial rains continue to fall until October or November (Friedmann and Smith 1950, 1955). In the hills of the Indian state of Coorg, lying around 12° north latitude across the Western Ghats of India, the peak of the breeding season comes somewhat earlier than in America at about the same latitude. Here the number of nests increases sharply from the beginning of the year to

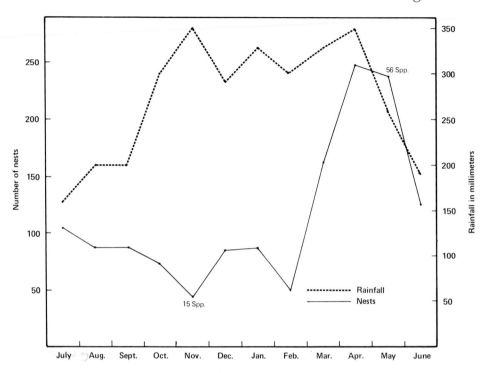

FIGURE 6. Rainfall and nesting in the vicinity of Buitenzorg, western Java, 6° 30′ to 7° 15′ S. lat., 650 to 5,000 feet (200 to 1,500 meters) above sea level. The broken line shows the average monthly rainfall in millimeters. The continuous line shows the distribution by months of 1,394 nests of 100 species. Notice that the breeding season is at its height toward the end of the heaviest rains, rather than at the beginning of the wet season, as in figure 4 (redrawn from Voous 1950).

a maximum in the dry month of March, falls to a low point in June, then rises to a minor peak in September. This early breeding is apparently timed to avoid the long-continued, drenching rains of the southwest monsoon in June and July. The minor nesting season in September comes in the sunny, steamy interval after the monsoon rains (Betts 1952).

A striking exception to my earlier generalization —that in rainy tropical lands most birds nest as the wet season begins—is provided by the abundant records from Java summarized by Voous (1950). On this island, around 7° south latitude, the rainy season comes chiefly from October or November to March or April, thus including the months when localities in the northern tropics, such as Costa Rica, Trinidad, and Venezuela, are driest. But, surprisingly, the peak of breeding in Java is from March to June, just as in Central America. Thus, the Javanese birds nest chiefly as the rainy season ends rather than as it begins, and also when the days, south of

the equator, are becoming slightly shorter rather than slightly longer (fig. 6). In Borneo, so near the equator that differences in day length are negligible, a rather similar situation prevails. Here the main breeding season begins in December when the monsoon deluges are heaviest, reaches its peak in March, when rainfall is diminishing, and continues until June or July. The reason for this appears to be that insects become most abundant in the wettest months when many trees renew their foliage (Fogden 1972). This may also explain the puzzling situation in Java.

Although Java, at 7° south latitude, has a single rainy season, Guyana, at about the same distance north of the equator, has two. There, in the Mazaruni district, the year is divided into a short dry season from February to April, the principal rainy season from early May to early or mid-August, the long dry season from August to December, and the short rains in December and January. Nevertheless, the annual march of nesting differs only slightly

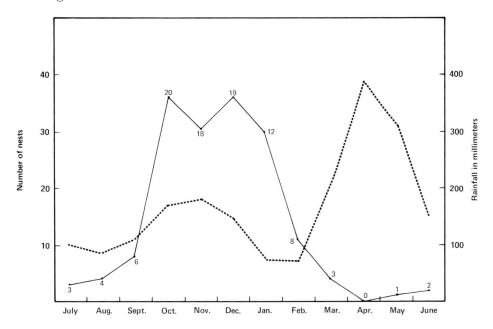

FIGURE 7. Rainfall and nesting at Amani, Tanzania, 5° S. lat., 3,000 feet (915 meters) above sea level, a region of heavy evergreen forest with two rainy seasons. The broken line shows the average monthly rainfall in millimeters. The continuous line shows the months in which eggs were laid in 168 nests. The numerals along this line give the number of species represented in the monthly totals (redrawn from Moreau 1936*b*).

from that in southern Costa Rica with its single rainy season. In Guyana, the number of nesting birds increases rapidly from February to March, continues high in April and May, falls through June to a low point in July, rises to a minor peak in September, declines again in October and November, and in December begins the slow ascent that culminates in March and April. Thus, in Guyana, the principal breeding season starts at a time of increasing rainfall and declines as heavy rains continue, and a secondary nesting season occurs in a dry period that is relieved by occasional showers (T. A. W. Davis 1953). Likewise in cloud forest and neighboring clearings in the western Andes of Colombia, at 3½° north latitude and 6,500 feet (2,000 meters) above sea level, A. H. Miller (1963) found by far the greatest nesting in the wet season of March, April, and May, after which the number of breeding birds declined through the drier months from June to September, to increase only slightly in the second rainy period, which occupied the last quarter of the year.

In tall, dense evergreen forest 3,000 feet (900 me-

ters) above sea level on the East Usambara Plateau, at 5° south latitude in what is now Tanzania, Moreau (1936*b*, 1950) found a single-peaked breeding season in a region with two rainy seasons. After a hot, dry period in the first quarter of the year come long, heavy rains and falling temperatures in April and May. These are succeeded by the cool dry season from June to September, which gives way to the short rainy season from October to mid-December, when the temperature rises and a few weeks of wet, quite variable weather occur. It is with this short wet season that the breeding season is associated (fig. 7). From September to October the number of species with nests rises steadily, and breeding continues at a high level into the warm dry season of January and February. During the heavy rains of April and May, no nests could be found in the forest, and few in neighboring clearings. In the Congo, in contrast to most tropical regions, nests appear to be rather evenly distributed throughout the year, with no distinct breeding season for most groups of birds (Moreau 1950). We still lack adequate studies in Amazonia.

66

Nesting Seasons

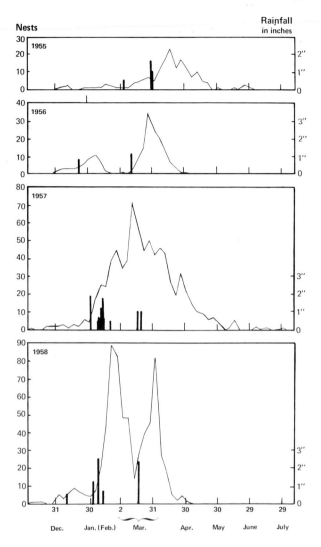

FIGURE 8. Rainfall and nesting on the Santa Elena Peninsula, southwestern Ecuador, 2° 15' S. lat., 0 to 300 feet (0 to 90 meters) above sea level. The graphs show the number of nests found in each five-day period from 1955 to 1958. The vertical bars show the important falls of rain, in inches, in this extremely arid region with unreliable precipitation (redrawn from Marchant 1959).

In the more arid parts of the tropics, the dependence of nesting upon the meager and unreliable rainfall is most striking. The Cape Verde Islands, lying in the Atlantic between 15° and 17° north latitude, have a long, severely dry season when, except in irrigated areas, the vegetation at lower altitudes dries up. All the small land birds, including species

that in Europe breed in the spring, delay nesting until rain comes in August, September, and October. If rain does not fall then, they mostly fail to nest (Bourne 1955).

One of the most thorough studies of breeding in a very dry region was made on the Santa Elena Peninsula of southwestern Ecuador, from 1954 to 1958, by Marchant (1959, 1960). Here, at 2° south latitude, in a region lying near sea level, the year is divided into two periods: a dry season from about May to November, when the air is kept cool by the almost continuously overcast sky and the only precipitation is the garúa, or fine mist, and a warm wet season from about December to May, when rain may fall but sunshine increases. The showers are undependable, in some years bringing as much as 45 inches (1,143 mm) of rainfall and in others scarcely 1 inch (25 mm).

The date and amount of breeding of nearly all the resident birds are controlled by these fickle showers. Each significant rain from January to April brings an upsurge of nesting, the volume of which depends on the quantity of rain (fig. 8). Accordingly, the peak of breeding may come at any time from mid-February to mid-April, and the length of the main nesting season may range from six weeks in a very dry year, such as 1956 when only 3.46 inches (88 mm) of rain fell, mostly on two days, to three and a half months, as in 1957 when 15.47 inches (393 mm) fell from early February to early April. Some birds, such as the Long-tailed Mockingbird, may start to build their nests before any significant rain falls, but these usually remain empty until a shower arrives; or, if the birds lay before the drought is broken, their sets of eggs are smaller than those in the rainy periods. Sometimes a rain triggers nesting efforts that fail because other showers do not follow and the country dries up, reducing the food supply and causing birds to abandon their nests, sometimes leaving their young to starve. In favorable years, however, some birds, including the Chestnut-throated Seedeater and Southern House Wren, may rear as many as three broods. In arid districts like this, the influence of rainfall on the nesting of tropical birds becomes unmistakably apparent.

Tropical Marine Birds

At high northern latitudes, cool summer seas, rich in dissolved oxygen, teem with vegetable and animal life. This sea life supports the myriad marine birds that breed at this season, often on forbidding cliffs that rim the continents and islands. At lower latitudes, where the area of water greatly exceeds that of land, the shores tend to be lower with fewer cliff ledges where colonial sea birds can nest in comparative safety. Instead, they mostly congregate to breed on small islands thinly scattered through immense expanses of ocean. In the warm, oxygen-poor water surrounding these tropical isles, birds do not find food so plentiful as in colder seas, and the quantity appears to fluctuate unpredictably. Around some of these tiny islands there seems to be no regularly recurring season of maximum abundance, as in northern seas and many terrestrial habitats. This uncertainty of the food supply, coupled with the huge numbers of birds that gather to nest on the few favorable bits of land, creates problems that tropical marine birds have met in unique ways.

On the more favored of these tropical isles, marine birds of one kind or another can be found nesting in every month; no general nesting season is recognizable, as in most regions inhabited by terrestrial birds. Some species, such as petrels and Sooty Terns, visit the island only to breed, whereas others, including Black Noddies and Brown Boobies, are based on the island throughout the year, returning to rest and sleep after each foraging expedition over the sea. Indeed, raising a single nestling binds some species, including frigatebirds, to the island for a year or more.

Some of these marine birds, such as the Fairy Tern on Ascension Island in the tropical South Atlantic and the Christmas Shearwater on Christmas Island in the central equatorial Pacific, have an annual cycle, with a peak of egg laying in the same months of successive years (Dorward 1963, Schreiber and Ashmole 1970). Others, however, regularly lay at intervals of less than one year. The first of such species to receive wide attention was the Sooty or Wideawake Tern, which on Ascension Island starts to nest at intervals of approximately 9.6

months (Chapin 1954, Ashmole 1963a). On Christmas Island, however, this same pantropical species has peaks of laying spaced about 6 months apart. Birds that successfully rear a chick lay again a year after they laid the egg from which this chick hatched; those that try to reproduce and fail lay again after an interval of 6 months. Since failures are frequent, many individuals lay sometimes in June and sometimes in December or January. On this same island, the Phoenix Petrel also has two breeding seasons, with maximum egg laying in November and December and a minor peak from April to June. In this species, however, all the individuals appear to lay at 12-month intervals, and different populations are involved in the two breeding seasons (Schreiber and Ashmole 1970).

Since Chapin's original account of "Wideawake Fair" on Ascension, a number of other cases of non-annual breeding have been discovered on this and other islands (table 2). On Ascension, the Brown Booby lays at approximately eight-month intervals, although the White Booby appears to have a one-year cycle (Dorward 1962). When Yellow-billed Tropicbirds nest successfully, their layings are separated by about ten months; however, if they fail, they lay again after only five months or a little more. Peaks of laying occur at intervals of about nine months. For the larger Red-billed Tropicbirds, the respective intervals are twelve and nine months, and maxima of laying come every eleven months (Stonehouse 1962). In the Galápagos Archipelago, Audubon's Shearwater nests throughout the year, but individuals lay on the average every nine months, an interval that seems not to be affected by the success or failure of the preceding attempt to rear young, although the intervals following failure are more variable (D. W. Snow 1965b). In the Swallow-tailed Gull of the same archipelago, the interval between layings by a successful breeder is nine or ten months, while those that lose their chicks have a shorter interval of seven or eight months (Snow and Snow 1967, Harris 1970).

Any regularly maintained cycle of less than twelve months will, over the years, result in laying in every month. Such a cycle can hardly be controlled by the movements of the sun or the march

TABLE 2
Periodicity of Nesting of Tropical Marine Birds

Species	Average Length of Individual's Cycle[a] Months	Interval between Peaks of Laying[b] Months	Island	Reference
Phoenix Petrel	12	6	Christmas[c]	Schreiber & Ashmole 1970
Christmas Shearwater	12	12	Christmas	Schreiber & Ashmole 1970
Audubon's Shearwater	9	—	Galápagos	D. W. Snow 1965*b*
Banded-rumped Storm-Petrel	12	6	Galápagos	Snow & Snow 1966, Harris 1969
Yellow-billed Tropicbird	5–10	9	Ascension	Stonehouse 1962
Red-billed Tropicbird	9–12	11	Ascension	Stonehouse 1962
Red-tailed Tropicbird	12	12	Christmas	Schreiber & Ashmole 1970
Brown Booby	8	8	Ascension	Dorward 1962
White Booby	12	12	Ascension	Dorward 1962
Great Frigatebird	12–24	12	Christmas	Schreiber & Ashmole 1970
Sooty Tern	9–12	9.6	Ascension	Ashmole 1963*a*
Sooty Tern	6–12	6	Christmas	Schreiber & Ashmole 1970
Black Noddy Tern	12	12	Christmas	Schreiber & Ashmole 1970
Black Noddy Tern	9–12	—	Ascension	Ashmole 1962
Fairy Tern	12	12	Ascension	Dorward 1963
Swallow-tailed Gull	7–10	—	Galápagos	Snow & Snow 1967, Harris 1970

[a] Longer intervals follow a successful nesting; shorter, a nesting failure.
[b] Approximate.
[c] All references to Christmas are to the island of this name in the Pacific, not that in the Indian Ocean.

of the ill-defined seasons in tropical seas, and in no instance does it appear to have been convincingly related to periodic events in the surrounding water. These nonannual cycles seem to be adaptive responses to an uncertain and often inadequate food supply and to scarcity of nest sites. In an ocean where food becomes abundant enough to rear a nestling only at unpredictable intervals, the probability that a bird's nesting effort will coincide with a favorable period will be greater the sooner after one laying it is able to lay again. Some, such as the Brown Booby, maintain an almost constant physiological readiness to seize the favorable opportunity (Simmons 1967).

Where there are not enough nesting sites to accommodate simultaneously the whole breeding population, the reproductive potential of this population will be increased if different pairs occupy the same site successively. On Plaza Island in the Galápagos, the Banded-rumped Storm-Petrel has two breeding seasons each year. Each individual appears to lay at annual intervals, but two different populations nest here. One does so in the warmer

season, the other in the cooler, and they occupy the same burrows alternately. On Pitt Island in the same archipelago, however, this petrel has only one breeding period, in the warm season (Harris 1969). On the tiny Salvage Islands, between Madeira and the Canaries, Banded-rumped Storm-Petrels appear to use the same burrows as White-faced Storm-Petrels, which are of similar size. They have solved the difficulty by nesting at different seasons—the White-faced in the first half of the year, the Banded-rumped in the second half (Lockley 1952).

On South Plaza Island in the Galápagos, where there are not enough niches in the lava rocks for all the Red-billed Tropicbirds that wish to nest there, these birds breed continuously throughout the year; but on Daphne Island, where the tropicbird population is better adjusted to the available nest sites, the species has a limited, well-defined breeding season, laying most of its eggs from October to December (D. W. Snow 1965*a*). On Ascension, the two tropicbirds, the Red-billed and Yellow-billed, have not adjusted the housing situation so well as the two storm-petrels have on the Salvages. Although their main breeding cycles are of different lengths, some individuals of both species try to nest in every month. The resulting competition for niches in the rocks causes the loss of many eggs and nestlings.

One may ask why any recognizable breeding seasons or peaks of laying persist in birds whose times of reproduction are not controlled by any recurrent external event, such as the march of the sun or the return of the rains. Since pairs lose their eggs or young at different stages and lay again after variable intervals, would not the different members of a population—even if they all started to nest at the same time and had the same basic internal rhythm —soon fall out of step? That is, would not laying be distributed randomly throughout the year with no evident periodicity?

Two factors help to keep the birds in a breeding colony more or less in step. One is social stimulation, which in such highly gregarious birds as Sooty Terns powerfully affects readiness to breed. Early arrivals on the breeding ground may delay nesting until enough gather and the whole group works up to a certain pitch of excitement, which stimulates late comers to begin more promptly. The other factor is fluctuation in the food supply. The one or two eggs of marine species are usually large in proportion to the size of the parent birds. Possibly an abundance of food is more critical when the female is forming her egg or eggs than later, when both parents help to feed a slowly growing chick. Until food becomes sufficiently abundant, laying can hardly begin; when it does, the females of a population may be simultaneously enabled to form eggs. On the other hand, a catastrophic failure of the food supply in the midst of the nesting season, as sometimes happens in tropical waters, may cause simultaneous abandonment of nests by whole colonies of birds, who might then be ready to try again after the same interval. Although such events may help to keep the birds synchronized, the persistence of a less-than-annual breeding periodicity, like that of the Sooty Tern, presents an ornithological problem that may take long to solve.

CHAPTER 6

The Timing of Nesting

We owe our knowledge of the breeding seasons of birds chiefly to the strenuous efforts of naturalists who have searched for nests in tropical forests, temperate-zone fields, Arctic tundra, remote islands, and suburban gardens, and who have diligently recorded the date and contents of each nest that they found. Information has been added by collectors who have examined the reproductive organs of the birds that they shot; but this quicker method of accumulating data is less accurate and less humane than the inspection of nests that are left undisturbed. It is largely the field workers, too, who can tell why certain seasons are most favorable for the nesting of different birds. But understanding of the physiological processes that bring birds into breeding condition at certain times of year has been contributed by the laboratory biologist and the experimenter. By collating the data from the field with that from the laboratory, we may attain deeper understanding of avian breeding seasons.

Breeding seasons and migration, two of the most perplexing fields of avian biology, present common problems. For many birds, migration precedes nesting. If we knew more about how wintering birds, in the tropics or far on the other side of the equator, time their departures for distant breeding grounds, we would have a better understanding of how

breeding seasons are regulated. The timing of migratory movements is no less perplexing than the navigational problems involved in long-distance flights. We know that birds can orient their journeys by the sun, stars, and perhaps even more subtle guides. They do not lack a compass or its equivalent; but accurate navigation also requires a chart that gives the latitude and longitude of the points of departure and destination. We do not know what these aerial voyagers substitute for this. The literature on both migration and the regulation of reproductive periods is vast and rapidly growing. That on the latter has been reviewed by Burger (1949), Thomson (1950), A. J. Marshall (1959), and Immelmann (1963, 1967), to whose summaries the following paragraphs owe much.

The Internal Cycle

Unlike man and his domestic animals, free birds are not, as individuals, capable of reproducing throughout the year. The study of the physiology of avian reproductive periodicity has largely centered on the male, whose organs undergo great changes in the course of the year and respond readily to experimental treatments. As the breeding season ends, the testes shrink to a small fraction of their former size, they seem almost to collapse, and their internal structure is profoundly altered. They must be reorganized or rebuilt before they are again able to secrete the hormones that induce sexual behavior and to produce the spermatozoa necessary to fertilize the female's ova. This process of reconstruction or regeneration may take from about six weeks to several months, according to the species and the external circumstances. Until it is completed, the bird does not respond to increased day length or other treatments intended to induce spermatogenesis or other aspects of sexual activity. Accordingly, the regeneration phase is often called the refractory period. It is followed by the acceleration phase, in which the sexual organs grow and become active, and ends in the culmination phase, in which nests are built, coition occurs, and eggs are laid.

This regular alternation of an interval of sexual activity with an interval of rest and restoration is the innate foundation of the cyclic or periodic character of avian breeding. In certain circumstances, the bird's breeding seasons may be controlled wholly by this internal rhythm, without modification by the environment. This seems to happen in those tropical birds that, in a continuously favorable environment, nest as species throughout the year. Each individual or pair has alternate periods of breeding and rest; but, since no external factor controls their reproductive efforts, the several members of a population fall out of step, and their eggs may be found in every month. Likewise, where conditions favorable for reproduction occur at unpredictable times, as seems to happen with capricious food supplies in tropical seas, the birds may follow their internal rhythm. Where this rhythm is unmodified by social behavior, the species breeds rather uniformly throughout the year, as in a number of tropical marine birds. Where social stimulation is strong, peaks of nesting come at less-than-annual intervals, as in the Sooty Tern.

An internal rhythm may be the cause of certain birds nesting at the same date in successive years of quite different weather. The Chestnut-headed or Wagler's Oropéndolas in a colony on Barro Colorado Island, Canal Zone, began to build their nests on January 8 in both 1926 and 1927. In the former year, they started thirty-three days after the dry season

began; in the latter, four days before the rains ended. Aside from the alternation of wet and dry seasons, there appeared to be, at 9° north latitude, no environmental variable pronounced enough to serve as a time giver. Accordingly, Chapman (1928, 1929) concluded that "each bird carries its calendar within itself." Similarly, a pair of Red-crowned Woodpeckers in Costa Rica laid eggs only a few days later in 1937, when the preceding months had been unusually wet, than in 1936, when these months had been quite dry (Skutch 1969*b*). Facts like these, which might be multiplied, suggest that an internal rhythm may, without external control, keep a bird on an annual schedule for at least a few years.

The strength of this internal rhythm is also demonstrated by the persistence with which some birds nest at their ancestral date when transported from the Southern Hemisphere to the Northern, or vice versa. When several Silver Gulls were taken from Australia to the National Zoological Park in Washington, D.C., they continued for two years to breed in November, as in their Southern Hemisphere homeland, but then changed their nesting time to the spring and early summer of North America. Strangely enough, descendants of these immigrants, twenty years later, reverted to the ancestral date of laying and tried to rear their chicks in snowy December (M. Davis 1945). Eventually, however, most birds transported across the equator seem to adapt their nesting to the season appropriate to their new environment. The Common Starlings and House Sparrows that homesick British colonists misguidedly took to New Zealand now nest in the months of the southern, rather than the northern, spring and summer. Some alteration of the time of nesting is an indispensable part of the acclimatization of any species that is transported, or spontaneously extends its range, to a region where the seasons differ from those in its original home.

A single species that breeds in the North Temperate Zone and winters in the tropics may pass this season on both sides of the equator in regions with quite different rainfall regimes. Yet all members of this species return to their northern breeding ground at about the same date. The most plausible explanation of this is that each individual is prompted to begin its journey by an internal rhythm that is set while it is in the north.

Regulation by Day Length

Although a strictly autonomous rhythm may, for a while, preserve its synchrony with rhythmic changes in the environment, sooner or later it is almost certain to fall out of step. Among the causes of this are the varying dates at which nestlings hatch, the different number of attempts that neighboring pairs make before they succeed in rearing a brood, and the retardation of physiological processes by adverse external conditions, such as excessive heat or cold or scarcity of food. Except in climates that are almost uniformly favorable for reproduction throughout the year, the bird that gets out of step with the march of the seasons will be severely penalized, and its stock may be eliminated by natural selection. To preserve its synchronization with the seasons, the bird must at some point in its annual cycle be subject to control by its environment. It must become sensitive to some external regulator, or time giver, some "proximate factor," that brings it into breeding condition at the appropriate season. It must be periodically set by its environment, just as a clock must at intervals be set by signals from the astronomical observatory.

The most dependable time giver for birds everywhere, except in the equatorial belt, is the duration of daylight, which is basically dependent only on date and latitude, although subject to some modification by cloud cover and topography—day dawns sooner on a clear mountaintop than in a neighboring overcast valley. Since W. Rowan (1925) demonstrated that Slate-colored Juncos could be brought into precocious breeding condition, even outdoors in the great cold of a Canadian winter, by prolonging their day by means of artificial lighting, the same has been found true of a number of other temperate-zone birds and even of some tropical species. It is not necessary to increase day length gradually, as happens in spring in the temperate zones; the light to which the birds are exposed may be increased abruptly by several hours and kept at this duration to the end of the experiment.

For several species of temperate-zone and some tropical birds, a schedule of 13 to 14 hours of light each day has been found effective in bringing male birds with shrunken testes into breeding condition (including the molt from eclipse into nuptial plumage in species that change color annually), while the control birds kept on short days remain sexually dormant. This is the day length used by poultrymen to make hens lay while winter days are short. The effective day length appears to be more important than the means by which it is reached. Burger (1949) induced spermatogenesis in Common Starlings by gradually reducing their day from 20.5 to 16 hours, yet the gradual increase from 6 hours of daylight to 9 hours had no effect.

Although sexual activation is achieved by a wide range of light intensities, if the period of illumination is adequate, certain intensities are more effective than others. Rollo and Domm (1943) found that 126 foot-candles caused Orange Weavers to molt into nuptial plumage more rapidly than did either stronger or weaker light. As to wavelength, yellow-red appears to be the region of the spectrum most effective in inducing sexual activation.

To respond to increased day length, the bird must have already emerged from its regeneration or refractory phase, for which certain temperate-zone birds appear to require days of considerably less than 12 hours. To prematurely lengthen the day of a temperate-zone bird in its refractory phase may prolong, perhaps indefinitely, the period of sexual quiescence rather than abbreviate it. This postbreeding refractory interval is, as A. H. Miller (1959b) pointed out, an essential part of the adaptation of the bird's breeding schedule to its environment. It prevents temperate-zone species from attempting to breed in autumn, when the day may be just as long as in spring, temperatures even more favorable, and food adequate, but when such an effort would perilously delay the parents' preparation for approaching winter. Furthermore, any young that they might produce would be unready to migrate at the appropriate time or to withstand the approaching inclement season. The refractory period prevents the useless expenditure of materials and energy that would ensue if sexual organs remained enlarged and active, secreting the hormones that induce singing, displaying, and courtship. Such economy may promote survival during migration and in the lean times that many birds experience after the close of the breeding season. The young of the year are prevented from premature attempts to breed—doomed to failure because of the lateness of the season—by a refractory period similar in length to that of the adults. Since they have never bred, this should be regarded as an interval of enforced delay of sexual maturation rather than of recovery, as in adults who have already reproduced.

In a number of common birds of North America and Europe, the post-breeding refractory period lasts three or four months and ends between late September and December, when days have become too short to induce breeding and will remain so until the following spring. In certain British birds, such as the Rook and Blackbird, the refractory period of some individuals ends early enough to permit spermatogenesis and even nesting in the autumn, if the weather is exceptionally mild.

In contrast to the young of related northern species, young Rufous-collared Sparrows, hatched in Colombia at 3½° north latitude, exhibited no post-juvenal refractoriness and by exposure to about 16 hours of daylight were brought into breeding condition when from three to six months old, although in their native home they normally reach this stage at six to eight months. Since this population of Rufous-collared Sparrows nests successfully throughout the year, precocious breeding is not, as in northern birds, doomed to failure by the climate (A. H. Miller 1959*b*).

More surprising is the discovery that adults of certain tropical species respond to abnormally prolonged days much as birds of the temperate zones do. Transported from the Colombian Andes to Berkeley, California, and kept in the natural daylight of 38° north latitude, adult male Rufous-collared Sparrows remained continuously in breeding condition for nine or ten months, through short winter days of 10 hours as well as long summer days of 15¼ hours, which seemed responsible for this abnormal prolongation of breeding. Females stopped laying when days were less than 11 hours but were stimulated by long days to continue laying far beyond

their usual reproductive period of four months (A. H. Miller 1965). Likewise, male Rufous-collared Sparrows from Costa Rica at 10° north latitude were brought into the reproductive state by exposure to 16 hours of daylight for a few weeks (Epple et al. 1972). Similarly, another tropical bird, the Black-faced Weaver, is sensitive to altered day length and can be brought into breeding condition by exposure to daily periods of illumination far longer than it would ever experience in its natural state (Morel et al. 1957, Lofts 1962). Males of a related tropical bird, the Orange Weaver, were induced to molt into nuptial plumage by artificially lengthened days (Rollo and Domm 1943).

The responsiveness to altered day length of certain tropical birds whose natural reproductive cycle appears not to be regulated by the photoperiod raises some interesting questions. Is this sensitivity an incidental consequence of their constitution, valueless to them at low latitudes, yet preparing the way for them or their descendants to invade higher latitudes, where responses to varying day length help birds to breed at appropriate times? Or is it, on the contrary, an inheritance from temperate-zone progenitors retained because, although no longer useful, it is at least innocuous? In the case of the Rufous-collared Sparrow, whose ancestors evidently dwelt in the North Temperate Zone where its closest relatives now live, and which as a species ranges into high latitudes in South America, responsiveness to long days may well be a vestigial character. We do not know how birds of exclusively tropical families, such as antbirds, manakins, or toucans, would respond to manipulated day length. Until we have this information, we shall remain ignorant of the true significance of this responsiveness now demonstrated to exist in a few tropical species.

Regulation by Rainfall and Other Factors

Although fluctuating day length is demonstrably a powerful proximate factor or time giver that brings temperate-zone birds into breeding condition at a favorable season, the situation in birds of low latitudes is more complex. The breeding seasons of tropical birds are, on the whole, adjusted, not to varying day lengths, but to the alternation of wetter and drier periods and their effects upon the abundance of food. The situation in northern Ethiopia provides evidence additional to that already given in chapter 5. On the Eritrean coast of the Red Sea, one of the hottest regions on earth, the scant rainfall comes mostly in the winter months, December to March, during which the great majority of birds of all kinds raise their broods. But across the mountains in central and western Eritrea, also around 16° north latitude, the rains fall in spring and summer, and the height of the breeding season is from April to July. Here the effect of rainfall completely overrides whatever influence the small fluctuations of day length at this latitude may have on the birds (K. D. Smith 1955).

The nesting of a single species may be timed by day length in one part of its range and by rainfall in another part. The Black-throated Sparrow inhabits the arid southwestern United States, the elevated plateau of northern and central Mexico, and the lowlands of Baja California. When possible, it nests during or soon after the rainy season. Between 20° and 26° north latitude in Baja California, this sparrow nests from October to June, during or following such slight rains as may fall, and on the central Mexican plateau it breeds from June to September, during the regular summer rains. However, between 28° and 38° north latitude in the interior U.S. deserts it breeds from April to June, regardless of precipitation, thereby avoiding both the cold winter and extreme summer heat (A. H. Miller 1960).

The first shower that falls after a long, severely dry period often has immediate and even spectacular effect upon birds. Gilliard (1959a) described how, in mid-April, birds of several kinds started to build their nests on the day after heavy rain broke a prolonged drought that had parched the Caracas region of Venezuela. Just beyond the Tropic of Capricorn in arid central Australia, where whole years may be rainless, showers fall at unpredictable times and may, for a short interval at any season, create conditions favorable for nesting. The onset of a shower after long, rainless months stimulates the birds to bathe, sing, court, feed their mates, and gather nest material. If the rain is adequate, or is followed by other showers that awaken the long-parched vegetation and tinge the landscape with verdure, nesting proceeds rapidly; but, if the drought

is soon renewed, nests that have been started may be abandoned (Immelmann 1963).

The foregoing and many similar observations show that the effect of rain upon breeding is different from that of increasing day length. Birds respond promptly to the rain; the lengthened day brings them into breeding condition only after weeks or months, according to the species. The photoperiod *prepares* the birds to breed, carrying them through the acceleration phase after they have completed the regeneration phase. The rain *releases* breeding activity in birds that have already been prepared, perhaps months in advance; it triggers the culmination phase. In regions where the advent of the wet season is fairly dependable, although the dry season may be severe, some birds start their nests before the rains that will create conditions favorable for rearing their young. In this case, rain is obviously not a time giver or regulator of breeding in the same way that day length is.

Although we have abundant evidence that the breeding seasons of small tropical land birds are adjusted to the regime of alternating dry and wet periods, rainfall is not, properly speaking, the ultimate factor determining the breeding season. The ultimate factor is the improved conditions, principally more abundant food, but in some cases also adequate building materials or better concealment for nests. The proximate factor or factors that bring birds into breeding condition at the proper time are more obscure. Probably, for many tropical birds there is no single dominating circumstance, like day length at higher latitudes, but a subtle interplay of influences modifying an essentially autonomous rhythm, some of which accelerate, while others retard, the attainment of the breeding state. Among these factors day length cannot be excluded. We know that at least some tropical birds are responsive to its alteration, and probably the majority of tropical land birds begin to nest while days are becoming longer, even though this increase in day length is slight. But this may be merely because the rains, to which their breeding season is adjusted, follow the sun in its annual swing back and forth across the equator. Day length seems never to be strong enough to overrule the influence, direct or indirect, of the rains. We do not know what effect increased humidity, apart from actual rainfall, has upon breeding (Moreau 1936*b*, Wagner and Stresemann 1950).

Doubtless autonomous or internal processes are largely responsible for timing the breeding of tropical land birds, although not to the same degree as in those tropical marine birds that have a nonannual breeding periodicity. What, then, binds the bird's internal cycle to the march of the seasons, so that it does not become maladjusted to its environment and nest at an unfavorable time? Probably, to synchronize the internal cycle with the external cycle, the former needs to be tied to the latter at only one critical point, and a number of points may be available for this setting. One suggestion is that the fixed point of the internal cycle is the molt that the bird undergoes each year. Black-and-white Manakins in Trinidad may hatch in any month from January to October, yet all of them, as far as is known, molt into adult plumage, a process taking about eighty days, from June to September of the year following their birth. Thus, the molt would synchronize the individuals of a population more closely than could be expected from their time of hatching alone (D. W. Snow 1962*a*). Similarly, in a subtropical locality of northwestern Argentina, most young Rufous-collared Sparrows began their postjuvenal molt at about the same time, although their ages ranged from one to nearly five months (King 1972).

In a later paper, the Snows (1964) pointed out that not only do nearly all individuals of a certain species, regardless of age, undergo their complete annual molts at about the same time, but also there is a strong tendency for different species with quite different breeding schedules to molt simultaneously. Even if we concede that the molt is the event that adjusts the bird's annual cycle to the seasons, we are left with our original problem, in a slightly altered form. We must now ask, not what environmental change prepares the bird to breed, but what determines the time of its molt. The Snows suggested that this might be the onset of the wet season, but this suggestion at present must be regarded as only a hint for future studies.

The shorter the season favorable for breeding, the more imperative it becomes for birds to start nesting promptly when suitable conditions arrive, or even

to anticipate them. Where the favorable period is long continued, the birds may wait to pair, build, and lay until food is plentiful enough for rearing their young, so that the proximate and ultimate factors tend to coincide. Where the season is short and its date unpredictable, as in dry central Australia, such delay might be fatal to the young who have not become self-supporting before the environment deteriorates so greatly that parents cannot nourish both their progeny and themselves. Here small birds have an unusually short refractory period and remain in almost perpetual readiness to breed, even if they seldom do so. They achieve sexual maturity and pair at extraordinarily early ages: Male Zebra Finches take mates when from ten to fourteen weeks old, females at nine to eleven weeks; one female laid her first egg when only eighty-six days old.

By duetting and group singing, engaging in courtship ceremonies, and nest building even outside the breeding season, the mated pair of these Australian birds maintain close bonds, synchronize their sexual cycles, and hold themselves ready to nest as soon as the indispensable and unpredictable rain arrives. Even individuals in the midst of the molt start to nest, in some cases suspending this process until they have raised their brood, but more often continuing to renew their feathers, perhaps at a reduced rate, while they breed—a rather unusual situation. By such a concentrated effort, these inhabitants of an inhospitable land do their best to replenish, during the brief favorable interval, a population that may have dwindled alarmingly in the long period since breeding was last possible (Immelmann 1963).

The Female's Decisive Role

In the foregoing account we have given attention chiefly to the males, whose sexual responses to various conditions are more pronounced, and have been more thoroughly investigated, than those of the females. Incidentally, we have noticed that female birds are also responsive to increased day length, as the domestic hens that continue to lay in midwinter when their day is artificially prolonged, and the Rufous-collared Sparrows who continued to breed longer in the long days of a California summer than in the tropical mountains whence they came.

Nevertheless, the female's response to lengthened

days differs greatly from the male's; he can thereby be brought into the breeding state, with production of spermatozoa, even in the absence of all the other conditions necessary for successful reproduction, but she does not form eggs in such circumstances. After a refractory period similar to that of the male, the female's shrunken ovary can be stimulated by long daily illumination to develop slightly, but it stops growing while the eggs are still rudimentary, as has been demonstrated in the Tricolored Blackbird, White-crowned Sparrow, and other birds (Payne 1969*a*, R. A. Lewis and Orcutt 1971). To form eggs without the possibility of hatching them and rearing the young would be a waste of nutrient materials, which she avoids by halting their development at an early stage and waiting for appropriate conditions. The known exceptions to this are certain domesticated breeds.

Favorable weather and adequate nutrition promote the further development of the eggs. Equally necessary are the presence of a mate, his songs and courtship displays, a territory, a site for the nest, and materials for building it. Only when everything is ready for their reception does the female complete the formation of her eggs. Even when her nest is finished, disturbance by man or some other animal may cause her to abandon it and postpone laying until she has completed another. To the male bird, who is often prepared to inseminate his mate long before she is ready for him, falls the coarse adjustment of the nesting season; to the female, more sensitive to the actual situation, belongs its fine adjustment. The event by which the start of the breeding season is usually dated, the laying of the first egg, depends wholly on her. In the culmination phase of the breeding cycle, the female controls events.

CHAPTER 7

The Forms and Sites of Nests

With the exception of a few aberrant species, including certain cuckoos and cowbirds, that foist upon other birds the care of their eggs and young, birds cannot become parents until they have made, or otherwise acquired, a receptacle for their eggs. Such a container is known as a "nest." This term may, I believe, be applied to any receptacle of the eggs that has some feature to distinguish it from the surroundings, even if it be no more than a hollow in the top of a rotting stub or an unlined depression that the birds have scraped in a sandy beach. It is questionable, however, whether we should designate as a nest the spot where a goatsucker lays its eggs on leaf-strewn ground or a flat roof, without any preparation for their reception, and with nothing to distinguish this spot from the surrounding surface. Similarly, the place where an auk incubates its egg on the bare ledge of a seaside cliff and the part of a bare branch on which a Fairy Tern deposits its egg hardly deserve to be called nests.

The function of holding the eggs is common to all the immense variety of nests that birds use. The simplest nests do no more than this, but many have additional functions, such as conserving the warmth of the eggs and nestlings, concealing the occupants, shielding them from rain or hot sunshine, or holding their enemies aloof. In addition, some nests are used as dormitories by one or both parents, with or without their fledged young. Relatively few nests serve all these purposes, and the actual form of any nest can often be understood by considering which of them it serves.

Types of Nests

Nests may be divided into those that are found or sometimes stolen, those that are excavated, and those that are built. A few birds in the most diverse families find the abandoned, or sometimes newly finished and occupied, nests of other species and appropriate them for their own use; but in many instances these opportunists add a little material or make certain alterations before they lay their eggs.

Excavated nests are usually carved into wood, as with woodpeckers and barbets, or dug into a bank or even level ground, as in many kingfishers, motmots, jacamars, bee-eaters, and a few swallows. But they may be carved into a hard termite nest, in the manner of a number of parrots, kingfishers, trogons, puffbirds, and jacamars (plates 9 and 10); in the globular papier-mâché nests of tree ants, as is the habit of the Rufous Woodpecker; or even into a papery wasp nest from which the stinging builders have been ousted by the birds, as in the Violaceous Trogon (plate 11). Some birds prepare a nest of soft

PLATE 8. Colonial Magellanic Penguins looking out of their nesting burrows on the coast of Argentina (photo from William Conway, New York Zoological Society).

materials in the chamber they have dug, but these species usually belong to families some of whose members build without excavating, as in the American flycatchers. Families composed almost wholly of excavators, such as woodpeckers and kingfishers, rarely line a hole or burrow that they have hollowed out but lay their eggs on its hard floor. Some ground-nesting birds, such as the Horned Lark, dig a depression in the soil to receive their well-made cups of vegetable materials. The slight hollows scraped into sandy ground by certain terns and shore birds might be included among excavated nests, but these birds do so little work that I am inclined to include them among those that find their nests.

The largest and most interesting class of nests contains those that are built by the birds themselves. These display such an endless variety of shapes and materials that it is difficult to describe all the types. To most people, no doubt, the term "bird's nest" calls up a picture of a simple, bowl- or cup-shaped structure, and this is certainly the most common sort in vegetation everywhere. These open nests might be divided into those that are supported below, as in many finches and thrushes, and those that are attached by their rims to twigs and hang between the supports, as in many vireos and antbirds. Some open nests are so shallowly cupped that they are best described as slightly concave platforms, such as those of certain pigeons and rails.

PLATE 9. Terrestrial termitary in which Rufous-tailed Jacamars raised a brood, beside a pasture in Venezuela. The entrance to their nest is at the center.

tain wrens and other birds construct a more completely enclosed nest, of which many variations are found in the tropics.

Still more elaborate are the pensile nests, which are usually covered above. The tropics afford a vast and fascinating variety of these, including swinging pouches, retort-shaped structures with tubular entrance ways, and greatly elongated masses of entangled materials with shallow niches in the middle (plates 18, 19, 20, and 21). Finally, there are nests, made largely of sticks by members of the ovenbird family in the tropical and South Temperate portions of the American continents, so large in relation to their small builders, so elaborate, and so well enclosed that they can best be described as avian fortresses or castles.

Although the great majority of nests are built and

PLATE 10. Termitary in which Massena Trogons nested in the Caribbean rain forest of Honduras. After the birds left, the termites filled the cavity they had made.

Some of the most elaborate of nests are the covered structures (plates 14, 15, 16, and 17). These are far more common in tropical and subtropical regions where the builders are permanently resident than at higher latitudes where the majority of the birds are migratory and cannot afford to devote so much time to building. The simplest of the covered nests are domed or oven shaped, such as those of the North American wood warbler called the Ovenbird, the Common Meadowlark, and many tropical species of diverse families. By extending the roof forward and downward in front of the doorway, thereby forming a vestibule, or antechamber, cer-

PLATE 11. Vespiary in which Violaceous Trogons reared three young, in the Térraba Valley of Costa Rica.

PLATE 12. Rock Cormorants build open nests on remote coasts and islands (photo from William Conway, New York Zoological Society).

occupied by a single pair of birds, several pairs of the black cuckoos called anis may join in making an ample open nest of sticks lined with green leaves. In this nest the several cooperating pairs lay their eggs in a single heap to be incubated by all the males and females of the group. The polygynous male Red-billed Buffalo Weaver of tropical Africa gathers thorny twigs to build a number of chambers so close together that when he covers them thickly with a bristling array of spiny sticks they form a single mass sometimes six feet long by four feet high (plate 22). According to the size of this compound nest, it may contain from three to six or even eight closed compartments, each of which is lined with grasses and acacia leaves by a female who lays her eggs and raises her brood in it. The Sociable Weavers of arid regions of Africa join forces to build in an isolated tree a huge domed roof of grasses and weed stems, which may attain four or five yards in diameter and weigh hundreds of pounds (plate 23). Often several such masses of varying size are built in the same tree. The underside of each mass is honeycombed with holes, each the entrance to a separate chamber in which a pair of weavers nests. A single colony may contain as many as three hundred chambers (Collias and Collias 1964*a*).

Unique in a family of birds that prefers to nest in holes, the Monk Parakeet of Argentina builds in a tree an elaborate apartment house of interlaced sticks. A large one may contain a cartload of material. Each of the several apartments, occupied by

PLATE 13. Nest of the Yellow-thighed Manakin. In tropical forests many nests are barely large enough to hold the eggs and young.

PLATE 15. Roofed nest of the Yellow-faced Grassquit, in a potted fern on a Costa Rican porch.

PLATE 14. Roofed nest of the Ruddy Crake, hidden low in a grassy field in Honduras.

PLATE 16. Roofed nest of the Slate-throated Redstart, on a steep, forested slope in the Guatemalan highlands.

PLATE 17. Roofed nest of the Vermilion-crowned Flycatcher in a surprising situation—beneath a rail on a tramline trestle over a Honduran stream.

a single pair, consists of a nest chamber entered through an antechamber that opens to the outside. The striped-breasted brown Palm-Chat, a starling-sized bird of Hispaniola, builds a somewhat similar structure of interlaced twigs, usually in a royal palm but sometimes in another tree. An exceptionally large nest that we saw in the Dominican Republic was a mass of twigs about ten feet high by four feet in diameter surrounding the crown shaft of a tall palm tree. It was impossible to learn the number of inhabitants of this huge nest, but at a much smaller nest I counted twenty-five birds flying back and forth, bringing sticks. The trunk of a palm that supports a Palm-Chat apartment house may be riddled with holes made by several pairs of Hispaniolan Woodpeckers, which also nest in a community.

Principles of Safety

The first step in the establishment of a nest is the selection of its site, which must be carefully chosen if the future occupants are to have a fair chance of escaping the thousand perils that will beset them. Every situation that human ingenuity might choose for safety and concealment has been used by birds —and many more that might never occur to us. The highest boughs of the tallest trees, the narrowest crannies in their trunks, the densest tangles of vines, the thorniest shrubs, burrows in the ground, rock-bound islets in a stormy sea, niches in forbidding cliffs—each has been chosen by some bird as its nest site. Add to these the interior of a wasp, termite, or ant nest, crevices in the sides of some larger and

PLATE 18. Globular nest of the Paltry Tyranniscus, a diminutive American flycatcher, built in an orchid plant dangling high in a tree at the edge of a Costa Rican forest.

PLATE 19. Pensile nest of the Southern Bentbill flycatcher in a Panamanian forest.

more powerful bird's nest, nooks and crannies about the dwellings of men, the labyrinth of a spider web —and still we have not enumerated all the sites chosen by birds. Some float their nests on still water, others sew them up in leaves, and still others hang them in mid-air from a slender dangling vine or swinging root of an epiphyte.

For the safety of their nests, birds depend on four principles: invisibility, inaccessibility, impregnability, and invincibility. Each of these is only relative; a bird that achieved perfection in one or more of them might cover the earth with its progeny.

Invisibility is generally attained by placing the nest amid dense concealing vegetation or by hiding it away in a crevice or cranny. Sometimes it is achieved by concealing coloration. Ground-nesting

birds, such as goatsuckers and snipes, do not always place their eggs beneath the thickest cover; often they select a rather open space and depend on the perfect assimilation of their plumage with the background, coupled with complete immobility, to escape detection while they incubate the eggs or brood the nestlings. Were some of these birds less alert while sitting on their nests, we might often step on them before we saw them. Small, neutrally tinted arboreal birds also rely largely on their cryptic colors to safeguard them while they incubate and brood. These birds often make nests barely large enough to hold their eggs. One of the most difficult to discover of all the nests that I have found was that of a Scarlet-thighed Dacnis, a slight hammock about as broad and deep as the cupped palm

PLATE 20. Pensile nest of the Sulphur-rumped Myiobius, a small flycatcher of tropical American forests. An apron or visor shields the entrance on the right.

PLATE 21. Nest of the Black-cowled Oriole, hung beneath the leaf of a "travelers' palm" by fibers passed through perforations.

of a child's hand, completely covered on the lower side by pieces of green living fern fronds. Situated in the midst of a parasitic vine in the crown of a tree with dense foliage, this nest was so well concealed that it took me several hours to find it, even with the nestlings' loud cries to guide me to it.

In Africa, some bishop birds (*Euplectes*) arrange loose screens of grass above their globular nests among grasses, either as camouflage or to shield them from strong sunlight in shadeless country (Crook 1963).

Inaccessibility to prowling animals is found by sea birds on small, isolated islands, often encircled by forbidding, wave-lashed cliffs. On these islands, seafowl were practically immune from terrestrial predators, until man began the careless dispersion

over the earth of cats, rats, pigs, goats, and other destructive quadrupeds. Nonetheless, gulls and other winged predators may take their toll of eggs and young. To place their eggs and nestlings beyond the reach of voracious birds as well as quadrupeds and snakes, petrels and shearwaters not only breed on islands but also seek additional safety by hiding their nests in burrows or crevices in rocks. At least one bird creates the island on which it nests. The Horned Coot, which lives high in the bleak Andes of Bolivia and neighboring parts of Argentina and Chile, piles up small stones and pebbles in the shallow water of a lake to form a mound that may exceed a yard in height and contain over a ton of rock (fig. 9). On this islet, above the level of the water and beyond reach of terrestrial predators, the coot arranges a nest of aquatic vegetation and lays its eggs. Farther north in the Andes, where lacustrine plants are more abundant, the Horned Coot builds nests of vegetable materials alone, but similar in form and placement to those made of stones (Ripley 1957a, b). Burrows in high vertical banks and niches in lofty cliffs also enjoy a high degree of inaccessibility. Especially safe are the high ledges deep in lightless South American caves where Oilbirds or Guácharos nest.

Oropéndolas and related birds of tropical America hang scores of long, pouchlike nests from the outermost branchlets of a single lofty tree, where their noisy, crowded colony becomes the cynosure of the neighborhood (plate 24). Unless the nest tree has a clean, smooth trunk and a crown that stands free of surrounding trees, with no vines leading up from the ground, it will almost certainly be invaded by snakes or other enemies and devastated, as I once saw. If it fulfills these conditions, it is hardly accessible except to winged predators, which as a rule do not pillage enough nests to jeopardize the existence of the colony. To escape predation, the yard-long nests of the Northern Royal Flycatcher, attached to slender twigs or vines dangling above a woodland stream, must, I believe, achieve almost perfect inaccessibility, at least in respect to the wingless nest robbers common in such localities

PLATE 22. Nest of the Red-billed Buffalo Weaver, containing several chambers, each occupied by a single female. The polygamous male owner is at the lower right (photo from N. E. and E. C. Collias).

PLATE 23. Many-chambered nests of the Sociable Weaver in a camel-thorn acacia in arid South Africa. The bottom of each great mass of straws and weed stems is honeycombed with chambers, each occupied by a single breeding pair (photo from N. E. and E. C. Collias).

(plates 25, 26, and 27). Without this inaccessibility, the slowly developing nestlings, hatched from eggs that require three weeks of incubation, would seem to have hardly a chance of flying unharmed from their conspicuously swinging nursery (Skutch 1960).

Although birds that depend upon invisibility for the safety of their nests often hide them amid dense foliage, those that rely upon inaccessibility sometimes seem to deliberately seek exposed sites. This is especially true of birds that build hanging nests entered through a downwardly directed spout. The Sulphury Flatbill of tropical America attaches its blackish, retort-shaped nest to the end of an exposed twig or to a dangling vine. In tropical Africa, the Red-headed Weaver goes even further, spending much time snipping leaves not only from the supporting slender branch but also from surrounding twigs (Crook 1963). The Strange Weaver has a

similar habit (Collias and Collias 1964a). Such defoliation, which makes the nest stand out conspicuously, evidently makes it harder for slender tree snakes to insert their heads into the freely hanging, downwardly directed spout. The thin, naked twigs give the serpents little support; they have been seen to fall while trying to enter such nests. Likewise, it must be difficult for wingless predators to reach nests of the Black-throated or Altamira Oriole that hang conspicuously from telephone wires in Mexico.

Thorns often help to make a nest difficult to reach. The Buff-breasted Acacia Warbler of Africa builds its little purse-shaped nest amid twigs studded with so many long, sharp thorns that a man can hardly touch it without being severely pricked. The Superb Glossy Starling improves the safety of its nest's natural setting in a spiny tree by collecting many thorny twigs to form a bristling barricade

PLATE 24. Clustered pouches of the Montezuma Oropéndola hang like gigantic fruits from a tall, isolated tree in the Caribbean lowlands of Honduras.

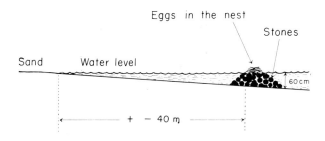

FIGURE 9. Nest of the Horned Coot on an island of stones that it has piled up in the shallow water of Laguna Verde, 14,100 feet (4,300 meters) high in the Chilean Andes (from Ripley 1957).

along the approaches, an arrangement that should hold aloof most climbing and creeping enemies (van Someren 1956). The White-headed Buffalo Weaver similarly covers the branches leading to its roofed nest with thorny twigs (Collias and Collias 1964a).

The sharp spines that protect a nest are not an unmixed blessing to the breeding birds. Sometimes one projects into the nest cavity and seems to jeopardize the eggs or nestlings, while rarely a parent suddenly darting from the nest is impaled on a spine, as Belt (1888) saw happen to a "yellow and brown flycatcher" in Nicaragua. Most birds that nest in thorny plants seem to do nothing about interfering thorns; but Red-backed Shrikes, before they start to build, try to bite away the thorns and other projections from the branches at their nest site (Kramer 1950). A nest of a Black-throated Sparrow in a cholla cactus was so closely surrounded by needle-sharp spines that the parent seemed to have difficulty avoiding them. Each time that it approached or left, the bird paused to snip off the ends of some of the thorns (Bent et al. 1968).

Impregnability, at least sufficient to ensure the perpetuation of the species, is achieved by the South American horneros or ovenbirds in their fortresses of hardened mud, often built on a post or tree bough with little attempt at concealment. So, too, the mansions of interlaced twigs made by castle-builders and their relatives seem strong enough to keep out some would-be predators; yet slender enemies, such as snakes, manage to insinuate themselves into the inmost sanctum, and some animals

tear a gap in the walls. Woodpeckers and barbets carve their nest chambers in wood sound enough to resist predators lacking powerful claws or bills; yet at times a snake will manage to enter, a strong mammal to enlarge the doorway, or a long-billed toucan to extract eggs or young. The cavity that Yellow-bellied Sapsuckers carve into the decaying heart of a living aspen trunk often resists the attempt of a principal enemy, the raccoon, to enlarge the doorway and reach the nestlings (Kilham 1971a).

One of the best examples of impregnable nests is provided by the hornbills of the Old World tropics. When ready to lay, the female enters her nest hole in a tree and, with her mate's help, closes up the doorway with a strong wall of clay or other mate-

PLATE 25. Nest of the Northern Royal Flycatcher. Hanging above a Honduran mountain torrent, it was not immediately recognized as a bird's nest.

the least successful and the one least employed by birds. Only birds exceptionally strong and bold can depend on it for the safety of their progeny. The larger and fiercer birds of prey are able to defend their nests against many assailants, and men have sometimes paid dearly for their attempts to reach them. The Tropical Kingbird often places its open nest in a bush or low tree with little regard for shade or concealment. Were this flycatcher as stout clawed and strong as it is stouthearted and confident, all would be well with it. No bigger than a thrush, it is only a catcher of small insects. Despite its spirited onslaughts on enemies, many of its nests are overtaken by disaster. Nevertheless, these sanguine kingbirds are successful in rearing their offspring often enough to make them one of the most universally distributed and abundant small birds

PLATE 26. Nest of the Northern Royal Flycatcher. The boy's finger indicates the shallow niche where eggs and nestlings rest.

rials, leaving only a slit barely wide enough for him to pass food in to her. Self-immured, she remains in the nest cavity, fed by her attentive partner, until her nestlings are half-grown, or in some species until they are ready to follow her into the open. Probably few small-climbing animals attempt breaking into a nest defended from within by a bird that can wield its huge and powerful beak through a gap in the protecting shield. Tropical American counterparts of the hornbills, but not closely related, the similarly huge-billed toucans do not close the doorway of their tree cavity, where, while incubating, the male and female exchange places many times a day. Instead of using their beaks to defend their citadel, they scramble into the open when danger threatens.

Invincibility is, of the four principles of safety,

PLATE 27. Nestling Northern Royal Flycatcher resting in its niche, in the usual orientation.

over a vast and varied territory stretching from southern Texas to northern Argentina.

Protective Nesting Associations

Other birds, by no means invincible in their own strength, increase the safety of their nests by placing them close to an abode of some more formidable creature which, by defending its own home, may incidentally drive away some of the birds' enemies. These associates of nesting birds include stinging insects, more powerful birds, and men. In the tropics of both hemispheres, the nests of certain kinds of birds have been found close to occupied wasp nests too often for this propinquity to be attributed to chance. In America, several kinds of flycatchers, wrens, and icterids have this habit; in Africa, a number of weaverbirds and sunbirds (Moreau 1942c); in Australia, warblers of the genus *Gerygone*, the Double-bar Finch, and occasionally certain other small finches. In southern Australia at least one of the *Gerygone* warblers and some of the finches carry this practice beyond the tropics (Chisholm 1952a, Hindwood 1955).

We have good evidence that the birds build near an established vespiary rather than that the wasps seek the proximity of birds' nests. We do not know why the insects, some of which are intolerant of other disturbances, desist from attacking the building birds. Although occasionally an open nest is found close beside a vespiary, the birds that most consistently place their nests close to wasps build covered structures or else deep pouches. The situation in the tiny seedeaters of the genus *Sporophila* is highly instructive. The many species make slight, shallow, open cups at a distance from wasps, with a single known exception. The Dull-colored Seedeater of Argentina, which habitually nests beside an occupied vespiary of *Polistes canadensis*, constructs a substantial pouch about eight inches deep (Contino 1968). This appears to be a necessary precaution, because wasps that at ordinary times live peaceably enough close to birds may attack them if excited. When some children disturbed a nest of large black wasps in an orange tree beside my dwelling, the irate insects stung a twelve-day-old Scarlet-rumped Black Tanager that hitherto had rested unharmed in its open nest two yards away. The "yellow and brown flycatcher" who, as already mentioned, was impaled on a thorn, was attacked by wasps stirred up by its flutterings and stung to death, despite the efforts of Belt's party to rescue it. Two feathered nestlings of the Tropical Kingbird were attacked by an excited swarm of small black meliponine bees from a nearby hive, while the parent kingbirds looked on helplessly. Since these bees bite but do not sting, the nestlings survived the onslaught (Skutch 1954).

The wasps evidently derive no benefit from the proximity of the birds that they tolerate. Sometimes they will severely sting a person who attempts to reach the bird's nest; but many tropical wasps are surprisingly pacific, and the investigator may, with impunity, examine a nest only a few inches from their hive if he can reach it without shaking the supporting branch. We have few observations to show from what enemies the wasps protect their avian neighbors. In Ecuador, a small snake took two newly laid eggs from the covered nest of a Southern Beardless Flycatcher, within thirty-two inches of a small nest of wasps (Marchant 1960). Perhaps predatory birds would be driven off by the stinging insects; but certain birds, ranging in size from Summer Tanagers to Red-throated Caracaras, habitually plunder wasp nests for their larvae and pupae, regardless of stings.

Mammals might be held aloof by the wasps, some kinds of which will regularly attack a man through the hair of his head or beard, even when they might more readily sting his bare face and arms, thus demonstrating their readiness to attack furry animals. Many of the bird-wasp nesting associations are in thorny trees, whose spines, one might suppose, would themselves hold quadrupeds aloof. But this is doubtful, because I have watched squirrels climb with no harm to the fruits of the pejibaye palm (*Guilielma gasaepes*), a plant with extremely long, sharp, and closely set spines. In the tropics, as in the north, squirrels plunder many bird nests. Thorny trees are more numerous in dry than in humid regions, and bird-wasp associations appear to be more frequent in the more arid parts of the tropics than in heavily forested regions. The thorny trees of dry country often have sparse foliage, so that the nests of both birds and insects that they support are

FIGURE 10. Paired thorns of the bull's-horn acacia. These hollow thorns are inhabited by fiercely stinging ants that defend the small tree and the nests that birds often build in it. The ants go in and out through the orifice they have made near the end of the thorn on the right (redrawn from Belt 1888).

poorly concealed. Apparently, in such sites invincibility compensates for lack of invisibility.

A number of instances of birds building close to ant nests or in ant-infested trees have been recorded. In Africa, the nests of several kinds of weaverbirds and hawks were surrounded, and in some cases overrun, by stinging ants, which drove the human observer from the nest tree but seemed not to harm the birds (Moreau 1942c). In Nigeria, the Bronze-headed Mannikin regularly builds close to one or more nests of the large, fiercely stinging red ant *Oecophylla smaragdina*. In an area of parkland that supported ninety-five trees suitable for both the ants and the birds, nineteen trees contained ant nests and seven of these trees held twelve nests of the mannikin, whereas seventy-five trees without ant nests supported only five mannikin nests in as many trees (Maclaren 1950). In Australia, the White-throated Warbler, one of the species of *Gerygone* that often seeks the proximity of vespiaries, sometimes breeds in an ant-infested tree (Chisholm 1952a).

The bull's-horn acacias of tropical America have paired, hollow thorns inhabited by stinging ants that eat the protein bodies borne at the tips of the leaflets (fig. 10). These trees are the preferred nesting sites of several species of birds, whose nests can

be examined, as a rule, only at the price of some severe stings. Sometimes these acacia trees contain nests of wasps, which further increase the peril of any creature that molests them. The Banded Wren prefers to build its elbow-shaped nest near a vespiary in a bull's-horn acacia. When arboreal snakes or lizards are placed quietly in one of these acacia trees, the ants bite them so fiercely that they quickly leave (Janzen 1969). Despite the poor concealment offered by their coarse, open branches, ant-infested cecropia trees are sometimes chosen as nest sites, especially by Vermilion-crowned Flycatchers. The little Azteca ants that inhabit the hollow trunk and limbs of the cecropia bite but do not sting; they probably do not make such an effective guard for the nests as do the *Pseudomyrmex* ants of the acacias. Although ant-bird nesting associations are far less common in temperate regions than in the tropics, in Scandinavia, Durango (1949) found three of the covered nests of the Long-tailed Tit close to the red ants *Formica rufa*. In one case, the ants, swarming over the walls of the nest, held aloof squirrels, and the tits successfully reared a brood in it.

At times small, weak birds build their nests very close to the nest of a larger, more powerful bird, frequently a raptor. In Africa, a number of weaverbirds have this custom. Their nests may be clustered around or even attached to the much larger structure of the bird of prey. On an island in Lake Victoria, dozens of nests of the Slender-billed Weaver were hanging so densely from the nest of a Yellow-billed Black Kite that from the ground the kite's nest appeared to be made of lumps of tow. In North America, the nests of Cliff Swallows are sometimes clustered around the eyrie of a Prairie Falcon, and, in regions where trees are scarce, the Arkansas or Western Kingbird may build close by a nest of a Swainson's Hawk (Bent 1942). In Norway, Fieldfares often breed close to a pair of Merlins, who occupy a nest built by the Hooded Crow. This crow often eats the eggs and young of the Fieldfare, but it cannot do much damage to a colony situated around a Merlin's eyrie. Even when, like the Merlin, the raptor preys on birds, it does not as a rule hunt close to its own nest. Thus the small birds breeding around an eyrie are immune not only from the attacks of the raptor but also from those of the nest robbers that the raptor holds aloof from its own eggs and young. Yet another form of nesting association is that in which one or a few pairs of birds of one species breed in a large colony of more social birds, thereby, apparently, taking advantage of the mass attack that the latter direct against predators. On the coasts of Finland and Sweden, Tufted Ducks and Turnstones habitually nest in colonies of terns and Black-headed Gulls (Durango 1949).

Although a variety of birds, from storks to wrens, nest on or in the dwellings of men, they usually do so simply to take advantage of attractive nest sites that man's constructions offer. Some birds, however, appear to seek the proximity of man for the protection from predation that his presence affords. An African weaverbird so often establishes its nesting colonies in palm trees and banana plants in streets and beside houses that it has been called the Village Weaver. The Black Weaver shares this predilection to some extent. These and several other African birds apparently find in the proximity of man a measure of immunity from the attacks of hawks and monkeys, although not always from man himself.

Moreau pointed out that birds seeking some special protection for their nests, especially weaverbirds, often have recourse to two or more alternatives. Some, such as the Black Weaver, may build either close to human habitations or in trees or bushes overhanging streams, where their nests are relatively inaccessible. Others, including the Yellow Weaver, may build over water or close beside a hornet nest. Yet others, such as the Village Weaver, frequently establish their colonies close to man but may on occasion cluster their nests about that of a more powerful bird. This weaverbird has, within the last few centuries, become established on the West Indian island of Hispaniola, where it associates with wasps instead of the men or bigger birds that it prefers in its native Africa.

The ability of birds to take advantage of different protective agents in different localities poses some interesting psychological questions. Are they aware that stinging insects, or a more powerful bird, or man will protect their eggs or young from some of their enemies? If so, how was this knowledge gained? Or do certain species or races of birds include strains with different, genetically determined "blind" instincts to nest in special situations that afford protection?

Selection of the Site

For successful reproduction, the nest site must be chosen in accordance with one of these principles of safety or a combination of them. The actual selection may be made by the male, the female, or both together. In all those species of which the male establishes a territory before he wins a mate, he is responsible for the general locality of the nest, if not for its exact position. Exceptionally, as has been recorded of the Song Sparrow, Snow Bunting, Common Meadowlark, European Blackbird, and some other species, the female, who may not be familiar with the boundaries of the domain established by her mate, builds her nest beyond them. When this occurs, he may try to enlarge his territory to include his partner's nest.

Some males go further than this and not only pick a nest site but also begin or even complete a nest before the arrival of a mate. Among them are the Phainopepla of the southwestern United States, the White-throat of the British Isles, and the bishop birds of Africa. Often the male builds several nests, and the female, when she arrives, is able to select the one that most appeals to her. The Winter or European Wren may make half a dozen or more nests but leaves them unlined. The Northern House Wren places sticks in from one to seven boxes, thereby suggesting to his newly arrived partner that they are potential nest sites. In both of these wrens, the female, after choosing one of the nest boxes or unlined shells of nests, adds the lining of softer materials, rarely receiving help from her mate. The polygynous male Long-billed Marsh Wren also constructs a number of nests, which may be lined and occupied for breeding by one of his females, although sometimes each prefers to build her own from start. In tropical America, the male Yellow-faced Grassquit, whose closed nest is shaped much like that of the marsh wren, often begins to build. If his mate approves his choice of a site, she later helps with the work. As the nest nears completion, the male may lose interest and leave the lining largely or entirely to her, or he may continue to perform an equal, or even major, share of the construction until it is finished.

In woodpeckers, too, the male is often responsible for the selection of the nest site. In species that sleep singly in their holes, including the Hairy Woodpecker, Red-crowned Woodpecker, and Red-bellied Woodpecker (Kilham 1958*b*), the male, who is usually more industrious in carving dormitories for himself, generally has the newer and sounder chamber, which at the outset of the breeding season may be chosen to shelter the eggs and young. In the Yellow-shafted Flicker, the nest site may be chosen by either the male (Sherman 1952) or the female (Kilham 1959). In the Downy and Hairy woodpeckers, the female also at times chooses the site of the hole (Bent 1939, Kilham 1960). Even in woodpeckers that continue to lodge in pairs or family groups in the cavity where the young were reared, the male may begin to carve the hole that will become next year's nest long before his mate takes any interest in the matter. As the time for laying approaches, she helps to finish the hole, already far advanced. If it becomes necessary to start a new chamber late in the season, both sexes of many kinds of woodpeckers work about equally carving it out; it is then difficult for the observer to decide who has chosen the site.

One morning before sunrise, a male Gray's Thrush flew into a dracaena shrub beside our house and sat for about a minute amid the branches, revealing his sex by singing a few notes. Later in the day, he came occasionally to the same spot, where there was no sign of a nest, as he never brought any material to it. After this had continued for three days, his mate, whom I had not previously seen there, started to build in this spot, and he, departing from the custom of his kind, helped substantially. Although I have never known the male Scarlet-rumped Black Tanager to participate in building, he may, at least occasionally, select the nest site. At the end of March, one came repeatedly to sing amid the dense twigs of a Thunbergia shrub in front of my study window. After this had continued for several days, I first saw his mate in the shrub, where she rested at a certain point amid clustered stems while he sang nearby. On the following day, she started to build in the site that she had tested the previous day.

Another male who does not build, the Vermilion Flycatcher, may also select a spot for the nest, which he "visits about once a minute, sits on it with his splendid crest elevated, tail spread out, and wings incessantly fluttering, while he pours out a continuous stream of silvery gurgling notes, so low they can scarcely be heard twenty paces off" (Hudson 1920). In some other American flycatchers, the nest site seems to be selected by mutual agreement of the mates. Certainly at this time they act in closest concord, with much soft-voiced consultation as various inviting locations are examined and rejected, until at last one that satisfies them is found. One day I watched a pair of Tropical Pewees sitting side by side in the crotch of a small tree. On the following day, the female began to build in the exact spot where they had rested, carrying on all the work unaided by her partner, who was zealous in the nest's defense.

Although birds that build nests amid leafy vegetation generally choose a territory that provides a number of potential sites, birds that nest in holes, but cannot make them, first seek an adequate cavity and then establish a territory around it, thereby making sure that they have not claimed an attractive foraging area that lacks facilities for breeding. In many northern species, the male arrives first, finds a hole, and claims the surrounding territory before he wins a mate. When a female comes, he courts her by offering the nest site. When a female Pied Flycatcher approaches the hole or nest box of a male, he flies to the doorway, singing a long, excited, whispering song. If the female follows him, he enters; usually she stays outside until she has seen him go in a number of times. Other Old World flycatchers of the genus *Muscicapa* demonstrate the hole to the female by similar displays. Indeed, analogous rites are performed by fishes in which the male chooses the nesting place, commonly a hole, or builds the nest (Haartman 1956, 1957).

Significantly, the male European Redstart, a member of the thrush family, shows his prospective mate the hole by displays rather like those of the Pied Flycatcher. These demonstrations include flying to the entrance and clinging to it with his head inward and his rufous tail conspicuously spread in front, and entering it to show his black throat with contrasting white forehead in the doorway. At high intensity, the male redstart flashes his white frontal patch in and out of the hole at a rate too fast to count (Buxton 1950). The male Great Tit shows a female the nest hole by sitting in the doorway and pecking at its lower edge. The male Blue Tit calls his mate's attention to the cavity by a number of displays. In one, he clings in front, turns his head back and forth showing his white cheek patches, or puts his head inside and withdraws it over and over. In another display, the male Blue Tit flies slowly toward the hole with his fully extended wings beating rapidly through a small arc, so that his flight resembles that of a moth. However, in this resident bird the female is probably already aware of all the holes suitable for nesting that the territory contains (Stokes 1960).

In many male birds, the ritual of courtship includes acts symbolic of nest building: carrying material that is later dropped and going through the motions of shaping a nest, often in an inappropriate spot. Thus, in England, the male Reed Bunting squats on the ground and shuffles around, always with an eye on his mate, evidently suggesting building. In the tropical forests of Costa Rica, the elegant male White-tailed Trogon, even before he has attracted a mate, interrupts his mellow calling to cling to the side of a rotting trunk and dig at it with his bill, indicating a spot suitable for a nest cavity, which he will never carve without a female's help. How much weight such masculine suggestions bear in the final choice of the nest site is often difficult to learn. In the trogons, the voice of the male, who usually takes the lead in the task of excavation, is possibly decisive.

Sometimes birds start to build in inappropriate sites, even such as could not possibly support their nest, and discover their error only after they have done a little or even much work. I have known hummingbirds to begin a nest on the tip of a leaf that touched a neighboring upright stem, forming a sort of false axil. The unattached point of the leaf soon slipped down with the weight of the material placed upon it. Probably curious blunders of this sort are made chiefly by young birds starting their first nest.

CHAPTER 8

Crafts and Materials Used in Nest Building

In ancestral birds, male and female probably shared about equally the labor of home making. However, with increasing specialization in courtship and song, the male found himself more and more occupied with other business and left the work of building ever more to the female. At the present time, we find among birds every possible degree of variation from the closest cooperation between the sexes to building by the female alone, or more rarely by the male alone. The situation varies so from family to family, from genus to genus, and even from species to species, that to make sweeping generalizations would be dangerous, while to list every exception would be tedious. But certain trends are clearly evident.

Participation of the Sexes in Building

In nonpasserine families, building by both sexes is the prevailing mode. This is true of herons, pigeons, gulls and terns, cormorants, hawks, kingfishers, motmots, trogons, nonparasitic cuckoos, barbets, jacamars, puffbirds, and woodpeckers, to mention only a few. The most striking exception is provided by the hummingbirds, in which building by the male has not, to my knowledge, ever been recorded.

When the sexes cooperate in building, the usual system is that of more or less equal sharing, or turn and turn about, but some families follow special procedures. In many pigeons and doves, the female settles on the site of the future nest and arranges its materials with her bill, while her partner, flying to and fro, brings additional straws and twiglets and lays them beside her. Often he alights on her back while he deposits his contribution, as I have seen in the Blue Ground-Dove, Ruddy Ground-Dove, and White-fronted Dove. Later, while incubation is in progress, the female may bring material and lay it on the nest while her mate is taking his long spell on the eggs at the habitual hours, perhaps to make it a more adequate nursery for the nestlings soon to appear. Among the black anis, too, the female usually sits on the nest to receive and arrange the sticks and green leaves brought by her mate, but in her absence he may place the materials with his own bill. In these birds, also, additions are frequently made to the nest after the eggs have been laid. In herons and egrets, the male presents sticks to his sitting mate with a little ceremony; but, if she is not present, he may himself work them into the structure. In the tree-nesting Red-footed Booby, the male also passes a stick to his partner with a display (Verner 1961).

Among passerines, the male, as we have seen, may complete the nest without assistance from a female; or he may help her to build, or merely make an ineffectual show of helping, or do nothing but sit nearby and sing while she works, or stay out of sight and take no interest at all. In the single family of the American flycatchers, I have watched each of these degrees of assistance except the first; I know no male flycatcher that builds alone. Active building is done by the male Black-fronted Tody-Flycatcher, Slate-headed Tody-Flycatcher, Great Kiskadee, Torrent Flycatcher, and Yellow-bellied Elaenia. A great show of helping is made by the male of the big Boat-billed Flycatcher. He faithfully follows his mate back and forth while she gathers materials for the open cup high in a tree, or else he awaits her near the nest. Frequently he finds a twiglet or a length of fibrous root and carries it as he flies with her to the nest tree, but when she goes off again he follows, still carrying the piece. I have seen him bring the same root thrice and take it away as many times! Or while his hard-working partner comes and goes, he may perch near the nest for ten minutes or more, holding in his bill a suitable piece of material, which it never occurs to him to lay there or even to pass to his mate. I have never seen a male Boat-billed Flycatcher make a single tangible contribution to the nest that he will later faithfully guard.

In the related cotinga family, male tityras have the same unproductive habit of following their building mates back and forth with a bit of material, a dry leaf or twig, in their bills. I have watched a male Black-crowned Tityra, escorting his busy partner, come and go ten times with the same leaf. Rarely a male tityra will manage to drop a piece of material into the old woodpecker hole that contains the nest.

In some flycatchers, the male, who does not attempt to build, rests near the nest and greets his mate with a cheery twitter or chirp whenever she passes him bearing something for the growing structure. Such is the habit of the Vermilion-crowned and Gray-capped flycatchers. The male Sulphury Flatbill shows little interest in the wonderful fabric, shaped like a chemist's retort, that his mate is building, although later he helps her to feed the nestlings in it. Finally, in a group of flycatchers that includes some of the most accomplished of avian architects, the male neither builds nor pays any attention to the eggs and young. Among these are the Bentbill,

Ruddy-tailed Flycatcher, Sulphur-rumped Myiobius, and Oleaginous Pipromorpha.

Male birds who participate in building are frequently more eager to follow their mates than to advance the undertaking. In many of the multihued tanagers of the tropics, which fly two by two at all seasons, the male takes a large share in nest construction, finding material, placing it in the nest, and helping to shape it with his own body, bill, and feet. Usually members of a pair arrive together with contributions. First one, then the other, enters the nest to arrange what it has brought. If the female goes first and completes her work, then flies off without waiting for her partner to deposit what he holds in his bill, he will often follow her at once, carrying in the wrong direction the material that he would have added to the nest had she waited for him. Some of the diminutive tanagers called euphonias avoid this wasted effort by having the male always precede his mate to the nest when the two come together with material; he then waits for her to arrange her contribution and the two fly off together for more material. Gray-headed Tanagers are sometimes so impatient to add their contribution that one enters the nest on top of the other, causing the latter's prompt departure. Often, however, the one outside passes its material to its sitting partner for arrangement, a system followed by many birds.

Other male birds help their mates faithfully but seem incapable of understanding the finer points of nest construction. The Southern House Wren prefers a narrow nest cranny. If a pair has chosen a cavity too capacious for the eggs and young, the male eagerly helps his consort fill it with twiglets to reduce the excess space. But when enough twigs have been carried in and the female starts to fashion the nest proper with fine rootlets and fibrous materials, her mate stupidly continues to bring coarse twigs, often interfering with her work by blocking the doorway with his superfluous contributions, which at times she carries away to keep the passage open.

So, too, the male of a pair of Rose-throated Becards that I watched in Guatemala brought much material for the bulky globular nest, which hung high above the ground at the end of a long, thin branch of an alder tree. As in many tanagers, he was more eager to be near his mate than to finish the nest. Often he followed her to it with empty bill, or he flew off with her, still bearing his intended contribution, to avoid being left alone while he placed it. Nevertheless, he contributed a substantial share of the great and varied mass of materials needed for a nest that is huge in relation to its builders. Yet everything that he brought was added to the outside. After the female began to line the snug nest chamber, he continued to deposit his materials on the exterior, never entering. A male Bay-headed Tanager helped his mate to lay the foundation of her beautiful mossy nest, but he soon lost interest and left her to finish it.

Sometimes a male bird departs from the usual practice of his family or his species in the matter of nest building. Although in the wood warblers male participation is rare, the male Buff-rumped Warbler regularly gives substantial help to his building partner. In the troupial family, the males of the Melodious Blackbird and Bay-winged Cowbird are exceptional in helping to build, and in the case of the cowbird the male may do most of the work. Recently I watched a male Gray's Thrush bring much material to the nest, although the male's participation in building is rare in this species and rather unusual in the thrush family as a whole.

Edmund Selous once remarked that a male bird, singing near the nest at which his mate works alone, is to him a more pleasing spectacle than that of the pair working together. A charming sight it certainly is, yet to me not so delightful as that of the pair working side by side in closest harmony, the male perhaps so bursting with song that melodious notes escape him even while he approaches the nest with a laden bill. He seems so eager to help that we readily pardon the stupid blunders he at times makes, yet still more eager to be near his mate. To follow her he often carries away the billful of material that he has laboriously gathered for the nest. Even if the male neither builds nor sings but merely follows his consort back and forth like her faithful shadow, our hearts warm to his constant devotion.

Crafts Used in Building

In building nests, birds employ a fair share of the constructive crafts known to man, and at most or

all of them they anticipated us by many thousands of years. They are wood carvers, miners, potters or masons, basketmakers, thatchers, feltmakers, weavers, tailors, decorators—even at times thieves.

The outstanding examples of wood-carvers are the woodpeckers, which in Spanish-speaking countries are appropriately known as *carpinteros*. With sharp bills and heads constructed to withstand sharp percussions without injuring the delicate structures of brain, eyes, and ears, those most adept at their art chisel into fairly sound wood, carving out deep, smooth-walled chambers with neatly rounded doorways. Some of the barbets with short, thick, prong-tipped bills, which operate by biting or gouging rather than by pecking, hollow out of solid wood cavities as neatly finished as those of woodpeckers. A few titmice also excavate their nests in trunks or branches, but to be workable by their weaker bills the wood must be well softened by decay.

The miners include all those birds that dig shafts in banks of clay, loam, or sand, or even in rather level ground. They include kingfishers, motmots, jacamars, bee-eaters, and among the passerines a number of ovenbirds and swallows. South American ovenbirds of the genus *Geositta* are often called *mineros*, or miners. The tunnels of the larger kingfishers and motmots are neatly rounded borings that may reach two or three yards in length and rarely even more. At the inner end, which is often raised somewhat above the entrance to permit drainage and prevent flooding, is an enlargement, usually with a low vaulted roof, where the eggs are laid and the young grow up.

Birds that make their nests of mud or clay might be called potters, for the results they achieve sometimes remind us of the products of the ceramic art. But instead of pressing and molding a mass of plastic stuff into the desired shape, in the manner of the human potter, they build up their structures by bringing pellets of mud, permitting them to dry and harden, then adding more, until the nest is finished. Their method of working resembles that of masons, who lay stone on stone or brick on brick and bind them together with mortar. A bird nest of clay might be compared to an adobe house, with the difference that in adobe construction the bricks of clay

PLATE 28. Nest of Gray's Thrush built upon a leaf of a Panama-hat plant. In rainy Panama, small plants sprouted in the middle layer of mud to form an aerial garden.

are dried in the sun before being laid in the walls, whereas the avian mason pushes the pellets into place while they are soft enough to stick. Often more or less vegetable material is mixed with the mud or clay, as straw in bricks, and serves as binder. Among the master masons are Cliff Swallows and House Martins, whose roughly globular nests are fastened beneath eaves of houses or on sheltered faces of cliffs; phoebes, which attach their massive half-cups to piers and abutments of bridges or beneath overhanging rocks; and the several species of horneros or ovenbirds of South America, whose little houses of clay closely resemble the old-fashioned baking ovens widely used until recently in Latin America. A number of thrushes equip their open nests with a middle layer of hardened mud (plate 28), and at least one, the Speckled-breasted Scrub-Thrush of tropical Africa, makes its cup-shaped nest almost wholly of mud (van Someren 1956).

The basketmakers construct their nests of interlaced twigs. This method of working is widespread in the ovenbird family, in which the numerous species of *Asthenes* are called *canasteros*, or basketmakers. Another large genus, *Synallaxis*, contains the castlebuilders. They start a nest by making a deep bowl or basket of closely interlocked twigs, which might satisfy most birds as an adequate receptacle for the eggs and young. But these indefatigable builders continue to add sticks and other materials until their finished structure has lost all semblance of a simple basket and reminds one of a fortified castle. After the builders have covered over their brood chamber, they often roof it with a thatch of broad pieces of leaf, bark, and similar materials that shed rain. Doubtless they were thatching their nests in this fashion long before our ancestors began to build covered dwellings.

The feltmakers press and entangle downy or fibrous materials into a compact fabric without actually interlacing them. The exquisite little chalices of hummingbirds are made largely of vegetable down matted and kneaded together, often with the addition of lichens or moss to the outside. The Long-tailed Silky-Flycatcher builds a bulky open cup of felted beard lichens (*Usnea*). The bushtits employ much the same technique to construct a

PLATE 29. A few of the neatly woven pouches of the Montezuma Oropéndola that fell when an overloaded branch snapped off.

long, hanging pouch that is one of the marvels of bird architecture. By felting coarser fibrous materials, some of the smaller American flycatchers fashion pensile structures that in some cases closely resemble the woven nests of certain weaverbirds.

The weavers of greatest skill are members of the Old World weaverbird family and of the New World troupial family (plate 29). Not satisfied with simply entangling their fibers, these accomplished birds actually lace them through the fabric from side to side, often tying simple knots. Their constructions range from deep cups to woven pouches over a yard long or, among the weaverbirds, to more elaborate structures equipped with spoutlike entrance ways. Because of the very regular texture of its fabric, Dr. and Mrs. Collias (1964a) considered Cassin's Weaver the best of all avian weavers (plate 30).

The most renowned of the tailors is the Indian bird of this name, a member of the Old World war-

PLATE 30. Sleevelike nest of Cassin's Weaver, one of the most skillful of avian weavers (photo from N. E. and E. C. Collias).

bler family, which sews up its little nest in living green leaves that it draws together by cobweb passed through perforations in their tissue. Several African representatives of the family have a similar custom, and some of them use vegetable fibers in addition to cobweb to firmly bind the living leaves that enclose their nests, making them hard to distinguish from surrounding foliage. In Central America, the Black-cowled Oriole suspends its pouchlike woven nest beneath the broad leaf of a banana or some other plant by means of fibers that it passes through perforations it has made in the leaf tissue (plate 21).

A number of birds attach to the outside of their nests pieces of lichen, moss, delicate ferns, or other things that are not integral parts of the structure (plate 31). In some cases, these additions help as-

similate the nest to its setting and make it less conspicuous. Apart from any utility they may possess, the result is so pleasing to the human eye that we are tempted to regard them as adornments and apply the term "decorators" to the birds who so embellish their homes.

Nest Materials

The materials that enter into the construction of bird nests are no less varied than the sites in which they are placed, the forms they take, and the skills by which they are made. Perhaps the simplest method of classifying them is according to their origin. First we have the inorganic materials, which are not numerous. Mud or wet clay is frequently employed; pebbles, small stones, or hard fragments of dry clay

PLATE 31. Nest of the White-eared Hummingbird, a dainty open cup of vegetable down ornamented with lichens.

PLATE 32. Pistillate inflorescences of *Myriocarpa yzabalensis*, a small tropical American tree. They are much used by builders of pensile nests.

less often. The Ringed Plover lines the depression in the sand where it lays its eggs with small stones of uniform size. The Great Shearwater builds a little platform of stones to raise its eggs above the floor of the cave in which it nests (Lockley 1942). In the far north, Thick-billed Murres may place small stones under their eggs on a ledge (Tuck 1960). True to its name, the Rock Wren makes a foundation of pebbles for its nest in a crevice in a cliff and paves the approach to it with little stones. The Black Wheatear of Spain, which also lives in arid, rocky regions, has a similar habit, sometimes carrying into its nest crevice rock fragments more than half as heavy as itself (Richardson 1965). When a pair of Cliff Flycatchers nested on the sill of a permanently closed window in Brazil, they began by making a pavement of pebbles, bits of brick,

plaster, and the like, the largest about the size of a small walnut. After they had filled the whole corner of the sill, they built their shallow bowl of fine straws and feathers on this platform (Euler 1867).

Desert birds that perforce build in exposed, wind-swept places weigh down their nests with fragments of stone to prevent their being blown away. When the Desert Lark of the Near East nests under an overhanging stone, it builds around the exposed side of its cupped nest a semicircular wall of rock fragments about an inch and a half high—as high as the nest's rim. A nest in a more exposed situation under a shrub was completely encircled by such a stone wall (Orr 1970). In North America, the Horned Lark may lay a pavement of rock fragments and pieces of clay around the exposed side of its cup-shaped nest set in a depression that it digs in the ground beside a tuft of grass or clod of earth (Bent 1942). Adelie Penguins outline their nests with stones, the only material available to them on the bleak Antarctic continent (Levick 1914).

Materials of vegetable origin form the bulk of the nests of the majority of birds. Among these are coarse pieces, usually stems of various sorts, such as weed stalks, straws, twigs of woody plants, and lengths of flexible vines. Wiry roots, leafstalks, pine needles, and seaweeds are also frequently used. Papery or membranaceous materials include leaves used either green or more often dead, either whole or more or less reduced to lacy skeletons, and the epidermis of herbaceous stems, separated from the interior tissues by decay. Then there is a whole series of fibrous stuffs: bast fibers from the inner bark of plants; fibrovascular strands from decaying vines or weed stalks or the hearts of rotting palm trunks; fine fibrous roots; and delicate fungal filaments, resembling lustrous black horsehair, which in wet tropical forests creep over trunks and fallen branches and provide an abundant and much-used fiber for the birds of these regions.

Plants supply a great variety of downy materials, chiefly attached to their seeds, including thistle plumes, milkweed silk, cotton, down from the pods of the silk-cotton and balsa trees, and from a great many other seeds tipped with plumes or embedded in cottony stuff. A few birds gather the hairy or downy covering of leaves, usually after this has been separated from the inner tissues by leaf-mining larvae, since otherwise this covering is difficult for them to detach from the leaf with their bills. The brown scales, known to botanists as ramenta, which thickly cover the young fronds of certain ferns, are a coarser material of allied origin. These are often used by birds, especially in those tropical regions where tree ferns grow abundantly. Mosses and liverworts are valuable as filler or padding for bird nests and sometimes form their chief bulk; likewise, the fine capsule-stalks of mosses are an occasional lining for small nests, including those of wood warblers. Lichens, mosses, and delicate ferns have already been noted as forming a decorative encrustation on the outside of many nests.

Tendrils enter into the composition of numerous nests and are the chief ingredient of a few. The retort-shaped nest of the Strange Weaver, except for the inner lining, is made wholly of coiled tendrils, giving it an odd appearance. They are also the chief component of the similar nest of another weaverbird, the Red-crowned Weaver of the Congo (Collias and Collias 1964a). The extremely slight, open nest of the Rufous Piha in the Costa Rican forests is made largely of coiled tendrils; the building female spends little time bringing material but much time arranging the stiff, recalcitrant pieces into her little mat, barely wide enough to hold her single egg.

Flowers are used to line the delicate nests of certain small birds, including those of the Reed Warbler and Bearded Tit in Europe and the Common Bushtit in America.

Materials of animal origin are, as a rule, subsidiary rather than principal components of bird nests. Feathers or hairs are frequently used as lining. The feathers so employed are usually picked up as they are found, but eider ducks pluck them from their own breasts, continually adding to the nest lining while they incubate their eggs. The plumage evidently becomes loose at this time, as does the hair that a female rabbit pulls from her body to make her nest when she is about to give birth. The parent's habit of lining the nest with its own down is widespread among swans, geese, and ducks, but rare in other families. Among passerines it may be confined to the Superb Lyrebird, the female of which

PLATE 33. Nest of the Band-tailed Barbthroat hummingbird attached beneath a narrow strip of a frayed banana leaf that forms a roof above it. The three eggs were laid by two females. A fourth was found on the ground below.

embeds her single egg in down from her flanks.

Birds frequently gather hairs or wool from thorns or barbed-wire fences, where it has been snagged and torn from the animals that bore it. Less often, nest builders collect it from its source. Given a favorable opportunity, the Brown-headed Honeyeater or "Hair Bird" of Australia will pluck hairs from a horse while the rider sits in the saddle, from the back of a cow or a native bear, or from a woman's head (Mathews 1924). In the Falkland Islands, the Tussac-Bird or Blackish Cinclodes is similarly bold, sometimes tugging at hairs on a man's neck and ears (J. S. Huxley et al. 1948). The Galápagos Flycatcher likewise tries to procure hair directly from a human source (Lack 1947). When building its nest in a hole, the Tufted Titmouse has been known to gather hair for the lining from men, women, squirrels, woodchucks, and opossums. Goertz (1962) watched a titmouse alight almost three hundred times on a pair of opossums resting in a tree and pull an estimated fifteen hundred hairs from their backs, while the marsupials, becoming resigned to her depredations, scratched, licked their fur, and napped. When Gray Jays found an abundance of shed deer hairs in early spring, they gathered billfuls from the snow and stored them between the needles of a spruce tree, apparently for future use in their nest (Lawrence 1968).

Although spider silk usually accounts for only a small proportion of nest bulk, it is one of the most important of all nest-building materials. A very large number of birds of the most diverse kinds, but chiefly of the smaller species, employ it to bind together the finer materials of their nests and to attach their structures to supporting twigs and leaves (plate 34)—it is the universal "cement of bird architects." The silken egg cases of spiders—white, yellow, or brown—are often used to cover or ornament the exterior of the nest.

Swifts do not need cobweb to bind together their slight nests of sticks, for they possess a special salivary secretion that is an excellent glue. While they build, their salivary glands enlarge greatly, then shrink in size after the nest is finished. Some cave-dwelling swiftlets (*Collocalia*) construct their nests almost wholly of this secretion, which forms the edible bird nests of the Orient. Swifts, a highly special-

PLATE 34. A hermit hummingbird's nest viewed from the back, to show how it is fastened by cobweb to a strip of leaf.

ized family of birds—in many respects a law unto themselves—are among the very few birds that, like the eiders, supply from their own bodies one of the materials used in forming their nests. In this they resemble the many insects, spiders, and other invertebrates that form their egg cases wholly of their own secretions. House Martins mix saliva with the clay of which their nests are built, as do hornbills with the matter that seals the incubating female in her nest cavity.

The cast skins of snakes and lizards are sought by numerous birds for their nests, probably merely because they provide a light, pliable material for the lining. Although it has been suggested that these exuviae are placed in the nest to ward off serpents that destroy so many eggs and nestlings, or to frighten away nest-robbing birds, there are two reasons for doubting these explanations. Birds that gather snake sloughs often bring to their nests superficially similar materials, such as onionskin, scraps of thin paper, or colorless plastic; if the exuviae of serpents frighten nest-plundering birds, these should be still more terrifying to the usually smaller birds who collect them.

The cast skins of reptiles are sought chiefly by birds that build substantial nests in enclosed spaces, whether in old woodpecker holes, natural cavities in trees, burrows in banks, or chambers in elaborate edifices made by themselves. Thus, among wrens, both the Northern and Southern house wrens and Bewick's Wren use snakeskins; but, as far as I have seen, none of the many species that build covered nests in trees and shrubbery collect them. Among American flycatchers, the Great Crested Flycatcher and other species of *Myiarchus* that nest in holes gather sloughs, but none of the numerous kinds whose so varied nests I have found in the open do so. Among antbirds, the Black-faced Antthrush, which occupies a hollow stub, may bring snakeskin to its nest, but none of the many species that hang their nests in forked twigs is known to have this habit. The Rufous-breasted Castlebuilder lays a carpet of cast reptilian skins the length of the long, narrow hallway that leads to the nursery in its imposing mansion of sticks; its relative, the Slaty Castlebuilder, stuffs such skins into odd crannies about its somewhat simpler nest. Among the few birds that place snakeskins in open nests are the Western Kingbird and Blue Grosbeak.

Another substance of animal origin used in bird nests is cow dung. Grackles of several kinds and Melodious Blackbirds sometimes use it instead of mud for the middle layer of their bulky cups. An excellent binding material, it is a component of the clay walls of the house in which I write. The droppings of the birds themselves enter importantly into the nests of certain species. Excreta serve to widen the narrow rock ledge on which the Black Noddy nests and to bind together the feathers that these terns bring to it (Cullen and Ashmole 1963). The flimsy nest of Inca Doves becomes more substantial as the droppings of the tenants, especially of the nestlings, accumulate, so that it provides a more stable recep-

tacle for second and later broods than for the first. Accordingly, the young are more likely to be successfully reared on a nest that has already been used one or more times than on a new nest (Johnston 1960). The nest of the Oilbird on a ledge in a dark cave is composed largely of regurgitated matter (D. W. Snow 1961).

The manufactured articles used in bird nests include string, yarn, rags, paper, fragments of plastic bags, and cellophane. Rarely a bird incorporates in its own nest one built by some smaller species. The Great Kiskadee, one of the largest of the American flycatchers, sometimes carries off the whole nest of a Vermilion Flycatcher or of a little seedeater to add to its own bulky, roofed structure of straws and weed stems. Hudson noticed this habit in Argentina, and I have seen it in the same species in Honduras.

The materials of bird nests can also be classified according to the parts they play in the construction, whether as foundation, middle layer, or lining. The difficulty here is that the materials that in large, coarse nests form the lining may serve as foundation in finer ones. Nevertheless, a few rough generalities can be made. Sticks, coarse herbaceous stems, vines, large dead leaves, mosses, and liverworts are common foundation materials for the bulkier nests. For the middle layer, birds frequently use leaves, especially small, thin ones, or strips of larger kinds, such as banana and wild plantain; papery plant epidermis; slender flexible vines; vegetable down; and mud or cow dung. The inner lining may be composed of fine fibrous rootlets, bast fibers, down, the setae or capsule-stalks of mosses, wool, horsehair, or the fungal strands that are sometimes called "vegetable horsehair." A scattering of birds, including the Red-footed Booby, a number of hawks, cuckoos, pigeons, colies, castlebuilders, broadbills, swallows, titmice, and sometimes even shearwaters and toucans, line their nests with few or many green leaves. As these soon wither, fresh leaves are in many cases brought throughout the period of incubation and more rarely after the young have hatched. They serve to stain white eggs in open nests and make them less conspicuous, to freshen the nest in hot weather, and perhaps to diminish the number of insect parasites by means of the hydrocyanic acid released as they dry, as suggested by Johnston and

PLATE 35. Nest and eggs of the Buff-throated Saltator. The lining of fine material contrasts with the coarse pieces on the outside.

available, yarns and string, are the textiles employed for woven nests. For thatching, which only a few careful homemakers undertake, coarse, flattish vegetable materials, such as bark, broad strips of banana leaves, grass blades, and leaf stalks are piled above their nest chambers by castlebuilders. The Red or Greater Thornbird, a member of the same family inhabiting Paraguay and neighboring countries, covers its mansion of sticks with old, dry horse droppings, which become felted together and form an efficient roofing (Hudson 1920). Some of the weaverbirds that live in the rainier parts of Africa insert broad grass blades under the roofs of their covered nests to form ceilings that shed heavy rain. The Striated Weaver and Black-throated Weaver plaster the inside of their woven, roofed nests with mud, which after drying hard probably helps to shed water (Crook 1963). The dull-colored hummingbirds called hermits that dwell in wet tropical American woodlands have a very different method of keeping their nests dry. By the liberal use of cobweb, the female hermit fastens her open cup beneath the drooping tip of a palm frond or a broad strip of a banana leaf, which forms a green living roof impervious to rain.

Gathering the Materials

Materials so varied as we have listed, without naming all that birds use, are naturally gathered in the most diverse manners. The bill is the instrument almost universally employed to gather materials for the nest, but in rare instances the feet are used. It is not easy to understand how, with a bill already laden, a bird manages to pick up additional pieces without dropping what it already holds, and then, if it be a male, to sing with a full bill. Watching at close range a male Buff-rumped Warbler who sang profusely while gathering material for the nest that he was helping his mate to build beside a stream, I saw clearly that while singing he held the straws against his upper mandible, which could only have been done with his tongue, invisible to me. The trick of adding piece to piece when building is another application of the capacity that enables a bird to continue to catch insects with a bill already full, as many passerines do while feeding nestlings. In

Hardy (1962). Flakes of stiff bark are used by woodcreepers to line their nest cavities in tree trunks.

For binding and cement, cobweb is the material so widely employed that it is difficult to imagine how birds of many kinds could continue to build their nests if they succeeded in devouring all the spiders. Swifts, swallows, and hornbills, as has already been mentioned, use more or less saliva to bind their twigs or clay together. The Buff-breasted Acacia Warbler of Africa collects the juice of acacia trees to consolidate its felted nest (van Someren 1956). A hummingbird on the high Peruvian plateau, the Andean Hillstar, uses a gum, possibly regurgitated nectar, to attach its nest to a slight projection on a cliff face (Dorst 1956).

Vegetable fibers of various sorts, strips of grass or palm fronds, slender flexible vines, and, where the Hawfinch, larvae for the young are clamped be-tween the tongue and lower mandible (Mountfort 1957). Since the bill of the finch is very much thick-er than that of the wood warbler, it is not surprising that it employs different methods. A bird can sing freely with his tongue pressed tightly against a load of straws or food because he does not use the ter-minal part of this organ to modulate sound, as we do. Not all birds are able to gather things in an al-ready laden bill. Many bring no more than a single article of food to their nestlings at one time; others, notably pigeons, consider one straw or stick an ade-quate load to carry to the nest.

Twigs may be picked up from the ground; but strong-billed birds, including a number of jays, cuckoos, plantain-eaters, and some of the larger pi-geons and American flycatchers, break them from dead branches high in trees, often struggling to ac-complish this. But with its powerful bill, the Haw-finch clips twigs from treetops as easily as this could be done with sharp garden shears. The Swallow-tailed Kite, soaring on outstretched wings, dips to the crowns of the loftiest trees of the tropical forest and snaps off slender terminal twigs with its feet, without ever alighting. This evidently requires extremely sensitive tactile discrimination, for if the kite does not grasp hard enough it will not detach the twig, but, if it seizes an unyielding branchlet too firmly, it could be jerked out of the air, with disas-trous consequences to its long wings. To alight on its growing nest, the kite finds it convenient to take the stick in its bill, and it effects the transfer while soaring above the treetops. The kite's weak bill can hold a long stick only by grasping it near the center of gravity; the bird sometimes tries again and again to secure a firm hold, and often it drops its burden to the ground. While soaring on ascending air cur-rents, these kites also catch in their feet the flying insects on which they largely subsist. Whether for immediate consumption or to be taken to the nest, the insects are transferred to the bill while the bird continues on the wing.

The Chimney Swift, which like the kite gathers twigs from treetops for its nest, has been seen carry-ing them both in the bill and in the feet. Careful ob-servers insist that they are broken from the trees by the swift's feet. Probably, after being carried be-neath the body for a while, they are taken into the mouth as the bird approaches its nest, just as is true of kites. Swiftlets that nest in the total darkness of deep caves carry materials in their feet because they must leave their mouths empty to make the sounds that they use in echo location; but species that breed in situations with more light generally carry materials in their bills or even in their mouth cavities (Medway 1962). Although swifts collect nest material high above the ground, those other tireless fliers, the swallows, do not disdain descend-ing to the earth to gather in their bills the straws, pine needles, mud, or whatever else they need for their constructions.

Although, among birds as a whole, the bill is most often used for carrying, and after that the feet, some birds have peculiar methods of transporting things to their nests. The Great White Pelican of Africa carries vegetable nest materials inside the ca-pacious pouch, which it uses for catching fish; it flies to the nest with the pouch "bulging irregularly, like a sack full of rubbish" (L. H. Brown and Urban 1969). Hanging parrots of the genus *Loriculus* and lovebirds of the genus *Agapornis* are, along with the Monk Parakeets of southern South America, ex-ceptional among parrots in building nests instead of breeding in unlined cavities. Hanging parrots and lovebirds cut strips of bark, leaves, and grass; to take these materials to their nests, they tuck them between their contour feathers, which are equipped with special microscopic hooklets that help hold them in place. In captivity, hanging parrots cut neat, curving strips from newspapers (Buckley 1968). Monk parakeets shred the ends of twigs that they break from trees.

The Red-headed Weaver specially prepares the twigs that compose his hanging, retort-shaped nest. First he tears all leaves from a twig, dropping them to the ground. Next he uses his bill to loosen a fine strip of bark, which remains attached to the twig an inch or less from the point where he will eventually sever it. Then he wrestles with the twig, bending it to and fro until it breaks off. Finally, he carries the twig to the swinging nest, where he holds it under a foot while he attaches it to the structure by means of the loose tag of bark, which he ties with a knot (Crook 1963).

Since the fibrous strands in stems and leaves of herbaceous plants are commonly more resistant to decay than are the soft tissues in which they are embedded, they are separated out for the birds by this natural process. Nest builders need only find a weed or vine stem in just the proper stage of disintegration in order to have a supply of strong, clean fibers. But birds that weave their nests often prefer to work with fresh green strands. Standing on the midrib of a great banana leaf that has been frayed by wind, a hen Montezuma Oropéndola bends down and nicks the hard, projecting underside of the rib with the sharp tip of her bill, grasps a strand of the strong fibers, rips out a foot or two of its length, doubles the prize in her bill, and flies back to the crowded nest tree, often with the free end of the strand streaming far behind her. She also tears long, thin fibrous strips from living palm fronds. With these materials and lengths of thin vines, she weaves her admirable swinging pouch. Many of the African weaverbirds prefer to work with fresh green materials, which a bird gathers by nicking the side of a living grass blade or segment of a palm frond, seizing the severed end in the bill, and tearing off a narrow strip as it flies away. Orchard Orioles also build with green grass that later turns brown.

The Great Reed Warbler of Europe builds its nest largely of half-decayed reed and grass leaves, which, while clinging to an upright stem of the emergent vegetation, it fishes from the water of the swamp where it lives. This warbler attaches its open cup to erect stems of the reed *Phragmites*. To prevent their slipping down, the materials of the nest must be wrapped tightly around the smooth reeds while soft and pliable. Accordingly, when the warbler varies its procedure by gathering dry pieces, it dips them in water before taking them to the nest (Kluijver 1955). The Yellow-headed Blackbird, which builds a somewhat similar nest in a similar situation, likewise uses wet materials that it fishes out of the water of the marsh (Bent 1958). Other birds, including European Dippers and Mistle Thrushes, have also been seen immersing their materials in water to soften them.

One reason why many birds build most actively in the early morning may be that at this time their vegetable materials are moist from dew or the higher humidity of the night. Later on a sunny day, when the air becomes drier, dead leaves and other plant fragments become stiffer and resist being molded into the desired shape. Indeed, when a Winter Wren builds with dry materials that do not cohere, its nest may fall apart after a few days. Accordingly, this wren prefers to build during or just after a rain, and the same is also true of the Kittiwake and some other birds (Armstrong 1955). Birds that require plastic mud or clay, such as Cliff Swallows and horneros, may be forced to suspend operations if their sources dry and harden in a long rainless period. Sometimes an understanding friend will moisten the ground for them in dry weather.

In the humid tropics, where some of the greater trees support veritable gardens in the air—smaller trees, shrubs, vines, orchids, bromeliads, aroids, ferns and other herbaceous plants, mosses, liverworts, and lichens—birds may find a wide variety of materials for their nests without ever descending to the ground. Even rootlets are to be had high in the treetops. Some birds, including the Golden-bellied Flycatcher of the Costa Rican highlands, place their nests in crannies in the profuse covering of mosses and other epiphytes far up in the trees of the cloud-bathed forests. At high latitudes, nest builders are more dependent on what they can find on the earth's surface.

Bird nests, as already mentioned, are an important source of materials for other bird nests. Feathered builders are ever willing to save labor by drawing upon some other nest nearby, as the inhabitants of medieval Rome used the magnificent ancient edifices as quarries for the stone they needed for their own ruder constructions. Many birds will not lay again in a nest from which they have lost eggs or young, but they often build a new nest near it, levying largely upon the pillaged structure for their materials. Or any conveniently situated abandoned nest may be drawn upon, if its ingredients are of the proper sort, well preserved, and easily disengaged.

Building birds extract pieces not only from unoccupied nests but also from those actually under construction, or more rarely from those containing eggs, of either the same or a different species. Blue Tanagers, widespread in tropical America, are frequent offenders in this matter, pulling fibrous stuffs indiscriminately from the nests of all their building neighbors, if they find them unguarded. Montezuma Oropéndolas, building in crowded treetop colonies, try to snatch loose fibers from their neighbors' nests, or even from a bird alighting with a billful of newly stripped banana or palm strands. The poor victim of such effrontery is in a most embarrassing situation, for, if she opens her bill to protest her neighbor's shameless conduct, she will lose everything! If this were the constant practice of all the females in a colony of oropéndolas, rather than an occasional misdemeanor, their long pouches would never be finished. Probably most birds that breed in colonies, including weaverbirds, grackles, terns, herons, and penguins, are guilty of a certain amount of thievery. This propensity is heightened on barren islands where suitable materials are scarce. Where such pilfering is rife, the nest must be continuously guarded by one member of the pair from the time its foundation is laid.

In Panama, I observed Rufous-tailed Hummingbirds that were nesting unusually close together persistently tear the downy ingredients from each other's structures. Some nests with eggs were destroyed in this manner while the owners were absent foraging. One unfortunate hummer, more honest or less clever than her neighbors, continued to build for a whole month without having anything to show for her unremitting efforts; her nest materials were stolen almost as fast as she could gather them. At least twelve distinct beginnings of nests came to naught during her thirty-one of fruitless toil. This was an extraordinary situation, evidently caused by the concentration of breeding birds attracted by an abundance of flowering shrubs in the garden where these strange proceedings happened, coupled with a local scarcity of vegetable down (Skutch 1931).

One wonders why natural selection has not repressed theft of nest materials, especially in colonial birds, in which it would seem to reduce reproductive efficiency. However, this unsocial conduct appears to have certain compensatory advantages. Among oropéndolas and other birds with similar nests, it is not easy for a thief to detach a strand that has been properly woven into the fabric; it is chiefly the loosely dangling pieces that invite pilfering. Thus thievery discourages slovenly building. Among penguins, herons, and other birds whose nests are loose accumulations, rather than well-consolidated fabrics, thievery falls least heavily on pairs that learn from the start of construction to coordinate their movements and keep their nests constantly guarded. Such pairs will make the best parents, least likely to lose their chicks to marauding skuas or other avian predators. Thieves in a nesting colony might be compared to inspectors in a factory; they help to keep the breeding pairs at high efficiency by penalizing the careless and the lazy.

CHAPTER 9

The Art of Nest Building

Our survey of the forms and sites of nests, the principles of safety that govern their choice, the participation of the sexes in building, the materials and how they are gathered has prepared us to examine the actual process of construction. We shall see how birds, with only their bills and feet, accomplish things for which we need special equipment; how with similar organs and procedures, but different materials, they achieve very dissimilar results. The marvel of bird nests is that, without specialized tools to help them, birds finish constructions so useful and beautiful that they win our admiration.

The Tools and the Mold

The building bird's only tools are its bill and its feet. According to the kind of nest that it will make, the bill may be pressed into service as a chisel or drill for carving into wood, a pick for delving in earth, a shuttle for weaving, a needle for sewing, a trowel for plastering. It is difficult to predict from the bill's shape how it will be employed in building; often it must serve in various contrasting ways. Although the sharp bills of orioles and other icterids are obviously well adapted for weaving, the short, thick bills of many weaverbirds seem inappropriate for this activity; yet some of the latter weave nests not inferior to those of orioles and oropéndolas. The jacamar's long, slender bill appears to be just the instrument for weaving a delicate fabric, but in fact it is applied to the coarse work of digging a burrow in the ground or a hole in a termitary. The short, swollen bill of the Prong-billed Barbet, with three fine points at its end, is, surprisingly enough, a wood-carving tool. It is used to excavate a chamber that much resembles that of a woodpecker, in wood almost as firm as the latter can work with its sharp chisel. The broad, flat bills of American flycatchers, some of which are called "spadebills," seem to be admirably fitted for constructing clay nests, such as Cliff Swallows and horneros build; yet, with the exception of the phoebe's, they are not employed for working with mud. The fairly sharp bills with which horneros construct their clay ovens appear to be more appropriate for fashioning the felted structures that many flycatchers manage to complete with their seemingly so inadequate tools.

When one surveys the whole array of bills in relation to the nests that are built with them, it seems clear that bills have not in any considerable degree been modified by natural selection to make them more adequate instruments for nest construction. If forces were operating to modify bills for more efficient nest building, by the principle of adaptive convergence we would expect birds employing similar building techniques to have bills of much the same form, rather than of the contrasting forms that we actually find. Bills are adapted primarily for food gathering, and the bird must make shift to construct its nest with a tool that has been evolved for another purpose. Even the woodpecker's bill, so well fitted for carving its nest hole, probably evolved in the first place as a drill for extracting insect larvae from solid wood. A possible exception to the generalization that bills are not modified to become efficient building tools is provided by the Celebesian Starling. It has a heavy bill, somewhat like that of a woodpecker, that it uses for carving a nest chamber in solid dead wood. Since this starling seems to feed largely on fruit, its bill does not appear to be adapted to its foraging habits so much as to its mode of nesting; but we need to know more about its manner of life before we can reach a firm conclusion (Gilliard 1958).

Compared to the bill, the feet seem to take a minor part in nest building. Indeed, their role is likely to be overlooked by one who casually watches the construction of a cupped nest, whose opaque walls conceal the nether parts of the building bird. From time to time, however, the careful observer will see the builder's body rising and falling slightly with a motion imparted by the feet, which are used for arranging the materials on the bottom and sides of the concavity. This stamping, scraping, or kneading with the feet is, in fact, one of the most widespread and persistent of all the movements of building birds. It is found in many avian orders and possibly was inherited from reptilian ancestors. Turtles, for example, use their legs or flippers for digging holes in the ground into which they drop their eggs, then for covering them over with the loosened earth or sand. Megapodes build the great piles of fermenting debris in which their eggs are incubated chiefly, if not wholly, by kicking the materials together with their strong feet. Numerous shore birds, terns, and other birds make a nest by scraping a shallow depression in the ground with their feet. The Least Tern throws back the sand with only one foot, using the other to brace itself (Hardy 1957). A domestic hen in search of a nest sometimes pushes behind furniture in a dark corner of a room, or enters an empty box, where she scratches loudly

with her feet. Sometimes she is accompanied by a rooster, who squats in a promising spot and does the same.

Kingfishers, motmots, jacamars, and other burrowing birds loosen earth with their bills. Each time one of these excavators returns to its work at the head of the shaft, it pushes the dirt outward by alternate backward kicks of its feet. This sends out alternating parallel jets of earth, which follow the bird inward until it is lost from view in the darkness of the tunnel. It is difficult to learn how nocturnal petrels and shearwaters dig their burrows, but, by using an infrared viewer, Grubb (1970) watched a Leach's Petrel kick back the loosened earth with its feet, throwing it as much as one yard.

It is hard to imagine building procedures more different than those employed in digging a burrow in the ground and building a cupped nest in a tree, yet when we analyze them we find an unexpected similarity in the basic movements involved. Both kinds of construction entail the use of the bill to loosen the earth, or to fetch, place, and arrange the building materials. Both involve kicking with the feet, which in one case serves to throw out the loose earth, in the other case to arrange and compact the materials in the nest's concavity. Even tiny hummingbirds employ a kicking or kneading movement to consolidate the materials of their downy cups, and a similar operation is widespread among building passerines. Most exceptional in this order is the Bearded Bellbird, in which the female fashions her open cup of forked twigs with her feet and breast, using her very broad, short bill only for gathering and carrying material (B. K. Snow 1970).

Although the feet are largely or wholly responsible for making the simplest nests, such as scrapes in the ground, they have ceased to enter directly into the construction of some of the most advanced types of nests. As far as we can see, the carving woodpecker employs its feet only to cling to a vertical surface, while it uses its bill for breaking away particles of wood and its mouth for throwing them out of the cavity. Orioles and oropéndolas that construct long pouches carry on most of their weaving while hanging head downward in a lengthening sleeve. Although it is difficult to see just what they do in this enclosure, they obviously need their feet to cling in the vertical tube and all their weaving is evidently done with the bill. Possibly in some species, after the rounded bottom of the pouch has been woven in, the feet are employed to arrange the lining. Oropéndolas hold beneath a foot the leaves that they tear with their bills to form a loose litter in the bottom of their long pouches. Weaverbirds use their feet, especially when starting a nest, for drawing together the supporting twigs and for holding one end of a strand while they tie or entangle the other end with their bills, but apparently they do not employ their feet for shaping their structures.

In addition to tools and the skills to use them, avian builders require a form or pattern to control the size and shape of their structures. Smaller and simpler nests are molded to the builder's own body. Even in elaborate nests with extensive coverings and long entrance ways, such as those of castle-builders, essential features, such as the actual receptacle of the eggs and the doorway, appear to be fitted directly to the builder's body. The dimensions of the "kidney-shaped" nest of a weaverbird are determined by the reach of the builder while he stands on the foundation ring from which he works.

Procedures Used in Nest Construction

Among the simplest, least skillful, and apparently also the most primitive, of the procedures used in nest building are sideways throwing and sideways building. When a broody hen leaves her eggs, or is lifted from them and dropped on the ground, she often picks up leaves or straws and tosses them sideward and backward, sometimes letting them fall on her shoulders or back. As a Marbled Wood-Quail walked from her eggs on the floor of tropical forest for her long morning outing, she often picked up dead leaves and threw them over her back. Some fell on the roof of her covered nest and some in front of the doorway in its side, improving its already good concealment (Skutch 1947).

Similar behavior may be seen in many other birds that breed in scrapes or loosely constructed nests on the ground, including ostriches, rheas, emus, loons, geese, grouse, pheasants, cranes, and numerous shorebirds. As the bird walks away from its nest site, often at the end of a spell of incubation, it picks up whatever small, loose objects it finds in its path and tosses them backward to the right or left, therefore in the general direction of the nest it is leaving. Thus, according to the situation of the nest, grass, leaves, sticks, pebbles, or shells are gradually shifted in its direction. When these objects come within reach of the bird sitting on the nest, they may be crudely arranged by sideways building, the incubating bird taking them in its bill and laying them on either side of itself, or sometimes dropping them in front of its breast. If it faces in various directions, it may accumulate a substantial rim around the nest hollow.

The nests of species that follow this crude method of building vary greatly in size, sometimes becoming bulky if the immediate surroundings provide much material, but growing only slightly, or not at all, if little loose material is available near the site. Likewise, since material is thrown nestward chiefly as the bird leaves, the more inconstantly it sits, or the more often it is disturbed, the bulkier its nest is likely to become. In sideways builders, the nest is not made to a standard size, as in birds that bring their materials from a distance. When sideways throwing is supplemented by carrying materials to the nest in the bill, a step toward the construction of more elaborate nests has been taken (Harrison 1967).

The procedure for constructing more adequate nests depends largely on whether they are begun at the bottom, top, or middle. Nests placed on the ground, on a horizontal branch, or in a vertical fork are usually built from the bottom up. Open cups suspended by the rim, like those of vireos and many antbirds, are started at the top and built downward. The same is true of cups supported between vertical stems, like those of the Great Reed Warbler and Red-winged Blackbird, and likewise of the long, pensile nests of many American flycatchers and icterids. Many weaverbirds begin their nests in the middle, by constructing a vertical ring of fibrous strands between twigs or divisions of a palm frond. On one side of this upright foundation-ring they weave the covered nest. On the opposite side, they extend the fabric to form a visorlike projection, a porch, or even a long, hanging entrance spout (fig. 11).

As an example of a nest that is built upward from

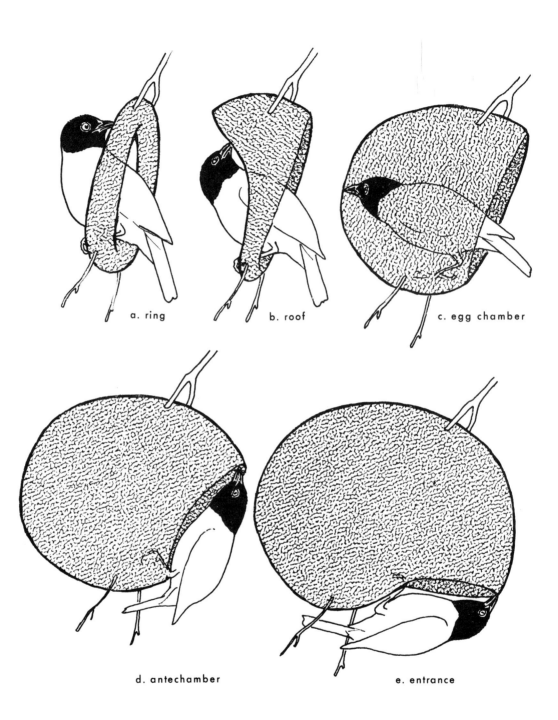

FIGURE 11. Stages in the construction of a woven nest by a male Village Weaver, including (*a*) initial ring, (*b*) roof, (*c*) egg chamber, (*d*) antechamber or vestibule, and (*e*) entrance (redrawn from Collias and Collias 1962).

the base, we may take that of the Bellicose or Lesser Elaenia. Some years ago, one of these small, severely plain gray flycatchers built on a slender, horizontal branch of a rose-apple tree in front of my window, where I enjoyed an exceptionally favorable opportunity to follow her procedure. The female, who worked alone, began by gathering many tufts of cobweb and wiping them from her bill onto the bark of the branch and twigs that provided lateral support. Then she collected fragments of dry grass inflorescences, bits of fibrous material, moss, and the like and attached them to the cobweb. From time to time, she brought more of this adhesive stuff to cover her growing accumulation and hold it more securely in place.

As the elaenia's nest grew and took the form of a ring with a depression in the center, she used her bill, feet, and body to press it into shape. She had two principal procedures. In one, she sat in the cup, pressed down her breast, and pushed the nest materials outward by rapid backward movements of both feet, often turning with a vigorous motion to face in different directions while she did so. This rotation about a vertical axis rounded the nest and helped to give its wall uniform thickness and height. In the second procedure, she sat in the nest and used her bill to arrange the materials, chiefly those on the nest's outer surface, pulling the loose ends inward and upward and then tucking them down into the rim. Thus the circular wall was built up as its constituents were pushed and pulled together from opposite sides by her feet and bill. The height appeared to be regulated chiefly by the base of the builder's tail, but also in part by her foreneck. I have seen the related Yellow-bellied Elaenia bend down her tail on the outside of her nest, apparently using it, as well as her bill and foreneck, to force the rim inward. By such procedures, many small, dainty nests are given an incurved rim that decreases the probability that the eggs will roll out when wind tosses the supporting vegetation.

As Kramer (1950) pointed out, the raking motion of the feet against the sides of the wall has an upwardly directed component that pulls down the posterior end of the body, thus pressing the underside of the horizontally held tail against the rim with a force greater than could be provided by the

bird's slight weight alone, although obviously the pressure that feathers can exert is limited by their fragility. The backward and upward kicking also pushes the bird's body forward and presses its breast against the nest, thereby helping to mold the wall. The Great Reed Warbler pushes so hard against the moist materials of her nest that her breast feathers become thoroughly wet, forming a dark breast patch that serves to distinguish her from her mate (Kluijver 1955). In some birds, the ends of the wings seem to aid the tail in smoothing down the rim to a uniform level. Birds that build deep cups sometimes push their breasts so far down into the concavities that their tails and wings point upward and they almost appear to be standing on their heads. However, I did not see the elaenia do this in her shallow structure. It seems not to be known—and it is a detail most difficult to observe—whether birds keep their toenails extended or turned inward while shaping a nest with their feet. If extended, the foot action would seem to be largely a kneading or raking; if turned inward, a stamping to consolidate the materials.

Apparently largely as a consequence of the elaenia's footwork, she had very little material in the bottom of her nest at a time when the sides were well advanced. I have noticed in other nests that are built upward, including oven-shaped structures resting on the ground, that the bottom is one of the last parts to be finished. It often remains bare until the soft lining is added as the nest nears completion. This elaenia exercised a good deal of discrimination in the choice of her materials; twice I saw her carry away pieces that, after being placed in her nest, proved to be too long or thick to be molded into the desired shape. She also flew up beneath her nest to pull away long strands that dangled untidily beneath it. She gave the impression that she knew exactly how she desired her finished structure to look. By no means all birds are so neat; some permit long, unsightly pieces to remain hanging below an otherwise shapely and compact structure.

In vireos and other birds that build open cups attached by the rim, the first step in construction is to wrap material around the supports (plate 36), often with the addition of cobweb to attach them more firmly. Strands are then looped from side to

PLATE 36. Nest of the Slaty Antshrike attached by its rim to the arms of a forked twig, like that of a vireo.

and female work together and choose for their nest site the end of a slender, drooping branch or dangling vine that hangs free of surrounding vegetation. The first step in construction consists in covering the point of attachment with fine vegetable fibers, cobweb, and other strands. When wrapping fibers around the thin supporting twig, the agile little flycatchers pivot their bodies completely around it. Soon they bring a variety of light, fluffy materials such as the pappi of composites and other kinds of down, shreds of vegetable epidermis and papery bark, fine grass blades, and often small withered flowers, all of which forms a loose mass that grows downward like an icicle. Then, arriving with a new contribution, they alight on the twig at the top of the elongated, tapering tuft and creep head downward over its surface, often at the same time circling it, tracing a descending spiral as they attach what they have brought.

The tody-flycatchers make no attempt to weave the fibers into a fabric but merely tangle them together until the accumulated mass is considerably larger than the birds themselves. Then, one at a time, the flycatchers cling to the side of this solid fluff of material and try to push into its midst. At first they succeed only in opening a slight depression in the side of the tuft, but little by little they force the fibers apart until the expanding cavity occupies the center of the mass. New materials are continually taken into the growing hollow, and additional pieces are fastened to the top and outside of the structure. When this central cavity, the future nest chamber, is first formed, the walls are thin and must be adequately lined with fibers and miscellaneous materials—a time-consuming task. Finally, an inner lining of small feathers is applied to the top, sides, and bottom of the chamber, until it is tapestried all around with feather down (plate 37).

Essentially the same procedures are used to build the more elaborate hanging nests of flycatchers, including the conical structure of the Sulphur-rumped Myiobius, in which the rounded chamber has a side opening that is covered by an apron (fig. 12), and the retort-shaped nest of the Sulphury Flatbill, which is entered through a downwardly directed spout. Each of these structures begins as an unbroken weft of entangled fibers, which is gradually

side, outlining the bottom. In this case, the base of the bird's tail cannot determine how high the rim will be raised, for the nest's rim is the fixed point of departure. But the base of the tail may control how far the bird pushes down the nest's bottom with the kneading or trampling movements of its feet. In deeper, more pouchlike nests in which the bird incubates with its tail and head turned upward, the depth must be otherwise controlled.

The long, hanging nests so common in tropical America and so diverse in form are made by two quite different methods, felting and weaving. The former method is followed by the flycatchers; the latter, by the orioles and their relatives. A species that constructs a felted nest is the Black-fronted Tody-Flycatcher, a diminutive yellow-breasted bird found from Mexico to Peru and Brazil. The male

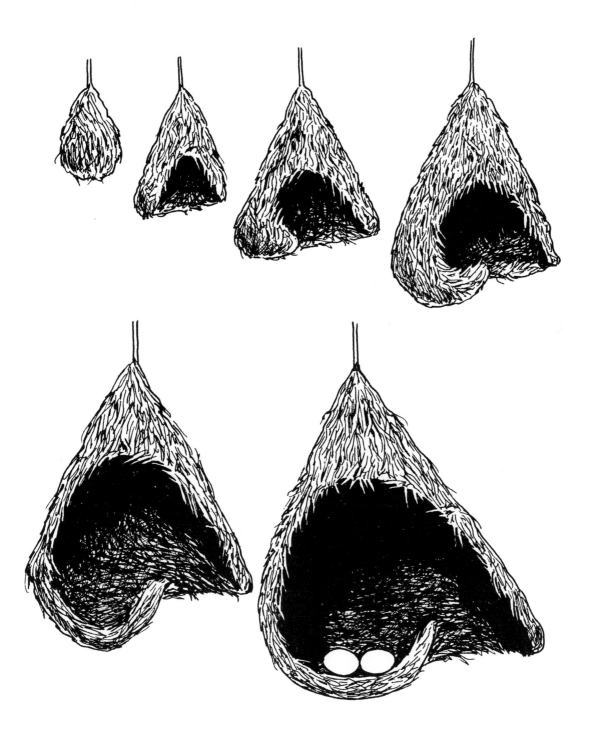

FIGURE 12. Stages in the construction of a felted, pensile nest by a female Sulphur-rumped Myiobius. A tangled mass of fibers (*upper left*) is hollowed out and spread apart, with the continual addition of new material to the wall, until it assumes the final form (*lower right*) (redrawn from Skutch 1960).

PLATE 37. Pensile nest of the Black-fronted Tody-Flycatcher, in an ornamental shrub in a Panamanian garden. Like other pendent nests of American flycatchers, it is made by felting rather than weaving.

forced apart into the desired shape, with the constant addition of fresh materials to maintain the thickness of the expanding walls. The tissue grows by intussusception, as botanists would say. New contributions are simply pushed into the mass, never woven through it. In these two species, the female builds alone.

Very different from the foregoing is the method of construction of the woven pouches of oropéndolas, caciques, and orioles. Here is no spreading apart of a continuous mass, but the successive addition of structural elements, each of which, from the first, is attached in the position it will occupy in the finished nest. In these birds the female, who alone builds, begins by wrapping fibers around, and looping them between, suitably situated twigs to form a more or less vertical ring, her future doorway. Fi-

bers torn from palm fronds or other large leaves, slender vines, and similar materials are worked into the ring until it becomes a sleeve, whose lower edge is extended by weaving. As the builder hangs head downward in the sleeve, hidden from view, the watcher sees her sharp bill push out here and there as she sticks a fiber through the fabric or pulls in a loose end—an operation not to be seen while flycatchers build their felted nests. The bottom is not closed off until the pouch reaches its final length—another point in which weavers differ from felters. Although it would seem much more convenient for the bird who has been working head downward to leave through the bottom of her nest as long as it remains open, she always turns around and climbs out through the doorway at the top. When a finished nest is examined, the long, strong fibers are found to be intricately interlaced and knotted together, forming a regular, even fabric, with meshes wide enough to admit air to the brooding female and her nestlings. These woven nests, together with those of some of the African weaverbirds, seem to represent the highest achievement of the nest maker's art.

Is Nest Building Innate or Learned?

We must now ask how birds acquire the ability to make the structures that so excite our admiration. To what extent are their building procedures innate and to what extent learned? There is no doubt that some of the basic movements involved in nest construction are inborn, and in some cases they are displayed long before they will be needed. When held in cupped hands, a seventeen-day-old Olive-backed Thrush "performed perfectly typical nest-shaping movements characteristic of adult females. The bird simultaneously kicked backward with both feet and forcibly thrust its breast against the side of the cup. The wings were held rather high on the back but not unfolded and the tail was rather depressed. The bird would perform a few rapid thrusts and kicks and then turn slightly in the cup and repeat these acts. It fell asleep after a few such attempts" (Dilger 1956).

A young Blue-throated Green Motmot, which was brought to me before it could fly and which I tried to rear, behaved most oddly one afternoon when it was probably somewhat over a month old. Dropping to the bottom of its box, it pecked at the paper covering the bottom and one of the walls, then kicked back rapidly with both feet, sounding a tattoo against the board. It was performing the two chief movements involved in digging a burrow, pecking to loosen the earth and kicking backward to throw it from the tunnel. Nestling European Cormorants, when only two weeks old, try to arrange the sticks of their nest with quivering movements of their bills, but in an ineffective manner. After two or three more weeks, they continue these movements until the twigs are firmly fixed in the nest (Tinbergen 1951). Rather surprisingly, hand-reared male Song Sparrows frequently made typical nest-molding movements when from one to three and one-half months of age, and they also carried grasses, although in this species the nest is normally built by the female alone (Nice 1943). Chicks of the Virginia and Sora rails and of the American Coot carried or pulled inward nest materials, or made other building movements, when from two to four weeks old (Nice 1962).

On Midway Island in the Hawaiian Archipelago, Black-footed Albatrosses begin kicking sand out of their nests as soon as they hatch. When storms blow sand into their nest hollows on the open beach, such activity may save their lives by preventing their being buried alive, which may happen if the storm is so prolonged that they become exhausted. As they grow older and wander from the nests where they hatched, the chicks may dig new nest depressions for themselves, up to one hundred feet from their natal spots and often in shadier places. Such nests are occupied only by day; at night the wandering chicks return to their original nests (Rice and Kenyon 1962).

These innate building movements must be refined and perfected by practice to result in effective nest construction. Unless they are performed in an adequate site, with appropriate materials, they will not produce a nest characteristic of the species. Although we have a dearth of pertinent observations, it is likely that each bird instinctively chooses for its nest a site like those its kind habitually uses. Caged Zebra Finches tended to select for their first nest the habitat, but not the same kind of container—a wire-mesh cup or wooden box—in which they were reared (Sargent 1965). Probably such innate or early acquired preference provides only a general orientation for the bird who is about to build for the first time, prompting it to start its nest on the ground, in a shrub, or high in a tree; but it does not provide guidance in all the fine points of site selection. The bird who builds in a bush must know, for example, just what configuration of branches is needed for the adequate support of its structure, and doubtless this knowledge is gained by experience, through trial and error. Birds about to build may test the quality of a prospective nest site by sitting in it, often repeatedly, and making shaping movements (Pokrovskaya 1968). One sometimes notices a bird starting a nest in a site that could never hold the finished structure. Although the past history of these birds is, unfortunately, unknown, probably in most cases they are youngsters experimenting with their first nests.

From the bewildering array of materials that many natural habitats provide, how does the bird choose just the proper ones for its nest at various stages of construction? Here, too, it seems that the bird must have certain innate or early acquired preferences that are refined by experience. In a very artificial situation, Sargent demonstrated that young Zebra Finches tended to select materials the color of the nest in which they were hatched, provided the color was not red. When young Ravens begin their very first nest, they are uncertain what material to use, and they pick up sticks that are quite unsuitable. With practice they learn to select twigs of the proper size and shape. A Jackdaw building for the first time will pick up a great variety of inappropriate objects, but it gradually learns to avoid those that will not serve its purpose. When grass was first presented to a Canary in a nest-building mood, it was promptly taken and carried to the nest pan in a cage. Here, however, the novice in building seems not to have been confronted by the necessity to choose among diverse materials (Thorpe 1956).

When we contemplate the vast variety of structures that birds build by means of long-continued series of a few simple, fundamentally innate operations, and the complexity of some of these nests, I believe that we must agree with Thorpe that "the bird must have some 'conception' of what the completed nest should look like, and some sort of 'conception' that the addition of a piece of moss or lichen here and here will be a step towards the 'ideal' pattern, and that other pieces there and there would detract from it." (Recall how the elaenia removed inappropriate and dangling pieces from her nest.) Without such an "ideal" or archetype to guide her, an oropéndola might go on lengthening her pouch indefinitely, or she might, on the contrary, close off the bottom prematurely. Yet oropéndolas' nests, despite some variation in length, fall within certain serviceable limits. One who has taken to heart the doctrines of John Locke may doubt that the "conception" or "ideal" is present in the callow bird's mind as a detailed visual image. What is probably present is a certain innate predisposition that determines the shape of an image gradually emerging as the bird continues to build—a blank form or empty matrix that is filled in the light of experience. Any development that does not conform to the predetermined model will be rejected; any that brings the growing nest closer to this innate model will be accepted and take the bird a step nearer to the complete visual image that it must possess to finish a nest that conforms closely to the ancestral type.

Little is known about whether birds improve their nests with practice in building. Canaries, whose imperfectly developed innate nest-building movements did not include the side-to-side weaving motion, tended to build diffuse, fluffy nests; but as they gained experience, their structures became tidier. The Serin Finch, however, can produce a virtually perfect nest at the first attempt (Thorpe 1956). Nice's (1943) observations on banded Song Sparrows of known age led her to conclude "that a Song Sparrow builds her first nest as quickly and expertly as she does her last, and that there was no evidence of improvement in building more secure nests, nor in concealing them more effectively. Nor was there evidence of some birds consistently surpassing others in these 2 prime requisites in nest building."

The aforementioned finches build simple open cups, and the situation may be different in birds that construct more elaborate nests. A Yellow-headed Blackbird, whose deep cup is attached to upright stems by strands wrapped around them,

built four imperfect nests before she succeeded in finishing one that was usable (Bent 1958). The ornithologist in the tropics, where elaborate nests abound, often finds imperfectly made examples that have been abandoned before completion, but unfortunately the history of their builders is almost never adequately known. The first stage in the construction of an oropéndola's woven pouch—the formation of a loop or ring of fibers from which the nest hangs—appears to be the most difficult step in the whole undertaking. In a single colony, some individuals start their nests without difficulty, whereas others have great trouble forming rings that will bear their weight, spending days on this preliminary stage of their work and trying again and again before they succeed. Chapman (1928) noticed this in the Chestnut-headed or Wagler's Oropéndola, as I did in the Montezuma Oropéndola. We both speculated that the females who had so much trouble, who often perched in the nest tree holding fibers as though they were uncertain what to do with them, were novices in their art, although we had no proof of this.

In a bird with a comparably elaborate nest, the Village Weaver, proof of the importance of experience in building has been supplied by observations in both field and aviary by Dr. and Mrs. Collias (1964a, b). Fledglings begin to manipulate a variety of materials soon after they leave the nest, and males build crude structures long before they are sexually mature. By practice, at first fumbling, they gradually achieve competence in the several skills involved in building, such as gathering material, tying knots and weaving, and finally constructing nests. The first nests completed by young males lacked the neatness and finish of the products of experienced builders (plate 38).

Although practice was indispensable for the development of a good building technique, the sight of an already finished nest, or the example of other individuals, was not. A male Village Weaver taken from the nest before his eyes were open and hand-raised in total isolation from other weaverbirds, gradually developed the capacity to build a typical nest. This definitely refutes an older view, favored by Wallace (1871), that birds acquire knowledge of the form and materials of the nest characteristic

PLATE 38. The loose, untidy nest built by an inexperienced young Village Weaver (*right*) contrasts with the neat, compact construction of an experienced adult (*left*) (photo from N. E. and E. C. Collias).

of their species by observing that in which they were hatched and reared. This view will appear a priori improbable to one who reflects that certain birds that hatch in the most elaborate nests, such as those of the castlebuilders, grow up in the dim interior where they have little opportunity to study the structure as a whole and after fledging do not return to it.

In addition to innate capacities and, at least in the case of the makers of more elaborate nests, a certain amount of practice, a bird requires adequate motivation to finish a nest with speed and efficiency. This depends largely on the state of its reproductive organs and the hormones they secrete, which in turn may be strongly influenced by social stimulation no less than by the wider environment. Internal and external factors, heredity and experience, are so intimately associated in the development of all the

more complex activities of animals that it is hardly possible to disentangle them. Improvement attributed to practice may be due, in greater or less degree, to the maturation of the bird's nervous system and the abundance of its hormones. The effort to disentangle these two factors often leads the experimenter to subject his animals to unnatural conditions that confuse what they are intended to elucidate.

Time Required for Building

The time taken to complete a nest varies enormously from species to species and with the circumstances in which each pair or individual builds. The same pair that starts early and works desultorily for weeks before the nest is finished is capable of accomplishing the same task in a fraction of the time, if the first structure is lost or pillaged and must be replaced later in the season. For this reason, it is often more entertaining and informative to watch birds build replacement nests than early nests. The observer is then more likely to find them at work, and the more concentrated activity makes it easier to decide whether both sexes engage in the task or only one. But replacement nests, especially those made hurriedly late in the season, are sometimes flimsier or less carefully finished than first nests.

The Rufous Hornero or Red Ovenbird of Argentina may, in favorable seasons, begin its clay nest in the autumn and resume its task whenever there is a spell of mild, wet weather during the winter. Some nests are finished early in winter, others not until the following spring, everything depending on the weather and the condition of the birds, for in cold weather, and when food is scarce, they do not build. The Greater Thornbirds of Paraguay may work all winter at their huge nests of sticks, sometimes devoting months to construction (Hudson 1920). In Surinam, a pair of Lesser Swallow-tailed Swifts started their long, sleevelike nest early in September and continued to build until the female laid eggs in it in the following March (Haverschmidt 1958).

A male Golden-naped Woodpecker started a new hole in June, after his young were well able to take care of themselves. He carved at it in a desultory fashion, chiefly in the evenings before he retired with his mate and their offspring into the old nest in the same trunk, which now served as the family lodging. The new hole was enlarged so slowly that by the following February it was still too small for the woodpecker to enter. As the breeding season approached, the female also took a share in the work, which proceeded more rapidly. By March, it was capacious enough for a single bird to sleep in. Apparently the woodpeckers were preparing to lay in it; but by a sudden change of plans, possibly caused by interference by a pair of Masked Tityras, they now began another hole, abandoning the one that had occupied their attention for so many months.

While certain Costa Rican birds that nest in holes or burrows may begin to excavate them half a year or more in advance of laying, no species, as far as I know, starts a breeding nest in the open so many months before it is needed. The reason for this seems obvious. In this humid tropical region abounding in termites, a nest composed of vegetable materials would deteriorate greatly during so long an exposure to the weather and destructive insects. Still, many birds of the humid tropics devote a number of weeks to building. A pair of Rufous-breasted Castlebuilders may begin their nest a month to seven weeks before the first egg is laid. Little Black-fronted Tody-Flycatchers may devote a full month to the construction of their cozy swinging nest; but, if the first nest is lost, the same pair is capable of completing a new one in half the time. A number of small American flycatchers, including the Northern Royal Flycatcher, Eye-ringed Flatbill, Brownish Flycatcher, and Sulphur-rumped Myiobius, take two or three weeks, and often more, to fashion their pensile structures, so various in design. A pair of diminutive Black-eared Bushtits that I watched in the high mountains of Guatemala required three weeks merely to apply the downy lining to their exquisite, lichen-encrusted pouch. In California, the Common Bushtit may spend as many as fifty days on a nest started early in the season (Addicott 1938).

Migratory birds, which each year make long round-trip journeys and raise large families to replace the losses inseparable from these hazardous voyages, could ill afford to devote so much time and energy to the construction of their nests. Doubtless it is for this reason that at high latitudes, where a large proportion of the birds are migratory, few kinds build nests comparable in size and complexity to those constructed by many birds in milder climates, where most of the breeding species are year-round residents. Nearly all of the few extratropical birds that make elaborate nests are resident where they breed rather than migratory. This is true of the various members of the ovenbird family in southern South America, whose nests of clay or sticks take so much time to build, and of the Long-tailed Tit of northern Europe, whose charming bottle nests may be built in as little as eight or ten days but may require two or three weeks—an interval that many a tropical bird squanders on a simpler structure (Lack and Lack 1958). Migratory birds, and even many that are permanently resident, often complete simple open nests in four or five days, and if pressed for time they will on occasion build them in a single day, as did a Village Weaver (Crook 1960).

In southern Costa Rica, Blue-diademed Motmots finish their long burrows in the ground many months before they will be used. In the wet season, mostly between August and October, the birds dig the tunnels in which they will not lay until the following March or April, and in the meantime they do not sleep in them. This procedure has two advantages. In September the soil is softer and easier to work than it will be in the dry months early in the following year just before they lay. During the long intervening interval, the earth kicked out into a heap below the mouth of the burrow will be covered by fallen leaves and sprouting seedlings; thus, it will be less likely to direct attention to the nest than freshly dug earth would be (Skutch 1964*b*). Another bird of the same region, the Buff-throated Automolus, likewise digs burrows in the wet months, long before it will occupy them.

The Origin of Nest Building

Selous (1927) developed certain interesting views on the phylogenetic origin of nest building: "Starting from the assumption that the earliest birds were ground-layers . . . I suppose that the sexual stimulus producing various frenzied movements in both sexes, and such movements (as also coition, to which they lead up) tending to become localized in some particular spot or spots, the eggs were often, as a natural consequence, laid in one of these." If the bird continued, during and after the laying of the eggs, to perform similar movements in the same spot, its presence upon the eggs would warm them, thereby hastening the development of the embryos. Since for most birds rapid development is advantageous (see p. 202), natural selection would strengthen this propensity and give rise to the habit of incubation. Selous continued: "The nest itself—or more strictly speaking, the lining of it—I suppose to have originated in that impulse which various birds seem to have, as a part or accompaniment of the sexual frenzy, to pick up from the ground sticks, leaves, or

other small objects—anything in fact that may be lying there—and drop or toss them down upon it again." These objects would, of course, be dropped at the place where the sexual activities were localized, where the eggs were laid, and where the birds tended subsequently to linger, because of its high emotional valence. Thus, in the view of Selous, the nest was originally the spot where the birds celebrated their nuptial rites, and the accumulation of materials at this spot began with the random billing of small objects under stress of great excitement. The prevalence of sideways throwing and sideways building among many birds that build the simplest ground nests lends some weight to this view.

Armstrong (1947) adopted a modification of Selous's theory. After giving a wealth of examples of birds picking up and billing nest materials in a wide variety of emotionally charged situations, he concluded: "Intense emotion, not necessarily wholly or exclusively of an erotic nature, stimulates movements which in the course of time may become fixed and adapted to foster the survival of the species by furthering the construction of a nest in which the eggs and young are safer than they would otherwise be."

It is well known that, when excited, birds (and other animals) make movements of no immediate use, and they are especially likely to perform out-of-context activities if for any reason they are prevented from carrying on the actions appropriate to the occasion. Thus, if a bird hesitates to attack a rival or an animal that threatens its eggs or young, or if its desire to incubate is thwarted, it may perform, often imperfectly, the movements of feeding, preening, bathing, or nest building, or it may even go to sleep. It is to be noted that these "displacement" activities are nearly always acts that enter largely into the bird's life and are useful to it; only now, because the appropriate responses are somehow blocked, they are performed out of context and to no purpose. The nervous impulse, deflected from its normal outlet, flows off through well-worn channels. If the sexual impulse is denied immediate satisfaction, perhaps because the partner is not ready, it is reasonable to suppose that this frustration would sometimes give rise to irrelevant movements. Such movements may be performed with inappropriate objects; in displacement feeding, for example, the bird may pick up such inedible things as pebbles or straws, only to drop them around the spot where it will lay its eggs.

Although these theories merit our consideration, the origin of nest building, like that of so many other things in this perplexing universe, is a subject fraught with tremendous difficulties. It is chiefly in birds that breed in crowded colonies, whose territories are in consequence reduced to little more than the nest site itself, that courtship and coition occur largely or wholly at the nest. In birds with larger territories, including the great majority of those of the woods and fields of continental areas, the nuptial rites are not performed at the nest site; it would be imprudent to perform them there, for this would unnecessarily draw the attention of predators to its position. Although the seaside and colonial-nesting birds on which Selous largely based his theory are mostly placed rather low in our taxonomic systems and are often regarded as relatively primitive, they are for the most part highly specialized in one way or another, and it is perilous to conclude from observations on them that ancestral birds performed their nuptial rites at a particular spot that in the course

of evolution became the nest. Possibly the habit of making a nest of some sort was derived from reptilian ancestors. Some contemporary reptiles, including certain turtles and lizards, dig holes to bury their eggs, which is a form of nest building at least as advanced as making a scrape in the sand, as many birds do; crocodiles accumulate heaps of mud and vegetation in which they lay their eggs. But, as far as I know, reptiles neither mate at the nest site nor manipulate materials with their mouths.

The preoccupation with materials that we sometimes witness in courting birds is, I believe, to be attributed to the association with the preexisting habit of nest building. The billing of nest materials and the shaping movements in which courting birds frequently indulge are symbolic of the whole complex process of which courtship is the prelude. Since birds have memory, the sight of a straw in the bill of a potential partner might suggest breeding to any bird who has had prior experience building a nest; it might even suggest nesting to a young bird who has not previously bred. All the emotions associated with reproduction, parental no less than sexual, seem to be stirred, at least vaguely, by the symbolic straw or stick.

CHAPTER 10

The Use and Maintenance of Nests

After a nest is finished, it may be abandoned before ever an egg is laid in it, used to raise a single brood, or occupied by a succession of broods, perhaps over a period of years, all depending upon the species of bird and what befalls the builders. Some kinds of birds normally build a new nest for each brood. A hummingbird's downy cup is often so flared out by the pressure of the nestlings' growing bodies, which sometimes burst it asunder, that it is no longer a fit receptacle for a second set of eggs. Many birds that rear several broods in a season will use the same nest for a second brood, if the first has been successful. The Mourning Dove has been known to lay five sets of eggs in the same nest in a single season. More rarely, a bird will repair and reoccupy a last year's nest, or it may choose the remains of the old structure as the foundation for a new one, as is a frequent practice of hummingbirds. An Eastern Phoebe used the same nest for more than twenty years, building up the rim a little at the beginning of each season (Sherman 1952). A number of raptorial birds also occupy the same nest year after year, each spring laying more sticks on it until, as in the case of the Bald Eagle, it becomes a huge pile that weighs tons and easily supports a man. Compare this with the tiny, frail nest of a hummingbird, or some of the honeycreepers, which fit into the palm of a child's hand!

As a rule, birds who have lost eggs or nestlings will not entrust another set of eggs to the same ill-fated receptacle. If they renest, they nearly always build a new structure in the same territory, often using the ravaged nest as a quarry from which they extract material for the new one. On the whole, this policy of not laying again in a pillaged nest is sound, for probably it was too exposed or too accessible. But frequently the replacement nest is in a site quite similar to that of the first and meets the same fate. And birds may be strangely blind in selecting the position of the new nest. Once I found a Boat-billed Flycatcher and Blue Tanager incubating in small trees growing close together. Both the flycatcher's and tanager's nests were plundered at the same time and doubtless by the same agent, probably toucans. The flycatcher then built a new nest in greater seclusion; the tanagers took possession of the flycatcher's abandoned nest, added lining to the interior until they had reduced it to the proper size, and then the female laid her eggs and started to incubate in this remodeled structure. As far as immunity to predation was concerned, it offered no advantage over the tanager's original nest.

Secondary Tenants of Nests

The foregoing is only one of innumerable examples that might be cited to prove that a nest has not necessarily outlived its usefulness when its builders desert it. Many kinds of birds are dependent largely on the abandoned holes of woodpeckers as receptacles for their eggs. These avian carpenters are the most serviceable of all birds in providing nests for their neighbors. The burrows from which kingfishers and motmots have brought forth their broods are often promptly claimed by Rough-winged Swallows, who for weeks have been patiently awaiting these vacancies. In tropical Africa, White-rumped Swifts occupy abandoned mud nests of swallows. The stout-walled clay houses of the Rufous Hornero or Ovenbird have many secondary tenants; and the massive mansions of sticks erected by their relatives, the firewood gatherers, thornbirds, and spinetails, are in great demand by neighbors less gifted as builders (Hudson 1920).

In Costa Rica, a bulky closed nest of a pair of Banded-backed Wrens, whose nestlings had been swallowed by a snake, was promptly occupied by a pair of Southern House Wrens. The last year's nests of thrushes and other birds are sometimes chosen by pigeons as the foundation of their own slight structures. The Solitary Sandpiper of North America regularly deposits its eggs in old nests of the American Robin and other birds that build in trees, and the Green Sandpiper of Eurasia has a similar habit. The Singing Silver-bill Finch of tropical Africa breeds in the closed nests of a number of weaverbirds, and in the same region the Orange-bellied Waxbill may adapt for its own use an old nest of a weaver or a warbler (van Someren 1956). One might continue for pages to multiply examples of this practice.

Other birds are not content to wait until the coveted nest has been abandoned by its makers. Occupied nests are captured by quiet persistence, clever strategy, or brutal fighting. When a tityra needs for her eggs the neatly chiseled hole in which a family of Golden-naped Woodpeckers pass their nights, she begins to fill it with bits of dry leaves and twigs during the day when the woodpeckers are not at home. When they return in the evening, the rightful owners patiently throw out the intruding material, for they want nothing more than a clean layer of wood particles on their bedroom floor. Next evening they must repeat this cleaning out—and the

next. Finally, they grow weary of this evening task and carve themselves a new hole nearby, leaving their former dormitory to the persistent tityra. The whole change of ownership may be effected without a struggle or a quarrel. I have known both the Masked and Black-crowned tityras to obtain possession of holes of the Golden-naped and related species of woodpeckers in this manner. Although the Masked Tityra fears the much bigger Fiery-billed Araçaris, she may dispossess these great-billed toucans of the hole in which they sleep by persistently filling it with litter.

The Piratic Flycatchers of tropical America like a comfortable covered nest but are unable to make one for themselves. Accordingly, they must steal it from some more gifted bird. Usually their victim is one of the yellow-breasted flycatchers of the genus *Myiozetetes*, such as the Gray-capped or the Vermilion-crowned, which build commodious domed nests with side doorways. More rarely, they take the swinging pouch of an oropéndola, the pensile, retort-shaped nest of a Sulphury Flatbill, or even the chamber that a pair of Violaceous Trogons has dug into the heart of a high wasp nest. Their strategy is simple but effective. While their prospective victims build, the pirates perch idly in surrounding trees, whistling breezily and watching patiently. As though aware that a premature assault would defeat their own purpose, they rarely become actively aggressive until the coveted nest has been completed and contains one or more eggs. Then they begin to annoy the builders, continuing their persecution until the latter chase them. While one member of the piratic pair leads off the pursuers, the other takes advantage of their absence to slip into the nest and carry out an egg, which it drops to the ground. This act is repeated until all the eggs have been removed. Since few birds will lay a second time in a ravaged nest, the Piratic Flycatchers then remain in undisputed possession, while their victims proceed to build a new structure nearby.

Other birds wage fierce, sometimes fatal, fights to gain control of a nest. Hudson vividly described the struggles between the Brown-chested or Tree Martin and Red Ovenbird for possession of the latter's oven of clay. Others have written of the conflicts between Common Starlings and Yellow-shafted Flickers, starlings and Common Bluebirds, and House Sparrows and martins for occupancy of a nest. The milder mannered feathered inhabitants of Central America do not often resort to such violence. The great majority of them make some sort of nest for themselves, and nearly all the rest depend upon patient waiting or bloodless stratagems to secure a ready-made structure for themselves.

The secondary tenants of bird nests are often animals other than birds. In Gabon, Africa, the kingfisher *Alcyon badia* and the woodpecker *Campethera nivosa* carve nest cavities in the arboreal termitaries of a species of *Nasutitermes*. The kingfishers' hole is completely sealed off from the termites' galleries; but the woodpeckers' chamber is not, with the result that sometimes the insects manage to close up the woodpeckers' doorway, causing the loss of their brood. After the birds leave, stingless *Trigona* bees may convert the chamber into a beehive, lining the walls with very hard resin to isolate it from the termites. When the bee colony degenerates, ants come in, devouring the honey and any remaining bee larvae. Finally, the termites displace the ants, closing up the opening and restoring the integrity of the termitary. The whole succession takes about two years (Brosset and Darchen 1967).

In covered nests of various New World birds, I have found mice, snakes, tiny marsupials called marmosas, and colonies of wasps and ants. These intruders do not always wait until the birds have left before they take possession. Mice close the original doorway with shredded material. The easiest way to ascertain what a covered nest contains is often by feeling with a finger, but the person who incautiously inserts a digit may receive a bite or sting from one of these invaders. A safer way to examine the contents of a covered nest is by inserting a small electric bulb attached by a cord to a flashlight and a tiny mirror pivoted on a bent wire.

Indeterminate Building and Nest Repair

No account of bird nests would be complete without reference to nest repair, which must be distinguished from the more widespread habit of renovation for another brood after an interval of neglect. A number of multiple-brooded birds refurbish their nests by adding a new lining before laying their next set of eggs, as I have seen in the Groove-billed Ani, Blue-and-white Swallow, Southern House Wren, Bay-headed Tanager, and Variable Seedeater, to mention only a few. Woodpeckers accomplish the same end by digging out the bottom of the nest cavity and leaving a bed of fresh, clean wood particles on its floor. In a sense, the birds that do this are building anew with the old nest as a foundation—repeating for a subsequent brood the normal cycle of building, laying, and incubation. Repair of a nest in the course of a single nesting, while it contains eggs or young, is an activity quite different in motivation and results.

After they have laid their eggs and started to incubate, many birds, especially among the passerines, cease to be concerned with the structural details of their nests. Building is for them a closed book, at least until they begin to prepare for another brood. This is unfortunate, for many a nest tilts over and spills out its contents, or else falls to the ground, when its builder might have saved it by making repairs that would require only the further application of skills that entered into its original construction. Thus I have watched the Montezuma Oropéndolas' long woven pouches gradually tear away from their supporting twigs, when by weaving in some new fibrous strands the birds might have prevented the disaster. Vireos' nests sometimes break away from the arms of the forks in which they are suspended. These nests are attached in part by cobweb; if the vireos would more consistently follow the example of the hummingbirds and daily bring fresh cobweb to strengthen their bindings, they might save their brood.

The British Nuthatch practices a simple sort of nest repair. If the cavity in a tree or wall in which the nest is built has too wide a doorway, the canny bird reduces the size of the aperture with a layer of clay, which blends with the surrounding bark or masonry. If later the clay cracks or falls away, it is speedily repaired (Coward 1928). The Pygmy Nuthatch has a somewhat similar habit; it calks with hair any seam or crevice that it finds in the walls of its nest hole. I do not know whether such stuffing is renewed if it falls out in the course of raising a family. Sometimes part of the clay wall of the Cliff Swallow's well-enclosed nest falls away while it is

PLATE 39. Rufous-breasted Castlebuilder about to enter its nest. The bird's bill points to the doorway, from which a narrow tunnel leads leftward through the mass of sticks to the chamber where the eggs lie on a bed of soft, downy green leaves.

still occupied, and then, whether there be eggs or young within, the swallows promptly replace the missing part with the same material. If fallen nestlings are placed for safety in a small box or tin by human friends, the parents soon cover most of the opening with a layer of mud. Since Cliff Swallows sometimes lay their eggs before they have finished building their nest, with them it is impossible to draw a sharp line between continued building and nest repair.

If a large segment—up to almost half—is cut away from the covered nest that a male Village Weaver is building, the bird soon repairs the damage. But if a far smaller piece is removed from the lower part of the ring, which is the first part of the nest to be woven and is in a sense its foundation, the bird usually abandons the nest, probably because the severed ring spreads apart when he stands in it and this impedes further operations (Collias

and Collias 1962). As long as a Baya Weaver's pensile nest with a long, descending entrance tube is occupied by a breeding female, the male continues to stitch thin strips to the outside. The wall may thus become a half-inch thick. This activity may have originated as a means of making the nest more impervious to rain, but it also serves to repair a nest severely damaged by thieving neighbors who tear strands from it while the builder is absent (Crook 1963). Male Baya Weavers, Village Weavers, and Yellow-backed Weavers may also strengthen the attachment of a nest containing eggs or young by weaving fresh green strips into it (Collias and Collias 1964a). To examine the contents of the long, tubular nest of a pair of Lesser Swallow-tailed Swifts, Haverschmidt (1958) made a small slit in the side, which the swifts always closed before the next inspection. These swifts also continued to add material to their nest during incubation.

PLATE 40. Rufous-breasted Castlebuilder's nest viewed from above, showing the doorway in the little pile of finer twigs (*center*) and the coarse thatch above the chamber where the nestlings are reared (*right*). Constant attention is needed to keep such complex nests in good condition.

Castlebuilders devote more attention to the maintenance of nests containing eggs or young than any other birds I know, for only by unremitting attention can they preserve their elaborate edifices in good order (plates 39 and 40). The sexes incubate by turns, and the partner not so engaged devotes much of its free time to taking care of their nest. Falling sticks are pulled or pushed into place; the coarser materials of the thatch are drawn up on the roof of the chamber; bits of cast reptile skin are constantly added to the collection; and every day fresh green leaves are brought and laid beneath the eggs. Often the eggs are neglected for short intervals while one or both members of the pair busy themselves attending to the structure that shelters them. Whenever I made a gap in the side of the nest chamber to examine the contents, the castlebuilders carefully filled it up with fine twigs. Often they continued to stuff fragments of reptile skin into the crevices, the better to obliterate all traces of the opening. The spot where the gap was made seemed to worry them; it continued to occupy their attention long after every sign of the breach in their castle wall had been effaced.

Repairing the nest, which occurs in castlebuilders, certain weaverbirds, and swifts, seems to be one aspect of what may be called indeterminate building. Birds, especially passerines, that build compact nests small in relation to themselves usually cease to add material after incubation has begun. I have never noticed the continuance of building after laying in antbirds, manakins, American flycatchers,

PLATE 41. A Clapper Rail adds a straw to its nest. Marsh-dwelling rails build up their nests to keep their eggs above rising water (photo by S. A. Grimes).

thrushes, wood warblers, finches, honeycreepers, or tanagers. But some nonpasserines that build quite simple nests in trees and shrubs, including pigeons and anis, may add materials during the course of incubation. The situation is different in species that construct nests that are unusually elaborate or bulky, or that require many billfuls of material. With some of these birds, building appears to become a mania, an end in itself, not just a means to produce a receptacle adequate to contain the eggs and young. Carrying to the nest many hundreds or even thousands of billfuls of material develops a habit too strong to be broken abruptly as soon as the eggs are laid. The impetus acquired during

weeks of laborious building carries over to the next stage, that of incubation, and at times, but less often, even into the period of caring for the young.

Becards, which build great globular nests of vines, fibers, and leaves; castlebuilders, with their relatively huge mansions of sticks; bushtits, with their lichen-encrusted pouches lined with a vast amount of laboriously gathered down—all continue actively to add to their structures until their eggs have hatched, and sometimes even in the intervals of attending the nestlings. Hummingbirds, whose small nests represent a more than ordinary measure of careful industry, bring cobweb, down, and an occasional lichen to them throughout the period of incubation. And birds that reduce the size of over-large cavities by laboriously filling them with leaves, twigs, or flakes of bark carry in more and more of these things as they return to their eggs after a recess for exercise and refreshment. Thus the Streaked-headed Woodcreeper brings pieces of stiff bark; the Tawny-winged Dendrocincla, lichens and fibrous bark; the Masked Tityra, fragments of dry leaves; the Southern House Wren, still more downy feathers to tuck beneath her eggs.

Birds whose nests rise above or float upon water are especially likely to continue building after they have laid, and often with greater necessity than in the foregoing species. Only by increasing the height of their structures can some of them prevent their eggs or young from becoming submerged (plate 41). When the water in its marsh rises, the King Rail hurriedly places more material on top of its nest in an effort to keep its eggs above the water level (Meanley 1953). Black-necked Stilts and American Avocets likewise build up their nests when water rises (Gibson 1971). After being released from incubation by the arrival of its mate, the Black Tern often places fresh pieces on its floating nest; thereby the structure's bouyancy is maintained even if the older materials become waterlogged from long submergence (Cuthbert 1954). For a similar reason, grebes often build up their nests during incubation. Adelie Penguins continue to bring stones to their nests during the incubation period; if the nest site is flooded by water from thawing ice, they may save their eggs by building the nest higher (R. H. Taylor 1962).

A curious instance of continued building was observed by van Someren in Kenya. While he watched a nest of the Blue-crested Plantain-eater, the parent brought sticks and built up the side toward him, apparently to shield its chick from his view.

Destruction of Nests by Their Builders

Although birds frequently tear apart nests, their own or those of other birds, to procure material for a structure they are building, only rarely do they deliberately destroy nests without this motive. A notable exception is found in the Village Weaver. Each male of this colonial, polygynous species builds a succession of covered nests—sometimes as many as twenty-three in a season—to which he tries to attract females by calling and displaying at each doorway. If no female accepts a nest after a week or two, the male then tears it down and builds a new one, for the female prefers fresh, green nests to old faded ones. He likewise demolishes used nests soon after the fledglings leave them. Since in crowded weaverbird colonies good nest sites are not overabundant, the advantage of such clearing of sites for new constructions that females will accept is obvious. Similar destruction of nests has been noticed in other weaverbirds (Crook 1960, Collias and Collias 1964a).

Another family in which deliberate nest destruction has been repeatedly witnessed is the cotingas. After nestlings of the Blue Cotinga disappeared prematurely from their treetop nest in Panama, Chapman (1929) watched the bereaved mother pull pieces from the nest and drop them. Similarly, after marauding Blue-throated Toucanets drove a nearly fledged Lovely Cotinga from its nest high in a great tree, its mother, who had evidently not yet found it on the ground far below, stood above the empty bowl and tore it apart. Although I could not fathom the bird's feelings, to me it looked like a gesture of heartbroken despair. Returning to a nest of the closely related Turquoise Cotinga that I had been studying, I found its materials scattered over the ground below with no trace of the two eggs that it had held. Since it is most unlikely that the open nest was destroyed by the predator that took the eggs, I attributed its demolition to the parent bird herself. On three occasions, the exceedingly slight nest of a Rufous Piha was destroyed soon after the single egg or unfledged young had vanished from it. Although I did not witness the demolition of these nests, I watched a piha scatter the materials of another nest, near which rested a recently fledged young bird. Since pihas and the other cotingas are thinly dispersed through forests with no lack of nest sites, their motives for dismantling nests are evidently different from those of weaverbirds.

A Sarus Crane, Scrub Jay, and Ring Ousel are among the birds that have been observed destroying their nests after the eggs had been taken or the birds otherwise disturbed (Armstrong 1947). Even more curious is the demolition of a nest by the young that grew up in it. Of this odd behavior, only a single example has come to my attention. As a nestling Red-footed Booby grows strong, it usually stands on a branch beside its nest of sticks in a tree and tears it apart, piece by piece, until nothing is left. At an earlier age, this nestling booby will catch sticks thrown to it, or try to take those that a parent has brought, and place them in the nest (Verner 1961).

CHAPTER 11

Eggs

The eggs of birds, perfect in form and varied in color, are among the most beautiful of small natural objects. Their attractiveness has caused them to be collected in vast numbers to fill the trays of museums and the cabinets of amateurs. The rarer kinds have brought prices comparable to those paid for semiprecious gems; their market value stimulated the activity of professional collectors and hastened the disappearance of some of the less common species of birds in densely populated countries like England. The study and minute description of these empty eggshells was the province of a special science, oölogy, which once ranked with taxonomy as the most intensively cultivated of the branches of ornithology.

With the recent trend toward the observation of living birds, egg collecting has lost much of its former popularity, but unhappily it is far from dead. Even from the aesthetic side, the zeal for amassing blown eggs was largely misapplied. A clutch of eggs in a cabinet drawer, no matter how carefully preserved, is not half so charming as a set of eggs lying in the nest that the parent birds made for them, embowered in verdure perhaps still spangled with dew. The pleasure of seeing a nestful of eggs is heightened by the difficulty of finding it, often by a long search in which the searcher matches his patience and ingenuity against those of the secretive parents.

If one leaves the nest undisturbed, he shares the birds' secret and may return again and again to follow the progress of their family. Moreover, old-fashioned oölogy did little to solve the many fascinating problems connected with birds' eggs—problems that require painstaking observation in the field, rather than the methods of the closet naturalist. And, despite all the collecting that has been done, the eggs of many kinds of birds have never been described. In tropical America, with its exceedingly rich avifauna, this is true of perhaps half of the resident species.

Laying

The interval between completion of the nest and appearance of the first egg is most variable, even in a single species. Some birds lay even before the nest is finished. Hummingbirds, for example, occasionally deposit their eggs in an inadequately lined receptacle, then continue to bring tufts of down to tuck beneath them; but hummingbirds continue this activity even after the nest is amply lined. Often birds lay the first egg on the day after they finish the nest. More often, two, three, or four days elapse before laying begins. In the Costa Rican highlands, I found that Slate-throated Redstarts, after building their nests in from three to five days, waited two to seven days before starting to lay. Birds that devote much time to the construction of an elaborate nest may not lay until many days after it seems adequate; however, such birds, as we learned in the preceding chapter, may continue to bring material not only in the interval between the virtual completion of the structure and the deposition of the eggs, but even while they incubate. Accordingly, it is difficult to determine this interval precisely. In a number of American flycatchers that build pensile or covered nests, it is not infrequently between one and two weeks.

Before laying an egg, some birds use the nearly finished or newly completed nest as a dormitory. The female Sulphury Flatbill sleeps in her hanging, retort-shaped structure for a number of nights, sometimes as many as ten, before her first egg appears. Woodpeckers often place their eggs in the hole where they have been sleeping for weeks or months; in species that roost singly, the female often lays her eggs in her mate's newer and sounder chamber. In the Guatemalan highlands in June or July, Blue-throated Green Motmots dig the long burrow where, if all goes well, they will incubate their eggs the following April. Throughout the intervening eight or nine months, the male and female sleep nightly in the unlined chamber.

Often we wish to know whether a certain activity at the nest, such as building, incubating, or feeding the young, is performed by the female, the male, or both together. They may be so similar in appearance and voice that the only way to learn their sexes with certainty, without opening their bodies, is to see which lays the eggs. To use this method, we must first give one member of the pair some distinguishing mark, such as a colored leg band or a paint spot on the plumage. Then we must watch, preferably from concealment, while an egg is deposited. If the observer has previously learned at what hour eggs are generally laid in the species in question, and what is the interval between the deposition of successive eggs, he will spare himself much tedious waiting.

Many species of birds have a more or less definite hour of the day for laying their eggs; although in this, as in other physiological processes, derangements sometimes occur. The numerous kinds of tanagers and hummingbirds that nest about my home

in Costa Rica usually lay their eggs early in the morning, from a little before to shortly after sunrise. Finches, too, are mostly early layers, although a few species delay until about midmorning. The considerable variety of American flycatchers often lay much later in the day than their neighbors the tanagers and finches, and they also deposit their eggs over a longer interval, from soon after sunrise until around noon. There is even considerable variation in the hour of laying successive eggs in the same nest. Birds that lay early, around sunrise, usually have a more definite time for this act than those that ordinarily delay until the day is well advanced. Thus Gray's Thrush, like the flycatchers, may deposit its eggs at almost any hour from soon after sunrise until near midday. In the north, the American Robin and Wood Thrush seem usually to lay in the second half of the forenoon and sometimes in the afternoon (Skutch 1952a).

Unlike most other passerines for which information is available, the little Orange-collared Manakins on our farm lay their eggs either late in the morning or early in the afternoon, chiefly between 11:00 A.M. and 2:00 P.M. Laying at times other than the forenoon appears to be more frequent in nonpasserine than in passerine families. The anis of the cuckoo family deposit their eggs in their communal nests usually around noon, although they may occasionally do so at any hour of the day. In northern Europe, the Common Gull lays at almost any time of day or night, and this is especially true of the first egg in a set. At least half of the second and third eggs were deposited between midday and 5:00 P.M. (Barth 1955). Likewise, King Penguins (Stonehouse 1960), Black-footed and Laysan albatrosses (Rice and Kenyon 1962), Gannets (Nelson 1966), and Wideawake or Sooty Terns lay their eggs at any hour of the day or night. In a sample of 136 eggs of the Wideawake, 4 percent were laid between 9:00 A.M. and noon, 57 percent between noon and 3:00 P.M., 34 percent between 3:00 and 6:00 P.M., 4 percent between 6:00 and 9:00 P.M., and the remaining 1 percent between 9:00 P.M. and 6:00 A.M. the next morning (Ashmole 1963a). The American Coot lays the earlier eggs of a set shortly after midnight, but the last 3 or 4 eggs are probably deposited around 4:00 or 5:00 A.M. (Gullion 1954). The few records

TABLE 3
Hour of Laying of Some Central American Birds

Species	Size of Set	Interval between Eggs in Days	First Egg	Hour of Laying Second Egg	All Eggs
Silver-throated Tanager	2	1(2)	5:00–6:15	5:30–6:45	5:00–6:45
Blue Tanager	2(3)	1	5:00–6:15	5:30–7:00	5:00–7:00
Scarlet-rumped Black Tanager	2	1	5:20–6:30	5:25–6:30	5:20–6:30
Yellow-faced Grassquit	2–3(4)	1	5:30–6:30	5:30–6:30	5:30–6:30
Southern House Wren	3–4(5)	1	—	—	5:30–6:40
Variable Seedeater	2(3)	1	5:30–6:30	5:30–7:00	5:30–7:00
Buff-rumped Warbler	2	1	—	5:30–7:00	5:30–7:00
Hummingbirds, 6 spp.	2	2	5:30–6:35	5:30–7:15	5:30–7:15
Buff-throated Saltator	2	1+	5:30–7:00	7:00–9:00	5:30–9:00
Black-striped Sparrow	2(3)	1+	5:30–7:00	7:00–9:00	5:30–9:00
Bellicose Elaenia	2	2(1)	7:00–10:00	7:00–10:00	7:00–10:00
Gray's Thrush	2–3(4)	1	—	—	7:00–11:00
Gray-capped Flycatcher	2–3(4)	2(1–3)	7:30–12:30	7:15–12:10	7:15–12:30
Yellow-bellied Elaenia	2	2	8:30–10:30	8:30–11:15	8:30–11:15
Anis, 2 spp.	4–5(7)	2–3	—	—	11:00–14:00
Orange-collared Manakin	2	2	11:00–14:30	11:00–14:30	11:00–14:30
Pauraque	2	2?	—	14:00–18:15	14:00–18:15

NOTE: With the exception of the anis, all observations were made in the same locality, the Valley of El General in Costa Rica. For the laying of the Pauraque, only 2 observations are available; for each of the other species, 6 to 41, a total of 262 for the 23 species in the table.

of laying of the Pauraque that I have been able to collect indicate that this crepuscular goatsucker deposits its eggs in the late afternoon, between 2:00 P.M. and sunset, before it becomes active (table 3).

The interval between the laying of successive eggs of a set may be one day, two days, or often more. It varies not only from species to species but even in the same set. For example, the second egg may be laid one day after the first, the third egg two days after the second. For songbirds and woodpeckers, especially those that produce large sets of eggs, the interval between eggs is usually a single day; for nonoscine passerines (Tyranni), including American flycatchers, manakins, antbirds, and ovenbirds, it is usually two days, as it is in hummingbirds. For large nonpasserines, the interval between eggs is often longer than two days. In the Australian Mallee-Fowl, the interval between eggs may range from two to seventeen days and is longest in unfavorable seasons (Frith 1962).

It is evident that if the interval is exactly twenty-four hours, all the eggs of a set will appear at the same hour of the day; however, if it consistently exceeds twenty-four hours but is less than forty-eight, each succeeding egg will be deposited later. In certain finches, such as the Black-striped Sparrow and Buff-throated Saltator, the second egg is usually laid about twenty-five or twenty-six hours after the first; so that if one finds the first egg around 6:00 A.M. he may expect the second between 7:00 and 8:00 A.M. the following morning.

These finches lay only two eggs in a set; but with species in which the number is greater, as in the Bobwhite and domestic hen, it is obvious that an interval between layings of somewhat over twenty-four hours will produce very considerable differences in the hour of depositing the eggs. Although hens may lay at any hour of the day and even at times drop their eggs from their roost in the night, most of their eggs are deposited in the forenoon and early afternoon. Typically, the first egg of a series is laid early in the morning, the second somewhat later on the following morning, and so on, until an egg is deposited rather late in the afternoon. Then the hen rests a day and begins a new cycle early on the second following day. When the interval between eggs is only slightly in excess of twenty-four hours, she may produce a number of them before she must skip a day in order to avoid laying in the evening or night; if the interval is about twenty-eight hours, only two eggs will be laid on consecutive days; many hens not selected for high egg production have a still longer interval and lay only on alternate days. Other things being equal, the most productive layers are, therefore, those whose eggs are produced with the shortest intervals. In some observations that we made on our hens years ago, the largest number of eggs laid without skipping a day was eleven, by a barred hen whose interval was usually about twenty-four and one-half hours, and once only twenty-three and three-fourths hours. Apparently few hens do better than this, but I have found one record of sixty-nine eggs laid in as many days (Turpin 1918, Atwood 1929).

Eggs usually emerge from the bird with the narrower end first, but occasionally they rotate through 180 degrees in the uterus shortly before laying and are expelled with the more rounded end foremost. This is more likely to happen as the bird grows older, probably because the tissues are stretched by repeated laying. While extruding the egg, the bird usually rises up and strains, from time to time looking beneath herself to see what has happened. Laying requires only a few seconds in cuckoos and cowbirds that try to foist their eggs on foster parents without being detected and possibly attacked. In most other birds it takes longer, usually at least a few minutes, and up to an hour or two in such large fowls as turkeys and geese. Many small birds seem hardly to be affected by the act of laying; but Fisher (1969) watched a Laysan Albatross who, after laboring about five minutes to expel her big egg, looked down at its glistening, mucous-covered surface, then sank down and lay flat upon it, eyes partly closed, as though exhausted by the effort. Her mate stood close beside her while she laid and examined the egg as soon as it appeared.

Bird nests are subject to many hazards and are often lost before the builder can lay the eggs in them, or while she is in the midst of producing her set. What does the bird then do with the eggs forming within her? If they are still incompletely developed, their growth may be suspended and they may be resorbed. Even if an egg is already enclosed in a shell, the female may be able to retain it for a day or so while she hastily builds a new nest. It is difficult to learn what a free bird who suddenly finds herself nestless does with a completely formed egg that she must lay, but I suspect that her predicament explains the presence of eggs in the nest of another species and of abnormally large sets of the same species, such as are recorded from time to time. Other alternatives are possible. Once I watched a female Blue-crowned Manakin, whose slight nest had completely vanished before she could lay, sit in the empty site and drop her first egg on the ground below. Two days later, I watched in vain for her to lay her second egg.

If a female Red-throated Ant-Tanager loses even one egg before her set of two or three is complete, she lays the remaining eggs in the pillaged nest and then deserts it (Willis 1961). The same happened when the first egg of a Wren-Thrush was removed (Hunt 1971). More tolerant of such loss, a Buff-rumped Warbler, whose first-laid egg promptly vanished, placed her second in the empty nest and proceeded to incubate it, as did a Silver-throated Tanager and Bellicose Elaenia. In similar circumstances, a Black-striped Sparrow failed to lay again in the ravaged nest; possibly she dropped her second egg on the ground at a distance.

Number of Eggs

Among the many birds that lay and incubate only a single egg at a time are albatrosses, fulmars, petrels, and shearwaters (Order Procellariiformes), Emperor and King penguins, tropicbirds, frigatebirds, flamingos, many auks, gannets, some tropical terns including the Noddy, Sooty, and Bridled, a large proportion of the world's pigeons, crested swifts, certain sunbirds of Africa and cotingas of tropical American forests. Two eggs are almost invariably found in the nests of the remaining penguins and pigeons, likewise in those of hummingbirds, many goatsuckers, thick-knees, manakins, antbirds, most honeycreepers and tropical tanagers. Four, or less often three, eggs are laid by the majority of shore birds.

In many species with sets larger than two, young birds nesting for the first time lay fewer eggs than older individuals, and they start to lay a few days later in the season. House Martins that build new nests lay, on the average, nearly one egg less than neighbors who recondition old nests, apparently be-

PLATE 42. Eggs of the Rufous-tailed Hummingbird in a soft, downy cup in an orange tree. Like many other tropical American birds, hummingbirds nearly always lay two eggs.

cause they devote metabolic resources to building that might have been applied to the formation of eggs (Lind 1960). Inclement weather and shortage of food often decrease the number of eggs laid, whereas a superabundance stimulates some birds to produce exceptionally large sets, as happens with raptorial birds in the far north in years when voles or lemmings become very numerous.

Although in pigeons, hummingbirds, goatsuckers, and some other groups the size of the set is largely independent of geographical position, many birds, particularly those that lay large sets, are more responsive to environmental factors. There is a strong tendency for the number of eggs in a nest to increase with latitude, so that in a single family, and often in a single genus or even species, the smallest sets are found near the equator and the largest in temperate and arctic regions. Thus, among the finches and tanagers that nest here in southern Costa Rica, the prevailing number of eggs in a nest is two, whereas in northern regions birds of these families lay much larger sets, often six or seven in the finches.

Likewise, habit of life influences the number of eggs in the same region. Precocial birds, whose young walk about and pick up food when a day or so old, can rear larger families than altricial birds, which for a number of days must bring nourishment to helpless nestlings. The largest sets, up to about twenty, are found among precocial families, such as ducks, pheasants, and quails, especially at high latitudes. Although there are records of still larger sets, they must be accepted with caution, because, when an unusually large number of eggs is found in a nest, it is probable that several females have laid in it. If all the eggs that one Mallee-Fowl lays at rather long intervals in a single incubation mound can be considered to constitute a set, it seems to hold the record for size, as a series of thirty-three eggs has been reliably reported. But the Mallee-Fowl neither sits on its eggs to hatch them nor pays any attention to its chicks (Frith 1962). In chapter 34 we shall consider more carefully the important implications of the number of eggs in a nest.

In species that raise several broods in a season, early nests often contain more eggs than late nests, although this reduction rarely occurs when the first brood consists of only two. One reason for the smaller size of later sets is that high midsummer temperatures at middle latitudes inhibit laying, as has been noticed in the Northern House Wren (Kendeigh et al. 1956). Since female birds of various species continue to feed the young of one brood while laying the eggs for a later brood, the smaller set may reflect the resulting drain on her metabolic resources. In the European Robin, Blackbird, and Yellow Bunting, however, second broods laid when the weather is warmer and the days longer often contain more eggs that do first broods (Lack 1954).

Many years ago, a curious ornithologist removed the eggs from a nest of a Yellow-shafted Flicker as they were laid, always leaving one so that the woodpecker would not desert, with the amazing and often-quoted result that she laid seventy-one eggs in seventy-three days, although the flicker's usual set consists of only five to nine eggs. Thereby the Yellow-shafted Flicker became the classic example of an "indeterminate layer," a term applied to a bird capable of laying an indefinite number of eggs in order to complete her set. Obviously, laying in such a bird is to a certain extent controlled by the stimulus provided by the eggs already present in the nest. But whether this stimulus is visual or tactile—the sight of the eggs or their contact with the naked skin of the female's brood patch—appears not to be known. When the stimulus reaches a certain intensity, provided by a number of eggs that is evidently a few short of the full set, no more of the rudiments present in the female's ovary are developed into complete eggs equipped with a shell.

Since a bird like the flicker is trying, perhaps unconsciously, to complete a set of a definite size, or perhaps one within certain limits, it seems likely that, if additional eggs were placed in her nest after she had laid the first, these would inhibit the maturation of her own eggs and she would lay fewer than usual. At least, this result might be expected in a bird that produces large sets; however, if the normal number is only two or three, there would hardly be time for the added eggs to inhibit the maturation of the eggs already far advanced within the bird. This method of demonstrating indeterminate laying has seldom been successfully employed. When Black-headed Gulls were given wooden models of gulls' eggs before they had laid any of their own (they would have eaten the eggs of other Black-headed Gulls), they often proceeded to incubate and did not lay at all. The onset of incubation inhibited the maturation of the eggs that they would otherwise have laid (Tinbergen 1958).

Among the few birds that have been demonstrated to be indeterminate layers by removing eggs as they are laid are the Wryneck, a relative of the flicker, which in one experiment laid sixty-two eggs consecutively; the House Sparrow, which laid up to nineteen on consecutive days; and the Gentoo and Adelie penguins, which may produce a third egg if one or both of the usual set of two are removed (D. E. Davis 1955, R. H. Taylor 1962). The Black-headed Gull may be induced to lay seven eggs, al-

though its normal set consists of only three. The Northern House Wren also appears to be an indeterminate layer, but only early in the season and to a limited degree.

The majority of birds are determinate layers. They are usually able to replace lost eggs but do so in a different manner. If they do not desert the nest from which eggs have been taken, they incubate the reduced number with no attempt at replacement and raise a smaller family. But if their nest is emptied, or if they abandon their remaining eggs because of the disturbance, they as a rule soon build a new nest and lay another set of eggs. They may do this repeatedly until either they succeed in rearing a brood or the breeding season ends. More rarely, they lay a new set in the nest that has been pillaged. In the far north, Thick-billed Murres replace only those eggs that are lost before they have been incubated for five days (Tuck 1960). The single egg of many petrels and albatrosses is scarcely ever replaced, and some penguins fail to lay again if their egg is lost. Parents so bereaved spend the breeding season as members of the "unemployed class," or else they return to sea and do not again attempt to breed until the following season. Thus they are the most determinate of layers. The reason for this unusual behavior appears to be the great length of the period needed to rear a single brood; hatched late from a replacement egg, the young bird might not be ready to leave the nest when the breeding season ends.

Some Physical Properties of Eggs

From the tiniest hummingbirds to the Ostrich and the even larger extinct flightless species like the moas and *Aepyornis*, birds exhibit a tremendous range in size, and this is matched by differences in the dimensions of their eggs. The eggs of the smallest hummingbirds are less than one-half inch in length by little more than one-quarter inch in thickness. Those of some of the smallest passerines, such as the kinglets and bushtits, are not much larger than this, although they are relatively broader. A large egg of the Ostrich measures about six by five inches. Such an egg weighs about 1,600 grams (3½ pounds); an egg of one of the smaller hummingbirds about 0.3 gram. Hence the Ostrich's egg weighs more than five thousand times as much as the hummingbird's.

Not only does the size of the egg vary with the species of bird that lays it; but there also are variations within each species and even among the eggs laid at various times by the same individual. Large individuals usually lay bigger eggs than do smaller ones of the same kind. The eggs of young birds breeding for the first time are generally smaller and lighter in weight than those of older individuals. There are likewise more or less regular differences in the size of successive eggs of a single set, but the variations are of a different character in different species. Thus in the domestic hen the eggs produced in a single cycle (see p. 137) tend to become progressively smaller and lighter, with the exception of the very first egg laid by a young pullet, which is often smaller than its immediate successors. In the Adelie Penguin, the first egg of a set is longer, broader, and heavier than the second and, if it occurs, the third egg (R. H. Taylor 1962). The last two eggs of a Northern House Wren's set are longer, broader, and heavier than the early eggs; but in the Song Sparrow no regular increase or decrease in size within a set was detected. Contrary to the situation in the Adelie Penguin, first eggs of the Rockhopper and Macaroni penguins are much smaller than second eggs in the sets of two. The small first egg tends to be neglected by the parents; although capable of producing a chick, it rarely does (Warham 1963).

The eggs of precocial or nidifugous birds, which hatch at a relatively advanced stage of development, are typically larger in relation to the parent than the eggs of altricial or nidicolous birds, which are blind and helpless at birth. For example, the Common Meadowlark and Upland Plover are birds of about the same size, and both nest on the ground, but the eggs of the precocial plover are very much bigger than those of the altricial meadowlark. There is tremendous variation in the ratio between the weight of a female bird and that of the eggs she lays. As a rule, the larger the bird, the smaller the fraction of her weight that her egg represents. The single egg of the big Emperor Penguin weighs only about 1.4 percent of the weight of the parent. The weight of a single Song Sparrow's egg is about 11 percent of that of the parent, and her set of four or five eggs weighs about half as much as she does. When a set of eggs is large, as in a number of ducks, rails, quails, and titmice, its total weight may exceed that of the bird who lays it. To produce so much organized material in the course of a week or two is certainly a very remarkable feat of metabolism.

Eggs are typically elongate bodies with one end broader and less pointed than the other and with the greatest transverse diameter nearer the blunt than the sharp end. But there are many exceptions to this rule. The minute eggs of hummingbirds are narrowly elliptical, with scarcely any difference between the two ends; and the same lack of noticeable polarity is exhibited by the larger and relatively broader eggs of many bigger birds. Some eggs, including those of certain owls, kingfishers, motmots, and quails, are so short as to appear almost spherical, and those of some turacos are practically spherical. At the other extreme are the long, strongly tapering or pyriform eggs of murres or guillemots, with the thick end broad and blunt, the thin end very much narrower. When rolled slowly, these top-shaped objects tend to describe circles rather than to move in straight lines. This is an obvious advantage for an egg that rests on a narrow rock ledge high on a seaside cliff, with at most a rudimentary nest to contain it. As incubation proceeds and the embryo develops, the enlargement of the air cavity at the broader end of the egg causes the center of gravity to shift toward the sharper end, with the result that the larger end rises and the egg rolls in smaller circles, so that it is less likely to fall off the rock shelf or into a crevice from which the bird cannot retrieve it. The much smaller eggs of a number of sandpipers, plovers, and related birds are often almost as strongly tapered; in this case the advantage appears to be that when arranged with the narrow ends inward they fit together much more compactly than would be possible with more rounded eggs of the same size, thus permitting the relatively small parents to cover them more effectively. When the set of eggs is large, as in many gallinaceous birds and ducks, this advantage of a strongly pyriform egg is lost and the eggs are usually rounder.

The shells of eggs consist almost wholly of calcium carbonate in the form of calcite crystals, with from 1 to 3 percent of organic matter. The shell is perforated by many minute pores, either simple or, as in the Ostrich egg, branched. The mouths of these pores on the outer surface of a hen's egg are visible with a hand lens as minute pits. They permit the diffusion of indispensable oxygen through the shell to the embryo developing within.

Eggs exhibit great differences in the texture of their shells. We commonly think of bird eggs as fragile and handle them with great care, but some are remarkably strong. Large birds, especially those that treat their eggs rather roughly, must enclose them in stronger shells than are needed by dainty birds that weigh less than an ounce. As a small boy, I was deeply impressed by the difficulty of breaking a Guinea hen's egg. Those of francolins are also amazingly sturdy. In Australia, the Black-breasted Buzzard preys upon the eggs of the Emu and some other large birds, including the Giant Crane and bustards, all of which nest on the ground. The strong shells of these eggs, especially those of the Emu, resist the buzzard's bill, so to break them the bird seizes a small stone, rises to a height of ten or twelve feet, and drops the stone on them. After smashing the shells in this fashion, it feasts on their contents (Chisholm 1954). This and the Egyptian Vulture's similar habit are among the few recorded instances of the use of a tool or instrument by a bird (see p. 37).

Doubtless it would be advantageous to many birds, at least of the larger kinds, to lay eggs with thicker shells, which might protect them from some of the smaller and weaker animals that prey on them. But shell thickness is limited by the ability of the enclosed chick to chip and break its way out when ready to hatch. In a murre's egg, the shell gradually increases in thickness from 0.4 to 0.6 millimeters at the broad end to 0.6 to 0.8 millimeters at the narrow end; this end needs to be stronger because it is in contact with the rock surface, whereas the broad end must be thin enough for the hatching chick to break through (Tuck 1960). The shells of megapodes' eggs are exceptionally thin; the parent birds rarely touch or move the eggs while they lie in the incubator mounds, so they need not be strong.

Some eggshells are smooth and glossy as though polished, whereas others are rather rough. Among the latter are the eggs of the Chestnut-winged Chachalaca, whose whitish shells are indented with many shallow, roundish pits that vary considerably in size and in addition bear embossed flakes of a pure white, limelike substance, sometimes more than one-eighth inch in diameter. The density of pits and flakes may vary greatly even on eggs in the same set. Most curious are the eggs of the anis, covered all over with a thick, soft, pure white layer that is easily scratched off, revealing a pale blue inner layer, harder and more resistant than the chalky exterior. The eggs of boobies likewise have a soft, chalky white covering over a harder, bluish or blue green shell; in the King Penguin, removal of the chalky surface reveals a pale green color that intensifies as incubation proceeds (Stonehouse 1960). Relatively few eggshells exhibit such pronounced contrast between their inner and outer layers.

As everyone who has eaten an egg knows, within the shell are two contrasting parts, the centrally located yolk and the almost colorless albumen, or "white," that surrounds it. Together they fill the interior of the egg, except at the blunt end where a pocket of air, which serves for the respiration of the developing embryo, is constantly renewed by diffusion through the porous shell. This air space is situated between the inner and outer layers of the thin, tough shell membrane that surrounds the albumen.

The relative size of the yolk varies greatly. In birds that hatch in a helpless state and are fed for some days in the nest—such as songbirds, woodpeckers, and pigeons—the yolk accounts for only 15 to 20 percent of the whole egg. It, no less than the albumen, is almost completely absorbed by the chick before it hatches. In nidifugous birds—such as geese, ducks, and domestic chickens—from 35 to 50 percent of the egg's weight is accounted for by the yolk (Heinroth and Heinroth 1959). When these eggs hatch, much yolk remains enclosed in the abdomen of the chick and serves as a reserve of nourishment during the first days of life in the open, when the bird might have difficulty picking up enough food for itself. This internal store of food enables it to live for several days without eating; but the rate of absorption of the yolk has been found to be about the same whether the chick eats well or starves. Nidicolous nestlings, which exhaust their yolks before they hatch, do not need this nutritive reserve, for their parents feed them adequately very soon after they escape their shells.

The Colors of Eggs and Their Significance

It is above all the colors of eggs that make them beautiful and present perplexing questions to the field naturalist. Many eggs are pure white, and some of these are most attractive, especially those of woodpeckers with highly glossy, translucent shells that seem to be made of the finest porcelain. Many kinds of birds lay eggs with uniformly colored shells. Both white and tinted eggs are frequently marked with flecks, spots, scratches, blotches, and scrawls, usually of some shade of brown, ranging from reddish brown to black. These markings are very often confined to a wreath around the thicker end of the egg, or form a cap on this end, or at least are more concentrated on this part of the egg than elsewhere. In some species, however, the mottling is rather uniform over the whole surface, and exceptionally it is concentrated on the sharper end. Frequently some of these spots or blotches are paler than others on the same egg, often of a color described as "pale lilac"; the fainter color of these markings is due to the fact that the pigment was deposited early and then covered by the outermost layers of the shell, which mask its intensity. Other variations in shade are caused simply by varying concentrations of the pigment that they contain.

A number of pigments have been isolated from eggshells, including red-brown, reddish yellow, bright yellow, and blue, the first two probably derived from hemoglobin and the last two from bile (Newbigin 1898). Yet I am aware of no eggs that are bright yellow; this pigment is combined with blue to produce green, as in the eggs of the Emu.

The brightest eggs that I have seen in their natural setting are those of the Great Tinamou of the tropical American forests. These exceptionally lovely eggs, the size of a hen's egg, are intense turquoise blue with a high gloss. Other species of tinamous lay eggs that are green, vinaceous purple, or deep gray, and equally glossy. In other families of birds, buff or brownish eggs are common, and some are pink or faintly reddish. The blackest eggs yet discovered were found by several ornithologists in the Guianas and neighboring regions. These small, black or purplish black eggs were attributed by them to the Blue Honeycreeper. But this identification is certainly incorrect. We now have abundant evidence that from southern Mexico to Brazil this honeycreeper lays whitish eggs with brown spots. Moreover, the honeycreeper builds a shallow, cuplike nest, while the black eggs were always found in a deep, thin-walled pouch composed of blackish fibers, with which the dusky eggs blended. The identification of the bird that lays the blackest known eggs continues to challenge ornithologists.

Except in the case of uniformly colored eggs, those in the same set are rarely exactly alike; often their markings differ greatly, some being much more heavily speckled or marbled than others. Sometimes the eggs within a single race or subspecies show amazing color contrasts; for example, those of the Tawny-flanked Prinia in the same locality of southern Africa may be either white, pale rufous, or olive green, with various hairlines and mottling (Thomson 1964).

It is far easier to admire the colors of bird eggs than to discover how they help the eggs to escape hungry predators and produce living chicks. It is obvious that we must consider this problem in relation to animals that seek their food by sight rather than by scent or some other sense, by day rather than by night. That their color patterns make eggs very difficult for us to detect is most obvious in the case of heavily pigmented eggs that rest on the ground in slight, open nests or in none at all, such as those of many shore birds, terns, and goatsuckers. But at times the sitting birds themselves blend with their background even better than their eggs. Thus the eggs of the Pauraque do not match the brown fallen leaves on which they rest nearly as well as do the parent birds, as I have seen on numerous occasions (plates 43 and 44). But since the two parents keep their eggs almost constantly covered, and, if driven up, the movement of their departure inevitably betrays the position of their eggs, it is doubtless more important that the birds be protectively colored than that their eggs be inconspicuous.

At the other extreme, it seems clear that when a

PLATE 43. Eggs of the Pauraque, on leaf-covered ground beneath a Central American thicket.

bird nests in a long, dark tunnel in the ground, in the manner of many kingfishers, motmots, and bee-eaters, the color of the eggs hardly affects their safety; hence there is little need to mask their shining whiteness by the deposition of pigments in their shells. Indeed, whiteness may be advantageous because it makes them dimly visible to the parents, so that they are less likely to be broken or left uncovered by the incubating bird. When Holyoak (1969) experimentally darkened the pale eggs of hole-nesting Jackdaws, they were broken more often than normal eggs. The chief predators on nests in long tunnels are apparently snakes and small mammals that do not depend primarily on sight to lead them to their food.

Although nearly all the nonpasserine birds that nest in burrows or deep holes lay white, unmarked eggs, many passerines that nest in such locations, perhaps in the very cavities that the nonpasserines

have earlier used, lay heavily pigmented eggs. (In some instances, however, their eggs are also immaculate white, such as those of certain swallows and thrushes.) These departures from the rule that hole-nesters lay white eggs are sometimes explained by the fact that the passerine species that breed in cavities often belong to families or genera most of whose members build nests in the open, as is true of American flycatchers and thrushes. On this view, the pigmentation of the eggs of these so-called secondary hole nesters is an ancestral character, which has been retained simply because, if of no present value, it is at least not detrimental, and no selective pressure has been at work to eliminate it. But perhaps it is too hastily assumed that the coloration of these eggs is without utility. Although pigmentation may be valueless in a very dark chamber, in a dimly lit hole, such as those occupied by passerines, a color pattern that breaks the outline of the egg or reduces its whiteness makes it very difficult to distinguish, as anyone can be convinced by peering into such cavities without the aid of artificial illumination. In dimly lit cavities and covered nests, the protective value of pigmentation is sometimes more obvious than in open nests. Strangely enough, although both Short-billed and Long-billed marsh wrens build cozy globular nests with side entrances, the eggs of the former are pure white, while those of the Long-billed are dull brownish, finely spotted with darker shades of brown.

The greatest puzzle is presented by the very diverse coloration from pure white to bright blue, with or without marbling or spotting, exhibited by the eggs laid in open nests built amid foliage by the majority of passerines and those of some other orders. These eggs rarely match the color of the nest lining; they seem to vary in shade without rhyme or reason. However, by a painstaking analysis, Lack (1958) was able to show some correlation between the nest site and the coloration of eggs in the thrush family. Whitish eggs are found in holes; speckled whitish or blue eggs, in shallow holes or in niches or on ledges; blotched whitish or blue eggs, in a nest in the fork of a bush or tree; and so forth. These correlations suggest that in this family egg color does affect egg safety. On the other hand, Cott (1954) proved by experiment that the eggs of many birds,

PLATE 44. The incubating female Pauraque is less conspicuous than the eggs that she covers (cf. plate 43).

especially of the smallest passerines, are distasteful to hedgehogs, ferrets, cats, and men; often the taste is bitter. Since poisonous, nauseous, or otherwise unpalatable insects and other organisms are often conspicuously colored to warn would-be predators, it is certainly not impossible that the bright colors of certain distasteful eggs are examples of warning coloration. However, we are not sure whether any eggs are distasteful to snakes, many of which prey insatiably on bird nests, or whether these reptiles pay any attention to colors. So difficult is the problem of the coloration of eggs in open nests!

A bird watcher rarely sees eggs, except those of ground nesters, before he has found the nest that contains them. Once he has discovered an open nest, he seldom has trouble seeing the eggs it holds, even if he must raise a mirror above it. Snakes appear to find nests by watching the parents, and, after locating a nest, they seem to have no difficulty detecting its contents, whatever their color. Toucans, jays, and other birds that eat eggs apparently often discover nests as they hunt through trees and shrubs for insects and fruits, and doubtless they frequently notice nests below themselves, with the contents exposed to their view. The same applies to the monkeys of tropical forests. With reference to these marauders, which are believed to see colors much as we do, protective coloration should be of some value to eggs in open arboreal nests. But I find it difficult to believe that many such eggs are camouflaged well enough to escape the keen vision of predators. My impression is that eggs in open nests in trees and bushes are, on the whole, protectively colored only to a minor degree. I believe that, if close assimilation to nest lining substantially improved their chances of escaping detection by predators, such eggs would match the lining much more often than they do, just as eggs laid on open ground blend into their substratum.

The eggs of pigeons and doves, which usually lie on nests that are scarcely more than open platforms, seem more in need of a dark hue than those of most arboreal birds, yet they are nearly always gleaming white. However, the two parents alternately keep them almost constantly covered, even spending much time on the first egg before the set is complete. As far as I have seen, the pigeon sits closely,

not leaving until aware that it has been discovered by an approaching man or other potential enemy. Hence concealing coloration for their eggs is less important than in birds that leave theirs exposed for a considerable part of the day while they forage. The importance of constant coverage of the eggs of pigeons was demonstrated by Murton and Isaacson's (1962) observations on the Wood Pigeon in England. Early in the season, when the parents' search for food takes so much time that the eggs are often exposed, more are lost to predators than in July and August, when abundant grain enables them to satisfy their hunger more quickly and, consequently, to keep their eggs covered continuously.

Exceptional among pigeon eggs are those of the Ruddy Quail-Dove, which are buff rather than white. This dove builds its slight platform a yard or two above the ground in tropical rain forest, an environment in which predation on eggs is particularly intense, and perhaps the tint of its eggs is to be attributed to extremely high selection pressure. But in these same forests the Great Tinamou lays its three or four vivid blue eggs on the ground, often at the base of a tree. These are so conspicuous that even in the dim light of the undergrowth they seem to almost strike one in the face. Probably the neutrally colored parent tinamou keeps them rather constantly covered; but the bird's shyness and the prompt loss to predators of every nest that I have discovered have prevented a thorough study of this matter. On the ground in neighboring thickets, the Little Tinamou lays two lovely vinaceous purple eggs, far less conspicuous than the blue eggs of the larger species, and it leaves them exposed for the greater part of each morning.

Nests are larger and therefore potentially more conspicuous objects than the eggs they contain. Hence it is more important that they blend protectively with their setting than that the eggs blend with the nest lining. There is an inverse relationship between the conspicuousness of nests and the protective coloration of the eggs in them. When the nest is well concealed, in a hole or burrow or amid dense vegetation that it matches in color, the eggs need little protective coloration. When the nest is exposed, a mere scrape in the ground or built conspicuously amid vegetation, the eggs require some pigmentation to make them less visible, unless the parents keep them constantly covered or are strong enough to hold enemies aloof.

This inverse relationship is well illustrated by the Northern Royal Flycatcher, whose yard-long nest of brown vegetable material hangs conspicuously, although rather inaccessibly, above woodland streams in tropical America. Somewhere near the middle of the irregular, elongated mass is a shallow, inconspicuous niche that holds two reddish brown eggs, darker in color than those of any other flycatcher that I know. The first of these remarkable structures that I saw I did not immediately recognize as a bird's nest, and when at last the behavior of the attendants convinced me that it was a nest, I had difficulty finding the dark eggs. Scarcely any other arboreal bird lays eggs with such good procryptic coloration; but the nest that holds them is so conspicuous that, if it were not difficult of access for all but winged predators, these eggs would, I believe, rarely escape destruction (see pp. 88–92).

The Bird's Ability to Recognize Its Eggs

If the great diversity in appearance of bird eggs is not easily explained by the principle of concealing coloration, perhaps the differences in color and pattern are of value in enabling birds to recognize their eggs, so that each parent will attend its own and not those of another bird. On the whole, a bird's strong sense of locality and its attachment to the chosen nest site are sufficient to prevent confusion of this sort, even when the ability to recognize its own eggs is lacking. In certain special circumstances, however, this ability may be important. When I studied Great-tailed Grackles nesting in a crowded colony in coconut palms, I was impressed by the fact that the blue eggs were so variously marked with dark scrawls and spots that each female might have distinguished her own from those of her neighbors. I made no experiments to test this point; but in the case of murres it has been demonstrated that in some, but not all, colonies the parents recognize their own eggs by the great diversity of color patterns. Since these sea birds breed on crowded ledges without nests to keep their eggs in definite spots, the value of this capacity to recognize their eggs individually is evident. Colonial-nesting Royal Terns

likewise recognize their single eggs by the great diversity in color (Buckley and Buckley 1972).

A bird's ability to recognize its eggs, or at least to distinguish those of its own species from eggs of other species, should discourage the habit that certain birds have of dropping eggs in the nests of other birds to be attended by the foster parents. In the parasitic cuckoos of the Old World, the eggs often resemble those of the fosterers in coloration and size. This assimilation could hardly have arisen unless there were, in the first place, a strong tendency for each parasitic species, or each strain within a species, to confine its attention to a single kind of host, and, second, some ability of the hosts to recognize foreign eggs in their nests along with a tendency to throw out or abandon these intruding objects. But in the American cowbirds, it has not been demonstrated that eggs of the parasite tend to resemble those of the host. On the contrary, the cowbirds drop eggs of the same appearance into the nests of birds whose own eggs vary greatly in coloration and size. Only a few of the numerous victims of the cowbirds appear to detect the imposition. American Robins and Gray Catbirds regularly throw out the foreign eggs, often quite promptly, and a number of other birds do so at times. More often, birds cover the cowbird's eggs, sometimes together with their own eggs, with a new lining and lay again on the higher floor. If repeatedly imposed upon, the Yellow Warbler may continue this process until it has a nest several stories high. Painted Buntings, Field Sparrows, and Yellow-breasted Chats often abandon the nest that the intrusive cowbird has profaned. But a great variety of birds accept the foreign eggs and hatch out young cowbirds that they raise, usually at the price of some or all of their own offspring (Friedmann 1929, 1963).

Not only do birds of many kinds apparently fail to distinguish their own eggs from those of other species, but sometimes they also accept quite dissimilar objects. Crows are among the more intelligent birds, yet they have been known to incubate a golf ball! King Penguins will incubate tins, bottles, or stones of size and shape similar to their eggs. One penguin sat for three days with a cylindrical tin, six by three inches, sticking out from beneath it like a ventilating shaft (Stonehouse 1960). At times birds actually prefer eggs different from their own. Ringed Plovers incubated artificial eggs with black dots on a clear white ground in preference to their own light brownish eggs with darker brownish spots. Both Oystercatchers and Herring Gulls neglected their own eggs to try frantically to cover painted models twice as long and broad, on which they could hardly balance themselves. Herring Gulls, whose eggs are olive brown with dark spots, also incubated models of the natural size painted bright blue, bright yellow, or bright red, although the latter color was least attractive, and they accepted rectangular and cylindrical artifacts colored in the pattern of their eggs. If the rectangle or cylinder had rounded edges the gulls would continue to incubate it, but if the edges were sharp they became restless and soon left (Tinbergen 1953). The Laughing Gull can distinguish its own eggs from artificial eggs of the same size painted to resemble them, yet it will accept and incubate artificial eggs of various colors, red being the least attractive to it (Noble and Lehrman 1940). Mourning Doves, whose white eggs were painted with a variety of bright colors, continued to incubate until they hatched but would not sit on eggs with cracked or slightly punctured shells (McClure 1945).

In all these experiments, as well as in the responses of birds to eggs of parasites, two different aspects of behavior appear to be involved: a bird's ability to distinguish its own eggs from foreign eggs or other objects, and its acceptance of the latter for incubation. Such tolerance of strange objects in the nest is certainly not proof that a bird fails to notice the difference between them and its own eggs. Birds of different species differ widely in their reactions to intruding objects.

How Birds Move Eggs

Most small birds appear unable to carry their eggs without transfixing them with their bills or taking advantage of a break in the shell into which they can insert their lower mandibles. By this means they can remove pierced eggs from the nest without spilling out the contents, which might attract ants or other animals and jeopardize the remaining eggs. But they seem unable to return, even to low nests, eggs that have been accidentally knocked out and

perhaps lie unbroken in full view. Rails, however, have been repeatedly seen to lift fallen eggs into a nest with their bills, and to remove eggs and young from a disturbed nest, carrying them in the same manner.

The Chuck-will's-widow and Common Nighthawk have been reported to transport eggs that have been touched by people to a safer spot, carrying them one by one in their capacious mouths for considerable distances; but the evidence for this is not beyond question (Ganier 1964). In New Hampshire, Kilham (1957), a careful observer, twice saw a Whip-poor-will fly up to a branch carrying an egg with its legs and feet. However, few people have witnessed this method of carrying eggs. Even in the goatsucker family, the parent's usual way of moving its eggs is to pull them toward itself with its bill (plate 45) or else push them over the ground beneath itself, by means of its feathers and possibly its feet (Weller 1958). Once, when I found fire ants swarming over the eggs of a pair of Pauraques, I transferred them to a safer spot on the ground a short distance away. The parents persisted in moving them back to the original site, with the result that, as soon as the hatching chicks pierced their shells, the ants entered and destroyed them.

Many other ground-nesting birds can move their eggs by rolling, and they commonly employ this method to return eggs that have been knocked or blown out of the nest. When a tern sees one of its eggs lying a short distance from the nest scrape in the sand, it seems undecided whether to sit on the two eggs that remain in the nest or on the truant egg. Often it compromises by settling in an intermediate position facing the single egg. Then it reaches out and with the lower side of its bill pulls the egg beneath its breast, incidentally bringing it closer to the nest. Not feeling at ease here, the tern soon returns to its nest, but presently it is lured by the sight of the exposed egg to move closer to it and tuck it beneath its breast once more. Each time this process is repeated, the egg is brought nearer the nest site, until finally it lies among the other eggs. This appears to be the accidental outcome of the competition for the tern's attention between the truant egg and the nest site with the remaining eggs (N. Marshall 1943, Tinbergen 1953).

PLATE 45. Common Nighthawk moving an egg by pulling it toward herself with her bill (photo by M. W. Weller).

In certain gulls, shore birds, and geese, eggs are retrieved in a more efficient and purposive manner; the parent walks backward toward the nest, pulling the egg along with the underside of its bill. The Black-headed Gull retrieves with its bill eggs placed a yard from the nest. The Ringed Plover uses wings, breast, and feet as well as its bill to shift its eggs. Certain quails also move their eggs by rolling, sometimes employing this method to shift them from a nest that has been disturbed, or from an exposed site, to a more secluded spot. In Africa, a Black-breasted Quail shifted her set over a distance of four yards by this method (van Someren 1956).

The Pheasant-tailed Jaçana lays its four eggs in a loosely made nest on the broad surface of a floating water lily leaf. When the decay of the leaf beneath the nest, or a rise in the water level following heavy rain, threatens the eggs with submergence, the male jaçana, who alone attends the nest, moves them to a safer spot on another lily pad, sometimes as much as fifty feet from the original site. Just how he accomplishes this feat is not known; but when an experimenter placed some of the eggs at a distance from the nest, the jaçana promptly brought them back by repeatedly going from the nest to an egg, each time settling at a point just short of the egg and pulling it beneath himself with his bill, much as a tern retrieves its eggs (Hoffmann 1949).

Unlike other species of albatrosses, the Waved Albatross of the Galápagos Islands has no recognizable nest or even a fixed nest site. It lays its single egg on any flat piece of ground, in the open or under a bush. For reasons that are not always clear, possibly because it dislikes a neighbor in a crowded colony or has been disturbed by a human intruder, the incubating albatross may move away, shuffling along with its egg held loosely between its brood pouch and the ground. In this fashion it may shift its egg for distances up to 125 feet (40 meters) in a few days. Eggs so moved may break or fall into a crevice from which the parent cannot retrieve them, but possibly the habit has certain advantages to compensate for substantial losses (Harris 1973).

CHAPTER 12

Patterns of Incubation

In reproducing by means of eggs in which the embryos develop outside their own bodies, birds follow the method that is by far the most common throughout the animal kingdom. They follow it more consistently than the other mainly oviparous vertebrates, for in both fishes and reptiles are numerous species that retain their eggs within the body until they hatch; but this system, known as ovoviviparity, has never been found in any bird. Although the majority of animals that lay eggs give them no further attention, many, from crustacea and spiders to fishes, frogs, and lizards, guard them carefully, sometimes attached to their bodies, or even in their mouths. Birds, however, attend their eggs more consistently than any other class of animals. The only species that regularly fail to do so either foist their eggs on other birds to be incubated or bury them where they will be warmed by solar or volcanic heat.

Incubation, which involves the application of heat to promote the embryo's development, sets birds apart from all those cold-blooded animals that merely guard their eggs. It is one of the most distinctive activities of birds, and outside this class it is known to occur only in the egg-laying monotremes among the mammals, and among reptiles in the Indian Rock Python, in which the female's temperature may rise as much as 10° F (5.5° C) above the surrounding air while she embraces her eggs in her coils. Without some external and fairly constant source of heat, bird eggs fail to hatch, and the only chicks that come into the world without heat supplied by their natural or foster parents are warmed either in man-made incubators or in the far more ancient ones prepared for their eggs by the mound-birds.

Not only is incubation one of the most characteristic of avian activities, but it also is one of the most interesting. One who casually glances at a domestic hen or some other bird sitting motionless for long periods on her eggs may exclaim, "Oh, how dull!" But as soon as we begin to think perceptively about this seemingly so boring occupation, a host of questions arises. What are the arrangements for transferring heat from the bird's body to the eggs? Is incubation performed by the male or female parent or by both? What determines which sex incubates in the several species and families of birds? How does the incubating parent combine with this duty the self-maintenance that it cannot neglect? How long do the eggs take to hatch? What determines the length of the incubation period? How long will a bird continue to sit on eggs that fail to hatch?

At What Stage Does Incubation Begin?

Some birds, including pigeons and doves, keep

PLATE 46. King Cormorants incubating in a crowded colony on the Argentine coast. Gulls in foreground (photo from William Conway, New York Zoological Society).

their nests rather constantly covered by one or the other parent from the time the first egg is laid. Yet they are guarding the egg without applying much heat to it, as seems evident because all the eggs hatch at about the same time, despite the fact that they were laid over a period of a day or more. Other birds actually begin to incubate before their sets are complete, with the result that hatching is spread over one or more days. If the set of eggs is large, or the interval between the laying of successive eggs is long, the first of the brood to hatch may be considerably older and stronger than the last, as is particularly noticeable in many owls and hawks. Ducks and gallinaceous birds usually delay effective incu-

bation until their sets are complete, so that all the chicks or ducklings will hatch and be ready to follow the parent to the feeding ground at about the same time. Still other birds only gradually warm up to their task and do not achieve full-time incubation until some days after they have laid the last egg. For some, the beginning of effective incubation at a certain point in the laying period has important consequences for the future brood of young; in other species, the start of incubation seems to be capricious.

A Synopsis of Incubation Patterns

Years ago, I studied the home life of a pair of big

Ringed Kingfishers nesting in a deep burrow in the loamy bank of a Guatemalan river. I wished to learn, among other things, how they divided their time on the eggs. Hour after hour, I sat among young canes on a sandy floodplain, with my eyes fastened on the hole in the bank across the channel. Only at long intervals, at most twice a day, did I see a kingfisher enter or leave the tunnel, and then it often darted in or shot out so swiftly that I failed to distinguish the color of its breast, where alone the plumage differed in the two sexes. In the first days, when I learned little except that the kingfishers incubated continuously for many hours, my study was making such great demands on my time and patience that I was tempted to abandon it without having reached firm conclusions. But my strong belief that the activities of birds are rhythmic rather than random buoyed me up and encouraged me to persevere until I proved that the kingfishers regulated their attendance on their eggs according to a simple schedule.

In subsequent years, I have studied the incubation habits of many other birds, nearly always finding that they follow a more or less definite schedule, and I have searched published reports for information on this subject. Although all the members of a species rather closely follow the same pattern of incubation, which in a number of instances prevails throughout a family, in the whole class of birds these patterns have proved to be amazingly diverse. Almost every scheme that one can imagine for keeping a set of eggs intermittently or constantly covered has been adopted by one species or another. Some birds sit continuously through a long incubation period, whereas others rarely stay in their nest as long as twenty minutes on fair days. Some birds incubate alone and others with the help of a mate. When both members of a pair share incubation, sometimes the female attends the eggs through the night, sometimes the male, and sometimes they alternate at nocturnal incubation. Although when first surveyed the diversity of birds' incubation habits seems utterly bewildering, careful analysis reveals seventeen well-defined types, the relationship of which becomes clear when we arrange them in outline form:

I. Both parents incubate
  A. They incubate simultaneously in separate nests
    1. *Red-legged Partridge Pattern*
  B. They incubate alternately in the same nest
  Either sex may cover the eggs by night
    2. The male and female replace each other at intervals of about twenty-four hours, so that the same parent does not often sit for the whole of consecutive nights—*Ringed Kingfisher Pattern*
    3. They replace each other at intervals of much more than twenty-four hours, so that the same parent may sit for several consecutive nights—*Albatross Pattern*
    4. They replace each other at intervals of less than one day, sometimes in both the night and the day—*Herring Gull Pattern*
  The female covers the eggs by night
    5. The male takes one long session each day—*Pigeon Pattern*
    6. The sexes alternate on the nest several times a day—*Antbird Pattern*
  The male covers the eggs by night
    7. The female takes one long session each day—*Ostrich Pattern*
    8. The sexes alternate on the eggs several times a day—*Woodpecker Pattern*
II. Only one parent incubates
  A. Only the female incubates
    9. She takes each day one recess that is usually long—*Quail Pattern*
    10. She takes several or many recesses each day—*Hummingbird Pattern*
    11. She sits continuously for many days, fasting—*Golden Pheasant Pattern*
    12. She sits continuously for many days but is fed by her mate—*Hornbill Pattern*
  B. Only the male incubates
    13. He takes one long recess each day—*Tinamou Pattern*
    14. He takes several or many recesses each day—*Jaçana Pattern*
    15. He incubates continuously for many days, fasting—*Emperor Penguin Pattern*

III. More than two birds incubate in the same nest
  16. The eggs are laid by only one female—*Bushtit Pattern*
  17. The eggs are laid by two or more females—*Ani Pattern*

The *Red-legged Partridge Pattern* (1) is exemplified by few birds. The female Red-legged Partridge lays two sets of eight to twelve eggs in different nests in close succession; she incubates one set while her mate takes charge of the other (Goodwin 1953). In the Sanderling, either the male alone or the female alone incubates a set of four eggs on the Arctic tundra; and apparently they take simultaneous charge of two sets laid by her (Parmelee 1970). Temminck's Stint has a similar habit (Lack 1968). Considering all the hazards to which nests on the ground are exposed, the advantage of not putting all the eggs in one basket is obvious. It is not difficult to imagine how this unusual system could evolve from incubation habits like those of the Bobwhite, of which either the cock or the hen may take exclusive charge of the nest, and more rarely they take turns on the same set of eggs. In captivity, they may even sit side by side in the same nest (Stoddard 1946); but I have found no well-authenticated example of consistent simultaneous daytime incubation of the same set of eggs by free birds of any species. Like the Bobwhite, the male Gambel's Quail may take over a set of eggs and hatch them if the female dies (Kendeigh 1952). In a number of woodpeckers, Australian grassfinches, barbets, swifts, titmice, and the Blue-throated Green Motmot, both sexes sleep in the nest with the eggs, but it is hardly possible to learn whether at this time each incubates part of the set.

In the *Ringed Kingfisher Pattern* (2) the male and female incubate for alternate periods of about twenty-four hours, so that the cycle repeats itself every forty-eight hours. The male and female Ringed Kingfisher relieve each other in the morning, and each interrupts its long period of responsibility for the eggs by a single outing in the afternoon. Red-footed Boobies usually replace each other on the nest in the evening, and each member of the pair incubates for twenty-four hours without relief or food (Verner 1961). In the Diving Petrel of New Zealand, the parents as a rule replace each other in the burrow every night, and each apparently remains continuously until relieved by its mate (Richdale 1943). Black Noddies on Ascension Island also take incubation shifts of about twenty-four hours (Ashmole 1962). In the swiftlet *Collocalia maxima* sessions are very long and one parent may remain on the eggs for over thirty hours, with the change-overs usually taking place at night (Medway 1962). Apparently, the swiftlet also follows this incubation pattern.

The *Albatross Pattern* (3) is followed chiefly by sea birds, especially albatrosses, shearwaters, petrels, and penguins. Petrels and the smaller shearwaters find it safe to visit their nest, in burrows or crevices on small islands, only at night, and preferably when there is no moon. Under cover of darkness they can escape the voracious skuas and Great Black-backed Gulls that might devour them in the daytime. Many of these birds gather their food at sea far from their nests—in the Manx Shearwater, up to six hundred miles, according to Lockley (1942). Penguins must often trudge many miles between an inland breeding colony and the water where they forage. Although the Diving Petrel's shifts of twenty-four hours on the nest seem long, they are far shorter than those of many other members of the order Procellariiformes and of certain penguins. Some intervals of continuous attendance at the nest without taking food are as follows:

King Penguin, 1 to 22 days, up to 35 when the partner failed to return (Stonehouse 1960)
Adelie Penguin, males up to 40 days, females up to 21 (The longest periods of attendance were by males who had been present at their nests before the eggs were laid and then remained, incubating and fasting, after females went to sea for food.)
Rockhopper Penguin, 10 to 19 days (Warham 1963)
Little Blue Penguin, 1 to 12 days (Richdale 1951)
Laysan Albatross, 1½ to 38 days, up to 52 days when the partner failed to return (Rice and Kenyon 1962)
Black-footed Albatross, 5 to 39 days, up to 49 days when the partner failed to return (Rice and Kenyon 1962)
Buller's Mollymawk Albatross, up to 24 days

(Richdale 1951)

Manx Shearwater, 1 to 26 (average about 6) days (Lockley 1942, Harris 1966)

Sooty Shearwater, up to 13 days (Richdale 1944*b*)

White-faced Storm-Petrel, 3 to 9 days (Richdale 1943)

Leach's Petrel, 1 to 6 days (W. A. O. Gross 1935, Wilbur 1969)

Yellow-billed Tropicbird, 3 to 4 days, sometimes less (Stonehouse 1962)

Wideawake or Sooty Tern, 89 to 157 hours; mean of nine records, 132 hours or 5½ days (Ashmole 1963*a*)

In the *Herring Gull Pattern* (4) the continuous sessions of each partner are usually much shorter than in the preceding types. In the Herring Gull (Tinbergen 1953) and Semipalmated Plover (Spingarn 1934) the male and female replace each other at intervals of several hours, and at times less, during both day and night. In the Sandhill Crane, the sexes cover the eggs alternately for periods ranging from about one-half hour to nearly eleven hours, and either sex may occupy the nest by night, apparently with no change-overs in the dark. Other cranes have similar incubation patterns (Walkinshaw 1965). The Erect-crested Penguin's spans of incubation range from two minutes to thirty-three hours, so that on some nights both parents will take turns on the eggs, whereas on others only one will cover them. The parent off duty often rests beside its incubating mate, and here it sleeps (Richdale 1951). The Brown Booby seems also to follow this pattern, since the majority of its sessions on the eggs last from one to twelve hours, although occasionally it may incubate for more than thirty hours continuously, with change-overs usually occurring in the morning but sometimes by night. The sessions of the White Booby average substantially longer (Dorward 1962). In the American Coot, the female sometimes incubates through the night, but more often the male is in charge of the eggs from nightfall until dawn. By day the two replace each other at intervals of from about one-half hour to five hours (Gullion 1954). The only instance known to me of a passerine of which either sex may incubate by night is the Cape Wagtail, of which Skead (1954) found some-

PLATE 47. Male Slaty Antshrike incubating two eggs in the undergrowth of Panamanian rain forest—an example of the Antbird Pattern of incubation.

times the male, but more often the female, sleeping on the eggs. This irregularity becomes more understandable in the light of Moreau's (1949) observation that in the African Mountain Wagtail both parents slept on or close beside the nest.

The *Pigeon Pattern* (5) seems to be of quite regular occurrence throughout this great family. The male pigeon or dove goes on the nest about the middle of the morning, but sometimes earlier or later, and sits, usually without interruption, for six

PLATE 48. Male Plain Antvireo incubating two eggs in Costa Rican mountain forest. He sat calmly while the camera was set on a tripod only a yard away.

to nine hours. In the middle or late afternoon, his mate comes to replace him and remains on the nest until he returns next morning. The gleaming white eggs are rarely left uncovered for more than a few minutes. A similar pattern of incubation is followed by a number of trogons, which nest in chambers that they carve in rotting trunks or termitaries, including the White-tailed, Citreoline, Black-throated, and Massena trogons. But the largest and most splendid member of the family, the Quetzal, follows a schedule that calls for less prolonged sitting.

The *Antbird Pattern* (6) is widespread among birds, but no family appears to follow it more consistently than the large tropical family for which it is named (plates 47 and 48). As in the Pigeon Pattern, the female incubates through the night, but each day she intercalates at least one session between those of her mate, and in many representatives of this category the two partners replace each other much more frequently. Single daytime sessions range from a few minutes, as in certain Old World flycatchers, to five or six hours, as in some antbirds.

This pattern of incubation is followed by the Quetzal, the smaller kingfishers, jacamars, gnatcatchers, some Old World warblers, and the Rose-breasted and Black-headed grosbeaks. A number of species, in which the sexes are alike and it is not feasible to determine which covers the eggs during the night, probably also belong to this group, among them certain woodcreepers, ovenbirds, swallows, and vireos.

Among passerines, incubation and brooding are the nesting activities least frequently undertaken by males, who as a rule do not incubate unless they also build the nest and feed the young. A curious exception is found among wagtails; males of both the Yellow and Japanese wagtails incubate about a quarter of the time, although they take no part in building the nest, and at least the latter fails to brood the nestlings (S. Smith 1950, Haneda and Shinoda 1969).

The *Ostrich Pattern* (7) is just the reverse of the Pigeon Pattern; the female takes one long session on the eggs each day but the male is in charge during the remainder of the day and through the night. In the South African Ostrich, the cock incubates from late afternoon or evening until the following morning, one of the well-camouflaged hens through the remainder of the daytime (Sauer and Sauer 1966). Much the same schedule is followed by the Large Pin-tailed Sandgrouse (Marchant 1963), Namaqua Sandgrouse (Maclean 1968), and Hudsonian Godwit (Hagar 1966). In the White-whiskered Softwing, a puffbird, the female enters the burrow in the leaf-strewn floor of the tropical forest in early morning and sits until midday, an interval of six to eight hours. A short while after her departure, the male goes into the burrow and sits until the following dawn (Skutch 1958b). Rufous-winged and Pale-billed woodpeckers also follow this pattern.

The *Woodpecker Pattern* (8) is the counterpart of the Antbird Pattern. In most species of woodpeckers for which information is available, the male occupies the nest cavity by night, and by day he and his mate replace each other a number of times. In a few woodpeckers, including the Golden-naped and at least some of the minute piculets, both parents sleep in the nest with the eggs and young, but indirect evidence suggests that the male covers the eggs through the night, as in less sociable woodpeckers (Skutch 1969b). This pattern is followed by the anis, at least when a pair nests alone instead of joining other pairs in a communal nest, and probably by other species in the cuckoo family. The American Coot sometimes conforms to this pattern.

Turning now to birds of which only the female incubates, we begin with the *Quail Pattern* (9), in which the female sits constantly, with the exception of a single recess each day. The Marbled Wood-Quail leaves her covered nest on the ground in tropical woodland early each morning for an outing that generally lasts from one and a half to three hours (Skutch 1947). The Bobwhite, however, usually takes her recess in the afternoon, neglecting her eggs for 1 or 2 hours if the weather is cool and showery to as much as 7 hours if it is warm and fine

PLATE 49. Female Rufous-tailed Hummingbird incubating two eggs in a bougainvillea vine.

afternoons may cause the Bobwhite to take her outings in the forenoon. The Helmeted Curassow of
northern South America likewise belongs here
(Schäfer 1954*b*). Some incubating domestic hens
leave their eggs once daily, but, if they are promptly
fed, their absence is likely to be much shorter than
that of free birds who must forage for themselves.

The *Hummingbird Pattern* (10) is widespread
among small and middle-sized birds of which only
the female incubates. The number of her absences
each day may range from two or three by very constant sitters to seventy or more by those that are
restless. With the exception of a few species in which
the male has been reported to share incubation,
hummingbirds quite generally follow this pattern
(plate 49). It is also the prevailing pattern in the
great order of passerine birds, but many exceptions
—species in which the male helps to incubate—
have been discovered. It seems, however, to be followed invariably by manakins, cotingas, American
flycatchers, wrens, wood warblers, honeycreepers,
American orioles and their allies, and tanagers
(plates 50, 51, and 52). It is also widespread among
gallinaceous birds. The hen Ring-necked Pheasant
takes one, two, or rarely more recesses each day,
each usually lasting from thirty-six to seventy-eight
minutes (Kessler 1962). Black Grouse, Hazel
Grouse, Willow Grouse, and Capercaillies take each
day two to five recesses that last from about ten to
forty minutes (Semenov-Tian-Shansky 1959). The
hen Rufous-vented Chachalaca of Venezuela usually leaves her eggs three or four times a day (Lapham 1970), and a Chestnut-winged Chachalaca that
I watched in Costa Rica did so twice a day (Skutch
1963*b*). This pattern is also exemplified by the Redhead Duck, of which the incubating female takes
recesses by night as well as by day (Low 1945).

At times the comings and goings of the incubating
female are so regular that it seems proper to speak
of the "rhythm of incubation." A Collared Redstart
that I watched for a morning in the Costa Rican highlands incubated as follows (intervals in
minutes):

Sessions	27	30	28	28	29	29	28	29
Recesses	9	9	7	8	12	9	11	13

Sometimes, however, the female's movements are
most irregular. One of the most erratic that I have

PLATE 50. Female Ruddy-capped Nightingale-Thrush
incubating two eggs in a mossy nest in Guatemalan highland forest—another example of the Hummingbird Pattern of incubation.

(Stoddard 1946). The difference between the Bobwhite and Marbled Wood-Quail may be due to climate; in the region where I studied the wood-quail,
rainy afternoons are frequent through much of the
year but rainy mornings are rare. A series of rainy

PLATE 51. Female Crimson-backed Tanager incubating two eggs in an ornamental shrub in Panama. She conforms to the Hummingbird Pattern of incubation.

PLATE 52. Female White-collared Seedeater incubating two eggs in a small, spiny palm in Honduras—another example of the Hummingbird Pattern of incubation.

watched was a Streaked Saltator, who in four hours attended her eggs as follows:

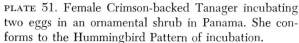

Sessions	27	108	7	31	
Recesses	7	8	10	16	30

This saltator's mate, who came to call her from her nest at random times, seemed to be largely responsible for her unpredictable behavior. In the regularity of their movements, most of the birds that I have watched while they incubated fell between these extremes.

In the *Golden Pheasant Pattern* (11), the incubating female sits continuously for many days, fasting. This seems to be true of the Argus Pheasant (Delacour 1951–1952) as well as the Golden Pheasant (Goodwin 1948a). The Common Eider Duck departs slightly from this pattern in that she may leave her nest every second or third day for an in-

terval of ten to fifteen minutes, during which she drinks but does not eat (Tinbergen 1958). Doubtless further studies will show that this type is not uncommon among large birds.

In the *Hornbill Pattern* (12), the female sits constantly throughout the incubation period but is well nourished by her mate. In these huge-billed birds of the Old World tropics, the female enters a hole in a tree about the time she begins to incubate. The doorway is then walled up with clay brought by the male or with a plaster composed of rotten wood, the indigestible chitinous parts of insects, and other matter that she finds within the cavity. She helps to build the wall that will confine her. In the plaster seal she leaves a narrow slit, through which her hard-working mate passes the food for her and likewise for the nestlings after they hatch. Waste is ejected through the aperture. The female hornbill remains incarcerated not only throughout the long incubation period, but also until the young are well grown or, in some species, until they are ready to fly. She takes advantage of this long period of enforced inactivity to molt and renew her plumage; if forcibly removed, she may be unable to fly. A female Silvery-cheeked Hornbill stayed in her nest for 108 days (Moreau and Moreau 1941).

I am aware of no other family in which a continuously incubating female is sustained by her mate, but an approximation to this pattern is found in a number of other birds, in which the female is so well fed by her partner that her absences from the eggs are infrequent or short, or both. Among them are owls, parrots, certain jays, Cedar Waxwings, goldfinches, siskins, and crossbills.

The *Tinamou Pattern* (13) is the counterpart of the Quail Pattern. Among the male tinamous that ordinarily take a single long recess each day are the Highland Tinamou (Schäfer 1954c), Little Tinamou (Skutch 1963a), Slaty-breasted or Boucard's Tinamou (Lancaster 1964b), and Brushland Tinamou (Lancaster 1964a). In a region of generally clear mornings and afternoon downpours, the Little Tinamou goes for a long outing each forenoon; but the Slaty-breasted and Brushland tinamous usually neglect their eggs for several hours in the afternoon while they forage. On some days these tinamous omit the excursion and incubate continuously. One or more females supply these male tinamous with from one to ten eggs in a single nest, which after laying they quite neglect; apparently they go off to produce more eggs to keep other males busy. When a male Bobwhite undertakes incubation, he follows this pattern.

The *Jaçana Pattern* (14) corresponds to the Hummingbird Pattern, with the difference that the male, rather than the female, attends the nest, taking a few or many recesses each day. In addition to the Pheasant-tailed Jaçana of eastern Asia (Hoffman 1949), the Ornate Tinamou of the high Andes, departing from the usual practice of his family, follows this pattern (Pearson and Pearson 1955).

The *Emperor Penguin Pattern* (15) is the counterpart of the Golden Pheasant Pattern. After laying her single egg, the female Emperor Penguin de-livers it to her mate and marches off to find food in the sea. He places the egg on his tarsi or feet and covers it with a fold of skin at the base of his abdomen. Then for about sixty days he stands upright on the snow, in the frigid night of the Antarctic winter, incubating his egg without taking food. In tempestuous weather, he huddles with other males of the colony, still carrying his egg, in order to conserve heat. About the time the eggs hatch, the plump females return to claim and feed the chicks, not necessarily their own, and the emaciated males trudge over the ice to feed and recuperate in the water (Stonehouse 1953). In New Zealand, the male Kiwi incubates two or three eggs in a burrow that the female has helped him to dig, and he may remain at his task as long as a week without going for food. The male Emu of Australia incubates the eggs laid by one, two, or three females in the same nest, which he rarely leaves (Kendeigh 1952).

In all the foregoing patterns of incubation, the eggs are ordinarily warmed by only a single parent or a single pair, although, as in some of the tinamous, Ostrich, and Emu, more than one female may have laid them. We now turn to patterns in which more than two individuals incubate in the same nest.

In the *Bushtit Pattern* (16), named for the minute bird that builds a cozy pensile nest in scrubby vegetation in the southwestern United States, additional individuals help the parents to incubate and later to feed and brood the nestlings (Addicott 1938). In the Ostrich, the dominant hen sometimes permits one of the subordinate hens to cover the eggs that all have laid, and rarely two hens may incubate simultaneously in the same nest. At a Costa Rican nest of the Acorn Woodpecker, a female and four males were incubating, replacing each other every few minutes, and at another nest at least two males and one female were taking turns on the eggs. At a third nest, two birds of each sex were feeding nestlings. Similar observations have been made on the Acorn Woodpecker in California. Unfortunately, the inaccessibility of all these high nests made it impossible to learn the number of eggs they held (Skutch 1969b). Possibly in some of them two females had laid, in which case they would fall under the following heading.

The *Ani Pattern* (17) is named for the long-tailed

black cuckoos of tropical America that have been repeatedly mentioned. When several pairs unite to build a bulky open nest of sticks lined with green leaves, all the cooperating females lay their chalky white eggs together, so that it seems impossible that any could distinguish her own. Incubation is performed by both the males and females of the flock, sitting in succession rather than two or more together; at night a single male incubates. After the young hatch, all the parents feed them, without favoring their own. In the Groove-billed Ani these communal nests are attended by two, three, or rarely more pairs (Skutch 1959*b*); in the Smooth-billed Ani, up to five females may be in attendance (D. E. Davis 1940*a*).

Once I found two female Blue Tanagers incubating in a nest in which both had laid. The dominant female sat when she wished and the less forceful one while the first went for food. This arrangement is most exceptional in the species and the tanager family as a whole. In a number of instances, two birds of different kinds have been found incubating in a nest in which both had laid, often sitting side by side or one above the other. Since such examples of interspecific cooperation are of sporadic occurrence, it does not seem necessary to establish another "incubation pattern" to accommodate them. We shall return to these strange associations in chapter 28.

A careful examination of the foregoing patterns of incubation reveals that almost every one has its counterpart, or negative image, in which the roles of the sexes are reversed. Thus the Ostrich Pattern is the reverse of the Pigeon Pattern, the Woodpecker Pattern of the Antbird Pattern, the Tinamou Pattern of the Quail Pattern, the Jaçana Pattern of the Hummingbird Pattern, the Emperor Penguin Pattern of the Golden Pheasant Pattern. The exceptions are the first four patterns, in which the roles of the male and female are so equal that no reversal would be significant; the last two patterns, in which more than two birds participate in incubation; and the Hornbill Pattern. This last is the only one for which a counterpart is needed to make the outline perfectly symmetrical; but I have no information on any species in which a female regularly feeds her incubating mate, even if he is not immured in the nest.

The life histories of a great many birds are still unknown, and possibly an example of this missing pattern will some day be discovered.

Numerous species for which we have detailed observations on incubation cannot be assigned to a pattern because of the difficulty of distinguishing the sexes. In the Blue-diademed and Broad-billed motmots, for example, one parent incubates throughout the forenoon and the other from early afternoon until the following dawn. Accordingly, these motmots follow either the Pigeon Pattern or the Ostrich Pattern, depending upon whether the female or the male incubates through the night. In these and other birds that nest in burrows or deep holes, it is scarcely possible to distinguish the sexes by seeing which member of a marked pair lays an egg.

The Evolution of Incubation Patterns

In rheas, Emus, kiwis, and tinamous, incubation and care of the young are performed largely or wholly by the male. When we recall that these birds are generally held to be primitive, we are tempted to conclude that long ago, when birds were evolving from their reptilian ancestors, the female after laying her eggs went off and forgot them, leaving to the male all their subsequent care. We may support this speculation by pointing to a number of fishes and frogs, including stickle-backs, fighting-fish, pipe-fish, and the nurse-frogs of Europe, in which the male alone takes charge of the eggs and young. We may suppose that, as birds developed psychically and the affective bond between male and female grew stronger, the latter stayed to help her mate at the nest, giving rise to the system that is now most common throughout the class of birds. Then, as the males in many species became more ornate, developed elaborate songs, or became greatly preoccupied in the defense of territory, they desisted from incubation, although they might still help with the care of the young. Finally, many males ceased to do this, too, giving rise to the situation that we find in many pheasants, hummingbirds, manakins, birds of paradise, and certain members of other families.

One difficulty with this view is that not all the species in which the male assumes most or all of the parental duties are regarded as primitive; some are placed fairly high in the evolutionary tree. Some of these birds—such as the hemipodes among the cranelike birds (Gruiformes), the painted snipe and jaçanas and phalaropes and certain sandpipers among the ploverlike birds (Charadriiformes)—belong in orders in which incubation by both sexes is the prevailing method, and which also contain species in which only the female attends the nest. In the groups just mentioned, the female's failure to attend the nest seems to be a secondary development; so that, if we suppose that sole attendance by the male was the primitive system, we must make the further assumption that in some evolutionary lines the female acquired the habit of attending her eggs but afterward lost it. We can avoid this complication by accepting the view that in ancestral birds the two parents shared incubation. The widespread occurrence of this system at the present time points to the soundness of this view; and, if we require an analogy among the cold-blooded vertebrates from which birds have descended, we find it in certain fishes, as in those cichlids of which the male and female alternate in the care of the eggs and hatchlings in a manner that reminds us strongly of birds.

Although, as is to be expected, the scanty fossil remains of ancient birds throw no light on this interesting question, it is probable that in ancestral birds incubation was shared by both parents. This system is still widespread throughout the avian class and is exhibited by at least some species in most of the existing orders. For reasons that we shall presently consider, some birds have departed from it in the direction of sole incubation by the male and others in the direction of sole incubation by the female. It is noteworthy that our present knowledge of the incubation habits of birds lends no support to the view that exclusive incubation by the female was the primitive method, as it does to the view that exclusive incubation by the male was primitive. Evidently, since birds were birds, most male parents have done more than simply fertilize the eggs.

One factor that might lead to abstention of one of the sexes from incubation is an unbalanced sex ratio. The mechanism of sex determination makes it probable that at the time of fertilization equal numbers of eggs will have male and female potentialities. But one sex may be less resistant than the other, so that more individuals die before or after hatching. If one

sex is more brightly colored than the other, if it calls or sings more frequently, or is less careful to remain concealed, it may suffer more from predation. This should, in many species, cause the heavier mortality to fall upon the male; but the female sitting on a low nest is exposed to dangers that a nonincubating male avoids. As Nice (1937) demonstrated in the Song Sparrow, 10 percent more females than males died between April and June. Finally, if one sex takes a year or two longer to mature than the other, it will be less numerous in the breeding population, even if the two sexes are equally represented in the total population. In all these ways, and especially in the last mentioned, an excess of breeding males or females might arise (see table 1, p. 2).

We know too little about the sex ratios of free birds, but some of the consequences of the preponderance of either sex seem clear (Mayr 1939). If one sex is substantially more numerous than the other, strict monogamy will deprive many individuals of the more abundant sex of parenthood. The abandonment of monogamy should give every normal individual an opportunity to breed and considerably increase the annual production of young, which is a critical factor in the survival of some species.

If males outnumber females, the latter may give employment to all of the former if each distributes her eggs between a number of nests, leaving the males to attend them. By refraining from incubation and giving much of her time to feeding, she might greatly increase the number of eggs that she laid in a season. Perhaps at first, in response to persisting parental impulses, she would remain to help the last of the males for whom she had provided eggs; but this male should be as capable of hatching them alone as the others are. With further evolution, the female's unnecessary participation in parental duties would disappear. Thus would arise the situation found in the Variegated Tinamou (Beebe 1925) and the Pheasant-tailed Jaçana, in which the males incubate small sets laid by the less numerous females. Harder to explain is the situation in the Highland Tinamou and Emu, in which several females deposit their eggs in the same nest and a male takes charge of them. In these species, it would seem to be more efficient for each female to attend her own family;

but we have much to learn about the habits and ecology of these birds.

When there is a shortage of sexually mature males, polygyny increases reproductive efficiency by permitting all the females to breed. In California, L. Williams (1952) found that the sex ratio of Brewer's Blackbirds fluctuated from year to year. The greater the scarcity of males, the more mates each one had, and these males helped their several consorts to feed the young. But, if the excess of females became considerable and persistent, such dispersion of the male's effort might prove unprofitable, leading to his complete withdrawal from parental occupations other than guarding the breeding colony, as in oropéndolas, caciques, Great-tailed Grackles, and other birds of similar habits.

The Influence of Environment on Incubation Patterns

When only one parent attends the eggs, they are, in most species, left exposed while it seeks food and water, or even rests and preens. These periodic absences are often long enough for the eggs to cool to a temperature close to that of the surrounding air. In mild weather, they suffer no harm; indeed, eggs of many kinds will hatch even after night-long chilling when the embryo is well formed. But long interruptions of incubation retard embryonic development unless the ambient temperature approaches that of the parent, and in freezing weather such exposure may be fatal. In the far north and high on tropical mountains, many birds breed when the weather is far from mild and even when the temperature falls below the freezing point. Must not these birds make special arrangements to keep their eggs continuously protected? Must not both sexes incubate, even if they belong to families in which incubation by a single parent is the rule?

When I began to study birds in the high mountains of Central America, I asked myself these questions. Even at seasons when nights are frosty and penetratingly cold, the sunny, windless days are usually mild at altitudes as high as 10,000 or 12,000 feet (3,000 or 3,600 meters). But when clouds envelop the heights, rain falls, or cold winds blow, as frequently happens while birds are nesting on tropical mountains, eggs and naked nestlings chill rapidly

if left uncovered. In several nests, I found dead nestlings that seemed to have succumbed to exposure. Yet I could discover no instance of a male who responded to this situation by taking turns on the nest, if his nearest relations in other regions did not do so. The males of hummingbirds, honeycreepers, wood warblers, wrens, thrushes, and other such families failed to incubate in the cool highlands, just as they did in the warm lowlands. In the high Andes, however, a male Sparkling Violet-ear Hummingbird may, at least occasionally, help his mate to incubate and brood (Schäfer 1954a).

In northern coniferous woods, crossbills and siskins often incubate while the country is blanketed with snow, so that they can feed their nestlings while the seeds of last year's cones are still available. The small eggs must be almost constantly covered, for on many days an exposure of more than a few minutes would result in their freezing. One method of keeping them constantly warmed would be for the male to sit on them while his mate goes for food. But incubation by the male is exceptional in the finch family and seems never to occur in the subfamily (Carduelinae) to which crossbills and siskins belong. So conservative are birds in their incubation habits that, instead of adopting this simple solution, these nesters in snowy woods solve the problem by the male's feeding the incubating female so liberally with regurgitated food that she need leave the nest only at long intervals and is rarely absent as much as ten minutes. The closely related goldfinches often breed in midsummer in the North Temperate Zone, or at middle altitudes in the tropics, yet they follow the same pattern of almost constant sitting by the female, while her mate provides for her. Possibly this practice arose in ancestral goldfinches that nested under more rigorous conditions.

Another bird that breeds in freezing weather, the Thick-billed Nutcracker of northern Europe, faces a problem somewhat different from that of crossbills and siskins. In autumn, these members of the jay family gather many hazel nuts and bury them in their territories. They depend on these stores for winter food and for nourishing their young the following year, but to use them they must breed early, before warm days cause the nuts to germinate. The nutcrackers have marvelous memories for the exact locations of their caches, but each individual knows only where its own nuts are hidden. Accordingly, to use her own resources, the female must leave the nest and find them, and while she is absent her mate incubates. He may do so as much as two-thirds of the daytime (Swanberg 1956). The Clark's Nutcracker of the Rocky Mountains sustains itself largely on the seeds of last year's pine cones while it nests, often when the woods are still laden with snow. In this species, too, the male takes turns on the eggs, but at one nest he did so only about one-fifth of the day (Mewaldt 1956). The males of both kinds of nutcrackers develop incubation patches as large as those of the females, which is unusual in Oscines. In most crows and jays, only the female incubates, usually receiving much food from her mate and at times also from their helpers.

The Emperor Penguin is another bird that departs from the incubation habits typical of its family in response to environmental conditions. The male of this species, as has been told, incubates continuously in midwinter in Antarctica, although in other penguins the sexes share incubation. The Emperors incubate in this most inclement season because they are exceptionally large penguins and take long to mature; if they laid their eggs in spring, their young would not be ready to go to sea the following summer, when open water extends near the breeding ground. But in winter the sea ice stretches far from the land, and the Emperors would lose so much time walking back and forth to the open sea where they forage that the male and female do not try to relieve each other during the two months' incubation. In other penguins, such as the Adelie, the female goes to sea after laying and her mate stays on the eggs until she returns some days later to replace him. The male Emperor seems to have prolonged this first long session to cover the whole incubation period.

We have found a hummingbird, two species of nutcrackers, and a penguin that depart from the incubation habits typical of their families to permit successful nesting at low temperatures. Undoubtedly, similar alterations of basic incubation patterns induced by the environment will turn up when more birds that breed in inclement regions are carefully studied. On the whole, however, birds adhere stubbornly to their ancestral patterns of incubation, which change less readily than certain other habits in adapting to new conditions.

The Influence of Coloration and Nidification on Incubation Patterns

A century ago, in an essay on bird nests, the great evolutionist Alfred Russel Wallace (1871) wrote: "When both sexes are of strikingly gay and conspicuous colours, the nest . . . is such as to conceal the sitting bird; while, wherever there is a striking contrast of colours, the male being gay and conspicuous, the female dull and obscure, the nest is open and the sitting bird exposed to view." Wallace, who was deeply interested in the subject of protective coloration, supposed that there was a close correlation between the form of the nest and the color of the female parent. He seemed to be unaware of the prevalence of incubation by males in many avian orders. Today, when we know vastly more about the breeding habits of birds than was known in his time, the problem appears more complicated. We must consider three factors: the form of the nest, the coloration of both parents, and their participation in incubation. How do these variables influence each other?

In most of the numerous species of wood warblers that breed in the United States and Canada, the male is more brilliant than the female, the nest is open, and she alone incubates. A roofed nest is built only by the Ovenbird, a terrestrial species in which both sexes are dull in color to match the ground litter over which they walk, and again only the female incubates. When, in the Guatemalan highlands, I discovered that the bright red Pink-headed Warbler built a covered nest much like that of the Ovenbird, I eagerly set up my blind to watch it. The male was no more brilliant than his mate and he had no excuse for not sitting on the eggs. Nevertheless, he failed to do so. Later, I found the same to be true of the Slate-throated Redstart, Collared Redstart, Chestnut-capped Warbler, and Black-cheeked Warbler. In all these fairly bright wood warblers the sexes are distinguishable only by the male's song, the nest is oven shaped, and the female alone incubates. Likewise in the more soberly attired Buff-rumped Warbler, only the female, clad exactly like her mate, incubates in the roofed nest.

In this family—whether the nest is open or covered, whether the parents are both dull or both bright or one is much more brilliant than the other—only the female incubates. Neither the form of the nest nor the color of the birds has any evident influence on the incubation pattern. But perhaps the covered nest has permitted the females of many tropical species of wood warblers, especially in the large genera *Myioborus* and *Basileuterus*, to become as brilliant as their mates.

In the Scarlet Tanager and Summer Tanager of the United States, the female is much less brilliant than the male and only she incubates, in an open nest. In tropical America, there are many small gem-like tanagers, especially in the great genus *Tangara*, of which the sexes are about equally brilliant. Yet their nests are open cups, and only the females incubate. Strangely enough, in another group of tanagers, the euphonias, the female is usually much duller than the yellow and blue-black male, the nest is a cozy covered structure with a round doorway in the side, and again only the female sits in it. Hence one need hardly ask whether males of the Scarlet-rumped Black Tanager and other species of *Ramphocelus* take turns in the open nests that their less colorful mates build. They, at least, are so eye-catching that they seem to have a valid reason for not incubating, although they do help to feed the nestlings.

The American orioles of the genus *Icterus* include species in which the females are much duller than the males and species in which both sexes are clad in the same uniform of black and gold. Some of these orioles build cups that fail to conceal the sitting bird, whereas others weave deep, pensile pouches that completely enclose the parent. The more brilliant females may construct either open cups or pouches. Whatever the form of the nest, the males abstain from incubation.

In the great family of American flycatchers, the sexes can rarely be distinguished by their appearance. Flycatchers' nests are most diverse—some open, some roofed, others pensile with the opening at the side or bottom. But whatever the shape of the nest, only the female incubates in the species that have been studied. In the large family of antbirds, however, the sexes often differ in plumage, the male being more deeply colored although he is hardly brilliant. The nest is often an open cup, and both sexes invariably incubate, as far as we now know.

We might continue to multiply facts of the same kind. It is evident that brilliant plumage, which might jeopardize the eggs, is not the only or the chief reason why male birds refrain from incubation. In most cases, we must seek elsewhere the explanation of their failure to incubate. It is likewise evident that the different coloration of the two sexes in many species cannot be explained solely by their participation in incubation. Matrimonial habits seem to be far more effective in determining whether the male and female will be alike or different in plumage. In nonmigratory birds that remain mated and on their territory throughout the year, the sexes are much more likely to have similar plumage than in migratory species that dissever pair bonds at the end of the breeding season. In the wood warbler, tanager, and American oriole families, for example, the sexes of the constantly paired, nonmigratory species more often wear the same bright colors than do those of the migratory species, in which the male, if brilliant, usually has a duller mate. In these migratory species, the brighter male often molts into duller plumage in the autumn, whereas in sedentary species of which the sexes are equally brilliant, both nearly always wear the same colors throughout the year. Even within the tropics, the sexes are more often alike in species that do not flock than in the more gregarious members of the same family. But in this matter it is impossible to make generalizations to which exceptions cannot be found. Doubtless, factors other than matrimonial habits are not without weight in determining whether the sexes will wear the same or different colors.

It would be wrong to conclude from the foregoing discussion that the coloration of the incubating parent does not affect the safety of the nest. When we studied the colors of eggs, we found that those in dark cavities require no concealing hues, those laid in very exposed places have great need of cryptic coloration, whereas eggs in cup-shaped nests amid clustering foliage are protectively colored only to a minor degree. The same seems to hold for the incubating parents. If while sitting they are wholly screened from view within a burrow or an enclosed nest, their coloration will not affect their safety at this time, although if brilliant they must exercise extreme caution not to betray the nest's location by their approach or departure. If the parent incubates on open ground or in a very slight nest in a tree or bush, it is usually protectively colored. But when a female sits in a cup-shaped nest amid screening foliage, concealing plumage seems to be of minor importance, and in many instances she may resemble her brilliant mate.

Although a bird's participation in incubation is not primarily determined by its coloration, its failure to sit may leave it free to acquire such gorgeous plumage that its approach to the nest would endanger the eggs or young. As we learned earlier, abstention from incubation—and then from all parental duties—was the first step toward acquisition of the profuse ornamentation of many pheasants, hummingbirds, cotingas, manakins, and birds of paradise.

The Influence of Song and Territorial Defense on Incubation Patterns

In the whole avian class, song is most highly developed and most freely used in the great group of Oscines, or songbirds. One of the principal functions of song is the proclamation and defense of territory. Many songbirds are territorial, and in a large proportion of them the male does not incubate. Might we conclude, then, that his release from this duty has been caused by his absorption in singing and defending his territory?

Since our effort to account for the loss of the incubation habit by male birds in many groups has thus far been rather unprofitable, we are tempted to clutch at this attractive hypothesis, but I doubt that it will afford the explanation we require. Territorial boundaries are usually settled before incubation begins, and thereafter their defense may make relatively slight demands on the male's time. Moreover, the female in some species helps to drive intruders from the territory without seriously neglecting her eggs. And in many territory-holding species, the male shares incubation yet manages to defend his domain.

Even irrepressible songfulness is not incompatible with the male's participation in incubation. Some birds that sing well and loudly continue to do so while covering the eggs. Once I watched a male Rufous-browed Peppershrike pour forth his far-carrying song for two hours while he sat in the open nest. In the related vireos, the male not infrequently sings while he incubates. The males of the Rose-breasted Grosbeak and Black-headed Grosbeak take a large share in incubating and frequently sing while they do so—so loudly that Weston (1947) depended upon the male's voice to guide him to nests of the latter. Who would expect to find these boldly attired, tuneful grosbeaks incubating, when in their family incubation by the male is exceptional, and even males of the most soberly attired sparrows rarely cover the eggs? In other families, such as the crows and jays, tanagers, and honeycreepers, males that are almost or quite lacking in song fail to incubate.

That songfulness is not incompatible with incubation is further attested by the fact that a number of females sing from their nests in response to their mates' voices or even when they are not heard. Among these songful females are the Cardinal, Yellow-tailed Oriole, Melodious Blackbird, and Highland Wood-Wren. One might suppose that such unrestrained songfulness would so often betray the nest to enemies that it would have been suppressed by natural selection. It seems absurd that a bird who has taken great care to hide its nest should so recklessly call attention to its location. Often, while observing a nest from concealment, I have watched the parent approach with the utmost wariness and hesitation, yet at the same time excitedly repeating loud notes that would seem to cancel all of its caution. And nestlings of numerous kinds fail to preserve the silence that seems indispensable for their safety. Not only do they receive their meals noisily, but while waiting for their parents many of them, especially among the nonpasserines and nonoscine passerines, also practice the simple but often sweet songs of the adults. We must conclude from this that the principal enemies of many nesting birds find eggs and young not by using their ears but by means of other senses. Probably these enemies are snakes and nocturnal mammals.

The outcome of our attempt to explain why some groups have deviated from the prevailing avian system of incubation by both sexes of monogamous pairs seems to be this: If breeding individuals of one sex are present in substantially smaller numbers than those of the other sex, the abandonment of monogamy, and of all attendance at the nest by the less numerous sex, is understandable because it permits full participation in reproduction by the more numerous sex, and this may be advantageous to the species. Otherwise, it is difficult to understand why either sex should refrain from attending the nests, especially when we recall that the brilliant colors and profuse adornments that unfit certain males for incubation were in all probability acquired *after* rather than before they ceased to attend their nests. Apparently, changes in incubation habits have arisen as random variations. In many instances, these changes conferred no advantage on the species but were not eliminated by natural selection because, at least, they did not decrease the species' reproductive efficiency below the critical point.

The incubation habits of many birds are not such as seem most efficient, as when in a rigorous environment one parent is obliged to keep the eggs warm without its mate's help. Perhaps, in migratory species that breed at high latitudes, the necessity to establish a territory, win a mate, and rear many offspring to replace high annual losses has placed a premium on the division of labor. Accordingly, the male is largely responsible for winning and defending a territory; the female, for incubating. But this specialization of functions seems unnecessary in the tropics, where both birds and humans manage to survive without being so efficient as those of more rigorous northern climates. With this exception, the accidental loss of the incubation habit by many male birds appears, on the whole, to have been more profitable to bird watchers than to the birds themselves. To them, it appears to have given no considerable advantage; for us, it has provided a vast variety of phenomena to discover, to ponder, and to enjoy.

CHAPTER 13

The Activities of Incubating Birds

The preceding chapter sketched in broad outlines the temporal pattern of incubation, as performed by a single parent or by two or more sitting by turns. Now we must fill in the details, considering how birds replace each other on the nest, how they warm the eggs, what other care the eggs require, and what is done with them while both parents are absent from the nest.

Nest-Relief Ceremonies

Even when both parents share incubation, they may fail to keep their eggs constantly covered. One may leave the nest before its partner comes to replace it, or either of them may interrupt its session to forage or drink and then return to the neglected eggs, as happens every afternoon in the Ringed Kingfisher and more frequently in certain other birds. In some species in which both parents incubate, notably toucans and ovenbirds, the eggs are less constantly covered than in many other species in which a single parent incubates.

Frequently, however, each of the incubating partners stays at its post until the other comes to relieve it, so that the nest is constantly, or almost constantly, attended. In certain species, the change-over is usually accompanied by a more or less elaborate ritual, involving posturing, presentation of material, or production of special sounds by one or both members of the pair. A distinguished ornithologist wrote that there is "generally" such a ceremony on these occasions. He happened to be a student of sea birds, and nest-relief ceremonies are indeed frequent among the larger birds and especially those that nest in colonies, as many sea birds do. In crowded colonies, the incubating bird must distinguish its returning mate from the many other individuals in the vicinity and relinquish its nest only to the former.

Among the smaller land birds, nest-relief ceremonies are so rare that in many years of watching I have seldom seen such rituals. Often the incubating partner quietly leaves the nest when it hears or sees its approaching mate. Sometimes, even when the mate comes silently, the bird on the nest detects its approach and quits the nest before the human watcher has become aware of it. A Spotted-crowned Woodcreeper, incubating in a deep hole in a trunk, would leave the nest when the mate alighted against the trunk but not when a bird of another kind did so; it evidently recognized its mate by the sound of its feet striking the wood. One of the very few displays that I have seen among land birds at the change-over was given by a diminutive White-flanked Antwren. When coming for his turn at incubation, the black male revealed his usually concealed white shoulder-bands as he alighted beside the open nest in the forest undergrowth. He did this when he arrived whether his mate was present or the nest was unattended (Skutch 1969*b*).

The Pied-billed Grebe has a simple nest-relief ceremony. In Wisconsin, the male and female replaced each other on their floating nest at intervals that usually ranged from twenty minutes to one hour and were rarely longer. The oncoming partner approached by swimming beneath the surface until it was beside the nest. After a quick look around it would submerge again, perhaps to reappear on the opposite side of the floating structure. It might repeat this several times before it hopped upon the nest, but at other times it shot out of the water a foot or two away and landed lightly on the nest's edge. Then it might touch the bill of its departing mate—the only ceremonial act (Deusing 1939).

Another marsh dweller, the Least Bittern, has a more elaborate nest-relief ceremony. Each partner incubates continuously for several hours on a platform built amid cattails, or some other emergent vegetation, about six to twenty-four inches above the water. The female is more attentive than the male, at least in the daytime. When one partner comes to relieve the other, the one on the nest raises its crown feathers and utters a low, hoarse *gra-a-a*. The oncoming bird, especially if it be the male, erects the feathers of its crown and often also of its body. Whatever its sex, the newcomer shakes its open, down-pointed bill from side to side, making a slight rattle. This act seems to be very important in nest relief, although it also occurs when a bittern returns to an unattended nest. In one instance, a male continued to rattle for at least ten minutes before his mate slipped from the nest and flew away. Sometimes the bittern brings a stick or a piece of marsh vegetation when it comes to take its turn at incubation (Weller 1961).

Among herons and egrets, a stick is often brought by the bird coming to replace its mate on the eggs. Usually its arrival is the occasion for a greeting ceremony in which both partners engage. In the Black-crowned Night-Heron, the male and female address each other by stretching their lowered necks forward until their slightly parted bills are close together, while the crown feathers are erected and the long nuptial plumes among them point upward or even forward. The duration of this display de-

pends on the incubating partner's readiness to leave (Allen and Mangels 1940). In a crowded heronry, the incubating bird's reluctance to leave without a display prevents its departure on the approach of any chance passer-by.

In the Willett, the oncoming partner, let us say the male, approaches the terrestrial nest by walking through the long grass, invisible except when at intervals he momentarily stretches up his long neck, periscopelike, above the grass-tops for a reconnaissance. He cautiously circles the nest and, if nothing excites his mistrust, crosses the little clearing on the edge of which it is situated. When directly in front of his mate, he bows low, with his bill pointing toward the ground near her breast, and utters his ringing cry. Then he deliberately walks onto the nest, stepping over his mate. As his breast touches her, she darts out beneath him and flies away, accompanying the flashing of her wings with a clear *pill-will-willet*. In this way, each partner replaces the other without exposing the eggs, even for an instant, to cold or the eyes of a marauder (Vogt 1957).

At a nest of the Stone-Curlew when incubation is well advanced, the incubating bird of either sex picks up a pebble as its mate approaches to relieve it and offers it to the latter with a bow. If the newcomer fails to accept the pebble, the outgoing bird deposits it between the two eggs before leaving the nest (Armstrong 1947).

In the Least Tern, the male of some pairs incubates nearly as much as the female, whereas at other nests he sits little but often gives a fish to his mate. As he brings an offering to his incubating partner he calls *keedee-cui, keedee-cui* very rapidly, and while she swallows it he holds his neck stretched outward, head erect, bill pointed up at an angle of forty-five degrees, and wings halfway elevated. He may then replace her on the eggs. Even if he brings no fish when he comes for his turn at incubation, he gives this display, which seems to be identical with one that is used in courtship (Hardy 1957). In the Black Tern, the change-over is more rapidly effected, without the presentation of a fish and with no display. Sometimes, when the sitting bird is reluctant to leave, the returning partner alights and pushes against it gently with its breast until it rises. Before departing, the tern who has just been relieved may remain at the nest for a minute or so, gathering bits of submerged vegetation from the surrounding water, often with a great flapping of wings, and tossing them about the mate who has just settled on the eggs (Cuthbert 1954).

In one way or another, the manipulation of nest material often figures prominently in the change-over, especially in birds that are indeterminate builders. Sometimes an incubating Herring Gull relinquishes the eggs at the least indication that its mate wishes to cover them. At other times, however, the sitter is reluctant to go. Then the other may approach the nest and utter the mewing call. If this is without effect, the gull may stand beside its sitting partner and wait, perhaps giving a few more mews, or it may go to fetch some nest material, often losing half the load when it opens its bill to call as it approaches the nest. These activities may sooner or later stimulate the sitting bird to rise; but if they fail, a gull eager to incubate may force its way onto the eggs. The two partners then try to push each other from the nest in a struggle that may last for several seconds, during which wings may occasionally flash (Tinbergen 1953).

The Razorbill lays a single egg on a bare ledge of a seaside cliff with no sign of a nest. At the change-over, which occurs rather frequently, the oncoming auk, cawing loudly, tickles the vibrant, upstretched throat of the partner in charge of the egg. After relinquishing it to the newcomer, the outgoing bird picks particles of earth or debris from the ledge and throws them back, beneath itself, or to either side, or perhaps it swallows some (Perry 1946). This pointless activity is probably inherited from distant ancestors that built nests. Far from the northern cliffs where auks breed, ovenbirds and woodcreepers often bring a contribution for the nest when coming to take over the eggs; and sometimes a departing woodcreeper bears a flake of bark in its bill. In these passerines, the carrying of material at the change-over appears to have no ritual significance. It can certainly have none in becards and hummingbirds, of which the female, who incubates with no help from a mate, frequently brings something for the nest as she returns to her eggs.

In penguins, change-over is often accompanied by a striking ceremony. As a Yellow-eyed Penguin ap-

proached his incubating partner, she broke into an "open yell." He ran up with arched back and beak to the ground. Then both put their heads together to perform a hearty welcome ceremony, in which a great volume of sound issued from their widely opened mouths as they faced each other, standing erect close together. After several less intense displays of mutual affection and three repetitions of the "welcome," the female resumed her position on the eggs, then rose to relinquish them to her mate. On another occasion, when the male was loath to give up the eggs to his partner, who had just come in from the sea, she hurried him off with a peck on the tail. When the male of another pair refused to budge from the eggs, the female inserted her head and neck beneath him and continued to push until she was covering the eggs with her mate resting upon her. The two penguins remained in this position until the observer left them. Although it has been claimed that in some species the incubating bird treats the incoming bird as a potential enemy until it has identified itself by an appropriate signal, this is not true of the Yellow-eyed Penguin, for in this species each partner recognizes the other at a distance, by its voice, or as soon as it marches into view (Richdale 1951).

Ruddy Ground-Doves, nightjars, avocets, and other birds have also been seen to push a mate from the nest when it refused to change over. At a nest of the Groove-billed Ani attended by several pairs, when one individual wishes to take over the eggs from another who has perhaps been incubating for only a few minutes and is reluctant to go, the newcomer quietly sits beside the other. As a rule, the incubating bird then leaves in a minute or two, with no sign of resentment at being so soon displaced. In other species, however, the newcomer simply leaves if its incubating partner refuses to make way for it, as I have seen in the Black-hooded Antshrike, Mountain Trogon, and Collared Trogon. The stubborn sitter may then have to remain at its post for a long while before it receives another offer of relief.

The importance of the nest-relief ceremony in certain groups of birds was well illustrated when a male Black Stork mated with a female White Stork at the Schönbrunn Zoo in Germany. These two species have different greeting ceremonies: the black species does not, like the white one, frequently *klapper* with its bill; instead, it moves its neck up and down and from side to side with a peculiar whispering. Although these two birds had been mated for years, the difference in their greeting ceremonies caused continual misunderstanding when they nested together. Similar confusion arises when Rock Doves mate with Wood Pigeons (Armstrong 1947).

As the time of hatching approaches, King Penguins, which earlier repeated a courtship ceremony at the change-over, perform regurgitation movements as they replace each other on the egg. While the egg is hatching, the oncoming parent may deposit regurgitated food upon it (Stonehouse 1960).

Although we have considered nest-relief ceremonies only in connection with incubation, in many species these rituals begin before the eggs are laid and continue while the parents brood their young. If in some birds these ceremonies are necessary to ensure recognition of the oncoming partner and cause the incubating bird to relinquish the nest, in other species such displays may be only a spontaneous expression of emotion when mates who have been separated for hours or days come together again—a renewal of the courtship rites that helps to hold the pair together. Only birds that nest in protected situations, such as small islands and seaside cliffs, or those strong enough to repel enemies, such as colonial-nesting herons, can afford to indulge in the more elaborate of these ceremonies. For small birds that breed in the forests and fields of continental areas teeming with predators, the more silently and discreetly the change-over is effected, the greater the chance of survival.

How Eggs Are Warmed

Plumage is a poor conductor of heat, and if an incubating bird sat with the feathered surface of its abdomen in contact with the eggs, these would remain at a temperature considerably below that of the parent's body, especially in cool weather. For the more efficient transfer of heat, the parent sheds the down feathers, and more rarely contour feathers, from part of its ventral surface before it begins to incubate. The skin thereby exposed then becomes richly supplied with blood vessels and also greatly thickened, so that heat flows freely from it to the eggs against which it is closely pressed. These featherless, edematous areas are known as "incubation patches" or "brood patches."

Many birds, including passerines, woodpeckers, pigeons, owls, hawks, grebes, petrels, albatrosses, tinamous, and penguins, have a single incubation patch in the center of the breast or belly between the ventral feather tracts. A number of birds have two lateral incubation patches, with or without a central patch between them; this is the condition found in gulls, shore birds, pheasants, grouse, rails, cranes, and auks. In the murres, which lay only a single egg, one of these paired patches soon disappears (Tuck 1960). Gulls have a separate incubation patch for each of the three eggs.

In the Banded-rumped or Madeiran Storm-Petrel, the denudation of the incubation patch occurs between forty and twenty days before the single egg is laid (Allan 1962), but in most small birds this happens later. In several kinds of finches, the first step in the formation of the incubation patch, the loss of all the down feathers from the area, occurs from six to four days before the first egg is laid. These feathers fall out spontaneously in an interval of about twenty-four hours. Then, while eggs are being laid, the blood vessels increase in size and number. Soon after this process begins, the skin starts to thicken and continues to do so during much of the incubation period. While the young are being brooded, the skin of the incubation patch gradually returns to its normal state, but the feathers are not replaced until the succeeding molt (R. E. Bailey 1952, 1955).

Ducks and geese lack true incubation patches, but they pull feathers from their breasts to line their nests, thereby making this surface more permeable to heat. In Ostriches, rheas, Emus, cassowaries, pelicans, boobies, and their allies, incubation patches are absent. Boobies and Gannets warm their eggs with their completely webbed feet. The Gannet holds its single big egg closely between the toes of both feet, almost completely embracing it with the broad webs, which become hot with suffused blood while incubation is in progress. By this means it keeps the egg at a temperature that compares favorably with that of eggs warmed by an incubation patch. To withstand this treatment, the Gannet's egg has a shell unusually thick for its size, which becomes very dirty, sometimes black and shiny, in the course of incubation (Nelson 1966). The Red-footed Booby, which incubates its single egg in an arboreal nest lined with green leaves, may stand on it with the inner part of the web of each foot stretched around it, or it may place both feet fully on the nest with the egg resting upon them and nestled amid the abdominal feathers between its legs (Verner 1961). Other boobies, such as the Brown and the White, manage to hatch two eggs by placing a webbed foot over each, with the outer toe on the ground, while the parent supports its weight on its tarsi, flat against the ground (Dorward 1962). Although the Gannet lays only a single egg, it also can hatch two with its feet, if a second egg is given to it. Lacking incubation patches, tropicbirds do not warm their single eggs with their feet, like other pelecaniform birds, but amid their dense ventral plumage, not in contact with the skin (Howell and Bartholomew 1962a).

While incubating the single egg on the Antarctic ice in midwinter, the Emperor Penguin holds it on his tarsi, covered by a fold of skin at the base of the abdomen. In penguin species that incubate two eggs, these are held upon the webbed feet and covered with the abdomen, where a bare incubation patch develops on the otherwise completely feathered body; unlike other birds, penguins do not grow plumage in definite tracts that alternate with bare spaces of skin. Similarly, a murre, incubating on a rock ledge, spreads the webs of its feet beneath the egg while it applies the incubation patch to its upper side (Tuck 1960). The male Pheasant-tailed Jaçana extends his wings beneath his four eggs to press them against his body and prevent their touching the wet material of his floating nest (Hoffmann 1949). The Razorbill pushes one wing between its single egg and the bare rock of the seaside cliff on which it breeds (Perry 1946).

As a rule, in those species that develop incubation patches, these are confined to the sex or sexes that attend the eggs. In grebes, petrels, pigeons, woodpeckers, and many other families, both sexes incubate and both have patches. In tinamous, phalaropes, jaçanas, and some sandpipers, incubation patches are confined to the males, who alone attend the eggs. In owls, hawks, and hummingbirds, they

are found only in the females. In passerines, the situation is puzzling. With the exception of social parasites, like cowbirds, the females throughout this order incubate and develop incubation patches, as far as is known. Male passerines may or may not incubate, and, strangely enough, the possession of an incubation patch does not coincide with their attendance of the eggs. Among jays, incubation patches have been found in male nutcrackers and magpie-jays (Amadon and Eckelberry 1955); in the former, the male incubates, but in the latter he apparently does not. They have also been recorded in the males of a few species of cotingas and American flycatchers (D. E. Davis 1945, Parkes 1953), although in neither of these families is the male known to incubate. The male Yellow Wagtail, who performs a minor share of the incubation, develops a bare patch that effectively warms the eggs (S. Smith 1950). Yet males of Common Bushtits, Wren-Tits, Rose-breasted Grosbeaks, and Black-headed Grosbeaks regularly take turns on the eggs but lack incubation patches. Although these attentive males cannot incubate so efficiently as their mates with incubation patches, their time on the nest is not wasted, for their presence over the eggs, like a blanket, at least retards the loss of the warmth that the female has imparted to them. That incubation by birds lacking incubation patches is not ineffective is proved by the above-mentioned examples in which such a bare area is lacking. Some of these birds hatch large sets of eggs, which they could hardly warm with their feet, as boobies do with their one or two eggs.

The temperature threshold below which embryonic development stagnates varies from species to species, and, to a much smaller degree, even in the same egg at different stages of development. In the Northern House Wren, it fluctuates between 61.3° and 66.6° F (16.3° and 19.2° C), being highest in the middle of the incubation period. In the domestic hen, a bird of relatively recent tropical ancestry, the threshold for development is much higher, 80° F (26.7° C). As the temperature of the egg rises above the threshold, the rate of development increases, although the embryo does not necessarily die if it falls below this critical temperature (Kendeigh 1963).

Measurements at the nests of a number of species

showed that the temperature of the egg's upper surface, in contact with the parent's incubation patch, was not over 1.8° F (1° C) below the parent's body temperature, which in most birds lies between 100.4° and 107.6° F (38° and 42° C). Sometimes the difference between body temperature and incubation temperature was much less than 1.8° F. Within the egg is a rather sharp temperature gradient from the top, in contact with the warm parent, to the bottom, resting on the nest lining or on the ground. In fairly large eggs, including those of the domestic hen, domestic goose, Yellow-eyed Penguin, Willow Ptarmigan, and Lesser Black-backed Gull, the difference between the top and bottom of an egg at full incubation temperature is about 12° to 16° F (7° to 9° C). Even in the tiny house wren's egg in a well-lined nest, the bottom may be 16° cooler than the top.

The stage at which eggs reach full incubation temperature varies widely among birds. From what has already been said about the development of the incubation patch in passerines, we would expect that this temperature would be attained by the time the set of eggs is complete; and Kendeigh's measurements show this to be true in the house wren. Apparently because of the slow development of their incubation patches, other birds do not apply full heat to their eggs until some days after they begin to incubate. In the Yellow-eyed Penguin, this delay is about fifteen days, more than one-third the incubation period of forty-two days. In the Goshawk, half the incubation period of forty-one to forty-three days had elapsed before maximum incubation temperature was reached; and in the European Sparrow-Hawk, twenty-two of the thirty-nine to forty-two days of the incubation period were required to develop full incubation temperature. In the Ruddy Turnstone, the temperature of the incubated eggs gradually rose from about 86° F (30° C) at the beginning to 100.4° to 104° F (38° to 40° C) on the eighteenth day, less than a week before the eggs hatched. The long incubation periods of some birds are evidently caused by this great delay in attaining full incubation temperature (Farner 1958).

Caring for the Eggs

The principal task of incubating birds is to sit, or more rarely to stand, upon the eggs to keep them warm; but certain other attentions promote successful hatching. In the first place, the eggs must be carefully adjusted to the bare incubation patch or patches, where alone they will receive full benefit of the heated skin. After settling on the nest, perhaps after first arranging its eggs with its bill, the parent slides its body slightly from side to side, bringing its eggs into contact with its bare skin. If you permit a broody hen to rest on the palm of your hand, you may often feel her warm abdomen sliding sideways across it, with the same movement that she uses to adjust her eggs. At intervals, the incubating bird repeats this action. After a heavy shower, a male Pauraque, incubating on the ground, made these sliding movements four times in forty-five minutes; and in sunshine another male Pauraque did so ten times in ninety minutes. Usually, however, the movements were much more widely spaced.

From time to time, the incubating parent rises in the nest to bend down its head and shift the eggs with its bill, an activity often called "turning the eggs." By marking eggs of the Herring Gull, Tinbergen (1953) demonstrated that they are in fact turned, so that now one side, now another, is uppermost. There was no definite order in these revolutions, and in the course of incubation every side was sometimes on top. Hazel Grouse turned their eggs thirty-three times in a day; Black Grouse, twenty-five; and Capercailles, from twenty-two to thirty-two times (Semenov-Tian-Shansky 1959). The yolk of an egg revolves freely within the shell, so that however a freshly laid egg is turned, the side with the germ, which is lightest, comes to the top, where it will receive most heat from the incubating parent. This rotation evidently continues through at least the earlier stages of embryonic development. Turning the egg, which is followed by rotation of the yolk, is sometimes deemed necessary to prevent the yolk from sticking in a fixed position relative to the shell, as might happen if the egg remained too long with the same side uppermost. If this sticking should occur, and the egg were accidentally turned with the embryo downward, the latter would remain at a lower temperature and develop more slowly, if at all. However, I am aware of no direct observations on this point.

The eggs of some birds seem never to be turned, yet they hatch successfully. As soon as an African Palm-Swift lays an egg, she glues it with saliva to her nest, to prevent its falling from the narrow shelf fastened to the vertical surface of a hanging palm frond. Apparently the eggs remain attached at the same point throughout the incubation period (Moreau 1941). Crested swifts also fasten their single eggs to their tiny cupped nests. While a hermit hummingbird sits in its nest fastened with cobweb beneath the drooping tip of a palm frond or a hanging strip of a banana leaf, it throws its head upward and backward until the crown almost meets the rump and the bill is almost vertical. Doubled up in such cramped quarters, it is unable to touch its eggs with its bill, although it might move them with its feet.

Aside from these occasional movements to turn the eggs or adjust them to the incubation patch by sideward sliding, birds sit inactive in their nests for long intervals. Sometimes they preen, and they may even call or sing, as is most likely to happen when a vocally gifted female hears her mate singing nearby and answers him (p. 166). Indeterminate builders may interrupt incubation to tidy their nests or bring more material. If small ants or other insects invade the nest, the incubating bird may rise up and spend considerable time picking them off; sometimes it even leaves the nest to hang beside or below it and pluck the troublesome creatures from the sides or bottom.

At intervals, especially on a warm afternoon, the incubating bird may drowse, but those that I have watched have rarely kept their eyes closed for more than a minute or two in the daytime. The eyelids of an incubating Pauraque are in almost continuous motion, up and down, now narrowed to a mere slit, now opened until the eye is half-exposed, rarely fully open in bright light, seldom wholly closed for more than half a minute. The Common Potoo, another crepuscular bird, likewise keeps its eyelids in restless movement in the daytime. At nightfall, most diurnal birds turn back their heads, push them into the feathers of their shoulders, and sleep in their nests, some so soundly that they are not easily wakened, others so lightly that a slight disturbance drives them from their eggs. Hummingbirds and pi-

geons sleep on their nests, as when roosting, with head forward and bill exposed. The plumage of sleeping antbirds is so relaxed that their shape is lost and the nest appears to be filled with a handful of loose feathers.

Keeping the Eggs Cool

Most incubating birds most of the time need to keep their eggs warm, but sometimes they must take measures to prevent their own bodies or their eggs from becoming overheated. This is particularly true of birds that nest on open ground in warm, sunny weather. Often such birds sit facing away from the sun. Except in the early morning, a Common Nighthawk nesting on a flat roof of a building in Missouri rotated in a clockwise direction throughout the day, keeping her tail directed toward the luminary (Weller 1958). A Pauraque incubating on a hilltop faced westward all day; in the forenoon he was exposed to the sun, which at that season passed almost directly overhead, and in the afternoon he was in the shade of neighboring woods. In the Kalahari Desert, Double-banded Coursers, nesting on strongly insolated, open ground, likewise incubate with their backs toward the sun throughout the day (Maclean 1967). On bright afternoons, a Long-billed Starthroat Hummingbird, who had built her shallow nest in a shadeless dead tree, sat with her tail directed sunward.

To cool themselves while sitting in strong sunshine, many birds open their mouths and pant. A few, including cormorants and their allies (Pelecaniformes) and goatsuckers, employ "gular flutter." Opening their capacious mouths widely, they rapidly vibrate the skin of their throats, which promotes evaporation from the exposed mucous membranes and effectively lowers body temperature. Strongly insolated birds also raise their head and back feathers, giving the air freer access to the skin, and rise up in their nests, permitting their naked legs and feet to dissipate heat. Thus cooling themselves while they shade their eggs, they keep the latter at a temperature substantially below that of the heated substratum. By shading her eggs, a Common Nighthawk kept them at the safe temperature of 114.8° F (46° C) while the gravel roof on which they lay was at 141.8° F (61° C). Maclean's observations attest to the importance of such protection: When Double-banded Coursers were kept from their nest for more than fifteen minutes on bright days, the strong sunshine killed the embryos by overheating. When the air temperature was above 95° F (35° C) on a sunny day, the coursers were more reluctant to leave their eggs, and much more eager to return, than when the day was cooler, as though aware that their unhatched progeny were in greater peril. In arid southern Africa, where the temperature at ground level may soar above 131° F (55° C), Ostriches must shade their eggs to prevent overheating.

To cool their eggs and prevent excessive water loss, some birds wet them. Least Terns nesting on sunny sandbars interrupt long sessions of incubation to fly to the river and dip into it. Returning, they may hover a few feet above the nest before alighting, while drops fall from their wet plumage upon their eggs (Hardy 1957). After African Skimmers had been kept from their eggs by a short inspection, the birds began to change places at their nests on the strongly insolated volcanic sand at intervals of about one minute. The partner so relieved would fly low over the water, dip its feet four or five times, then return to the nest and gently push off its mate, who then went to dip its feet in the same manner. After eight or nine such sorties, the skimmers settled down to steadier incubation. Whether the footwetting was done for the parents' own comfort or to cool the eggs quickly was not clear, but without much doubt it helped to reduce the temperature of the eggs (D. A. Turner and Gerhart 1971). Still another method is followed by the Egyptian Plover, which regurgitates water over its eggs more or less completely buried in sand (Thomson 1964).

Hiding the Eggs

The expedient of covering the eggs when the parent leaves the nest would seem to prevent the loss of many an egg, yet birds rarely take this simple precaution. Among the few that do are grebes, which carefully cover their eggs with loose nest materials whenever they quit them. Alarmed by a noise, a Pied-billed Grebe pulled pieces of vegetation over its eggs with three or four short thrusts of its bill, the performance taking less than five seconds. Then it slipped off the nest backward and vanished underwater. The eggs, however, were not completely hidden, so after five minutes the bird returned to cover them more thoroughly, then dived again. Sometimes these grebes covered their eggs as an anticipatory measure. Hearing a noise, the incubating bird would hastily pull material over them, then continue to sit above the covered eggs. If not further alarmed, the grebe uncovered its eggs and resumed normal incubation (Deusing 1939).

Before they leave, many kinds of ducks, geese, and mergansers cover their eggs with the downy feathers usually so abundant in their nests or with other materials. During the period of laying at temperatures close to freezing, the Rock Ptarmigan of the Arctic tundra carefully covers her incomplete set with fragments of vegetation, but after incubation begins she leaves them exposed during her recesses. The air is now warmer, and the originally richly colored eggs have faded to less conspicuous hues. White-tailed Ptarmigan and Red Grouse also cover their eggs when they leave their nests (MacDonald 1970). On the bleak Peruvian puna, the male Ornate Tinamou pulls loose feathers over his eggs before leaving them. In warm Central American lowlands, the male Slaty-breasted Tinamou tosses fallen leaves and twigs backward toward his eggs as he walks away for his afternoon outing, although they often remain imperfectly covered. Little Tinamous and Great Tinamous permit their glossy, highly colored eggs to lie quite exposed during their absences.

From the time she lays the first egg on arid, nearly barren ground in South America, the female of both the Least and Gray-breasted seedsnipes covers her set with soft nest lining when she leaves (Maclean 1969). The covering procedure of Kittlitz's Sandplover is different from that of most birds. Rising from the nest, a mere scrape in the sand, as a man approaches, the sandplover of either sex straddles its two eggs and with alternate movements of its legs kicks inward, pushing the sand from the rim toward the center. At the same time, it rotates, usually in a clockwise direction, throwing in the sand from all sides until the eggs are completely buried (plate 53). Six seconds suffice for the parent to cover the eggs, then the bird runs, crouching, away. If, suddenly frightened from the nest, the sandplover flees leaving its eggs exposed, it may return, bury

PLATE 53. Kittlitz's Sandplover standing above the two eggs that it rapidly covered with sand when disturbed while incubating. Now it will run away in a crouching posture (photo from William Conway and Joseph Bell, New York Zoological Society).

them in sand, then run away again. Returning to its hidden nest when all appears safe, the bird first probes the sand with its bill, then squats over the eggs and kicks the sand backward with its feet. It rises, rotates above the eggs, and resumes kicking, until finally the eggs are uncovered and it proceeds to incubate them. The eggs are freed of sand by much the same movements as were used in making the original nest scrape (Conway and Bell 1968). A related species, the White-fronted Sandplover, also of southern Africa, likewise covers its eggs with

sand when driven from them by an approaching man, but neither so consistently nor so thoroughly as Kittlitz's Sandplover (Hall 1960).

Concealing the eggs beneath nest material is by no means restricted to ground-nesting birds; even those that breed in holes and burrows sometimes do so. Flesh-footed and Short-tailed shearwaters often block the mouth of their subterranean tunnels with grass or other vegetation. The reason for this is not clear, since they breed on islands where nest-robbing mammals are usually absent (Warham 1958a, 1960).

After their father ceases to brood in the short burrow in the floor of the tropical forest, nestling White-whiskered Softwings sleep behind a screen composed of the loose fragments of dead leaves that line their chamber, which each evening they somehow raise up from its floor (Skutch 1958*b*). During the period of laying, the Pygmy Nuthatch keeps her eggs covered with feathers, moss, and other nest materials when she is absent from the hole in a tree most of the day, but after incubation begins she neglects to cover them when she leaves for her brief recesses (R. A. Norris 1958). Hole-nesting titmice of a number of species have the same practice.

The Masked Tityra usually nests in a woodpecker's hole or some other cavity high in a decaying tree dangerous to climb. When, after many years, I found a low tityra's nest that could be reached without great risk, I never saw the eggs. The doorway was too narrow to admit my hand, but on frequent inspections of the hole with a small mirror and electric bulb, I always found the eggs completely concealed amid the litter of small dead leaves that the tityra carries into her nest cavity. This was true whether I chased the bird from the hole or she left spontaneously. During their first ten days, the two nestlings rested so consistently beneath a blanket of leaves that I enjoyed only infrequent and partial glimpses of them. For the next five days, they lay with their heads exposed but their bodies covered by the litter. Only during their final two weeks in the hole did the young tityras sit above the leaves with their bodies fully visible when I peeped in with a light.

Among the very few birds that equip their closed nests with hinged doors are the penduline-tits. The Cape Penduline-Tit builds a pendent pouch of downy materials so compactly felted that it sheds a heavy rain. On the side of the oval structure is a spoutlike projection that looks like an entrance but is sham, as the base of the spout is walled up. Above this false doorway is a flap of material hinged at the bottom, closing the true entrance. To open this door, the tiny bird alights on the spout, seizes the free upper edge of the flap with its bill, and pulls it down. After climbing in, it turns around and closes the door. To leave, it pushes the flap out, emerges, then nudges the flap upward until the lips of the opening stick together. Not satisfied with a simple door, after entering the nest the penduline-tit pulls the floor of the entranceway up against the ceiling, thus blocking the tube for its full length. When leaving, it accomplishes the same end by leaning into the false entrance and prodding against the floor of the entrance. Like the related true titmice, the penduline-tit buries its eggs in the nest lining during the laying period (Skead 1959).

Although only a minority of ground- and hole-nesting birds cover and conceal their eggs while absent from their nests, this practice is even rarer among species that build open nests in trees and shrubs. Among the few that do so is the Bald Eagle, which before leaving its huge nest may spend five or ten minutes carefully pulling loose materials over its eggs (Bent 1938). Since, as we have already noticed, the eggs in open arboreal nests, far from being camouflaged, are often white, blue, or some other color that contrasts, rather than blends, with the lining, one might suppose that the simple expedient of covering would have evolved frequently, if it saved an appreciable proportion of such nests from predation. The virtual absence of covering in cupped nests in trees and shrubs strengthens my belief that cryptic coloration of eggs is of little value because, when a nest is discovered by a predator able to reach it, the eggs are doomed in any case. The nest itself must provide safety for its contents, by invisibility or inaccessibility, as explained in chapter 7.

CHAPTER 14

The Constancy of Incubation

Birds that incubate in accordance with certain of the patterns described in chapter 12 sit with great constancy, keeping their eggs continuously, or almost continuously, covered. Among these are the species that conform to the Ringed Kingfisher, Albatross, Pigeon, Ostrich, Golden Pheasant, Hornbill, and Emperor Penguin patterns. Great variations in constancy, however, are not inconsistent with certain other schedules of incubation, including the Herring Gull, Antbird, Woodpecker, Quail, Hummingbird, Tinamou, Jaçana, Bushtit, and Ani patterns. What causes these differences in the lengths of separate sessions and in the proportion of the day that the eggs are warmed?

To investigate this problem, we must first define constancy. For purely diurnal birds, which are most easily studied and show great fluctuations in attentiveness, the measure of constancy of incubation is the percentage of a bird's active day that it spends on the eggs. If the observer has made a record of the incubating bird's goings and comings from dawn to dusk, the simplest procedure is to add all the sessions of incubation, divide this sum by the total time elapsed from the bird's first departure in the morning to its return to the nest for the night, and multiply the quotient by one hundred. But perhaps one who can only watch for a few consecutive hours on different days has timed a number of sessions and

recesses and desires an index of the bird's constancy. In this case, he may compute it by the formula $C = (S \div S + R) \times 100$, where C = constancy of incubation, S = average length of the sessions timed in full, and R = average length of the recesses or absences timed in full during the same observation periods. This formula compensates for the different number of sessions and recesses that the watcher is likely to record during several observation periods, each of which lasts only a few hours. If it happens that he has timed equal numbers of sessions and recesses, he can compute with the sums rather than the averages and get the same result. This method of calculating constancy of incubation is applicable when a single parent is on and off the nest a number of times each day.

To gather the data necessary for these computations, various methods are used. Sometimes I have watched continuously from dawn to dusk, from concealment when there was the least suspicion that my unconcealed presence would influence the behavior of the birds I studied. Often I have made a record that seemed equally sound by watching all one afternoon and all the following morning, thereby including all hours of the day. Two or more observers, replacing each other at intervals, can make much longer records with less fatigue. Others prefer to use one of the ingenious devices that have been

TABLE 4
The Hummingbird Pattern of Incubation
Incubation by the Female for One Full Day[a]

Species	No.[b]	Sessions in Minutes Range	Average	Recesses in Minutes Range	Average	Constancy %
Scaly-breasted Hummingbird	35	<1–103	14.3	<1–23	6.0	70
White-eared Hummingbird	49	<1–24	9.5	1–17	4.9	66
White-crested Coquette	37	<1–78	13.4	<1–22	5.7	70
Tawny-winged Dendrocincla	10	10–96	45.8	11–51+	29.3	61
Buff-throated Woodcreeper	3	111–181	144.7	25–34	27.5	84
Yellow-thighed Manakin	7	29–108	65.1	6–21	14.0	82
Orange-collared Manakin	8	17–258	71.4	3–25	12.2	85
Rose-throated Becard	30	3–38	12.1	2–19	9.3	57
Bright-rumped Attila	5	63–111	93.2	13–60	38.3	71
Streaked Flycatcher	13	15–72	30.3	8–18	12.6	71
Northern Royal Flycatcher	19	4–52	22.4	3–33	13.6	62
European Wren	17	24–49	33.0	6–14	11.0	74
White-breasted Blue Mockingbird	28	8–42	20.8	1–23	7.1	75
American Robin	31	6–45	20.6	1–10	5.6	79
Long-tailed Silky-Flycatcher	36	4–68	15.1	1–6	3.5	81
Cinnamon-bellied Flower-piercer	22	4–62	19.1	4–24	10.5	65
Shining Honeycreeper	9	14–390	69.0	7–23	14.3	83
Pink-headed Warbler	24	13–35	20.1	4–13	8.3	71
Chestnut-capped Warbler	9	27–70	44.6	16–35	23.3	66
White-vented Euphonia	16	12–53	30.2	4–26	12.5	71
Speckled Tanager	14	20–77	37.8	3–23	11.4	77
Scarlet-rumped Black Tanager	16	8–102+	29.3	5–32	11.8	71
Black-faced Grosbeak	7	37–123	60.4	22–40	31.6	66
Orange-billed Sparrow	6	14–102	77.7	29–52	39.1	67

[a] The record for each species was made either during a single dawn-to-dusk watch on a typical day, or during two watches that together covered all hours of consecutive days. The record for the European Wren is from Armstrong (1955) and that for the American Robin from Schantz (1944). The others were made by the author, sometimes assisted by his family.
[b] The number of sessions is given. The number of recesses was usually one greater; less often, the same.

developed for recording automatically, usually with the aid of electrical apparatus, the arrivals and departures of the incubating birds. In a bird box, a trip lever in the doorway can be made to record on a revolving drum each entry and exit; in an open nest, a thermocouple among the eggs will indicate, by rising and falling temperatures, when the parent arrives and when it leaves. By such means, one readily accumulates a much greater mass of data than he is likely to gather by direct observation, but he misses many intimate details. In any case, some careful watching of the nest is necessary for the interpretation of the mechanically made record.

The Range of Constancy

Table 4 gives some samples of the incubation behavior of female birds that follow the Hummingbird Pattern. Most of the records are my own, but a few are from published sources. From these and many similar records that I have examined, it is evident that most small birds that incubate without help from a mate keep their eggs covered from 60 to 80 percent of their period of diurnal activity. This is true not only in tropical regions where daylight lasts only twelve or thirteen hours but also near the Arctic Circle when daylight is continuous and the incubating bird is active for sixteen or seventeen hours (Weeden 1966). Even when the two parents alternate on the nest, as in those that follow the Antbird and Woodpecker patterns, they may fail to keep their eggs more continuously covered (table 5). We might call 60 to 80 percent the normal range of constancy of incubation (Skutch 1962a).

An examination of table 4 suggests a distinction

TABLE 5
Antbird (A) and Woodpecker (W) Patterns of Incubation
Incubation by the Male and Female during One Full Day[a]

Species	No.	Sessions in Minutes Range	Average	No.	Intervals of Neglect in Minutes Range	Average	Attendance %
Quetzal (A)	{ 3 { 2	46–211 15–242	130.0 } 128.5 }	6	6–29	14.3	88
Rainbow-billed Toucan (A?)	14	4–86	32.9	12	2–44	14.7	70
Yellow-shafted Flicker (W)	{ 8 { 6	11–100 6–132	31.8 } 50.1 }	10	2–10	4.8	92
Golden-olive Woodpecker (W)	{ 2 { 3	82–118 51–297	100.0 } 146.3 }	0	—	0	100
Red-crowned Woodpecker (W)	{ 4 { 6	22–105 2–80	62.0 } 57.7 }	5	1–12	5.2	96
Golden-naped Woodpecker (W)	{21 }22	5–39 2–51	17.0 } 13.4 }	25	1–7	3.4	89
Olivaceous Piculet (W)	{ 5 { 5	2–76 7–89	44.4 } 55.2 }	3	17–35	27.0	88
Streaked-headed Woodcreeper (A)	{ 8 }11	6–37 5–57	16.4 } 26.9 }	17	4–41	16.9	60
Rufous-fronted Thornbird[b]	{20 }16	5–33 1–49	15.7 } 22.8 }	23	1–12	3.4	90
White-flanked Antwren (A)	{ 4 { 2	3–174+ 98–140	83+ } 119.0 }	5	11–94	35.6	76
Chestnut-backed Antbird (A)	{ 3 { 2	45–95 92–136	69.3 } 114.0 }	5	1–39	18.0	88
Spotted Antbird (A)	{ 3 { 3	44–217 36–164	112.7 } 109.0 }	4	8–22	13.5	92
Long-billed Gnatwren (A)	{ 5 { 4	14–95 60–90	67.0 } 79.8 }	0	—	0	100

[a] For all except one species, the alternating sessions of the male and female are given in consecutive lines, those of the male above. At the nest of the thornbirds, the two partners were easily distinguished, but their sexes were not determined with certainty. The male and female toucans could not be distinguished and their sessions are lumped. Attendance was calculated as the percentage of the birds' active day that the nest was occupied.
[b] Both sexes of the thornbird sleep in the nest at all seasons.

between constancy and patience. Such birds as the Thrush-like Manakin and Tawny-winged Dendrocincla sat very patiently, but because of their long absences they achieved low constancy. The White-eared Hummingbird sat impatiently, rarely remaining on the nest longer than fifteen minutes, but because she soon returned from her excursions, her constancy was high. The small manakins—Blue-crowned, Yellow-thighed, and Orange-collared—incubated with both patience and constancy; a number of small flycatchers, leaving their nests after brief sessions and remaining away about as long as they sat, showed neither patience nor constancy.

The upper limit of normal constancy, about 80 percent of the active day, is set by the incubating bird's own needs; most birds of small or medium size evidently require at least 20 percent or one-fifth of the daylight hours to find enough food for themselves while they incubate. The lower limit of 60 percent constancy is apparently determined by the developing embryo's need of heat. If the parent heats an egg for less than three-fifths of the day, embryonic development might be retarded or the egg might fail to hatch, unless the day is very warm or the nest exceptionally well insulated. Among the birds that incubate for less than 60 percent of the day are becards, members of the cotinga family whose bulky nests, with only a small opening in the side or bottom, retard the cooling of the eggs. Nevertheless, their incubation periods of about eighteen days are long for birds the size of sparrows. Small insectivorous birds, including some of the American flycatchers, may also incubate for less than 60 percent of sunny days when volitant insects are active, and their incubation periods are likewise often long.

Constancy of more than 80 percent is shown chiefly by birds that receive food from their mates. If the male feeds the female very generously, as in certain hawks, owls, hoopoes, parrots, jays, siskins, crossbills, goldfinches, and waxwings, she may achieve a constancy well over 80 percent, or even sit continuously, as in hornbills. Even if not fed by her mate, a female may incubate with unusually high constancy if she has close at hand an abundant supply of nourishing food, such as ripening grains. In tropical American forests, female manakins sit with steadfastness quite unexpected in birds so tiny, often remaining on their eggs two or three hours at a stretch and attaining a constancy of slightly more than 80 percent. They sit with most of their bodies above, rather than in, their shallow nests, depending on their greenish cryptic coloration and immobility to escape detection in their exposed position. The less often they approach or leave a nest, the less likely they are to draw an enemy's attention to it, and they satisfy their hunger with the small berries that are usually abundant when they breed.

Even fairly large birds that take only one or two recesses each day, such as tinamous, guans, and quails, often incubate with a constancy of 80 percent or less. When both sexes share incubation rather equally, neither need sit for much more than half the day in order to keep the eggs constantly covered. Those that follow the Ringed Kingfisher Pattern or Albatross Pattern may, indeed, sit fasting for a whole day or even many days together, but then they enjoy a long interval for foraging, resting, and recovering their strength. Birds as diverse as Song Sparrows, pigeons, and Yellow-eyed Penguins have been found to gain weight during the incubation period. It is most important for these and other birds that feed their young in the nest with food brought from a distance to come through the incubation period with no loss of strength, for they have strenuous days ahead.

Birds that attend their eggs with a constancy of well over 80 percent, without receiving food from a mate, pay for their devotion by losing weight and vigor. The female Common Eider Duck, who during four weeks of incubation may leave her nest only ten or fifteen minutes on every second or third day, to drink but not to eat, sometimes becomes so emaciated that she is forced to abandon her eggs and stagger to the water in order to preserve her own life. She may permanently desert her nest even when the eggs are on the point of hatching (Tinbergen 1958). Domestic hens usually leave their nests for a short while each day, for food and water, but nevertheless they grow thin. Three broody hens, who generally ate one-fifth of their usual ration and took more water than food, lost from 4 to 20 percent of their body weight. A cock who, in an experiment, was given no more than one of these hens ate, died when the hen passed three days without breaking her fast, while another cock and an active hen lost from one-quarter to one-third of their weight on the broody hen's diet (Wood-Gush 1955).

The abstemiousness of the broody domestic hen is evidently an inheritance from wild ancestors, since her human attendants usually provide an abundance of food that she can eat in a short time. Pheasants, which stick to their eggs even more pertinaciously, must lose more weight than incubating domestic hens. It is noteworthy that the birds who make these heroic fasts hatch out nidifugous chicks, which leave the nest soon after they are dry and pick up their own food under parental guidance. This method seems less work for the parent than bringing food to helpless nestlings and likewise allows her more time for eating. It is doubtful whether parents who nourish their young in or near the nest could feed them adequately if they lost so much weight while they incubated. As far as I know, the only such bird that fasts throughout the incubation period is the male Emperor Penguin, who loses a great deal of weight during more than two foodless months. But at the end of this period, a female, fat from feasting in the ocean, arrives to take charge of his newly hatched chick, while he trudges seaward to break his long fast.

Factors That Influence Constancy

Size: Large birds sit fasting on their nests for periods that would be fatal to small ones. They can do this because their metabolism is less intense than that of small birds, in consequence of the fact that, as solid bodies become larger, their volume increases as the cube of their dimensions, whereas their surface area increases only as the square of their dimensions. Accordingly, when two bodies of similar shape, but different sizes, are at the same temperature higher than that of their surroundings, the smaller body will in a certain interval lose more heat per unit of volume than the larger. The body temperatures of birds of different orders vary little, whence it is evident that small ones will need to produce relatively more heat and will burn their fuel more rapidly. Moreover, these small birds have less capacity for storage, whether of fat in their tissues or undigested food in their alimentary canals. In European titmice, Gibb (1954) found that the smaller the species, the more time it devotes each day to feeding.

Birds exhibit a very great range of size, the largest being thousands of times as heavy as the smallest. Although, on the whole, large birds incubate more steadily than small ones, there is by no means an invariable correspondence between a bird's size and its constancy. Among passerines, the diminutive manakins, only three or four inches in length, incubate more steadfastly than most larger species, including many that are ten or twelve inches. Among woodpeckers, Olivaceous Piculets, only three and a half inches long, frequently remain in their tiny nest holes for over an hour at a stretch and sometimes for nearly two hours. In sharp contrast to this, in thirty-eight hours of watching at five nests of the Golden-naped Woodpecker, a bird nearly seven inches in length, the longest session of either sex lasted only fifty-one minutes, and shifts on the eggs exceeding forty minutes were rare. Among trogons, the male and female of most species each take one

turn on the eggs in the course of twenty-four hours; but in one of the largest members of the family, the Quetzal, each parent generally takes two turns in the same interval. Thus the Quetzals' separate sessions in the nest are considerably shorter than those of smaller trogons, although they, too, keep their eggs almost constantly covered. In the pigeon family, which exhibits a great range of size, the same schedule of shifts on the nest is followed by all the species for which we have information, from the largest to the smallest. Size is obviously only one of the factors that influence the length of a bird's separate sessions and the constancy it achieves.

Food: Another important influence is the bird's dietary habits. Here we must notice two distinct factors that have different effects on the incubation pattern. Some birds digest their meals more rapidly; others take longer to fill their stomachs. A bird who gets hungry after one-half hour of sitting, but can find enough food in five minutes, will, despite its more frequent comings and goings, cover its eggs for a larger part of the day than one who sits for an hour at a stretch, then devotes one-half hour to foraging. Among the birds least steadfast in incubation, in regard to both the length of their sessions and their total constancy, are swallows, flycatchers, and other species that subsist largely on small flying insects. When tiny insects and spiders are supplemented by copious draughts of nectar, as in hummingbirds, the more sustaining fare permits more constant incubation. A diet that includes much fruit seems more favorable for steady incubation than one of small insects alone, possibly chiefly because the fruit is more easily gathered. A meal of fish or squids sustains birds on their nests for surprisingly long intervals. Birds whose mates feed them liberally are not averse to almost constant sitting.

Type of Nest: The nest itself may strongly influence the activities of the bird who incubates in it. In the middle of a warm day, the interior of a thick-walled, well-enclosed nest may become so uncomfortably hot and stuffy that the parent skimps incubation, as happens in bushtits and becards. Instead of incubating, the parents may bring additional material. Castlebuilders also neglect their eggs to put their elaborate house in order. To compensate for this neglect, the thick-walled nest retards the cooling of the eggs.

Number of Participants: The number of birds that participate in incubation influences the time that each spends in the nest, but the effect is not in all cases the same. When the two parents replace each other several times a day, as in the Antbird and Woodpecker patterns, the long outing that each enjoys while its mate sits permits it to fortify itself well with food, with the result that when it returns it can remain on the nest a long while. Where enemies abound, as in the tropical forests where most antbirds live, these long sessions are further advantageous, for, the less often a bird approaches its nest, the less likely it is to betray its position. On the other hand, the absent partner's eagerness to return may reduce the length of its mate's sessions, as often happens in Golden-naped Woodpeckers.

Once I watched four or five Acorn Woodpeckers incubating in a high, inaccessible hole in a great charred trunk. In nearly twelve hours, they took 108 sessions, which averaged only five minutes. The longest turn on the eggs was seventeen minutes; and once they replaced each other three times in slightly over a minute. Since woodpeckers' holes are usually high and have strong walls, secrecy is not so necessary as at the low, open nests of antbirds.

Temperament: The fairly obvious factors that we have already considered are inadequate to account for all the variations in constancy. Intimacy with a variety of species reveals that some birds are stolid and restful, others mercurial and restless; and these differences in temperament affect their behavior while incubating. Toucans and trogons live in the same forests, feeding largely on fruits and insects. Both are colorful and both nest in holes. Trogons sit for long periods and cover their eggs almost constantly; the larger toucans change over frequently and fail by a good deal to attend their eggs continuously. What is the reason for these differences? Trogons are staid, dignified birds, perching upright for long intervals. Toucans are restless, always on the move; with their enormous bills and angular postures, they often remind one of feathered clowns. These contrasts in temperament of trogons and toucans are reflected in their manner of incubating.

When, early in my ornithological studies, I discov-

ered what long hours Ringed Kingfishers spend on their eggs in a dark burrow in a river bank, I regarded them as martyrs to parental duty and was sorry for them. But one day I saw a female alight in a streamside tree with a large fish dangling crosswise in her heavy bill. For the next two and a half hours, by my watch, she rested almost motionless there, without eating her fish. I could discover no burrow from which my presence might have been keeping her; her long immobility seemed to be a spontaneous expression of her angler's temperament. Thereafter, I no longer pitied the kingfishers.

In some birds, devotion to the nest overcomes temperamental restlessness. Jays are as active and excitable as toucans, but they sit on their eggs for hours at a time, receiving food from their mates and often also from helpers. Jacamars, which remind me of overgrown hummingbirds, seem to live at the highest tension, and their sessions of an hour or two in their dark burrows are longer than I would expect of such volatile creatures. Even in birds of a single species, we find differences in the rhythm of incubation that can hardly be attributed to external factors and appear to be an expression of individual differences in temperament.

Stage of Incubation: Sometimes a bird who at the beginning of the incubation period is shy and difficult to approach while on the nest will permit us to come close, perhaps even to touch it, as its eggs near the point of hatching. This increased devotion suggests that it is now incubating with greater constancy. To test this inference, we must spend long hours watching from concealment, or else use a mechanical recorder. Contrary to our expectation, the records of a number of species fail to demonstrate a rise in assiduity toward the end of the incubation period: increased stanchness in the face of apparent danger is not an indication of more continuous sitting when the bird is undisturbed.

Some birds, including anis, bushtits, and Scarlet-rumped Black Tanagers, gradually increase the time on their nests in the days following the laying of the last egg of their sets. Others, including pigeons, doves, and such passerines as the European Goldfinch, American Robin, Yellow Warbler, and Bananaquit, may sit with practically full constancy from the day their sets are complete. Thereafter, the time spent on their nests fluctuates somewhat from day to day, with rarely an increase and perhaps as often a decrease, as the date of hatching approaches.

Influence of the Weather

One cause of these day-to-day fluctuations in the constancy of the same individual bird is the weather. When a shower starts, birds often hurry to their eggs and cover them until it is over, thereby protecting them from chilling. If the rain continues long, they may leave for food while it still falls, but not until after they have sat longer than they habitually do on fair days. A number of birds that I have studied took their longest recorded session while rain fell. Of forty-two sessions of a Bellicose Elaenia, only two exceeded twenty-seven minutes; these lasted thirty-two and sixty minutes and were taken in the rain. A Golden-masked Tanager who in rainless weather was not seen to sit for more than twenty-seven minutes remained on her eggs fifty-one minutes beneath a heavy shower. An Orange-billed Nightingale-Thrush sat continuously for fifty-six minutes while rain fell slowly, although in ten hours of rainless weather her longest session was only twenty-one minutes. Many similar instances have been recorded for other birds.

The effect of prolonged rainfall may differ from that of an hour-long shower. By bringing to the surface worms, larvae, and other small invertebrates that live in the soil or in the litter that covers it, protracted rainfall often makes food easier for ground-foraging birds to find. Hence, they shorten their recesses and thereby increase their total time on their eggs, as I have seen in the White-breasted Blue Mockingbird. A long-continued, hard rain may have just the opposite effect on birds that catch small insects in the air or amid foliage. By beating down the insects and making them harder to find, the downpour causes the bird to devote more time to foraging, so that its constancy decreases.

In some tropical regions, including that where I write, mornings are clear through much of the year, including the season when most birds breed, followed by afternoons of long-continued rain. Some birds adjust to this situation by foraging mostly in the forenoon and middle of the day, then incubating steadily through the afternoon. A Band-tailed Barb-

throat Hummingbird, who earlier in the day had been taking periodic recesses, settled on her nest at 2:07 P.M. and incubated continuously until nightfall. A Piratic Flycatcher, who earlier in the day had followed the usual flycatcher practice of leaving the nest rather frequently, had retired by 2:35 P.M. on a rainy afternoon and remained until it grew dark. Another individual was absent only three minutes between 3:17 P.M. and dusk.

Years ago, I noticed that Rough-winged Swallows covered their eggs in riverside burrows more constantly in cool, cloudy weather than in warm, dry weather. The effect of temperature on incubation is, however, best investigated by the use of automatic instruments for recording both the bird's sessions and recesses and the daily march of temperature, if this is not available from a nearby meteorological station. In a number of species it has been demonstrated that as the air becomes warmer the bird incubates less constantly. Kendeigh (1952) found that when the thermometer stood below 70° F (21.1° C) Barn Swallows were on the nest 80.5 percent of the time, but as the temperature rose to above 85° F (29.4° C) their constancy fell to 31.6 percent. The American Robin's time on the nest fell from 78.1 percent at 58° F (14.4° C) to 60.7 percent at 83° F (28.3° C). In the Netherlands, Kluijver (1950) learned that Great Tits, when incubating their first brood, reduced their time on the nest by eight minutes per day for each 1.8° F (1° C) that the temperature rose, but for the second brood the temperature effect was more pronounced, each rise of 1.8° F causing a reduction of fifteen minutes per day in the time the titmouse spent on her eggs.

The effect of temperature on the incubation of nonpasserines has been less often investigated. Low (1945) found that a Redhead Duck incubating late in the season left her nest more often and spent fewer hours on her eggs than did ducks incubating first layings early in the season, and he attributed this reduced constancy to the higher temperature prevailing when the duck renested after losing her eggs. The single daily outing of an incubating Bobwhite lasts only an hour or two on cool and showery days, but it may be prolonged to as much as seven hours in fine, warm weather.

Exceptions to this temperature effect sometimes

occur. The Rufous-sided Towhees studied by J. Davis (1960) showed no consistent response to changes in temperature. One individual sat more constantly on cool than on warm days, but another did just the reverse.

When an incubating bird decreases its constancy as the temperature rises, it may make this adjustment in various ways. It may curtail its sessions; it may prolong its recesses; or it may do both together, in which case the effect will be most pronounced. Barn Swallows and Great Tits made this double shift as the thermometer rose. The American Robin shortened her sessions but made no consistent change in the length of her absences. In the Gray Catbird, the recesses also changed little with variations in temperature, but the sessions first lengthened, then became shorter, as the thermometer rose from 63° to 79° F (17.2° to 26.2° C) (Kendeigh 1952).

The influence of temperature on incubation is shown not only by the different constancy on colder and warmer days, but also by hourly variations on the same day. Many birds spend more time on their eggs in the cool of the morning and evening than in the middle of the day, when the air is usually warmest. I have noticed this especially in small American flycatchers, which incubate less patiently as the sun rises higher into the sky, insects become more active, and flycatching yields larger returns. At such times, an insect that blunders temptingly close may entice the bird from her nest to snatch it up.

The most impatient incubation that has come to my attention was by a female Black-faced Weaver in Senegal, where the thermometer may soar to 104° F (40° C) at noon. In forty-one minutes, she was seen to enter and leave her roofed but loosely constructed nest twenty-one times. In other parts of Africa, the eggs of this highly colonial weaverbird may be left unattended when the air temperature by day differs little from normal incubation temperature, but at night the female covers them (Collias and Collias 1964a). When the nest is exposed, dangerously high temperatures may increase the parent's attentiveness, not to warm the eggs but to prevent their becoming overheated by shading them (pp. 175–176).

The utility of these changes of constancy with variations of temperature is obvious, because in cool weather the eggs need more heat from the parent than in warm weather. In unseasonably cold weather, this admirable adjustment unfortunately breaks down, particularly in small birds that feed chiefly on flying insects, which then become scarce. The parent may be obliged to neglect its nest for long periods in order to find enough food to keep alive; it may even permit its eggs to become thoroughly chilled, as happens when swallows and Eastern Phoebes are overtaken by a late snowstorm after they have begun to breed. Such wintry weather scarcely affects incubating siskins, jays, and other hardy birds that are generously fed on their nests by their mates.

Although air temperature influences the constancy of incubation of many birds, the temperature of the eggs themselves may not. Captive Ring-necked Doves were given artificial eggs equipped with tubes through which a current of water could be passed. The doves continued to incubate whether the eggs were cooled to 24.8° F (−4° C) or heated to 143.6° F (62° C). On very cold eggs, the birds fluffed out their plumage and shivered; on hot eggs, they opened their mouths and tried to cool themselves by gular flutter. These experiments refute the suggestion that birds incubate to relieve the irritation of inflamed incubation patches by applying them to the cool, smooth surface of eggs (Franks 1967). It would be interesting to repeat this experiment with birds that incubate intermittently instead of keeping their eggs continuously covered, as do pigeons.

The Male's Influence on His Mate's Constancy

When the male bird does not incubate, he may influence the constancy of his mate's attendance on the eggs in various ways. If he feeds her abundantly, as occurs in hornbills and goldfinches, he may enable her to sit far more continuously than she could do if obliged to find all her own food. The more often a male Pied Flycatcher feeds his mate, the more constantly she covers her eggs. When a male was removed, his mate lengthened both her sessions and her recesses; but the latter increased more than the former, with the result that the time she devoted to incubation fell from 79 to 58 percent of the day. Despite her increased foraging, she lost weight (Haartman 1958).

Some males call their mates from the nest to feed them; others, who bring little or no food, summon the female from her eggs to accompany them in foraging. If the male guards the nest, without sitting, during his mate's absences, his too prompt return to assume this duty may send her off much sooner than she would otherwise have gone. A female Streaked Saltator sat far more steadily while her mate remained beyond sight and hearing for two hours than when he came periodically to guard. Once he was responsible for her departure only seven minutes after she had returned to her eggs. An Orange-billed Nightingale-Thrush whose mate never came near the nest took longer sessions than a neighboring female, whose mate through much of the day returned, after she had been sitting for ten minutes or so, to resume his guardianship, which was the signal for her to start her recess.

At the end of the female's outing, her mate often escorts her to the nest. Apparently the female usually takes the initiative, returning when she has satisfied her hunger, and her mate simply accompanies her. But from ancient times men have believed that the male bird may drive his partner back to her task if she has dallied too long away from her eggs. Plutarch, in the essay in which he compared the intelligence of land and water animals, remarked that the male pigeon takes upon himself part of the female's duty, in warming the eggs and feeding the young, and, if she happens to stay too long from the nest, he chastises her with his bill until she returns to it. The driving of female pigeons and doves by their mates has frequently been observed, but the male's motive seems to be to remove his consort farther from a rival rather than to force her to return to her eggs. In my own experience, if a female dove is late in returning to her eggs in the afternoon, the male will continue to cover them rather than go off and chase her back. Indeed, I have never seen in any kind of bird an unequivocal instance of a male compelling his mate to return to the nest. If he is so solicitous for the welfare of his brood, why does he not go to cover the neglected eggs himself?

Although the driving of the female to her eggs by the male is rare among birds, we should not too hastily conclude that it never occurs. A male Yellow-shafted Flicker in Kilham's (1959) aviary savagely attacked his mate when, after a long wait for her, she failed to come for her turn on the eggs. In the open, this female would doubtless have flown afar to avoid her irate partner. Among free birds, males seem more often to coax than to drive their negligent mates to the eggs. While a prisoner of war in Germany, Buxton (1950) made a thorough study of the nest life of the European Redstart. He saw the male, who in this species does not help to incubate, entice the female back to her eggs when she stayed off too long. Since the male redstart chooses the nest cavity, often before a female arrives in the spring, and then, by simple displays, persuades her to enter and build there, it is not improbable that he should use the same display to induce her to return to the nest at a later stage in the breeding cycle. Another male redstart performed the enticement display to bring his mate back to a nest with young, after an owl had frightened her away.

A male European or Winter Wren who was, at times, more active in feeding the nestlings than his mate, chased her back to it when she was disquieted by Armstrong's (1955) presence. It is probably significant that in these three species the male usually selects the nest site, and in the wren and flicker he takes an important part in building the nest, although only the flicker shares incubation. Apparently, the male is not concerned about the cooling of the eggs so much as about the continuance of the pair bond, which in these three birds is largely dependent on the female's acceptance of, and continued attendance on, the nest that he has chosen.

Although I have never seen a male bird drive his mate to her nest, once I watched a male Black-fronted Tody-Flycatcher who seemed to try persistently to coax the female back to her pensile nest. She had neglected it when the eggs became spotted with vermilion paint in my effort to mark her for identification. When she delayed overlong in the open, the male sometimes flew toward the doorway, as though to enter, but veered aside just in time to avoid striking the nest. He was performing alone his part of the ceremony with which he, like a number of other small birds with closed nests, often escorted her as she returned to the eggs. After she had definitely abandoned them, he spent much time near the nest, calling her with a measured *tick tick tick*, but in vain. No other male bird that I have watched has appeared to try so hard to recall his mate to her duty. Male tody-flycatchers do not incubate, but they help to build the nest, the site of which is evidently chosen by both members of the constantly mated pair.

In other cases, the male seems to restore the confidence of his mate when she has been frightened from her nest. Once I watched a Black Phoebe incubate in a substantial cup of dried clay attached beneath a great rock overhanging a rushing tropical torrent. Since the stony shore provided no spot to set a blind, I watched unconcealed. When I sat on a rock only thirty-five feet from the nest, the female phoebe would not return to her eggs until her mate came and stood below the nest, hovered beside it, or perched on its rim. In the latter case, a low, sweet trill was audible above the river's roar as she settled on the eggs in front of him. When I watched from a more distant point, the female often returned to her nest while her mate was beyond sight. It was evident that his presence restored her confidence when I was too near (Skutch 1960). Similarly, when a female Red-winged Blackbird was frightened from her nest, the arrival of her mate, who was drawn by her cries of alarm, would cause her to return promptly. If he did not soon come, she would fly around the nest, scolding, until he arrived. In the egg-laying period, a female Red-wing incubates irregularly, and this seems to disturb her mate, who may perform symbolic nest-building movements near her, or more often near the neglected nest, apparently trying to induce her to return to it (Nero 1956).

Abnormal Incubation and Brooding

The pattern of incubation that prevails in each species or race is inherited rather than determined by mutual agreement of the partners, and, as is usual in such cases, it is followed rather consistently by all individuals of the species. But, as in other aspects of living things, aberrations sometimes occur. Occasionally one finds a nest that departs from the specific pattern, either as to the participation of the sexes in incubation or the time of day when they occupy the nest. Such aberrations seem more likely to arise, or at least to be recorded, in captive than in free birds.

When a female domestic pigeon undertook to incubate alone a set of eggs fertilized by a male with whom she did not pair, she sat almost continuously. Except for brief absences for food and drink, she remained covering her eggs through the middle of the day, when the male would ordinarily have taken charge of them (Goodwin 1947). In contrast to this, when two male Barbary or Ring-necked Doves paired, built a nest in an aviary, and were given the egg of another pigeon to incubate, they sat side by side upon the nest all day, but at night they left the egg exposed while they roosted on a perch (Neff 1944). The female domestic pigeon departed from the hereditary pattern by incubating through the middle of the day, but the male Ring-necked Doves spoiled their chance of raising a brood by blind adherence to this pattern.

The sex that does not normally incubate the eggs or brood the young by night scarcely ever does so in response to exceptional circumstances. When a female Common Mockingbird, the mother of four nestlings a few days old, was killed by a cat, her mate fed them faithfully, but he neglected the maternal role of brooding them through the night, with the result that they died of exposure (Laskey 1936). When a male Common Bluebird's mate, who had been incubating for nine days, was killed by a cat, he did not himself undertake the feminine role; instead, he promptly sought another partner, who hatched four of the six eggs, despite two days' exposure to chilling temperatures while he was absent searching for her (W. J. Hamilton, Jr., 1943).

Nocturnal incubation by female woodpeckers is rare, at least in the many species that always sleep singly when adult. But Sherman (1952) noticed that, for some obscure reason, on a few nights a female Yellow-shafted Flicker, instead of her mate, covered the eggs. On at least one occasion, a male Field Sparrow brooded nestlings a few days old, a most unusual occurrence (Walkinshaw 1944). Almost unique in the annals of ornithology is the case of a male Tree Swallow whose mate died three days after she started to incubate, and who then took full charge of the nest, hatching all five eggs on the fifteenth day (Kuerzi 1941). Hardly less surprising is the behavior of a drake Wood Duck who, when in almost full eclipse plumage and unable to fly, undertook to incubate a set of eggs that his mate had earlier deserted. Although his daytime sitting was less prolonged than that of a normally incubating female, he sat in the nest box through several nights, and on leaving the nest he carefully covered the eggs with the down that the duck had provided for them. He continued to incubate even after one of the spoiled eggs burst (Rollin 1957). As told earlier (p. 154), the male Bobwhite may undertake full charge of the eggs, attending them day and night if his mate dies, and the male Gambel's Quail may do the same. But since, even if he does not lose his mate, the cock Bobwhite may assume all parental offices, perhaps this behavior should not be considered abnormal in this species.

Of all the aberrations of incubation, sitting persistently in an empty nest is the most curious and "irrational." As everyone who has raised chickens knows, when a hen becomes broody she sits stubbornly in a nest, even if deprived of all eggs, and it is often most difficult to get her up, so that she will resume laying. I have only once encountered this curious behavior among free birds. For over two weeks, a Gray-headed Tanager "incubated" in a nest that, as far as I saw, had never contained an egg. By night she slept in the empty nest, and by day she sat almost as constantly as a normally incubating female of her kind. The rare New Zealand rail known as the Takahe may also sit as though incubating in eggless nests; one pair did so by turns for seven or eight weeks (G. R. Williams 1960). A number of similar instances, involving the Rook, Hedge Sparrow, Long-tailed Tit, American Robin, Eastern Kingbird, and Green Woodpecker, were gleaned from earlier literature by Nice (1943). Some of these birds that "incubated" persistently in empty nests had lost their eggs; others apparently had never laid. Were they suffering from hallucinations or blindly carrying out a sequence of instinctive acts, the normal course of which had somehow been deranged?

Resistance of Eggs to Chilling

Most birds who incubate alone leave their eggs exposed for longer or shorter intervals each day, and even when pairs sit by turns, they may fail by a good deal to keep their eggs constantly warmed. Accordingly, eggs at any stage of incubation must be able to withstand a certain amount of cooling without loss of hatchability; the time they may remain uncovered without suffering harm will depend upon the temperature of the surrounding air. Thus we find birds that nest in tropical lowlands or the warm summers of the temperate zones, such as certain tinamous and quails, leaving their eggs exposed for several consecutive hours each day without disastrous consequences. The Ornate Tinamou adjusts to the climate on the bleak Andean puna by taking three shorter recesses daily, covering his eggs with loose feathers before he goes. Nevertheless, they become chilled during his absence. Similarly, in the far north, grouse and Capercaillies take two to five short recesses each day, instead of the single very long one of quails that inhabit warmer regions.

Sometimes a fright late in the day, or a nocturnal disturbance, causes a bird to leave its eggs exposed during all or most of a night; yet, at least in milder weather, they often hatch if incubation is resumed. Eggs appear to be more resistant to chilling than are recently hatched young. The less advanced the embryo within an egg, the more cold it can withstand. After a night of snow and cold rain in late May, I found a full set of four Field Sparrow's eggs stone cold and apparently abandoned; but, milder days returning, the parent resumed incubation and hatched them all. For sixteen hours, an incomplete set of two Dunlin's eggs remained covered with snow in an unattended nest, yet they hatched after the parents laid two more eggs and incubated them (Norton 1972). A normal Mallard duckling emerged from a shell that, early in the incubation period, had been cracked by freezing; it is not known how deeply into the egg the freezing had penetrated. Eggs of the domestic chicken (a bird of tropical origin) have hatched after being held for nine or ten hours at temperatures well below the freezing point, although such treatment greatly reduces their viability (R. J. Greenwood 1969). In the far north, newly laid eggs are often exposed to snow and freezing temperatures. To reduce the danger of loss by freezing, some Arctic birds start to incubate before their sets are complete.

The less frequent the arrivals and departures of a single incubating parent, or the longer the intervals between change-overs when both parents share incubation, the longer the eggs are likely to lie unattended if the bird is driven from its nest or prevented from returning at its usual time. Accordingly, it is not surprising that the eggs of birds who take very long sessions can withstand long periods of exposure. Many eggs of the Manx Shearwater, including some far advanced in incubation, hatched after lying neglected in the burrow for up to seven days. In eggs exposed to temperatures of 52° to 76° F (11° to 24.5° C) in a room, embryos remained alive for as long as thirteen days. Similar resistance to chilling appears to be widespread among the tube-nosed swimmers; eggs of several other species, including albatrosses and petrels, have contained living embryos after four to eight days of exposure (G. V. T. Matthews 1954).

CHAPTER 15

Incubation without Animal Heat

The small minority of birds that do not hatch their eggs with the heat of their own bodies fall into two divisions. The first consists of several groups of unrelated birds that drop their eggs into the nests of birds of other species and then give little or no further attention to their progeny. These brood parasites include a number of cuckoos widespread over the Old World and more sparingly represented in the New World, the cowbirds of the Americas, the parasitic weaverbirds and honey-guides of Africa, and the Black-headed Duck of southern South America. These birds do not concern us here; their eggs are incubated according to the pattern followed by the foster parents.

The other division of birds that fail to incubate is more homogeneous. With the exception of the Egyptian Plover, which buries its three eggs in the sand along the Nile and depends largely on the sun's rays to hatch them, it contains, as far as known, only the megapodes or mound-birds. This small family of gallinaceous birds, named for their conspicuously large feet, consists of seven genera and about thirteen species distributed from the Nicobar Islands in the Indian Ocean eastward to Samoa, and from the Philippines to southern Australia. They are largely terrestrial birds, mostly dull in plumage, and ranging in size from that of a small domestic hen to that of a turkey. Some have bare, wattled heads and necks.

No megapode is known to sit on its eggs. In the family as a whole the eggs are warmed by heat from three sources, alone or in combination: the sun, the fermentation of decaying vegetation, and the interior of the earth (that is, heat that escapes from volcanoes and hot springs). Even birds of the same species employ surprisingly different methods of warming their eggs in different localities. Mound-birds are opportunists with a strong aversion to unnecessary labor, quick to take advantage of any situation that will save them work; but it is not known whether the variations in procedure exhibited by the same species are innate or due to individual insight and initiative. Nevertheless, some of them, lacking access to such natural incubators as hot springs or sun-warmed beaches, work so hard to regulate the temperature of their eggs that birds that sit on theirs seem to have the easier time. Indeed, no bird succeeds in reproducing without considerable effort. Even the brood parasites must carefully study the activities of the birds on which they impose in order to deposit an egg in the nest at the proper time, upon which the whole success of their intrigue depends.

The largest mounds, which in exceptional cases

PLATE 54. Incubation mound of Common Scrub-Fowl in an open glade in monsoon forest in northern Australia (photo by H. J. Frith).

reach 35 feet in diameter by 15 feet in height (10.7 by 4.6 meters), are built by the widespread Common Scrub-Fowl, usually in the jungle just inland from the seashore (plate 54). Although a big mound is the accumulation of years and may be attended by several pairs simultaneously, it is nevertheless an impressive monument to the industry of a bird about the size of a domestic hen. Using its strong feet to scrape together the vegetable litter of which the structure is composed, it moves slowly backward, with its large toenails raking leaves and sticks behind it. As the bird works, it devours whatever fallen fruits, seeds, insects, worms, snails, or other edible objects its scratching exposes. If one of these mounds is excavated, the deep interior is found to contain completely rotted vegetable material, mixed with soil and quite cold. The surface layer, in the breeding season, is composed of freshly collected soil and leaves, many of which are still green, and this mass is likewise cool. The intermediate layer, from two to five feet below the surface, is actively fermenting and quite warm, at a temperature of 95° to 102° F (35° to 39° C). Here the eggs are laid singly, in burrows that the birds dig diagonally downward from the top for a distance of several feet, then fill with fresh leafy material to cover the egg. If the mound is old and large, the scrub-fowls may add fresh material and lay their eggs on one side only. Such annual increments may extend the pile to a length of 60 feet (18 meters). These big mounds containing much vegetable material are built chiefly in humid woods with much ground litter. In more open situations, scrub-fowls make their mounds largely of earth and depend chiefly on direct sunshine to heat them.

In addition to large mounds, scrub-fowls employ a variety of other methods to incubate their eggs, and sometimes several different procedures will be used in the same locality. If volcanoes or hot springs are available, they deposit their eggs in burrows that they dig in the crater walls or into ground heated by steam or warm water, carefully choosing spots at just the right temperature. They sometimes bury their eggs singly in sandy beaches exposed to hot sunshine. Some individuals deposit an egg in a deep fissure between two flat rocks heated by the sun, then cover it with an insulating layer of leaves. They may even lay an egg between the buttressing roots of a large tree and cover it with a heap of decaying leaves. Or they may dig holes in the ground, drop an egg into each, and build above it a small mound of organic matter, perhaps 6 feet in diameter by 2 or 3 in height (1.8 by 0.6 or 0.9 meters).

Another mound-bird that depends on solar heat to hatch its eggs is the Maleo of Celebes Island. From the forests of the interior where it dwells, it descends periodically to the seacoast to lay its eggs, sometimes walking as much as twenty miles to reach a beach of black volcanic sand that absorbs much heat. The eggs are laid in holes dug obliquely into the beach. According to some observers, each hole is the work of a single pair and contains one egg; according to others, several pairs join forces and lay their eggs in a pit that may be 4 or 5 feet in diameter by 1 or 2 feet deep (1.2 or 1.5 by 0.3 or 0.6 meters). Each huge egg is more than 4 inches long by nearly 2½ inches wide (10 by 6 centimeters). From it hatches a remarkably precocious chick, which digs its way out of the sand alone and immediately begins the often long journey to the inland forest. Other populations of Maleos bury their eggs in the soil near hot springs or volcanic vents, or in any clearing exposed to the sun.

In the shady depths of the forest, the Wattled Mound-bird or Brush Turkey constructs a mound composed wholly of decaying vegetation, scratched together from the ground, which may reach 12 feet in diameter by 3 in height (3.7 by 0.9 meters). Unlike the scrub-fowls, this species builds a fresh mound each year, as the breeding season approaches in August. The active fermentation of the vegetable matter may raise the mound's temperature to 120°F (49°C), and, since this is too warm for the eggs, the female waits until it cools to between 90° and 100°F (32° and 38°C) before she deposits them. Thereafter, the male takes charge. To test the temperature, he digs a hole in the mound and pushes in his featherless head and neck. If the pile has become too cool, he brings new vegetable material and mixes it with the old. If the fermentation of this generates too much heat, he opens the top of the mound to reduce the temperature.

The only mound-bird of arid country is the Mallee-Fowl or Ocellated Megapode of western and southern Australia. In a region where daily and seasonal variations in temperature are much greater than in the humid bush and tropical islands where most megapodes dwell, it faces a more difficult problem of temperature regulation, which it solves by skillfully combining heat of fermentation with that received directly from the sun. In the autumn, it digs a hole 7 to 15 feet in diameter by 3 or 4 feet deep (2.1 to 4.6 by 0.9 to 1.2 meters). This it fills by systematically scraping inward fallen leaves and other vegetable debris from distances up to 50 yards (15 meters), until the pile may rise a foot above

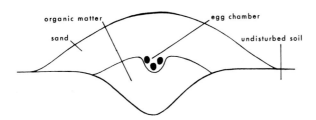

FIGURE 13. Diagrammatic cross section of a Mallee-Fowl mound (redrawn from Frith 1962).

the surrounding ground. After rain has wet this organic matter so that it will ferment, the bird piles loose, dry sand above it to a thickness of about two feet (fig. 13). Each mound belongs to a single pair. The female deposits the big eggs, in no regular arrangement, in a depression that the male makes in the top of the mass of moist vegetable matter. After each laying, he carefully fills the hole.

Early in the breeding season, the eggs are warmed chiefly by the fermenting vegetation, which may generate so much heat that the Mallee-Fowls are obliged to come daily in the cool early morning and open the mound to permit the excess warmth to escape (plate 55). After they have uncovered it to the required depth, they immediately begin to replace the sand by scratching it inward. With less intense fermentation, in combination with solar heat, the mound maintains the eggs at the required temperature of about 92°F (33.3°C) without the daily chore of cooling them. But, as summer advances and the sun rises higher in the midday sky, intense insolation tends to overheat the eggs. To counteract this effect, the Mallee-Fowls pile loose dry sand on top of the mound, sometimes to a height of four feet. Even with this protection, the eggs may gradually become too warm. To permit the excess heat to escape, the birds dig out the whole mound in the very early morning, scattering the sand widely until it cools, then kicking it back over the eggs before the sun rises high—a laborious operation that takes two or three hours.

At the season's end, when fermentation ceases, the birds depend wholly upon insolation to keep the incubator warm. Every day, after the sun has

risen high, they dig the soil away to within three or four inches of the eggs to allow the sun's rays to heat the sand above them. In the evening, the sand warmed by exposure to the sunshine is piled over the eggs to keep them at the proper temperature through the night. By installing an electric heating unit in the mound, Frith (1956, 1962) could raise its temperature unusually high or, by drying the vegetable matter so that fermentation stopped, cool it abnormally. The birds sensed these temperature changes, apparently by testing the sand with their tongues or the inside of their mouths, and worked hard to bring their incubator back to the usual 92°F. When undisturbed by an experimenter, they can maintain this temperature most of the time to within one degree. Just as the first efforts of birds that build elaborate nests may result in inferior structures (p. 122), so the earliest mounds of young Mallee-Fowls are so poorly made that they fail to hatch eggs. These birds must learn by experience to prepare effective incubators.

In one of these elaborate incubators, a single hen lays each year from five to thirty-three eggs, usually fifteen to twenty-four, which she deposits at intervals of from four to eight days, or rarely more. Unlike the eggs of the domestic hen and certain other birds, the eggs become progressively larger as the season advances, and the last to be laid may be 50 percent bigger than the first. Their incubation period varies widely, with the season and the mound's temperature, from forty-nine to ninety days; it averages about fifty-seven days. If not enough rain falls in the Mallee-Fowl's arid homeland to moisten the vegetable matter in the mound and start fermentation, the eggs fail to hatch or may not even be laid. The task of constructing the mound and regulating its temperature falls almost wholly to the male, who works hard for about eleven months each year to build and maintain it. His mate helps, especially toward the season's end after she has laid her last egg.

How did the megapodes' strange breeding habits arise, and what advantages do they have over more conventional methods of incubation to ensure their perpetuation and diffusion? When one recalls that certain reptiles, notably alligators, build mounds of mud mixed with vegetable matter in which to lay their eggs, he might infer that the megapodes have

PLATE 55. A female Mallee-Fowl (*left*) watches her mate test the temperature of their incubation mound by burying his head in the soil, during a routine check before dawn (photo by H. J. Frith).

simply retained the custom of their remote reptilian ancestors—a habit that all other birds have forsaken to incubate their eggs with the heat of their own bodies. But when we consider the taxonomic position of the megapodes—their close affinity to pheasants, grouse, and guans—we must reject this suggestion.

Those megapodes that hatch their eggs by means of the earth's internal heat or in sun-warmed beaches seem to enjoy the greatest advantage, for the parents' only task is to bury the eggs in the earth or sand, and, moreover, they need make no conspicuous mound that betrays their position to enemies. But only restricted localities offer such facilities, and, if we assume that the first megapodes to abandon the usual avian system of incubation followed this method, it is difficult to understand how the habit of building mounds of fermenting vegetation arose from it. On the other hand, mound making is a vari-ation of the widespread avian custom of nest build-ing. A bird that already laid her eggs in the heated compost of a mound might readily deposit them in warm sand or in earth heated by subterranean fires, if she happened to find such ground when her eggs were mature. Doubtless the mounds came first.

An obvious disadvantage of the mounds is their conspicuousness. Since some species use the same ones from year to year, they must become familiar to most egg-eating animals of the neighborhood. In many localities, they are well known to the aborigi-nal human inhabitants, certain of whom claim par-ticular mounds as private property, which is re-spected by their neighbors. Although the owners remove many of the big, thin-shelled eggs for eating, they have enough foresight to leave some eggs to hatch and to protect the birds that lay them. Ap-parently the aborigines have carried these valuable birds from island to island in their canoes, even though the megapodes were already widely distributed through the archipelagos of the western Pacific before the advent of man (Gilliard 1958).

The conspicuousness of the mounds is offset by the separation of the eggs in both time and space. Most gallinaceous birds lay all their eggs in a heap and do not begin to incubate until they complete their sets, which may be large. This exposes all the eggs to the same hazards, from the time they begin to accumulate until they hatch. As soon as a megapode's egg is laid in the heated medium, incubation begins. The next egg is laid from four to eight or more days later, in another part of the mound or beach, except in the Mallee-Fowl, which places its eggs close together, sometimes one touching another. Before the last of the numerous eggs is deposited, the earlier ones have hatched, and the amazingly precocious chicks, which can run soon after hatching and fly when only twenty-four hours old, have gone off to shift for themselves, never knowing their parents. This staggering of incubation

and distribution of risks seems to be the secret of the megapodes' success, the advantage that compensates for the obvious disadvantages of the mounds.

Nevertheless, it is probable that only in the Australasian region could such a system arise and persist. Doubtless it is not a matter of chance that Australia and New Guinea possess three of the most extraordinary groups of birds in the world—the megapodes, bower-birds, and birds of paradise. The first two are absolutely unique, while the last carries to supreme heights the tendency of males that remain aloof from the nest to develop lavish ornamentation and spectacular displays. The evolution of these remarkable families is evidently related to the paucity of predators in the region where they occur, and particularly to the virtual absence in the native fauna of placental mammals, which include so many highly efficient predators. The presence here of these fascinating birds suggests the glories that life might have achieved in far fuller measure if it could have avoided the horror of predation.

CHAPTER 16

The Duration of Incubation

How long do eggs take to hatch? If we desire only a rough approximation, the answer is easy. An old-fashioned farmer's wife could tell you that pigeons' eggs hatch in two weeks, hens' eggs in three weeks, and ducks' eggs in four weeks. If we need more accurate values, as we do to study the many interesting questions connected with the lengths of incubation periods, the answer is not so easy. As in many scientific investigations, our success will depend largely on our acumen in delimiting the intervals that we wish to measure. In the past, the incubation period was sometimes considered to be the time that elapsed between the laying of the last egg in a set and the hatching of the first chick or nestling; but this is often too short, for the first egg to hatch is likely to be the first that was laid, and it may receive more or less warming before the last egg is laid. If, on the contrary, we take the incubation to be the interval between the laying of the first egg and the hatching of the last chick, it will be too long.

The best recent practice is to consider the incubation period as the interval between the laying of the last egg in a set and the hatching of this same egg. To make sure when the last egg hatches, it is necessary to mark the eggs as they are laid, as can be done by numbering them lightly with a soft pencil on daily or more frequent visits to the nest during the laying period. Often it is not practicable to mark the eggs. Either they are very delicate, as in hummingbirds; or they are in a closed nest with a narrow doorway, where we can see them with a small mirror and electric bulb but cannot reach them without injuring the nest; or they are in a nest out of reach, where we view them by lifting up a mirror, thereby avoiding the disturbance caused by climbing or setting up a ladder. In this case, we consider the incubation period to be the interval between the laying of the last egg and the hatching of the last nestling, since observations on numbered eggs have shown that the last egg to be laid is nearly always the last to hatch. But, if any egg fails to hatch in an unmarked set, we cannot determine the incubation period with the desired accuracy, because the unhatched egg may be the last of the set.

An objection to this delimitation of the incubation period is that the parent may not begin to incubate promptly after the last egg is laid, or that it sits only sporadically during the first few days. To overcome this difficulty, Kendeigh (1963) developed a method of measuring the actual amount of effective heating that it takes to hatch a newly laid egg. Below a certain threshold temperature, embryonic development ceases; above this threshold, it proceeds at a rate proportional to the temperature, up to a certain optimum. Accordingly, the total amount of effective heating that an egg receives may be calculated by multiplying the difference between the actual and threshold temperatures by the number of hours that the egg remains at the higher temperature. For the eggs of the Northern House Wren, with a temperature threshold of 63° F (17.2° C), Kendeigh found the required amount of heat to be 5,867±144 degree-hours, or 32.9 kilocalories, during an incubation period of approximately fourteen days. Such measurements demand costly apparatus, easily accessible nests, and much laborious computation. Doubtless it will be long before this method is applied to many birds. For the present, we must be content in most investigations to regard the incubation period as a quantity that is conventionally defined as the interval between the laying and hatching of the last egg in a set.

Factors Affecting the Length of Incubation

Like the periods of gestation of mammals, the incubation periods of birds vary greatly, from about ten days in a few small songbirds to about eighty in the Royal Albatross (tables 6 and 7). Naturally, we are interested to know what causes this great variation. A first answer is likely to be that size is the principal determining factor (Worth 1940). To form a large chick requires more cell division and the organization of more material than to form a small one, and, other things being equal, this takes more time. No songbird's egg takes as long to hatch as that of the Ostrich, which is hundreds of times larger. But the eggs of the Northern Royal Flycatcher, no bigger than those of a sparrow, take as long to hatch as a domestic hen's eggs. Obviously, size is only one of the factors that determine the length of the incubation period, and probably not the most important one.

Another suggestion is that the incubation period is correlated with the taxonomic position of the bird. Bergtold (1917) supposed that the more primitive birds, those placed lower in our systems of classifi-

TABLE 6
Nest Type, Number and Size of Eggs, Incubation and Nestling Periods of Altricial Birds[a]

Species	Zone[b]	Nest Type[c]	Eggs No.	Size in mm	Incubation Period in Days	Nestling Period in Days
Manx Shearwater	L	B	1	60×41	51	70(62–76)
Great-winged Petrel	L	G	1	—	53	128–134
Leach's Petrel	L	B	1	33×24	42	66
Yellow-billed Tropicbird	T	H	1	54×38	41	70–85
Red-billed Tropicbird	T	H	1	65×45	42–44	80–100
Shag	L	H	3(2–4)	63×38	30–31	53(48–58)
Gannet	L	G	1	78×47	43–44	90(84–97)
Brown Booby	T	G	2	59×40	43–47	86–120
Ascension Island Frigatebird	T	O	1	—	43–46	180–200
Green Heron	TL	O	2–3	38×29	19–21	21–25
Bald Eagle	L	O	2(3)	74×57	34–35	72–74
African Crowned Eagle	T	O	1(2)	—	49	111(103–116)
American Sparrow Hawk	L	H	4–5	35×29	29	26–27
Common Puffin	L	B	1	63×44	40–43	47–51
Wood Pigeon	L	O	2	—	17	22
Mourning Dove	L	O	2	28×22	14–15.5	13–15
Ruddy Ground-Dove	T	O	2	23×17	12–13	12–14
Ruddy Quail-Dove	T	O	2	28×20	11	10
Black-billed Cuckoo	L	O	2–3	27×21	14	8–9
Groove-billed Ani	T	O	3–5	32×24	13	11
Screech Owl	L	H	4–5	36×30	25–26	32
Elf Owl	L	H	2–4	27×23	24	28–33
Oilbird	T	H	2–4	42×33	33–35	100–115
Common Potoo	T	O	1	41×32	33	51
Chimney Swift	L	H	4–5	20×13	18–19	25–32

Species	Zone[b]	Nest Type[c]	Eggs No.	Size in mm	Incubation Period in Days	Nestling Period in Days
Common Swift	L	H	2–3	—	19–20	35–56
Cave Swiftlet	T	H	2	17×11	21.5	36–42
White-cheeked Coly	T	O	4–5	—	12–14	17–18
Black-throated Trogon	T	H	2	28×22	18	14–15
Mountain Trogon	T	H	2	29×24	19	15–16
Amazon Kingfisher	T	B	4	32×27	22	29–30
Blue-throated Green Motmot	T	B	3	29×23	21–22	29–31
European Bee-eater	L	B	4–7	26×22	20	20–25
Red-and-white-billed Hornbill	T	H	2(3)	—	33–37	46
Rufous-tailed Jacamar	T	B	3–4	22×19	20–22	20–26
Blue-throated Toucanet	T	H	3–4	—	16	43
Olivaceous Piculet	T	H	2–3	17×12	14	24–25
Yellow-shafted Flicker	L	H	6–8	27×21	11–12	25–28
Golden-naped Woodpecker	T	H	3–4	—	12	33–37
Yellow-bellied Sapsucker	L	H	3–7	23×17	14	23–28
Band-tailed Barbthroat	T	P	2	15×9	18.5–19	23–25
Little Hermit Hummingbird	T	P	2	12×8	15–16	20–23
Scaly-breasted Hummingbird	T	O	2	—	17–19	25(22–29)
Rufous-tailed Hummingbird	T	O	2	14×19	16	19–23(27)
Spotted-crowned Woodcreeper	T	H	2	—	17	19
Buff-throated Automolus	T	B	2–3	27×20	20–21	18
Rufous-fronted Thornbird	T	P	3	—	16–17	21–22
Bicolored Antbird	T	H	2	24×18	15–16	13–15
Plain Antvireo	T	O	2	20×15	15	9(10)
Black-faced Antthrush	T	H	2	40×25	20	17–19
Orange-collared Manakin	T	O	2	21×15	18–20	13–15
Masked Tityra	T	H	2	30×21	ca. 21	28–30
White-winged Becard	T	R	3–4	20×15	18–19	21
Rufous Piha	T	O	1	31×22	25–26	28–29
Eastern Phoebe	L	H	5(3–7)	19×15	15–17	15–16
Gray-capped Flycatcher	T	R	2–3(4)	23×17	16–17	19–21
Least Flycatcher	L	O	3–4	16×13	15–16	14
Sulphur-rumped Myiobius	T	P	2	18×13	22	22–24
Skylark	L	G	3–4(5)	24×17	11–12	9–10
Horned Lark	L	G	2–5	22×16	11	10
Rough-winged Swallow	TL	BH	5–7	18×13	16(18)	18–21
Barn Swallow	L	H	4–5(7)	19×14	14–15	20–24
Blue-and-white Swallow	TL	H	2–4	17×12	15	26–27
Purple Martin	L	H	4–6	25×18	15–16	24–35
Blue Jay	L	O	4–5	28×20	17–18	17–21
White-tipped Brown Jay	T	O	2–3	35×25	18	23–24
Common Raven	LT	OH	4–6	50×34	20–21	35–42
Black-capped Chickadee	L	H	6–9	15×12	12–13	16–20
Black-eared Bushtit	T	P	4	14×10	15	17–19
Cape Penduline-Tit	L	P	5–6	—	ca. 15	23–24
Pygmy Nuthatch	L	H	5–9	15×12	15–16	20–22
Brown-headed Nuthatch	L	H	4–7	15×12	14	18–19
Yellow-breasted Forest Bulbul	T	O	2	25×17	14	18
White-throated Forest Bulbul	T	O	2(3)	24×16	11–12	16–18
American Dipper	L	H	4–5	26×19	16	24–25
Cactus Wren	L	R	3–5	24×17	16	19–23
Winter or European Wren	L	RH	4–8	17×13	15–17	15–21
Southern House Wren	T	H	3–5(6)	18×14	14–16	17–19
Long-billed Marsh Wren	L	R	3–8	16×12	13–16	14–16
Common Mockingbird	L	O	4–5	24×18	12–13	11–13
Donacobius	T	O	3	25×17	17	17–18
Gray Catbird	L	O	3–5	23×18	12–13	10–12

Species	Zone[b]	Nest Type[c]	Eggs No.	Size in mm	Incubation Period in Days	Nestling Period in Days
American Robin	L	O	3–4(6)	28×20	12–13	13(16)
Gray's Thrush	T	O	2–3(4)	28×20	12	15–16
Common Bluebird	L	H	4–6	21×16	13–15	17–18
Orange-billed Nightingale-Thrush	T	O	2(3)	24×18	13–15	14–15
Cape Robin	L	O	2(3)	24×17	15–17	15–17
Great Reed Warbler	L	O	4–5(6)	—	14–15	12
Graceful Warbler	L	R	3–5	—	11–13	13–14
Blue-gray Gnatcatcher	L	O	(3)4–5	15×11	15	11–13
Long-billed Gnatwren	T	O	2	19×14	17	11–12
Superb Blue Wren	L	R	3–4	16×12	13–14	12–13
Pied Flycatcher	L	H	5–7(9)	—	11–15	16–18
Paradise Flycatcher	T	O	2–3	—	12–13	11–12
Yellow Wagtail	L	G	3–6	—	12–13	10–12
Mountain Wagtail	T	H	2	—	14	15–16
Cedar Waxwing	L	O	3–5(6)	22×16	12–13	16
Long-tailed Silky-Flycatcher	T	O	2	26×17	16–17	24–25
Greater Long-tailed Pied Shrike	T	O	3–4	25×19	13–14	16–18
Lesser Red-winged Bush Shrike	T	O	2–3	—	12–13	16
Red-winged Starling	L	H	2–4	—	13–23	22–28
Violet-throated Black Sunbird	T	P	(1)2	19×13	13–14	16
Emerald Long-tailed Sunbird	T	P	1(2)	—	13–14	14–16
Golden-fronted White-eye	T	O	2–3	18×12	11–12	15
Red-eyed Vireo	L	O	3–4(5)	20×15	12–14	10–11
Yellow-green Vireo	T	O	2–3	20×15	13–14	12–14
Bell's Vireo	L	O	3–5	17×13	14	11–12
Gray-headed Greenlet	T	O	2	18×14	16	12
Yellow Warbler	L	O	4–5	17×13	11–12	8–10
American Redstart	L	O	4(5)	16×12	11–12	8–9
Slate-throated Redstart	T	RG	2–3	18×13	13–15	12–14
Buff-rumped Warbler	T	RG	2	21×15	16–17	13–14(15)
Yellow-tailed Oriole	T	O	3	24×18	14	12–13
Great-tailed Grackle	T	O	2–3	34×23	13–14	20–23
Red-winged Blackbird	L	O	3–5	25×18	11(12)	11
Swallow-Tanager	T	H	3	21×15	13–17	24
Blue Honeycreeper	T	O	2	19×14	12–13	14
Bananaquit	T	R	2–3	17×13	•12–13	15–18
White-vented Euphonia	T	R	3	17×12	15–17	18–20
Blue Tanager	T	O	2(3)	23×16	13–14	17–20
Scarlet-rumped Black Tanager	T	O	2	24×17	12(13)	12(11–13)
Silver-throated Tanager	T	O	2	21×16	(13)14	(14)15
Cardinal	L	O	2–5	25×18	12–13	9–11
Black-striped Sparrow	T	R	2	25×18	13–14	11–12
Variable Seedeater	T	O	2(3)	17×13	12–13	12–13
American Goldfinch	L	O	4–6	16×12	12–14	10–15
Red Crossbill	L	O	3–4(5)	20×15	12–14	18–24
Gray-headed Waxbill	T	R	4–5(6)	14×10	12–15	14–16
Barred Waxbill	T	R	5–7	—	12–13	13–16
House Sparrow	L	H	3–7	23×15	11–12	15(11–19)
Black-faced Weaver	T	R	2–3(5)	18×14	11–12	11–14

a The data for this table have been selected from the most careful studies; often several were consulted for the information on a single species. To avoid crowding the table, references are omitted, but most of those used are cited in the text and listed in the References.

b Abbreviations used to indicate climatic zone: T = tropical; L = higher latitude.

c Abbreviations used to indicate type of nest: B = burrow; G = on ground; H = hole or niche in tree, termitary, cliff, building, or some similar protected site; O = open nest amid vegetation; P = pensile nest; R = roofed nest, amid vegetation unless otherwise indicated.

TABLE 7

Incubation and Fledging Periods of Precocial, Subprecocial, Semialtricial, and Related Birds

Species	Nest[a] Site	Type	Eggs No.	Size in mm	Incubation Period in Days	Fledging Period in Days
Slaty-breasted Tinamou	C	G	10[b]	46×40	16	15–20
Brushland Tinamou	C	G	9[b]	48×38	19–20	—
Ostrich	C	G	15–50[b]	150×125	ca. 42	—
Kiwi	I	B	2	135×84	76	—
Adelie Penguin	C,I	G	2(3)	69×55	34	41–56[c]
Rockhopper Penguin	I	G	2(3)	—	33–34	67–71[c]
Yellow-eyed Penguin	I	G	2	—	44	97–118[c]
Royal Albatross	I	G	1	—	79	236
Laysan Albatross	I	G	1	109×69	64	165
Greater Flamingo	IC	G	1	90×56	28	75–78
Canada Goose	C	G	5–6	86×58	28	ca. 60
Mallard	C	G	8–12	58×42	27	ca. 56
Helmeted Curassow	C	O	2	110×63	34–36	—
Prairie Chicken	C	G	7–17	45×34	23	ca. 14
Ring-necked Pheasant	C	G	8–14	42×34	23	7–14
Sora Rail	C	O	10–12	32×23	16–19	36
American Coot	C	O	8–12	49×34	23–25	ca. 75
Killdeer	C	G	4	36×27	24–26	21–31
Spotted Sandpiper	C	G	4	32×23	20–22	13–16
American Avocet	IC	G	3–4	50×30	24	27
Herring Gull	I	G	3	72×51	27–33	45
Sooty Tern	I	G	1	50×35	28.5–30	56–60
Least Tern	C	G	2–3	31×24	18–20	—
Thick-billed Murre	C,I	K	1	80×50	33–34	18–25[c]
Namaqua Sandgrouse	C	G	3	36×25	21	30–42
Common Nighthawk	C	G,T	2	30×22	18–19	18–25
Whip-poor-will	C	G	2	29×21	19–20	14

[a] Abbreviations used to indicate situation of nest: C = continent; I = island in sea; IC = island in continental lake or swamp. Under "type" the abbreviations are: B = burrow; G = nest or scrape on ground; K = breeds on cliff; O = open nest amid marsh vegetation or (curassow) in undergrowth of tropical forest; T = on flat roof.

[b] Eggs laid by two or more females.

[c] Age of going to sea.

cation, have not been able to speed up their embryonic development as much as the higher, more advanced birds. One reason why penguins, petrels, albatrosses, Ostriches, Emus, and related birds have long incubation periods may be that their body temperature averages about 3.6°F (2°C) lower than that of the rest of the avian class (McNab 1966, Warham 1971). But these and other relatively "primitive" birds with long incubation periods are on the whole larger and lay bigger eggs than the advanced songbirds, so that to settle this question we must disentangle the effects of size, taxonomic

position, body temperature, and other possible factors. The shortest incubation period that I have determined is that of the Ruddy Quail-Dove of the tropical forest, a member of a family placed about halfway up the taxonomic ladder. Its eggs hatch in eleven days, which is less than the incubation period of many songbirds with smaller eggs. Evidently we should be wary of any single-factor explanation of incubation periods.

Still another possibility is that the length of the incubation period is influenced by the hazards to which the eggs are exposed. Let us suppose that the eggs of a certain bird require fifteen days to hatch, and that 30 percent of the nests are destroyed, mostly by predators, during this incubation period. This is an average loss of 2 percent per day. If the bird could reduce its incubation period to twelve days, its success in hatching its eggs might be as much as 6 percent higher. But the accelerated development would probably be carried over from the embryos within the eggs to the young in the nest, because in nidicolous birds we find a rough correlation between the lengths of the incubation and nestling periods. Accordingly, the gain in reproductive success by a reduction of three days in the length of the incubation period could be substantially more than 6 percent, since birds that lose many eggs to predators usually lose many nestlings to them. Thus, the more the nests of any species are exposed to predation, the more natural selection should shorten the incubation period, by eliminating the genotypes whose eggs develop more slowly. Or, to put the matter differently, birds with short incubation periods have a better chance of surviving than those with long incubation periods, and the advantages of rapid incubation will be greater in proportion to the hazards to which the nests are exposed.

This selection pressure to shorten incubation periods where losses of nests are great is, I am convinced, the most powerful of the many factors that affect their length. Let us return to the Northern Royal Flycatcher and the domestic hen, whose eggs take about the same time—about twenty-one days—to hatch, despite the tremendous disparity in size. The flycatcher's eggs are laid in a remarkable nest that dangles above a forest stream, beyond reach of most predators, and moreover they are cryptically colored to a degree unusual in their family. The domestic hen is descended from the Red Jungle-Fowl, whose eggs were laid on the ground in tropical lands teeming with egg-eating animals. The ancestors of the domestic hen were evidently subjected to far stronger selection pressure to reduce the incubation period than the flycatcher has been for countless generations. In general, birds that breed on small isolated islands, on cliffs, in burrows, or in otherwise inaccessible situations, and those like hawks and eagles that are well able to defend their nests, have long incubation periods. Birds that nest on or near the ground in continental areas with many predators have short incubation periods.

Although the ecological factor is undoubtedly weighty, we must not underestimate other factors. The egg's size is certainly important, and it is doubtful whether selection pressure, however strong or long continued, could reduce the incubation period of the domestic hen to twelve days, which is the period of many eggs the size of the Northern Royal Flycatcher's. To fully appreciate the difference in the rate of embryonic development of the hen and the flycatcher, we must take into account not only the size of the young that emerge from their eggs but also their condition. The hen's chick is able to walk and pick up its own food a few hours after it hatches; the flycatcher's offspring is a typical helpless passerine nestling, which for days rests inertly in its nest, fed by its parent. The difference between the two is much the same as that between a newborn calf and a newborn human baby: in a few hours the former can walk better than the latter will be able to after a year. Although the periods of gestation of the calf and the baby are the same, the calf's prenatal development is far more rapid, for it is much bigger in addition to being much more advanced at birth. The greater development within the egg doubtless explains why no precocial bird has achieved such a short incubation period as that of many altricial and semialtricial birds. Curiously enough, the latter birds have not only the shortest (by a narrow margin) known incubation periods, but also the longest, as exemplified by albatrosses that nest on islands where their eggs and helpless chicks are fairly safe from predation.

Another instructive comparison is that between the Ostrich and the Storm-Petrel. Ostrich eggs weigh up to about 3½ pounds (1,618 grams); the Storm-Petrel's egg weighs ¼ ounce (7 grams). Nevertheless, the incubation periods of these so different eggs are the same, about six weeks. Moreover, the Ostrich chick escapes from the egg at an advanced stage and is soon able to walk; the petrel's nestling remains in its burrow for nine weeks before it can fly. Why does the huge Ostrich develop so much more rapidly than the little petrel? The answer is that the petrel nests safely in a burrow on an islet with few enemies; the Ostrich lays its eggs on the ground on the predator-plagued African continent. Despite its great power and ability to defend itself and its nest, its eggs are subject to so many hazards that development has been greatly accelerated.

The same contrasts emerge but are less striking when we compare birds of the same family rather than those so distantly related as the Northern Royal Flycatcher and domestic hen or the Ostrich and Storm-Petrel. The South American Torrent Duck, an inhabitant of tumultuous Andean streams, nests in deep burrows in riverbanks, cavities in streamside trees, or crevices in forbidding cliffs. In such a protected situation, one duck hatched her eggs in forty-three or forty-four days, by far the longest incubation period recorded for any member of the duck family, including the much larger geese and swans (Moffett 1970). But the eggs of the Mallard, a bigger bird, laid on the ground where they are exposed to many dangers, hatch in only twenty-seven or twenty-eight days.

Among American flycatchers, eggs in hanging nests difficult to reach, including those of the Northern Royal Flycatcher and a number of other species, have incubation periods ranging from eighteen to twenty-three days; but eggs of about the same size in more accessible nests amid trees and shrubbery hatch in thirteen to eighteen days. Among tanagers nesting in the same locality, the largest belong to the genus *Ramphocelus*, the smallest to *Euphonia*, while those of *Tangara* are intermediate in size. Corresponding to this, *Ramphocelus* lays the biggest eggs, *Euphonia* the smallest, and those of *Tangara* fall between them in dimensions. Yet the eggs of *Ramphocelus* hatch in twelve days, those of *Tangara* in thirteen or fourteen days, and those of *Euphonia* in fifteen to eighteen days—the smallest eggs take longest to hatch. The explanation seems to be that *Ramphocelus* builds rather conspicuous open nests amid shrubbery at no great height; *Tangara* builds smaller, less conspicuous open nests usually considerably higher in trees; and *Euphonia* makes covered nests in nooks and crannies, often high in trees, where they are harder for predators to find.

The old notion that the incubation period is related to the bird's position in the taxonomic system—which aspires to reflect the evolutionary sequence—is not wholly unfounded. The flycatchers nesting in the Valley of El General in open or domed nests more accessible to predators than the pensile structure of the Northern Royal Flycatcher, generally take two to four days longer to hatch their eggs than do the tanagers, finches, thrushes, and other songbirds that build in the same trees and bushes, where their eggs, of about the same range in size, seem to be exposed to the same hazards. The suborder Tyranni, to which the American flycatchers belong, is classified just below the Oscines or songbirds. Other families of the Tyranni also have fairly long incubation periods. Although the slight nests of manakins are built in the undergrowth of the forest where they suffer heavy predation, their tiny eggs take from seventeen to twenty days to hatch.

The Evolution of Incubation Periods

It will help us to understand variations in the duration of incubation if we recall that vital processes, in both animals and plants, are often slow unless subject to considerable pressure to accelerate. The acorns of many species of oaks take two summers to mature, although others of about the same size in the same locality ripen in a single summer. In cycads, pines, and other gymnosperms, the whole process of fertilization and embryonic development is amazingly slow, for no apparent reason. Living in comparative security beneath the soil, certain cicadas take as long as seventeen years to complete a series of transformations no more complex than many another insect undergoes in a month or two. On remote islands with few or no predators, the eggs of birds take a surprisingly long while to hatch. In all these cases, and many more that might be cited, development proceeds at its own leisurely pace because it is under no great pressure to go faster. It is not the long incubation periods that demand explanation, but rather the short ones.

Environmental pressures can greatly accelerate the activities of living things. Competition for a place in the sun speeds the growth of plants in crowded vegetable communities. In many organisms, a short growing season makes rapid development and reproduction a necessity. After a good shower soaks the desert, herbaceous plants germinate, grow, flower, and set seed with amazing speed. In *Ephedra*, which flourishes in arid places, rapid growth of the pollen tube and prompt fertilization contrast strongly with the slowness of these processes in other gymnosperms (Coulter and Chamberlain 1910). In the brief Arctic summer, birds of many kinds pass through their reproductive cycles with a rapidity that contrasts sharply with the leisurely procedures of birds in the wetter parts of the tropics.

If any stage in the life history of an organism is particularly vulnerable, that stage is likely to be abbreviated in the highest possible degree. In birds, eggs and helpless young are far more exposed to predation than are adults, whose high mobility enables them to escape enemies. The hazards—such as weather but even more predators—to which the eggs of many birds are exposed have placed a premium upon their rapid development; the greater the perils that beset them, the shorter their incubation periods tend to be. Danger forces embryos to develop faster, just as fear prods a sluggish man to run.

The rate of development within the egg is often continued after hatching. Birds with short incubation periods generally have short nestling periods, whereas those with long incubation periods tend to have long nestling periods. Exceptions, of course, occur. In woodpeckers and toucans, the eggs hatch quickly but nestlings remain long in the nest cavity; in antbirds, many species have incubation periods long for such small birds, followed by exceptionally brief nestling periods. The positive correlation between the lengths of incubation and nestling periods is understandable when we recall that the same hazards that threaten eggs commonly threaten the young that emerge from them, so that in safe nests both embryos and nestlings can develop slowly, whereas in nests exposed to many perils it is equally important that both develop rapidly.

When subject to strong selection pressure for rapid embryonic development, the hatching of eggs can be accelerated in various ways. Perhaps the most important is the internal chemistry of the embryo, the genetically determined rate at which its organizing processes go forward at favorable temperatures. The reduction of the temperature threshold for development should also help, especially in cool climates, as it permits development to proceed, although at a diminished rate, even when the egg is

cooling, as during the absences of the incubating parent. This, however, is a matter of which we still know little. The prompt development of the incubation patch, as in songbirds, is evidently one of the prime factors in shortening the incubation period. The tardy attainment of full incubation temperature that we noticed earlier in certain nonpasserines (p. 174) seems to be one of the main reasons why their eggs take so long to hatch. Apparently selection pressure has not been strong enough to accelerate this feature—so necessary for rapid hatching—in penguins that nest rather safely in remote places or in raptors able to defend their nests. In the Ruddy Turnstone, a small ground nester, the tardy attainment of full incubation temperature is more puzzling. Likewise, the very long incubation periods of manakins are difficult to account for. However, it is evident that what natural selection can achieve in any lineage depends upon the incidence of favorable mutations, which arise by chance rather than in direct response to the needs of the species. In the long, hazardous gamble of evolution, some organisms have never received genes that might help them greatly and perhaps prevent their extinction.

The thermal insulation provided by the nest also affects the temperature at which eggs are maintained and, accordingly, the length of the incubation period. The long periods of some birds that lay their eggs directly on the ground, which is often wet and a good heat conductor, are doubtless to be attributed, at least in part, to this circumstance. The longest incubation period of a passerine that I have ever determined—one of the longest on record—is that of a Rufous Piha, a thrush-sized cotinga. Its single egg, laid on an exceedingly slight, unlined nest in a tropical forest tree, took twenty-five days to hatch. Passerines with short incubation periods commonly line their nests well, providing good insulation.

To what extent do birds realize their eggs' inherent capacity for rapid development? The temperature at which they keep their eggs is, of course, limited by their own body temperature and the efficiency of their incubation patches. In any event, incubation at temperatures much above those that birds actually apply to their eggs would not be helpful, as it causes abnormal development. But perhaps birds that warm their eggs intermittently could hatch them more promptly if they covered them more continuously, for example, if those that follow the Hummingbird Pattern would adopt the Antbird Pattern. To answer this question, several tests are available: the effects of different natural environments, the results of incubation at controlled temperature in an incubator, and the comparison of related species with different incubation habits.

Near sea level in Ecuador and Surinam, the incubation period of the Southern House Wren is fourteen days (Haverschmidt 1952, Marchant 1960). Between 2,500 and 3,000 feet (760 and 910 meters) in Costa Rica, the period is only exceptionally as short as fourteen days and a few hours, usually about fifteen days, and often sixteen days. At a nest situated at 8,500 feet (2,600 meters) in the Guatemalan mountains, the incubation period was seventeen days. These differences evidently result from the more rapid cooling of the eggs during the female's absences from the nest in the cooler air at increasing altitudes. Eggs of the Northern House Wren hatched in thirteen days when the average daily temperature was 78°F (25.6°C) but in sixteen days when it was 67°F (19.4°C), despite the fact that the female wrens sat more constantly at the lower temperature. The shortest incubation recorded for the house wren, thirteen days, was achieved by a female who covered her eggs only 43.6 percent of the day, but in a nest box exposed throughout the day to the full glare of the sun, which raised the temperature of the interior to 100° F (37.8° C) in the middle of the day (Kendeigh 1952).

In Iowa, Eastern Phoebes hatched their first brood, early in the spring when days were cool, in an average time of 17 days, but, in the warmer weather when second broods were incubated, the time was reduced to about 15.5 days (Sherman 1952). The average length of the incubation period of the Blackbird in Great Britain was 13.7 days in March, 13.3 days in April, 13.1 days in May, and 12.7 days in June (D. W. Snow 1958). In the European Robin, a similar but smaller reduction of the incubation period as the advancing season brings warmer days has been noticed (Lack 1948). Since all these birds follow the Hummingbird Pattern, the

longer incubation periods in cooler weather seem to be caused chiefly by more rapid cooling of the eggs during the female's absences, although, because nests do not provide perfect insulation even while the parent sits over them, the eggs may remain slightly cooler on cold days than on warm ones.

When eggs of the Bobwhite were placed beneath bantam hens, who take a daily outing of about an hour instead of the Bobwhite's several hours, they hatched in 22 to 23 days (average 22.5 days) instead of in 23 to 23.5 days, as when they are incubated by the quail (Stoddard 1946). Kept at a constant temperature in an incubator, the eggs of a number of birds have hatched in from six to twelve hours less than when warmed intermittently by the birds that laid them (Nice 1954). On the other hand, Skylarks' eggs required 13 to 14 days in an incubator, although the female Skylark, sitting intermittently, hatches them in 11 to 12 days (Bent 1942). Likewise, artificially incubated eggs of Bell's Vireo required 15 days, yet the parents, sitting alternately, hatch them in 14 days. Apparently, the temperature or humidity in the incubator was not so favorable as in the nest (Graber 1955).

The fact that the eggs of certain birds that incubate in the Hummingbird Pattern or Quail Pattern hatch more rapidly when days are warmer, or when more constantly warmed in an incubator or beneath some other bird, suggests that the parents do not take full advantage of their embryo's innate capacity to develop rapidly. If, by taking turns on the nest, the male and female together kept their eggs constantly covered, these might hatch more promptly, thereby increasing the eggs' chances of escaping predation. If this turned out to be true, we should have to ask why so many male birds, especially among the passerines, fail to incubate, when we would expect natural selection to favor the pattern that most reduced losses of eggs by abbreviating their time in the nest. Fortunately, to test this question we can compare certain birds that follow the Antbird Pattern with closely related species that follow the Hummingbird Pattern.

Among the vireos, both sexes of a number of species take turns on the nest and keep their eggs rather constantly covered, while in other species of the same genus only the females incubate, with frequent recesses. Nevertheless, available records fail to suggest that incubation periods in the first group are shorter than those in the second group (Skutch 1960). Among the finches, grosbeaks, and sparrows, where incubation by the female alone is the rule, the Rose-breasted and Black-headed grosbeaks are conspicuous exceptions. These grosbeaks keep their eggs covered almost continuously, the female sitting somewhat more than her mate, yet their incubation period is at least twelve days, as in many finches and sparrows of which the female incubates alone (Weston 1947). Doubtless the lack of a bare incubation patch on male vireos and grosbeaks prevents the time they spend on the nest from being as effective as it might be. Taking all the evidence together, it appears that birds that incubate with a constancy of 70 or 80 percent approach rather closely, but do not always attain, the shortest incubation periods possible for their eggs. Natural selection accomplishes much, but the notion that it consistently achieves the absolute maximum of adaptation or efficiency is a fallacy.

The situation appears quite different when the constancy of incubation falls below 60 percent. The Rough-winged Bank-Martins studied by Moreau (1940) in Africa incubated with a constancy of only 31 to 66 percent and took nineteen days to hatch their eggs, although other swallows that warm their eggs more constantly hatch them in fourteen to sixteen days. Since these martins nest rather safely in burrows, perhaps the long incubation period is not detrimental to them. One of the reasons why American flycatchers with pensile nests have such long incubation periods may be that they sit impatiently, but here again prolonged incubation is compensated by the inaccessibility of their nests. Although small flycatchers with open nests also incubate with low constancy, especially while the sun shines, their incubation periods are shorter. It is noteworthy that a female Bellicose Elaenia who sat with the unusually high constancy of 80 percent hatched her eggs in thirteen days and a few hours, while most of her neighbors of the same species incubated less steadily and took fourteen or fifteen days to hatch theirs.

The short incubation periods of woodpeckers, often no more than eleven or twelve days, are puzzling, for these birds nest rather safely in holes that they carve in trees and their nestling periods are long, sometimes three times as long as the incubation periods. Woodpeckers hatch in a very undeveloped state, yet they are only slightly less developed than passerines, and certainly no less than hummingbirds, whose much smaller eggs take considerably longer to hatch. In their dark cavities, woodpeckers seem unable to distribute food equally among their brood, with the result that the youngest and smallest often succumbs. I believe that woodpeckers' eggs have, so to speak, engaged in an evolutionary race to hatch first, with consequent greater probability of survival, and this has shortened the incubation period. In woodpeckers, the one-day interval between the laying of successive eggs has given later eggs a chance of overtaking earlier ones, if genetically better endowed for swift development. The longer interval between eggs of other birds in which nest mates compete strongly for survival—as in raptors, boobies, and skuas—virtually precludes that the embryos in the later eggs can overtake their older siblings.

In the less arid parts of the tropics, where birds do not need to replace such great annual losses as birds commonly must in high latitudes and where, moreover, they enjoy a longer season favorable for breeding, the whole sequence of nesting and the interval between broods tend to proceed at a more leisurely pace. The incubation periods of certain tropical birds are significantly longer than those of their closest relatives in the North Temperate Zone. The difference is especially pronounced in the wood warbler family, in which species nesting in the tropics generally have incubation periods of fourteen to sixteen days, instead of eleven or twelve days, as in the migratory species that breed in the north (Skutch 1954).

It is evident that incubation periods are influenced by many factors and are responsive to a variety of external conditions that affect the lives of birds, but for shortening them the most powerful factor has been the hazards, especially from predators, to which the eggs are exposed.

The Incubation of Unhatchable Eggs

If from infertility, chilling, or any other cause,

TABLE 8
Incubation of Unhatchable Eggs

Species	Length of Attendance in Days	Usual Incubation Period in Days	Reference
Adelie Penguin	49, 59, 63	30–39	R. H. Taylor 1962
Shag	65, 84	30–31	B. K. Snow 1960
Gannet	102	42–46	Nelson 1966
Black-crowned Night-Heron[a]	40, 49, 51	22–24	Noble & Wurm 1942
Wood Duck	62	ca. 30	F. Leopold 1951
Bobwhite	56	23	Stoddard 1946
Sarus Crane[a]	70–72	ca. 32	Walkinshaw 1947
Dunlin	33	22–23	Norton 1972
American Avocet	39	22–29	Gibson 1971
Mourning Dove	ca. 130	14–16	D. Luther in litt.
Smooth-billed Ani	24	13–15	D. E. Davis 1940*a*
White-tailed Trogon	51	ca. 17	Skutch 1962*b*
Yellow-shafted Flicker	30	11–12	Sherman 1952
Black-chinned Hummingbird	24	14	Bené 1946
Buff-throated Woodcreeper	30	ca. 18	Skutch MS
Blue-and-white Swallow	26	15	Skutch 1952*b*
Common Crow	26, 28, 32	16–18	Emlen 1942
Carolina Chickadee	24	12–13	Odum 1942
Blue Tit	25	13–15	Gibb 1950
European Wren	25, 26, 51	15–17	Armstrong 1955
European Robin	35, 48	13–15	Lack 1953
Common Bluebird	33	13–15	Thomas 1946
American Goldfinch	23, 25, 26	12–14	Berger 1968

[a] In captivity.

eggs fail to hatch in the usual time, a bird will usually continue to incubate them for a number of additional days (table 8). Most birds seem to remain faithful to their eggs for an interval at least 50 percent longer than their normal incubation period, and occasionally they continue to attend them for double the usual period. Even greater persistence in incubation has been recorded. A pair of White-tailed Trogons, whose eggs were spoiled by an invasion of ants that failed to pierce the shells, were still diligently warming them when their nest was broken up by some animal fifty-one days after incubation began, although their normal incubation period is about seventeen days, or one-third as long. The European Wren and European Robin have also on occasion incubated unhatchable eggs for three times their usual incubation periods. It is clear that, in the normal course of events, incubation stops because the hatching of the nestlings gives the parents something else to do, not because the latter have exhausted their impulse to sit on their eggs.

Eggs that fail to hatch in 20 percent more than the average period scarcely ever do; thus, the strength of the parents' urge to incubate provides a wide margin of safety.

An outstanding exception to the rule that birds persist in incubating eggs well beyond the usual time is found in certain pigeons, who will abandon a nest if hatching is delayed by as little as one day. They may even desert eggs in which the young are peeping and trying to break out of the shell only twenty-four hours too late. This strange behavior is caused by the formation in the parents' crop of the "pigeon's milk" on which the nestlings are nourished; if the young are not there to receive it, the parents become upset and neglect the nest. Most birds do not have this difficulty, because they do not prepare food for their nestlings until they see them. Some pigeons, however, attend unhatched eggs beyond the normal time (Heinroth and Heinroth 1959).

Many birds appear to be unaware of the condition of their eggs. They stubbornly continue to incubate long after the eggs are spoiled and incapable of hatching; even after they are so rotten that they are malodorous, as happened with a Wood Duck who attended her eggs for twice the normal interval (F. Leopold 1951). When only part of a set hatches, the unhatched egg or eggs are often left in the nest until the nestlings leave, even when the empty shells and all waste are promptly removed. A few birds, however, seem somehow to know the condition of the eggs and remove them if they are abnormal or infertile. After much squeaking and rustling by both parents inside their closed nest, one member of a pair of African White-rumped Swifts appeared in the doorway with an egg in its mouth. The bird threw the egg to the ground, where it was examined by Moreau (1942*b*), who found it to be without an embryo after seven days

of incubation. Once I watched a Black-cheeked Woodpecker throw from his nest hole an egg that was yolkless and seemed abnormally small. I have also seen an Acorn Woodpecker and an Olivaceous Piculet each carry an egg from its nest, but I was unable to examine them. If a woodpecker's nest is broken open or severely disturbed, the owners are likely to throw out their eggs (Truslow 1967, Skutch 1969*b*).

In addition to removing cracked eggs from their nests, Common Swifts sometimes eject fertile eggs that hatch after a human hand has replaced them. That the ejection is deliberate rather than accidental is suggested by the fact that the birds may repeatedly, and inexplicably, throw from the nest a hatchable egg that has been returned to it. Sometimes the egg is carried out in the swift's mouth (Lack and Lack 1952).

CHAPTER 17

Hatching

If you hold an egg that has long been incubated against your ear, you may hear low, rhythmic tapping within the shell. About the same time, weak peeps may be audible, proof that the chick has pierced the inner shell membrane with its bill and is now breathing air into its lungs from the air space at the blunt end of the egg, which has grown larger as the egg lost moisture during incubation. The obvious inference from these sounds is that the chick, now well formed, is pecking or hammering against its prison wall, to break its way out. The tapping, we suppose, continues until it pips the shell, producing a slight elevation or roughness with fracture lines radiating around it, the first visible sign that the egg is about to hatch. We must be wary, however, of the inference that the birdling within is pecking against its shell. Despite a good deal of painstaking research, scientists do not agree as to exactly how the chick ruptures its shell. Before considering this point, we must examine the structures that help the young bird to escape its egg.

The Egg-tooth and Hatching Muscle

Since the chick's bill is still soft, it might be injured by striking or pushing against the hard, calcareous shell if it lacked some special protection. This is provided by the egg-tooth, a whitish horny shield, often with a tiny sharp projection, that covers part of the upper mandible, frequently not at its very tip but at the subterminal point where the bill decurves sharply. In many birds, a similar but smaller shield covers the tip of the lower mandible. Egg-teeth were evidently inherited from reptilian and even earlier amphibian ancestors. In frogs of the genus *Eleutherodactylus*, which hatch in the adult form, the tip of the upper jaw bears a tiny, spinelike egg-tooth that splits open the eggshell. Similar evanescent egg-teeth are widespread among snakes and lizards; in geckos they are double. Crocodiles and tortoises have a pointed horny callosity on their snouts, instead of an egg-tooth. In some of these animals, an enzyme secreted by the embryo dissolves the shell from inside and facilitates the mechanical splitting (R. Mertens 1960). In birds, as far as I know, no such chemical action helps the chick to break its way out of the shell.

At the time of hatching, almost all birds bear an egg-tooth on their upper mandibles, although a few exceptions are known. Among the many that also have a similar shield on the tip of their lower mandibles are certain loons, rails, bustards, pigeons, shore birds, terns, auks, and hornbills. In woodpeckers, which hatch with the lower mandible conspicuously longer and broader than the upper mandible, this covering on the tip of the lower mandible

PLATE 56. Hudsonian Godwit chick escaping from its egg at 10:30 A.M. Its short, firm bill and one large foot are clearly visible. The nest, at Churchill, Manitoba, is well lined with leaves and grass (photo by J. A. Hagar).

PLATE 57. Same nest as in plate 56 at 4:15 P.M. on same day. Three chicks (one not visible) have hatched; the fourth hatched the following night. The two visible nestlings still retain their egg-teeth. Hudsonian Godwit chicks have egg-teeth on both upper and lower mandibles that drop off as soon as their bills are dry (photo by J. A. Hagar).

seems necessary to protect it. The slender bills of sandpipers, particularly in species like snipes and woodcocks that use the sensitive bill tip for probing in the ground, would seem to require special protection while breaking their shells. Accordingly, we find the ends of both mandibles, which are of about equal length, completely enclosed in protective caps (Jehl 1968b).

Egg-teeth begin to develop early in embryonic life and are fully formed at the time of hatching. After hatching, when they appear to have no further use, their fate varies widely in different groups of birds. Newly hatched sandpipers and plovers often shed their egg-teeth before they are dry, and nearly always before they are a day old. Domestic chickens and quails drop theirs after two or three days (Clark 1961). Auks and their allies retain their egg-teeth much longer, up to at least two weeks in the Razorbill and nearly a month in guillemots and puffins (Sealy 1970). In petrels and related birds the egg-tooth is also quite persistent, remaining on the Sooty Shearwater until it is two or three weeks old (Richdale 1945). Passerines, as I have seen in several tanagers and has been reported by Parkes and Clarke (1964) for other species, do not drop their egg-teeth entire as sandpipers and gallinaceous birds do but seem to absorb them slowly during the course of life in the nest; they may not completely vanish until after the young have fledged. Those of siblings in the same nest may dwindle at different rates.

Examination of the egg-teeth of a variety of birds should make us doubt that hatching chicks fracture their shells by a forward pecking movement. If this were true, the prognathous young of woodpeckers should have the larger shield on their lower mandibles, which alone would strike the shell, but actually we find it on the upper. In a bird whose upper mandible is, at hatching, as long or longer than the lower, the tooth should be at the very tip of the upper, but very often it is located slightly farther back, leaving the tip exposed. The situation in certain sandpipers, in which the hard shield embraces the tip of each mandible, appears to be somewhat exceptional. Moreover, chicks are so tightly packed in their shells that they hardly have enough free play to deliver an effective peck.

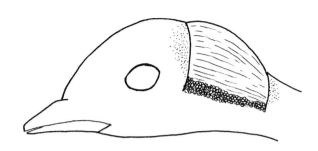

FIGURE 14. Diagrammatic lateral view of the "hatching muscle" (M. complexus) of the domestic chick at hatching (redrawn from Fisher 1958).

Megapodes differ from other birds not only in their incubating methods but also in the absence of egg-teeth on the hatching chicks. Studies of embryonic development have shown that the megapode does in fact have an egg-tooth at about the time that other gallinaceous birds hatch, that is, after about three weeks of incubation. But the megapode's egg-tooth is at best rudimentary and disappears during the several additional weeks that the chick remains within its exceptionally thin shell. The hatchling finally shatters its shell by vigorous movements of its legs and body. This vestigial egg-tooth strengthens the view that the megapodes, despite their unbirdlike methods of incubation, are not primitive birds but derive from a gallinaceous stock with more conventional breeding habits. Even embryos of certain marsupial mammals have vestigial egg-teeth, suggesting their descent from oviparous ancestors (Clark 1960, 1964).

The second structure that evidently helps the chick to break out of its shell is the *musculus complexus*, or hatching muscle, a paired muscle that extends broadly from the upper hindneck to the back of the head (fig. 14). In domestic chickens, grebes, gulls, and doubtless other birds that remain to be studied, it reaches its maximum size just before hatching, after which its volume decreases, although it persists into adult life and serves to elevate the head. Its size appears to be related to the thickness and strength of the shell; it is smaller, relative to embryo size, in gulls, and especially in grebes, whose almost constantly moist eggs have soft, chalky, easily broken shells, than in the domestic chicken, whose shells are firmer. At the time of hatching, this muscle is suffused and swollen with lymph from glands that lie beside it. Some investigators have thought that so much lymph would interfere with effective muscular action, but more probably it nourishes muscle fibers that have much work to do (Fisher 1958, 1961, 1962).

Methods of Hatching

The development and subsequent shrinkage or loss of the hatching muscle and egg-tooth run parallel courses, suggesting that they work together to release the chick from the egg. From the muscle's position, it is clear that its contraction will elevate the head but that it could not help deliver a forward peck. Neither the hatching muscle nor, in many kinds of birds, the position of the egg-tooth is compatible with the notion that the chick pecks or hammers at the shell until breaking it—even if there were space to deliver an effective blow. It is likewise doubtful that the chick by an upward twitch of the bill could strike the egg-tooth against the shell hard enough to fracture it.

The small amount of observational evidence on this matter has been gained by watching the movements of chicks of the domestic hen and Bobwhite in eggs that have been equipped with a "window." This, of course, interferes more or less with the normal behavior of the emerging chick. As Brooks and Garrett (1970) observed, just before pipping, a series of strong, convulsive movements passes through the chick's whole body, each causing it to expand once and then immediately contract. R. A. Johnson (1969), who called this movement the "surge-pip," noticed that in the Bobwhite it starts with a leg push and progresses upward through the trunk to the head, which is turned back and wedged between the thorax and the shell. Evidently the contraction involves the hatching muscle as well as others. By this movement, the bill and, especially, the egg-tooth are pressed against the shell. The tiny, sharp projection on the egg-tooth apparently catches in the shell membrane and, together with the pressure of the tumid hatching muscle against another part of the shell, serves to

hold the bill in position until the successive surges weaken and finally fracture the shell. Thus the shell is pipped or starred not by impact but by pressure intermittently applied. Finally, the chick pierces the shell with a small hole through which the tip of its bill is visible.

From this point onward, events move more rapidly and are somewhat easier to observe. I have sometimes held an egg in my hand while the imprisoned birdling completed the final stage of hatching. In the Groove-billed Ani and Buff-throated Saltator, representing the cuckoo and finch families, the chicks broke out of their shells in much the same way. When I began to watch, each had made a small gap through which the end of the bill protruded. At intervals the little bird drew back its head, which was turned sideward beneath a wing, then pushed it forward again. With each outward thrust, the culmen or ridge of the upper mandible pressed against one end of the perforation and often broke off a fragment of shell. Each rhythmic movement of the chick's head may well have been the visible portion of a surge-pip through the whole body. As it continued, the chick rotated slowly in the egg, propelling itself in a manner that I could not discover because so little of its body was visible to me. The direction of this rotation was such that the head moved backward, and the culmen was constantly applied to a fresh part of the shell, breaking away more of it and lengthening the ragged gap. When this opening extended from one-half to two-thirds around the egg, the struggles of the chick cracked the unbroken part of the shell in the same line, so that the blunt end fell off like a cap. Then the chick squirmed out into the palm of my hand, where it lay exhausted by its vigorous effort. After I began to watch, a saltator that had made a gap in its shell one-eighth inch in length, completed its emergence in only ten minutes.

Other birds vary this procedure. A Ruddy Crake that I also held in my hand while it hatched operated much like the saltator. However, instead of pressing its culmen against the edge of the lengthening gap in the shell, it used the point, equipped with a small egg-tooth, and the left side of the bill to chip off fragments. After the linear gap had been extended almost halfway around the larger end of the egg, the struggling little rail pushed off the cap and escaped about one-quarter hour after I found it with a small perforation in the shell. Instead of making a continuous cut like the foregoing birds, a Black Duck that C. S. Allen (1893) watched hatch in his hand punctured its shell at intervals of about one-eighth inch. Rotating in the egg, it encircled the larger end with twenty-five or thirty such perforations. It started around the second time, but, before this second revolution was half-completed, it tore loose the cap of the shell so that it could be raised like the lid of a box, an inch of unsevered shell membrane acting as the hinge. Then the chick pushed out, head, shoulders, first one wing and then the other, until it lay gasping with widely opened bill, its feathers plastered against its wet, slimy, nearly bare skin.

The method of emergence of the kingfishers and motmots that I have watched while they hatched is somewhat peculiar. Instead of pressing in the same place until it perforates the shell, or making a series of fractures in the same transverse belt, the chick moves its head about and cracks the shell in a number of spots scattered over an entire quadrant between the greatest circumference and the blunter pole. The cap that it finally pushes off is markedly asymmetric and is separated from the body of the shell by an oblique line, rather than the transverse line characteristic of passerines and many other birds.

In the Herring Gull, Tinbergen (1953) gathered evidence that the parents, who frequently turned their eggs through most of the incubation period, ceased to do so after pipping began. Probably the peeps of an imprisoned chick apprised them that it was about to hatch and prompted them to treat the hatching egg like a nestling, which is never deliberately turned over in the nest. Usually the first perforation made by a chick was at the top of the egg; far less often it was at the bottom; and even less frequently it was in some other position. This indicates that a hatching chick responds to gravity: When its bill points either upward or downward, its head is in the neutral position, with the gravity receptors in the two ears equally stimulated. In most of the pipped eggs, a line of cracks around the

blunt end led up to the hole, revealing how the little gull had turned within the shell until it was in the correct position for breaking through. As an American Flamingo's egg pipped, the parent stood over it and moved it with its bill until the opening was uppermost, thus giving the hatching chick access to the air (Bent 1926).

Although the American Woodcock has an egg-tooth at the end of each mandible and uses these to perforate the shell, afterward it proceeds in a manner different from that of most other birds. With the tip of its bill it seizes the rim of the rather large hole and pushes until its spinal ridge splits the egg longitudinally; the shell seems to rip, rather than to fracture. Willets hatch in a somewhat similar manner (Wetherbee 1962).

The young Ostrich breaks out of its huge egg in an astonishing fashion. Even before it pips the shell, the chick utters various social calls, most of which are pleasantly musical. With its strongly developed hatching muscle and bill ensheathed in a terminal egg-tooth, it breaks a small, irregular "window" in the shell's side, revealing its beak and one foot. By stretching its body and kicking downward with this foot, it often shatters the shell into many pieces. The thick shell breaks with a loud report that may frighten the incubating parent, who leaves the nest with every sign of alarm, although the chick's earlier calls should have prepared it for the event. The young Ostrich's efforts to hatch may drag on for hours, sometimes for two days, with intervals of repose. When finally it frees its head from the shell and stretches out its neck, it celebrates its release with a "call of triumph" (Sauer and Sauer 1966).

While an egg hatches beneath it, the parent bird often sits restlessly, perhaps, if it is inexperienced, because it is puzzled by the movements and increasing roughness of the object that for so long lay smooth and inert in contact with the bare skin of its incubation patch. At times the parent rises up to look or poke beneath itself, and it may even perch on the nest's rim to watch what is happening. It may answer the chick's peeps with low, encouraging notes. A female Gray-cheeked Thrush sang repeatedly while one of her eggs was hatching—evidently an emotional response to the event (Wallace

TABLE 9
Time Required to Escape from Shell

Species	Incubation Period in Days	Interval from First Fracture to Escape in Days or Hours	Reference
Ostrich	42	up to 2 d.	Sauer & Sauer 1966
Pied-billed Grebe	23–24	3.5 hr.	Deusing 1939
Adelie Penguin	30–39	<1–2+ d.	R. H. Taylor 1962
Laysan Albatross	64	2–5.5 d.	Rice & Kenyon 1962
Sooty Shearwater	ca. 56	4 d.	Richdale 1945
Diving Petrel	—	3.5 d.	Richdale 1945
Yellow-billed Tropicbird	40–42	1.5–2 d.	Stonehouse 1962
Torrent Duck	43–44	3–7 d.	Moffett 1970
Redhead Duck	23–24	16–18 hr.	Low 1945
Sora Rail	16–19	1–2 d.	Walkinshaw 1940
American Coot	23–25	12–76 hr.	Gullion 1954
Old World Golden Plover	27–28	2–6 d.	Williamson 1948
Sanderling	24–32	2.5–5 d.	Parmelee 1970
American Avocet	23–27	4–5 d.	Gibson 1971
Black Noddy Tern	34–36	3–4 d.	Ashmole 1962
Herring Gull	28–33	1–3 d.	Tinbergen 1953
Least Auklet	—	2–7 d.	Sealy 1970
Blue-throated Green Motmot	21–22	1–2+ d.	Skutch MS
Gray-capped Flycatcher	16–17	24–41 hr.	Skutch 1960
Ovenbird	12–14	15–20 hr.	Hann 1937
Scarlet-rumped Black Tanager	12–13	18–29 hr.	Skutch 1954
Silver-throated Tanager	13–14	12–16 hr.	Skutch MS

in Bent 1949). A male Black-hooded Antshrike refused to relinquish the nest to his mate while the first of their two eggs was hatching, although it is customary for antbirds to leave rather promptly when the other partner arrives to take charge of the nest.

Exceptionally, a parent helps a chick emerge from the shell by picking at the edge of the gap that the birdling has made. This assistance, which has been reported of birds so diverse as the Gray-cheeked Thrush, Wood Thrush, European Blackbird, Gray Catbird, Clark's Nutcracker, Prairie Chicken, Greater Flamingo, Gannet, murres, and cranes, is most likely to be given if the chick is weak and takes unusually long to break out of the egg. The parent Ostrich sometimes cracks the big shell with its breastbone and seizes the chick by the head to pull it out (Kendeigh 1952).

The time that the chick takes to break out of its shell varies greatly from species to species and even among the eggs in the same nest, but in general it is longer in birds with long incubation periods than in those with short periods. Many songbirds have hatched by the day after that on which the first minute fracture can be detected on the shell, perhaps by feeling with one's finger tips or lips even before it is readily visible. American flycatchers tend to take a little longer, from twenty-four to forty hours instead of fifteen to thirty. For birds larger than most passerines, two to four days frequently elapse between the first starring or cracking of the shell and final emergence. In the Least Auklet, the interval ranged from two to seven days (Sealy 1970). Torrent Ducks, with an incubation period exceptionally long for their family, took a week to escape, but Redheads chipped out of their shells in sixteen to eighteen hours (table 9).

Most of the hatching interval is taken up with

cracking the shell and making the first perforation, after which the chick has an easier time and works faster, as seems necessary, for now it is in a peculiarly vulnerable state. Ants, which seem powerless to harm an intact egg, may enter through the gap. The parent is then unable to remove them, as many birds do from their nests, and they may start to eat the chick, finally killing it, a tragedy that occasionally happens in the ant-infested tropics when hatching is delayed. A different peril confronts hatching grebes in floating nests: water may enter the perforation and drown them. Accordingly, they do not pip their eggs until shortly before they hatch, although two days before this they may begin to call loudly, warning the parents not to desert them. A Pied-billed Grebe escaped the shell only three and a half hours after it made the first fracture (Deusing 1939). In contrast to this, chicks of the Wideawake Tern may make a hole in the shell large enough for the bill to protrude as much as two days before they escape. On dry ground, they are in little danger of drowning, and apparently ants are not abundant in their colonies (Nice 1962).

Fisher (1961) noticed that eggs of chickens and ducks that do not hatch within twenty-four hours of becoming pipped usually fail to hatch at all. But, in other birds, successful hatching may be abnormally prolonged. Once I studied a nest of a Buff-throated Saltator who was frequently driven from her egg by a rival of her own species nesting nearby. The persecuted saltator's egg did not hatch until the third day after I noticed a rough spot on the shell. By this delay, the incubation period was prolonged to sixteen days instead of the usual thirteen.

Disposal of Empty Shells

Songbirds usually remove the shell as soon as the nestling has wriggled out. It may be either carried away or eaten for its calcium content. American flycatchers often take a little longer to remove the shell. Even hummingbirds frequently carry the shell away, although their bills seem ill fitted for this. Once I watched a Band-tailed Barbthroat grasp the larger part of a recently vacated shell in the tip of her long, slender bill, fly with it for a foot or so,

drop it, catch it in midair, carry it a foot or two farther, all on a descending course, then drop the white object to the ground. In general, altricial birds that keep their nests clean remove the empty shells; of those that neglect sanitation, some do and some do not. Although Ruddy Ground-Doves and Blue Ground-Doves permit their shallow, open nests to become fouled with droppings, they carry off the shells. Broad-billed and Blue-throated Green motmots also remove the empty shells, although they do not otherwise clean their burrows. Trogons and jacamars permit the shells, along with other waste, to litter the bottom of their closed nests.

In nidifugous birds, whose young abandon the nest soon after they hatch, shell removal seems less necessary, yet many species practice it. Among those that carry off the shells before the chicks leave the nest are certain grebes, rails, gallinules, terns, plovers, sandpipers, sandgrouse, and coursers. A Long-billed Curlew dropped the shells 360 feet (110 meters) away (Graul 1971). So strong is the Stilt Sandpiper's urge to remove shells from near its nest that it is easily caught in a trap baited with an empty shell (Parmelee et al. 1968). Among semialtricial ground nesters, several species of goatsuckers—including the Pauraque and Common Nighthawk—carry away empty shells (Bent 1940, Rust 1947). On the other hand, gallinaceous birds and ducks tend to leave the shells in their nests, or at most to push them over the edge onto the ground or into the water. For birds that lay more than one egg, the prompt removal of shells would be advantageous, because occasionally an unhatched egg slips into an empty shell in such a fashion that the occupant cannot escape through the double wall—a mishap that is most likely to occur when the brood is large.

The Hour of Hatching

Birds may emerge from their eggs at any hour of the day or during the night, presumably also at any hour. In some species, however, there is a marked tendency for hatching to occur during the night and forenoon, whereas in the afternoon far fewer eggs hatch than one would expect if the distribution of this event were random throughout the twenty-four hours. Of 119 eggs of the Yellow Warbler, only 9 hatched in the afternoon, the other 110 in the night or early morning (Schrantz 1943). Of 38 eggs of the Red-winged Starling of South Africa, 30 hatched between dawn and noon, only 2 in the afternoon, and the rest at night (M. K. Rowan 1955). Of 33 eggs of the Wideawake, 1 hatched between midnight and 6:00 A.M., 21 between 6:00 A.M. and noon, 11 between noon and 6:00 P.M., and none between this hour and midnight (Ashmole 1963a). Eggs of captive Mourning Doves usually hatched in the early morning or between 10:00 A.M. and noon (Nice 1922–1923). Of 26 hatchings of two closely related species of American flycatchers, the Gray-capped and Vermilion-crowned, 14 occurred in the forenoon, 11 at night, but only 1 in the afternoon. Of 93 eggs of eleven species of Costa Rican finches, tanagers, and wood warblers, 51 hatched in the forenoon, 27 during the night, but only 15 in the afternoon. If the hour of hatching were random, we would expect 23 eggs to hatch in the afternoon, the same number in the forenoon, and 46 during the approximately twelve hours of night (Skutch 1952a).

The simplest explanation of these facts is that the movements of the imprisoned chick breaking out of the shell follow a diurnal rhythm, just as many other activities of birds occur predominantly at certain hours rather than uniformly throughout the day. Apparently, in those birds in which hatching occurs chiefly in the night or forenoon, the chick works most actively to break its shell toward the end of the night and in the early morning, but rests later in the day and during the first hours of night. Perhaps, even before they hatch, diurnal birds form the habit of sleeping by night, when they are kept continuously warm in the egg, and becoming active by day, when in many species incubation is intermittent. The prevalence of hatching by day in Wideawakes is more puzzling, because these terns are active by night as well as by day. However, if the eggs of any species were regularly laid early in the morning, as is true of many birds, and the incubation period were a whole number of days, plus or minus no more than six hours,

the eggs would hatch during the second half of the night or in the forenoon, if the activities of the chicks within them were governed solely by their state of development rather than a diurnal periodicity.

Synchronous Hatching

One of the most exciting ornithological discoveries of recent years is that the social interactions of birds begin while they are still enclosed in their shells and they may influence each other's time of hatching. This has been demonstrated in the Bobwhite and Old World Quail, which, after they begin to breathe with their lungs, make a wide variety of sounds and vibrations. Whether being hatched beneath a parent or in an incubator, chicks in eggs that lie in contact with each other communicate by these signals. Apparently the most important is a rhythmic clicking, which is associated with the chick's respiratory movements and usually begins about twenty-four hours or less before it hatches (Driver 1967, Forsythe 1971). Clicks produced artificially at rates of one and one-half to about sixty times per second accelerate clicking and hatching in isolated eggs, but at rates between ninety and four hundred times per second retard emergence from the shell. By causing some chicks to work faster and others to slow down, these mutual interactions promote synchronous hatching of all the eggs in a nest (Vince 1969).

By placing side by side in an incubator pairs of Bobwhite's eggs, one of which had started incubation forty-eight hours after the other, R. A. Johnson (1969) caused them to hatch at nearer the same time. The eggs incubated first hatched in an average interval of 558.8 hours; those started later hatched after 528.8 hours and almost caught up with the former. Some of the earlier eggs were pipped as much as 33 hours before they hatched, but some started late hatched only 4 hours after they were pipped. The surge-pip movements that finally broke open a shell seemed to be particularly infectious by jarring adjoining eggs. When one chick begins this activity, its companions do the same and escape from the egg 20 or 30 minutes later.

Probably, future work will demonstrate that similar systems promote simultaneous hatching in ducks and other precocial birds with large sets of eggs. For such birds, the emergence of all the brood within a few hours or, at most, a day is of the utmost importance, since those that hatch early need food and often the parent leads them away, leaving retarded chicks to die in the nest. It is doubtful, however, that nidicolous embryos stimulate each other to hatch. With them, hatching is usually at rather long intervals if the parent begins to incubate while still laying, but more nearly simultaneous if she waits until her set is complete.

CHAPTER 18

The Newly Hatched Bird

After their eggs hatch, birds enter the most strenuous phase of their breeding activities. For a number of days they had little to do except sit passively warming their eggs and satisfy their own needs. Now the parents must still devote much time to sitting, for the newly hatched young require warming, but in addition they must find food for their offspring as well as for themselves, protect the young from enemies as far as they are able, and, in those species with the most perfect parental instincts, keep the nest clean. Scarcely any other animals work so hard for their young, or attend them in so many ways, or make such obvious sacrifices for them, as birds do.

Instead of feeding her offspring from her own mouth, in the manner of many kinds of birds, the mammalian mother continues to eat as usual, often increasing her consumption of food, then passively stands or lies while her young suck milk from her teats. Only a few kinds of mammals, especially carnivores like foxes and coyotes, bring food for their older offspring to eat. Although some kinds of mammals impart a degree of warmth to their young by huddling with them in the nest or den, mammalian parents seldom shield their offspring with their own bodies from cold air, beating rain, or burning sunshine. Some mammals clean their young by licking

up their wastes, but none gives such painstaking attention to sanitation as many birds do. Aside from birds, the method of feeding young from the attendant's mouth is common only in the social insects, including bees, wasps, ants, and termites. Some of these insects also give attention to the sanitation of the cells that contain their larvae. Being cold-blooded, they cannot warm their offspring, but they may cool them by fanning air over or through the nest with their wings, many individuals acting in unison. And, as everyone who has deliberately or accidentally molested a hive of bees or wasps or a nest of ants knows to his cost, many kinds defend their families with spirit. In the extent of their parental care, these social insects offer the closest parallel to birds that we find in the animal kingdom, apart from man.

The activities of parents of very young birds depend in large measure on the stage of development at which the latter emerge from their shells. Before proceeding to follow these activities, we must, therefore, briefly survey the different types of hatchlings. According to their condition when they hatch, birds may be classified as:

1. *Altricials*: helpless at birth, unable to leave the nest for a week or sometimes for months, during which they are wholly dependent upon their par-

ents for food and protection (all passerines, woodpeckers, hummingbirds, swifts, trogons, kingfishers, pigeons, parrots, etc.)

2. *Semialtricials*: able to leave the nest within hours, or at most a day or two, of birth but remain on or near it, where they are fed and brooded by their parents (gulls, terns, skimmers, goatsuckers, albatrosses, etc.)

3. *Subprecocials*: able to leave the nest as soon as they hatch, or at most within a day or two, follow their foraging parents and are fed from their bills (grebes, rails, cranes, guans, some pheasants, etc.)

4. *Precocials*: able to leave the nest, follow their parents, and pick up their own food soon after they hatch:

   a. Parents help the young to find food (Ostrich, pheasants, quails, etc.)

   b. Young find their own food (ducks, shorebirds, kiwis, etc.)

5. *Superprecocials*: wholly independent of parents from the moment they hatch, or at least as soon as they dry (megapodes and the Black-headed Duck, the only known examples)

These five classes of young birds are arranged in the order of their increasing maturity at hatching and decreasing dependence on their parents. The precocial habit is appropriate for birds that feed and nest on the ground or inland waters in continental areas where predators abound, and especially for those whose heavy flight makes it impracticable for them to alight on or close beside their nests. Such birds would wear a revealing path to their nests, and bringing food on foot would consume too much time. It is more expedient for their young to hatch able to follow their parents. Precocial birds that nest in trees, including certain ducks and pheasants, are generally closely related to ground nesters and have not become fully adapted to arboreal life. The precocial habit is well fitted to the needs of birds that breed on the Arctic tundra, the seashore, the plain, and other places with few woody plants.

The altricial habit evidently offers considerable advantages, for it is followed by the great majority of the birds of woodland and thicket, as well as by those that nest in the safety of small islands. At a stage of development when precocial birds are still

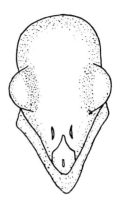

FIGURE 15. Head of two-day-old Golden-fronted Woodpecker viewed from above, showing the bulging, tightly closed eyes, the broad lower mandible with projecting corners, and the shorter, narrower upper mandible with slitlike nostrils and the prominent white egg-tooth at its tip (from Skutch 1969).

dependent upon the nourishment stored in their eggs, altricial birds hatch and can receive food from both parents, which should more than compensate for the lack of the larger reserves present in the relatively bigger eggs of precocial birds. Hence altricial birds should attain the power of flight, and the safety it gives, at an earlier age than precocials, and many do. But, as we saw in the case of incubation periods, the rate of development depends so much upon the safety of the nest, as well as the size of the birds, that valid comparisons are difficult to make. Some large altricials and semialtricials raised in secure retreats, remain flightless longer than any precocials except those that have completely lost this power. When they leave the nest, the most advanced altricials most resemble their parents; precocial chicks resemble them least.

Altricial or Nidicolous Nestlings

Altricial nestlings, so called because they are tender and need nursing, escape from their eggs in a more or less developed state, but always helpless. Far less attractive than soft, downy, bright-eyed precocial chicks, they often look like embryos prematurely escaped from their shells. This is particularly true of the grublike newly hatched hummingbirds and the underdeveloped hatchlings of woodpeckers (fig. 15) and their kin. Their eyes are tightly closed, their swollen abdomens grotesquely disproportionate to the rest of their bodies, their skin often imperfectly clothed or quite naked.

Sometimes an altricial nestling lifts up its head in a silent plea for food as soon as it has wriggled out of its shell. The widely gaping mouth that it then reveals often has a bright interior, the color of which is fairly uniform throughout a family: red in finches, tanagers, honeycreepers, and icterids; yellow or orange in vireos, thrushes, wrens, nuthatches, swallows, American flycatchers, and many other families (Ficken 1965). In wood warblers, the lining of the mouth may be either yellow, red, or yellow marginally with red or pink in the center.

In a few birds, the mouth cavity and tongue are marked with a pattern that contrasts with the ground color, making them more strikingly conspicuous. In a number of small African grassfinches, as in the viduine whydahs that foist their eggs upon them, the interior of the nestling's mouth is pinkish, with several darker spots, often black in color, forming a symmetrical pattern on the roof, and sometimes also on the tongue. In addition, there may be bars and crescents, especially on the floor of the mouth. The only New World passerines in which I have noticed similar, but simpler, patterns are the Donacobius, a South American member of the mockingbird family, which has the interior of the mouth dark flesh color with three conspicuous black dots on the tongue, and the Long-billed Gnatwren, which has a yellow mouth with two black spots on the tongue, as has the related Blue-gray Gnatcatcher, according to Root (1969). Striking buccal patterns are also found in the cuckoo family.

The mouths of certain nestling passerines, including thrushes, tanagers, and doubtless others, are equipped with inwardly directed projections or bristles that help keep the food moving in the proper direction. A nestling Gray's Thrush has long, prominent bristles at the back of the cleft palate and on the rear margin of the horny tongue; on the floor of the mouth these are shorter.

As though the nestling's gaping mouth were not already sufficiently conspicuous, its apparent size is increased by the presence of flanges, or projecting

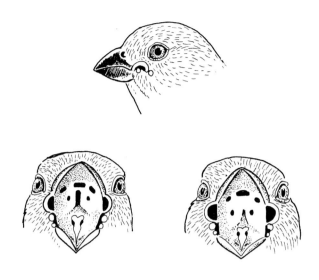

FIGURE 16. Heads of nestlings of the parasitic Pin-tailed Widow-Bird and the Barred Waxbill, its chief host, showing the prominent marks on the roof of the mouth and the tongue and the glistening "reflection tubercles" at the corners of the mouth. The markings of the parasite closely resemble those of its host (redrawn from Friedmann 1960, after a field sketch by J. P. Chapin).

folds of skin, at the corners (Clark 1969). These are usually orange, yellow, or whitish, and if the cavity is red they contrast strikingly with it. In recently hatched woodpeckers, the lower mandible, longer and broader than the upper, bears at the base very prominent projecting corners; in the larger species, these become great, swollen, white knobs. Nestling grassfinches and viduine whydahs have, in addition to the elaborate mouth patterns, knoblike "reflection tubercles" at the corners of their mouths (fig. 16). These remarkable bodies have a mirrorlike structure that is extremely efficient in gathering and reflecting light. Even in very dim illumination they glow so strongly that early observers conjectured that they were luminescent organs, somewhat analogous to those of fireflies; but this was disproved by placing them in total darkness, where they were invisible (Friedmann 1960).

The brightly colored mouth cavities of many altricial birds, often bordered with flanges or knobs, are believed to guide the parents to the spots where they must place the food they bring. In open nests with good illumination, such guidance seems somewhat superfluous, especially when one remembers that the eyesight of birds is excellent. Possibly they not only direct but stimulate the hard-working parents. When a nestful of drowsing finches or tanagers suddenly stretch up at a parent's approach and gape like so many suddenly opening red blossoms on swaying stalks, the effect is strikingly beautiful, and it may well increase the parental zeal of creatures so sensitive to color as birds seem to be.

In covered nests, such as grassfinches build, and even more in the deep holes of woodpeckers, the parents seem in greater need of visual aid for finding their nestlings' mouths, often while their own bodies block the doorway and cut off most of the light; this probably accounts for the reflection tubercles of the former species and the grotesque white knobs of the latter. These swellings at the corners of woodpeckers' mouths are sensitive and the parents of very young nestlings touch them with their bills to stimulate the tiny naked woodpeckers to rise up for their meals (Sielmann 1958).

Passerines that nest in dark holes tend to have more conspicuous oral flanges than those that nest in the open, doubtless because the parents have greater need of conspicuous marks to guide them to their nestlings' mouths. When I peered into the deep hollow stub in which Black-faced Antthrushes were nesting, the enormous flanges at the corners of the nestlings' mouths resembled great white eyes staring up at me from the gloom. But some birds that nest in dark chambers, including kingfishers and motmots, lack such guides to the nestlings' mouths. My experience in feeding a young motmot led me to believe that these nestlings simply grasp more or less blindly in the direction of the parent's bill until they seize the food. Such a boisterous method of taking meals contrasts strongly with that followed by passerines and other nestlings with conspicuously colored mouths that simply gape passively until the parent pushes in the food.

Nestlings that grow up in unlined chambers have special pads on their heels or tarso-metatarsal joints that, together with their swollen abdomens, bear the weight of the young birds. The pad, which protects each heel from abrasion as a nestling shuffles

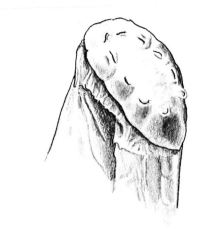

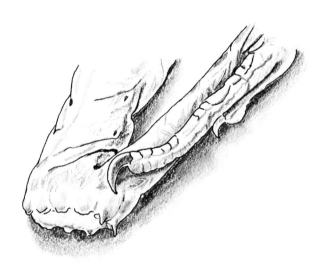

FIGURE 17. Heel pad of nestling Rainbow-billed or Keel-billed Toucan, viewed from below (*above*) and from the side (redrawn from Van Tyne 1929).

about on the hard nest floor, is a cushion of thickened skin broader than the leg (fig. 17). In some nestlings, such as those of toucans, it is grotesquely large in comparison with the tiny foot, which seems to be a mere appendage of the callosity. Indeed, the toes are less useful to the nestling until it is feathered and climbs up to look through its doorway. The surface of the pad that makes contact with the substratum varies with species: it is smooth in jacamars (plate 58), motmots, and nunbirds, which rest on

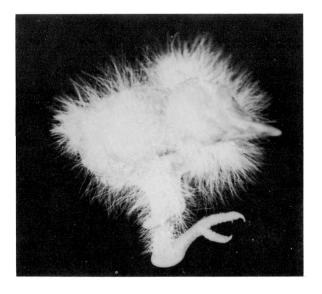

PLATE 58. Altricial nestling of the Rufous-tailed Jacamar, two days old. A thickened pad of skin prevents abrasion of its "heel" on the burrow's unlined floor.

earthen floors of burrows; roughened with low tubercles or papillae in burrow-nesting American kingfishers; and studded with spikes in trogons, toucans, barbets, and woodpeckers, which grow up in holes in trees.

Precocial chicks are covered at birth with ample soft down. The covering of altricial nestlings is more variable. Many that hatch in holes and burrows are quite naked, as is true of woodpeckers, toucans, hornbills, puffbirds, barbets, motmots, kingfishers, some parrots, and many others. Strangely enough, the Rufous-tailed Jacamar hatches with rather copious down, although it seems to need this thin, loose garment no more than do kingfishers, motmots, and puffbirds, which likewise grow up in subterranean burrows. Pigeons and doves and certain cuckoos at birth bear short, sparse bristles, usually yellowish or buff in color, rather than the tufted down that is more characteristic of altricial nestlings.

Most of the newly hatched passerine nestlings that I have seen bore scattered, plumose tufts of down, gray or whitish in color, especially on the top of the head and along the center of the back. Other passerines, however, are born as naked as woodpeckers and kingfishers. Even in the same passerine family, some species have natal down whereas others

lack it, as is true of vireos, Old World warblers, crows and jays, and American flycatchers. Indeed, this difference is found even in a single genus, as in vireos. It is difficult to correlate the presence or absence of natal down with the form of the nest, the environment, or other pertinent circumstances. This is what we should expect if down were of little or no importance to nestlings, for in this case its occurrence would not be controlled by natural selection but would be subject to random genetic variation. It is difficult to imagine how natal down so thin and sparse as that of many passerine nestlings could be of much service in keeping them warm, protecting them from insect bites, or making them more difficult to detect by breaking their outlines, as Ingram (1966) supposed.

Exceptional among passerines is the Bearded Bellbird, which after drying is covered with a thick coat of pale grayish white, hairlike down. Apparently this extraordinary vestment serves as a disguise; when disturbed, even at the age of one day, the nestling rolls up into a ball, with eyes, bill, and feet hidden beneath the down, so that it looks more like a small mammal, or a big hairy caterpillar, than a nestling (plate 60). In warm Trinidad, it is, from the first, brooded little (B. K. Snow 1970). Among nonpasserine altricial nestlings, a rather similar coat of down completely covers the newly hatched Common Potoo, which during the first fortnight of its life is so continuously brooded by its parents that it seems to have little need of this garment (Skutch 1970).

Some altricial nestlings hatched with little or no down acquire much of it early in life. Boobies, born nearly or quite naked, become well covered with soft, white down, which starts to grow out during their second week, when parental brooding diminishes and they are often exposed to strong sunshine or to rain. This downy garment is gradually supplanted by vaned feathers of adult type (plate 61). Likewise pelicans, cormorants, frigatebirds, and anhingas hatch naked and acquire their downy coverings early in their prolonged nestling lives; but tropicbirds, which also belong to the order Pelecaniformes, leave their eggs well clothed in soft, abundant down.

Black Swifts and Chestnut-collared Swifts, also

PLATE 59. Altricial nestlings of the Ruddy Ground-Dove in pinfeathers. Their slight nest rests on a bunch of bananas in a Panamanian plantation.

hatched naked, are thickly covered with gray woolly down by the time they are two weeks old. Since these swifts often nest on cliffs in cool, damp caves and ravines, even behind a curtain of falling water, this garment, lacking in most other nestling swifts, serves them well (C. T. Collins 1963, D. W. Snow 1962*b*). Other swifts that acquire a thick coat of postnatal down are African Palm-Swifts and crested swifts, whose nests give less protection than those of most other members of this family. Hatched quite naked in a deep burrow high in the Guatemalan mountains where nights are chilly, the Blue-throated Green Motmot is soon enveloped in long, soft down that springs from the feather tracts, so that when ten days old it has become a fluffy ball (plate 62). Motmots of warm lowlands do not develop downy cloaks (plate 63) but remain naked until they put on plumage resembling that of their parents (Skutch 1945). In addition to their natal down, which is confined to the feather tracts, Yellow-bellied Elaenias and Long-tailed Silky-Flycatchers acquire early in life fairly abundant postnatal or secondary down on the originally bare skin between these tracts. It consists of short, dense whitish tufts and assimilates these nestlings to their shallow, light-colored nests to a degree that can hardly be claimed for the longer, sparser natal down of most other passerine nestlings. Moreover, it helps to keep the nestlings warm, especially in the case of the silky-flycatcher that lives at high altitudes.

We have already noticed that, in newly hatched woodpeckers and the related barbets and toucans, the lower mandible is both longer and broader than the upper. In hornbills, jacamars, and kingfishers, the lower mandible is at least longer than the upper. The difference in length may amount to as much as ¹⁄₁₂ inch (2 mm) in the Amazon Kingfisher. It is not evident whether this bizarre, temporary condition is of utility to the young nestling; possibly, the lower mandible becomes longer simply because it has more room to grow inside the shell. In the kingfishers and toucans for which I have information, the upper mandible catches up to the lower in length from ten days to two weeks after the nestling hatches, but in the Golden-fronted Woodpecker the mandibles become equal in about eight days. Skimmers, which when adult have the lower mandible

PLATE 60. Bearded Bellbird, four days old, an exceptionally downy altricial nestling (photo by B. K. Snow).

much longer than the upper, hatch with the two mandibles of equal length.

Nice (1962) designated as semialtricial all those birds that when newly hatched are well covered with down but unable to leave the nest, including tropicbirds, herons, hawks, owls, turacos, and others. Some, like herons and hawks, are born with open eyes, while in owls the eyes are closed. According to this classification, we should have to include among the semialtricials the Bearded Bellbird, which is born with a dense downy coat but remains in the nest for thirty-three days, one of the longest periods on record for a passerine. According to present information, it would be the only passerine to fall into this category. To avoid this and other difficulties that arise from trying to establish too many categories, I have included among the altricials all birds that remain and are fed in the nest for at least the

PLATE 61. Young Brown Booby. Hatched naked, it grew a coat of dense white down that is being replaced by dusky feathers.

first week of life, and I have applied the term semialtricial to the following division.

Semialtricial Chicks

In general, semialtricials are the offspring of birds that lay their eggs on the ground, often in scrapes or hollows that they dig, rarely in well-constructed nests. To compensate for the hazards of life on open ground, they hatch in a less helpless state than most altricial nestlings, with an ample cloak of down, open eyes, and the ability to walk or at least hop away from their natal spot when a day old, if not sooner. Their precocity is revealed even before they escape their shells: in some, a warning cry from a parent who detects danger causes them to silence the peeps by which they announce their impending emergence. But for many days, food is brought to them by their parents, whom they cannot follow on foraging excursions.

Upon hearing the alarm call, Herring Gulls less than a day old leave the nest and crouch nearby. As days pass, their excursions from the nest become

PLATE 63. Altricial nestling of the Turquoise-browed Motmot, twelve days old, with feathers just beginning to emerge from the ends of the horny sheaths (cf. plate 62).

longer. Young skuas are equally ambulatory. On the second day after they hatch, Least Terns begin to wander from the nest and rarely return. Their parents, who often seemed distressed by their errant behavior, make a scrape in the sand and brood them where they find them (Hardy 1957). When disturbed, day-old Black Terns jump into the water and swim a few feet into the surrounding vegetation, to climb back into the nest when all is quiet. They keep their nest clean by swimming out a short distance to defecate (Cuthbert 1954).

Although most gulls and terns nest on the ground and are semialtricial, a few instructive exceptions are fully altricial. These include the Kittiwake, a marine gull that nests on tiny ledges on the faces of forbidding cliffs, and the Fairy Tern and Black Noddy, which choose similar sites on sea cliffs but may also nest in trees. Instead of wandering long before they have grown flight feathers, the chicks of these three species cling to their nests until they can fly well. Apparently this is not merely because the narrow ledges or branches on which they hatch make wandering perilous; their departure from the ground-nesting habit of typical gulls and terns has caused many subtle changes in their behavior, including their reactions to danger, which have been analyzed in detail by E. Cullen (1957), Dorward

PLATE 62. Altricial nestling of the Blue-throated Green Motmot, eleven days old. Hatched quite naked, in a burrow high in the Guatemalan mountains, it has grown a thick coat of down that is lacking in the slightly older Turquoise-browed Motmot of warm tropical lowlands.

PLATE 64. Semialtricial chick of the American Flamingo, caressed by its parent (photo from William Conway, New York Zoological Society).

(1963), and Cullen and Ashmole (1963). Kittiwakes
and Black Noddies build more substantial nests than
most gulls and terns. When threatened, the chicks
of both species, and likewise those of the Fairy
Tern, crouch rather than flee, which would be fatal
on their high cliffs. Young Fairy Terns have such
short legs that they could hardly run. Brown Noddies, which build bulky structures in trees and
bushes but in some localities nest upon the ground,
are less altricial than Black Noddies and Fairy
Terns; when frightened, the young leave the nest
even before they can fly.

African Skimmers, hatched on sandy shores that
may heat up to more than 140°F (60°C) on sunny
days, are at a very early age led by their parents to
the water's edge, where they spend the warmest
hours lying in the soft, moist sand, with a parent
guarding them. In the evening, when the water becomes rough and waves break on the shore, they
move inland toward the grass. They follow the parents over the beach, bathe and drink in the surf, and
scratch shallow resting places in the sand. When
alarmed, they stretch out motionless on the beach,
not stirring even if touched (Modha and Coe 1969).

Chicks of the much bigger Lesser Flamingo are
somewhat less precocious than those of gulls, terns,
and skimmers. For twenty-four hours after hatching,
they cannot stand. On the second day they can
stand, but not until the third day will they leave the
nest if alarmed. Even then they cannot climb up the
steep side of the mud nest to return but are brooded
by their parents beside it. At the age of five days,
the flamingo chicks easily leave and regain the nest.
When eight days old, they abandon their nests and
gather in vast hordes, numbering up to 300,000
young, in which they live, still fed by their parents,
until at the age of two or three months they can fly
(L. H. Brown and Root 1971).

Despite their exceptionally long period of flightless dependence on their parents, albatrosses are
semialtricial. It was earlier (p. 121) mentioned that
when only a few days old they build their own
scrapes beside that on which the parent is sitting to
guard them. Long before they can fly, albatrosses
wander rather widely over the beach.

Although various species of goatsuckers have
been credited with carrying their young, this is

PLATE 65. Semialtricial nestlings of the Pauraque, one
and two days old.

doubtful and certainly not the usual method of
changing the position of their broods. On the contrary, when the parent wishes to feed or warm its
young somewhere else, or to remove them from
danger, it settles in the preferred spot and utters a
special call, which draws them with impelling
power. I have known one- or two-day-old Pauraques
(plate 65) who progressed by hopping over the leaf-
strewn ground, to surmount apparently insuperable
obstacles, such as a fallen banana plant with huge,
slippery leaves (Skutch 1972).

It is difficult to draw a sharp line between the
several categories of young birds. Although kingfishers and motmots are hatched sightless and perfectly naked, if one of their long burrows is opened
at the rear when the chicks are only a day or two
old, they toddle forward into the entrance tunnel to
avoid the light. If the nest proper is the enlarged
chamber at the inner end of the burrow where they
hatched, they have already left it, just as a gull or a
tern of the same age leaves the nest; but their short
peregrinations are confined to the burrow, which
they will not abandon until they are fully feathered
and can fly well, four or five weeks later. From a
tender age, burrow-nesting puffbirds are fed at the
mouth of the tunnel, for after the eggs hatch the

PLATE 66. Two subprecocial chicks of the Red-fronted Coot (*left*) share a nest with a superprecocial duckling of the parasitic Black-headed Duck in an Argentinian marsh. The shell of the coot's egg in the center has just been pierced by the chick within it (photo by M. W. Weller).

parents do not enter except to brood. Naked, sightless White-fronted Nunbirds, raised in burrows over four feet long in the forest floor, make a pedestrian round-trip journey of about eight feet to receive each meal (Skutch 1972).

When alarmed, six-day-old anis, in pinfeathers and still quite flightless, leave their nest of sticks and scramble away through the branches of the supporting tree or bush, aiding their progress by hooking their bills over the twigs. Sometimes they lose their grip and fall to the ground, where they creep amid the herbage to hide. When all is quiet they return, if they can, to be fed and brooded in the nest for several days more. Other cuckoos have similar habits. Likewise, Hoatzin nestlings, alarmed

in their nest above the margin of a South American river, drop into the water, then use the claws on their still featherless wings, as well as feet and bills, to clamber up through the branches to their nest when it seems safe to do so. If frightened, wood warblers, sparrows, and other ground-nesting passerines "explode" from their nests when hardly able to fly, and, unlike cuckoos and Hoatzins, they rarely return to them. Although all these mobile nestlings seem properly to be called altricial, they show some approach to the precocial habit.

Subprecocial Chicks

Like the semialtricials, subprecocial chicks hatch with open eyes and an ample garment of down, able to leave the nest in a very short while. But they are from the first even more mobile than the former, able to accompany their foraging parents. Although the chicks cannot yet feed themselves, the parents deliver food to them where and when it is found, instead of carrying it to them at or near the nest, perhaps from a feeding area hundreds of miles distant, as in certain marine birds. This is the chief difference between the subprecocials and the semialtricials.

Typical representatives of this category are rails, coots (plate 66), and gallinules or moorhens. Usually the downy, open-eyed chicks slip from the nest within a day of hatching, and if disturbed they do so much sooner. Alarmed chicks of the American Coot drop from the nest when only fifteen or twenty minutes old, with their natal down still clinging to their moist bodies. Usually they drown, for they still have little buoyancy and may not be able to scramble back upon the platformlike nest, as chicks do when slightly older. Although so precocious in some respects, young rails, coots, and gallinules are fed from their parents' bills, at least until they are some weeks old. The mouths of these chicks are not brightly colored on the inside as in passerine nestlings, but the bills and heads in some species are remarkably brilliant. In the newly hatched European Coot the nearly bald crown is reddish. Above the eyes, the skin is tinged with blue. There are brilliant red papillae around the eyes and at the base of the bill, which is banded with vermilion, black, and white. The chick of the Purple Gallinule has a dull red bill, banded with black and dusky blue. The newly hatched Moorhen's bill is red with a yellow tip, and at its base is a poorly defined pink frontal plate.

The bright bill and head colors of young coots and gallinules evidently have the same function as the conspicuous mouths of passerine nestlings: to stimulate the parents to feed their young and probably also to provide a conspicuous target when they deliver the food. The different location of the colorful areas in passerines and in coots may be related to their different methods of taking their meals. Instead of holding up a gaping mouth, the coot adopts a begging posture, presenting its face to the parent,

then seizes the food with a grasping movement as the adult's bill approaches. While soliciting food, the chick may sway its brightly colored head from side to side, thereby making it more conspicuous. As the chick grows older and finds an increasing proportion of its nourishment on its own, the brilliance of its head colors and the intensity of the associated behavior diminish. In most species of rails, however, the chicks lack bright marks on their heads (Boyd and Alley 1948).

Among ducks, geese, and swans, the only species known to feed its young directly from bill to bill is the Magpie-Goose of Australia. Like young coots, the goslings have conspicuously colored bills and heads, as well as begging cries. However, at least in captivity, they can pick up food for themselves the day after they hatch and when four weeks old are quite independent (Kear 1963). Accordingly, the Magpie-Goose must be regarded as transitional between subprecocial chicks and true precocials. In a related family, Northern or Black-necked Screamers in the New York Zoological Park picked up food and placed it in a chick's mouth until it was at least ten days old, both parents feeding it in this way (J. Bell et al. 1970). With these exceptions, ducks and their allies all belong in the precocial category.

Like rails and coots, grebes leave the nest at a very early age—if disturbed, even before they are dry. They climb upon the back of either parent and lie more or less concealed amid the feathers, seeming to be brooded dorsally rather than ventrally, while they cruise around the marsh or pond. Although they continue to receive food from the parents' bills until fairly well grown, they at no time have bright bill or head colors. However, the crown stripes of certain species, often meeting to form a conspicuous V on the forehead, as in the Pied-billed and Horned grebes, may serve the same function as the head and bill colors of coots and gallinules.

Newly hatched chicks of guans, curassows, and chachalacas are covered with ample, soft, prettily variegated down and have open eyes. Within a day, sometimes only two or three hours, of hatching, they leave their nests in trees or bushes and very soon can climb and flit through the boughs and vine tangles of the heavy tropical vegetation where most of them live. Until they are well grown, their parents feed them directly from their bills. Once I watched a half-grown Chestnut-winged Chachalaca receive pieces of green leaf from a parent, then proceed to feed itself quite competently.

Some of the pheasants, including the Great Argus, Crested Argus, and Peacock Pheasant, pass food directly to their chicks, which during the first few days of their lives seem unable to pick food from the ground as do most members of their family. With the exception of these pheasants and the guan family, most gallinaceous birds are fully precocial. Other families with subprecocial chicks are the loons, cranes, limpkins, bustards, thick-knees, and coursers. Although many shore birds appear to be wholly precocial, oystercatchers sometimes feed their young directly from their bills. Chicks of the Common Snipe can pick up small items of food when newly hatched, but during their first week they receive nearly all their food from the bills of their parents, who may continue to feed them sporadically for at least two months (Tuck 1972).

Precocial Chicks

A precocial chick hatches with open eyes and a soft, downy coat. Its brain, muscles, and limbs account for a greater proportion of its weight than do those of an altricial nestling, but its digestive tract is lighter; it could hardly travel with its parents if burdened with such a relatively huge abdomen as nidicolous nestlings have. The best-known precocial bird is the domestic chicken, and after that the duck, but their methods of feeding are different. Domestic chicks receive much help from the hen, who scratches vigorously for her downy brood, sometimes sending a chick rolling if it gets in the way of her busily kicking feet. When she finds an edible morsel, she may pick it up, mandibulate it, then lay it down, perhaps pointing her bill at it, until one of the chicks, coming to her food call, swallows it. In the Red Jungle-Fowl, from which the domestic chicken is descended, a young rooster may help a hen to forage for her chicks and if she is killed may raise them unaided (Stokes 1971). In monogamous members of the pheasant family, such as the Bobwhite, the cock regularly helps his mate forage for

the brood; if she dies, he may take full charge of them. Although most pheasants and quails do not pass food to their young in the manner of parents of subprecocial birds, some of them at times hold it in their bills until a chick snatches it away. Turkey chicks mostly feed themselves under parental guidance, but when they are newly hatched the hen wild Turkey may place food directly into their mouths (Donohoe et al. 1968). Despite their downy coats, gallinaceous chicks are sensitive to cold and require much brooding until they are fairly well grown.

Ducklings are more independent. On the very first full day of their lives, some kinds display amazing stamina and vitality. They may hatch far from water, to which they must walk before they take their first meal. After jumping from their nest box five or six yards above the ground in Illinois, a brood of Wood Ducks was led by their mother to the Mississippi River by a circuitous route that offered the greatest amount of cover. After descending a high bluff and crossing a railroad track, duck and ducklings crossed the broad, swift river, swimming about three-quarters of a mile to reach the timbered bottomland, with lakes and sloughs, where they found safety and food (F. Leopold 1951). The energy for this hazardous journey was supplied by the yolk remaining from their eggs and now enclosed in their abdomens. A similar reserve of yolk enables newly hatched domestic chicks to withstand long journeys from the hatchery to the consignee, during which they receive no food.

Although the duck, and in some species also the drake, lead and protect the young brood, the parents do not make food available to them by scratching and calling, as quails and pheasants do. From the first, ducklings are quite capable of finding it for themselves. When only three days old, a Lesser Scaup duckling could dive, capture a minnow, and return to the surface in less than five seconds (Bartonek and Hickey 1969). Incubator-hatched Ruddy ducklings, turned loose in the marsh when only a day or two old, raised themselves without adult care, as did a brood of Tufted Ducks deserted by their mother when only a few days old. Next to megapodes, ducks are the most precocial of birds

PLATE 67. A superprecocial Mallee-Fowl chick emerging from the incubator mound in which it was hatched by a combination of solar heat and heat of fermentation (photo by H. J. Frith).

(Nice 1962). As we shall see, one species of duck is superprecocial.

Another large group of birds commonly classified as precocial includes the jaçanas, plovers, turnstones, sandpipers, avocets, stilts, phalaropes, seedsnipes, sandgrouse, and related birds, and also the kiwis. Although the young of all these families pick up their own food at an early age, future intimate studies may disclose that some of them are fed more or less frequently from the parents' bills, as has been observed in the Oystercatcher and Common Snipe. The young of tinamous may also be precocial. Although we have some fine nesting studies of these terrestrial birds of tropical and South Temperate

America, after the day-old chicks are led from the nests by their fathers they are so elusive that nobody seems to have watched them forage.

Superprecocial Chicks

When superprecocial chicks hatch, they are so well able to take care of themselves that they need no parental assistance of any sort. As far as we now know, all megapodes belong to this category. By constructing in the laboratory a replica of the Mallee-Fowl's mound—or rather half a mound—with a sheet of glass covering what would be the center of a complete mound, and placing eggs nearly ready to hatch in contact with this window, Frith (1962) was able to watch the emergence of a chick. Within two hours after the appearance of the first crack in one of the shells, the young megapode burst its egg asunder, forced its legs downward, wriggled its shoulders, and started to work its way upward through the several feet of sand and other material lying above it. Spasmodic efforts that continued for five or ten minutes, and might raise the chick from one to six inches, were succeeded by hour-long intervals of rest, while it remained in the same spot with slowly pulsing body.

From these experiments and observations in the field, Frith learned that the young megapode might take from two to fifteen hours to reach the surface of a mound. A condition necessary for successful emergence is the frequent stirring of the sand, either by the parents regulating the mound's temperature or, in the laboratory, by the attendant of the incubator. If the mound remains unworked for several days, the sand becomes so compacted that the chick might not have the strength to force its way upward.

As the chick pushes itself upward, its head is bowed forward and rests against its chest. Suddenly the back of its neck appears at the mound's surface. After the neck is free, the head quickly follows. The chick opens its eyes for the first time and rests briefly (plate 67). Then it resumes its struggles, freeing one wing and then the other. Soon the whole body follows. Temporarily exhausted, the young Mallee-Fowl may lie exposed on the surface for some time, an easy prey to predators; but more often it tumbles down the side of the mound and staggers to the nearest bush to collapse in the shade, where it re-

cuperates its strength after such prolonged exertion. Its recovery is swift: within an hour it can run firmly; after two hours it runs very swiftly and can flutter above the ground for thirty or forty feet. Twenty-four hours after its escape from the mound, it flies strongly. The natal down, which like other gallinaceous birds it developed during the first three weeks of embryonic life, has been succeeded during the exceptionally long incubation period by a more mature plumage that enables it to fly.

From the moment it bursts out of the shell, the Mallee-Fowl is wholly on its own. Even if the parents, working to regulate the mound's temperature for the benefit of later eggs, happen to see it emerging from the sand, they pay not the slightest attention to it. But it is well equipped to take care of itself. At dusk on the day of its escape, it ventures cautiously forth from the shade where it has rested and forages for its first meal, scratching the ground strongly with its feet and vigorously pursuing the insects that it stirs up. Although it may pass its first night in the open on the ground beneath a sheltering shrub, more commonly it struggles up through the branches. On the following nights, it roosts high. Its defense and escape reactions are already well developed. Two hours after emerging, a chick vigorously attacked a ruler pushed toward it. If surprised in the bush, it flies up to a cluster of leaves, grasps them, and hangs upside down, an effective way to escape any terrestrial enemy.

Aside from the megapodes, the only bird entitled to be classed as superprecocial is the parasitic Black-headed Duck of southern South America. It deposits its eggs indiscriminately in the nests of birds as various as other ducks, ibises, spoonbills, herons, storks, limpkins, screamers, gulls, and Chimangos—indeed, in the nest of almost any large bird that breeds in fairly dense marsh vegetation, without regard to the size of the host's eggs or how well their color matches that of its own white eggs. Its preferred hosts, however, seem to be two species of coots, the Red-fronted (plate 66) and the Red-gartered. To adapt its diet and method of receiving food to so great a variety of foster parents, a duckling would need to be extraordinarily versatile. Apparently, it does not depend upon any of them for nourishment, but only for brooding during the first day or two after it hatches. Then it swims off alone through the marsh, capable of finding all its own food, avoiding enemies, and keeping warm without being brooded. In common with most other brood parasites, adult Black-headed Ducks take no interest in their offspring (Weller 1968).

From the foregoing survey, it is evident that the only true nidicolous or nest-inhabiting birds are the altricials of our first category. All the rest, including semialtricials and the three types of precocials, are nidifugous or nest-leaving birds in greater or less degree.

CHAPTER 19

Brooding and Temperature Regulation

Just as the bird's development while in the egg can continue only within a certain range of temperatures and proceeds most rapidly at a temperature near that of its parent's body, so, after hatching, the chick grows well only at certain favorable temperatures. Excessive cold or heat may be fatal to it. Until it acquires the capacity to hold its body temperature almost constant amid the fluctuations of environmental temperature, it needs its parents' help. Although certain birds, notably the megapodes, can incubate their eggs without animal heat, no bird has perfected a method for warming its young other than with its own body. Perhaps this is the reason why megapodes are the only birds, except the social parasites, that never brood their young, although certain others do so very little. The amount of brooding that young birds receive depends largely upon how soon they develop homeothermy, so that these two subjects are most profitably treated together.

Brooding

Brooding is the continuation of behavior that has become well established during the preceding period of incubation and, as a rule, requires the formation of no new habits by the parent. It serves, first of all, to dry the hatchling's downy feathers that are still plastered against the skin by the fluids of the egg, until their filaments separate and, in many chicks, cover and protect the body. The brooding parent also warms the young and protects them from rain, wind, or direct sunshine, which may be fatal to tender birdlings. The amount of brooding that the young require varies immensely with the species and is affected by such factors as the weather, the protection provided by the nest, the rapidity of feathering, the age at which they themselves can maintain constant body temperature, and the presence of enemies from which their parents can shield them. Thus the young need to be brooded more hours each day, and for a longer sequence of days, in cold weather than in warm, in rainy weather than in dry, in exposed nests than in well-enclosed ones, in slowly maturing young than in rapidly maturing young.

That brooding is essentially the continuation of habits formed during the period of incubation is evident from the fact that the same parents that incubate the eggs brood the young. If both parents incubate, both take turns covering the young. If only one parent incubates, only this parent broods, even when, as often happens, the other parent feeds its offspring indefatigably and defends them with the greatest zeal. When the brooding parent (usually the female in passerines) dies, the mate will often continue to feed the nestlings and to clean the

PLATE 68. Barn Owl brooding downy young (photo by S. A. Grimes).

nest, but rarely brood them, with the consequence that, if the nestlings still require warming, they succumb to exposure and all effort is lost. Or if the male habitually broods by day but not by night, as in pigeons and doves, he may attend the motherless nestlings by day but neglect them through the night, with fatal results. There are a few records of brooding by males of species in which the male does not as a rule incubate, but this is an abnormality probably no more frequent than incubation by the male in species of which the female normally incubates without help.

In species of which only the female incubates and broods, especially in passerines, the male some-times discovers very promptly that the nestlings have hatched, and in an initial burst of activity he may bring much more food than they can eat. His brooding mate profits by this excess and need not seek so much for herself. In consequence, she may spend more time covering the newly hatched young than she devoted to her eggs while incubating. But in the following days, as the young grow bigger and need more food, she devotes less and less time to brooding. Usually there is a daily decrease in both the length of her separate sessions on the nest and the total time she spends there (table 10). As a rule, the parent continues to brood by night after she has ceased to do so on rainless days, but a hard

TABLE 10
Brooding of Altricial Nestlings[a]

| Species | Percentage of Time Brooded during a Morning at Age in Days: | | | | | | | Nestling Period in Days | Type of Nest[b] |
	0–1	2–4	5–7	8–10	11–15	16–21	22–29		
Black-throated Trogon	83	—	54	—	0	—	—	14–15	H
Blue-diademed Motmot	—	39	0	—	0	0	0	29–32	B
Rufous-winged Woodpecker	—	100	52	—	0	0	0	23–24	H
Golden-naped Woodpecker	90	—	72	—	14	—	0	33–37	H
White-crested Coquette	—	43	—	0	—	—	0	21–22	O
Black-hooded Antshrike	85	76	36	0	—	—	—	10–11	O
Chestnut-backed Antbird	80	80	28	0	—	—	—	10	O
Rufous Piha	75	—	64	—	64	47	0	28–29	P
Masked Tityra	—	—	18	—	0	0	0	28–30	H
Yellow-bellied Elaenia	69	—	61	18	—	0	—	17–18	O
Long-tailed Silky-Flycatcher	65	—	46	—	4	0	—	24–25	O
Tawny-bellied Euphonia	—	36	—	0	—	0	—	20	R

[a] The records are from a single nest of each species, which was watched for 4 to 6.5 hours continuously in rainless weather, always for the same interval on different mornings at the same nest. A zero indicates that no brooding occurred during the observation period; a dash, that the nest was not watched. The nestlings of the trogon, motmot, woodpecker, antshrike, and antbird were brooded by both parents; the others, by the female only. The nestlings of the hummingbird, piha, tityra, elaenia, and silky-flycatcher bore more or less down; the others were quite naked.
[b] B = burrow; H = hole in tree; O = substantial open nest; P = thin platform; R = roofed nest amid epiphytes.

downpour may send her to cover well-feathered young on the point of leaving the nest. One rainy afternoon, I watched a sixteen-day-old Yellow-bellied Elaenia, a small tropical flycatcher, whose mother brooded it much of the time to shield it from the downpour. Suddenly this young bird, who was being treated almost like a newly hatched nestling, left its sodden open cup and climbed up a neighboring leafy branch to pass the night alone amid the foliage. It never returned to its nest and, therefore, was almost certainly never brooded again.

Parents that use their nest as a dormitory (see chapter 30) may continue to sleep with their nestlings until the young fly forth and even after they become self-supporting; but usually it is impossible to learn how long the offspring are actually brooded, rather than simply accompanied, in these well-enclosed structures. Among these dormitory users is the Blue-throated Green Motmot, in which the parents continue to pass the night with the young until a few days before their departure from the burrow, or even throughout the whole of the month-

long nestling period. But the Blue-diademed Motmot, which inhabits lower and warmer regions and does not use its burrow as a dormitory, treats its young very differently. When the little motmots are only about six days old and still quite naked, the parents cease to brood them by day or by night. Another burrow-nesting bird of the tropical lowlands, the Buff-throated Automolus, whose tunnels are sometimes in the same banks as those of the Blue-diademed Motmot, behaves in much the same way, leaving its young alone through the night when they are only about a week old and still without expanded plumage. The naked young motmots and automoluses suffer no harm by remaining unbrooded in their burrows in the warm earth, and the advantage of this system to their species is obvious. Any nocturnal prowler that entered the mouth of the tunnel would have all its occupants trapped; but if the parents are elsewhere, they may live to rear another brood. In the highlands where the Blue-throated Green Motmots live, predators, especially snakes, are less numerous.

In contrast to these parents who leave naked

PLATE 69. Australasian Gannet trying to cover a downy young too big to be brooded (photo by John Warham).

nestlings unbrooded are others who seem to carry brooding to extremes (plate 69). In England, Rooks may brood their large nestlings to the day of their departure (Coombs 1960). Even in fair weather, the female Boat-billed Flycatcher sometimes broods well-feathered nestlings almost ready to fly, who are so restless that from time to time she leaves the nest to perch beside it. The same is true of White-fronted Doves. I have often found a parent sitting on the platformlike nest with the nestlings' fore-parts protruding from beneath it, or even with the young sitting beside the parent. White-fronted Doves attend their young almost constantly in this fashion until the latter fly away. One day I approached a nest on which a parent was sitting with

a nestling's head and shoulders projecting from beneath its breast. Suddenly the parent dropped to the ground and tried to lure me away by "feigning injury" in a most realistic fashion. While my attention was held by the adult's seemingly agonized struggles, the youngster rose from the nest and flew so well that it was soon beyond view in the surrounding thicket. The Boat-billed Flycatcher and White-fronted Dove seemed to be present primarily to guard their feathered young, and the brooding, which was obviously superfluous on warm days, appeared to be incidental.

When hot sunshine falls upon an exposed nest, the parent sometimes broods the nestlings in the normal way, but often it stands on the rim, perhaps with partly spread wings, shading the young from rays that would be harmful or even fatal to them (plates 70 and 71).

Altricial young, which we have been considering chiefly, are brooded in the nest where they hatch but scarcely ever, as far as I know, in some other spot after they have flown from the nest. The situation is quite different with semialtricial, subprecocial, and precocial chicks, some of which, as everyone who has kept domestic chickens knows, require brooding long after they leave the nest. Gallinaceous birds settle on the ground in a secluded spot and their chicks crawl beneath them. Bobwhite chicks are brooded until they are about two weeks old, by either the male or the female parent, and sometimes by both simultaneously, some of the chicks being covered by their father and some by their mother.

Aquatic birds face a special problem, as they cannot brood their young in the water or on sodden, marshy ground. To overcome this difficulty, rails, coots, and gallinules construct special brood nests, which are usually no more than platforms raised above the water in marsh vegetation or in the growth surrounding a pond; or they may lead their chicks to an old structure that they built for displaying or breeding but failed to use. When the eggs of the American Coot begin to hatch, the female covers the older, more active chicks on a brood nest that has been built for them, while the male takes charge of the hatching eggs and newly emerged young in the breeding nest. A family of coots from six to fifteen days old may be divided at night, each parent covering some of the chicks on a separate sleeping platform. One mother slept with her offspring on a brood nest forty-six days after the last of them hatched, but apparently she roosted beside rather than above them (Gullion 1954). Platforms, sometimes called "dummy nests," are built for brooding their chicks by the Sora Rail, Virginia Rail, Common Gallinule or Moorhen, and doubtless by many other members of this family.

Downy young Black Terns, when frightened, swim out from their floating nest, beginning when only a day old. Usually they soon return, either spontaneously or in response to a parent's calls. Sometimes, however, the alarmed parents hastily

PLATE 70. Masked or White Booby shading a downy chick in the characteristic drooped-wing posture (photo by G. A. Bartholomew).

construct a new floating nest and call their chicks to it, to feed and brood them there exactly as though they were in the nest where they hatched. Chicks only three days old may swim to a new nest 100 feet (30 meters) from their birthplace (Cuthbert 1954). As downy chicks of the Least Tern wander over a sandy beach, the parents repeatedly scrape depressions in which to brood them (Hardy 1957).

The young of certain gallinaceous birds are brooded on a perch in a tree or shrub, where they are safer than on the ground. Because these so-called game birds are so shy of their human persecutors, this habit has been observed in aviaries more often than in the wild. For the first three nights after it hatched in a zoo, a chick of the Ocellated

PLATE 71. Female Red-winged Blackbird shielding nestlings from sun (photo by S. A. Grimes).

PLATE 72. When cold or disturbed, murre chicks seek shelter under the wings of any nearby adult. This photograph includes examples of the bridled phase of the Common Murre on Funk Island, Newfoundland (photo by L. M. Tuck).

Argus Pheasant was brooded on the ground. On the next evening, the mother flew to a perch four feet high and called the chick, who climbed a specially prepared hurdle to reach her and was brooded on the perch. Two male chicks of another hatching were brooded on a perch by this female pheasant, one under each wing, until they were three-quarters her size and could fly well. A captive Red Jungle-Fowl hen brooded up to four chicks on a perch, beginning when the young were three or four weeks old. The habit of brooding on a perch has not been wholly lost by the domesticated descendants of the Red Jungle-Fowl, for I have seen domestic chicks huddle beneath their mother on a branch, when they were sent to roost in a tree before brooding had ceased. A captive Great Curassow, which nests in a tree, brooded her chicks on the ground for a few nights after they hatched, but thereafter they roosted on low branches by their mother's side (fig. 18). The female Emerald Dove of India also broods her young on a perch, one beneath each wing—the only instance of brooding away from the nest by an altricial bird that has come to my attention (J. S. Huxley 1941).

FIGURE 18. Female Great Curassow roosting at night with a chick under each wing (by A. E. Gilbert, from Delacour and Amadon 1973).

Development of Temperature Control

The primary function of brooding is to maintain the young bird's temperature at the point most favorable for its vital processes and development. Accordingly, it is supplementary to the young bird's own unconscious effort to preserve a constant body temperature. Newly hatched birds, like animals as a whole, may be divided into two classes: poikilothermal, or "cold-blooded," animals, whose body temperatures follow rather closely that of the air or water that surrounds them; and homeothermal, or "warm-blooded," animals, whose body temperatures remain nearly constant amid fairly wide fluctuations in the temperature of the surrounding medium.

The advantage of homeothermy is that it permits the animal to carry on its vital functions, to move and even to think, at the same rate in cold weather as in warm. That this is no small advantage is attested by the facts that the only fully homeothermal animals, birds and mammals, live successfully at high latitudes where cold-blooded amphibia and reptiles are rare or absent, and that in the temperate zones they remain active in severe winters when poikilothermal animals of all kinds are dormant. However, there is a disadvantage to homeothermy. To maintain a temperature different from that of the environment demands an expenditure of vital resources: of heat-yielding nutrients when the temperature is lower than that of the body, and principally of water for cooling by evaporation when it is higher than that of the body. Accordingly, the capacity to abandon homeothermy in certain circumstances is valuable to a number of mammals and even birds. By hibernating at a temperature not greatly different from that of the surrounding air they can husband their resources in severe weather. In birds, such torpidity occurs in the Poorwill, Lesser Nighthawk, and, to a minor degree, in swifts of several kinds—likewise in many hummingbirds on chilly nights, except while they are incubating eggs or brooding nestlings.

Homeothermy is always relative to external conditions. No animal that ever lived could maintain a constant body temperature indefinitely in all circumstances, for example, in boiling water or at absolute zero. A strict definition of homeothermy would stipulate the conditions. For example, it could be the ability to hold the body temperature constant within 3.6°F (2°C), while the animal is subjected for one hour to a temperature of 50°F (10°C) and for another hour to 113°F (45°C). But experiments on free animals can hardly aspire to the precision attainable in a physical laboratory; our information on temperature regulation in young birds has mostly been gathered in the endlessly variable circumstances of the field. And, however efficient their metabolisms, small birds cannot be expected to attain the thermal stability of large ones. Recently hatched birds differ greatly in the ability to maintain a fairly constant body temperature in the absence of brooding parents, but few, if any, can hold their temperatures constant in their natural environment as well as their parents, who are larger and wear a denser coat of feathers; if they could, brooding would be superfluous. A practical measure of the thermal efficiency of young birds is the amount of brooding they require.

In general, precocial birds develop temperature control at an earlier age than altricial birds, as they need to do in order to leave the nest and follow their parents, but there are exceptions. A classification of newly hatched birds based on their heat regulation, rather than upon how soon they leave the nest and how they eat, would be somewhat different from that in the preceding chapter. Among superprecocials, megapodes, which are never brooded, seem to be homeothermal from the day they hatch, and Black-headed ducklings appear to be so after their foster parents dry their down.

Of all precocial birds, ducklings appear to be hardiest; many already have efficient temperature control when only a day old. But species differ greatly, those that nest in the far north, such as the Common Eider, resisting cold much better than those of more southerly distribution, such as the Mallard (Koskimies and Lahti 1964). Many gallinaceous birds, despite their downy coverings, acquire the ability to regulate body temperature much more tardily, as one would infer from the long-continued brooding that they receive. Although the Ringnecked Pheasant is one of the hardier members of the family, thriving in regions with severe winters, its chicks are surprisingly sensitive to exposure. On the day after they hatch, their temperatures, even in a warm brooder, are about 102°F (39°C), six degrees lower than that of adults, and they do not attain the adult temperature of 107.6° to 109.4°F (42° to 43°C) until they are about two weeks old. Exposure for thirty minutes to the moderate temperature of 68° to 77°F (20° to 25°C) caused the body temperatures of pheasant chicks from two to four days old to fall 5.4° to 9°F (3° to 5°C), and those of week-old chicks dropped almost as much. In chicks a few days old, repeated half-hour exposures to this temperature at daily intervals did not, as one might expect, increase their resistance to cool air but actually reduced it. On each repetition of the test their temperatures fell lower and finally a number succumbed. Although their resistance to cool air improved with age, even when two weeks old their temperatures might fall 18°F (10°C) when they were kept for an hour at 68°F (20°C). Young domestic fowls are similarly affected by cool air (Ryser and Morrison 1954).

Another gallinaceous bird, the Capercaillie, acquires adult temperature and the ability to maintain it at about eighteen days of age, when it weighs four times as much as when it hatched. Before this time, cold, wet weather causes the loss of many chicks, as it does in quails and related birds, because the young cannot be brooded enough without dangerously reducing the time they devote to eating (Höglund 1955).

I have found no study of the development of temperature control in subprecocial birds. By the end of their second week, American Coots are seldom, if ever, brooded in the daytime, and when they are about twenty days old nocturnal brooding also ceases, which suggests that by this age they have developed efficient temperature regulation.

For semialtricial chicks, which are easier to observe because they are less mobile, we have a number of careful studies. Even before they hatch, gulls can maintain their body temperatures well above that of the ambient. At air temperatures between 66.2° and 82.4°F (19° and 28°C), newly hatched Western Gulls could hold their body temperatures almost as high as that of fully feathered individuals. At slightly lower temperatures, the downy gull chicks tried to keep warm by huddling together,

but still their temperatures fell (Bartholomew and Dawson 1952). When only a day or two old, California Gulls can maintain their temperatures well above that of the air, and this ability improves rapidly on the following days (Behle and Goates 1957).

Probably because of their smaller size and proportionately greater surface area, young Least Terns do not regulate their temperatures as well as gulls of the same age when exposed to cool air, but they have greater ability to prevent overheating when subjected to high temperatures (Howell 1959). This ability is of vital importance to young Sooty Terns that are hatched on open ground on sunny tropical islands. From the first day, the downy young can keep their temperatures down by vigorous panting, but they crawl into the nearest available shade, which is usually that of their brooding parents, who sit so steadfastly on their nest scrapes that they are easily captured by hand (Howell and Bartholomew 1962*b*).

Albatrosses regulate their temperatures at least as early as gulls do. As soon as the abundant natal down has become dry and fluffy, chicks of the Laysan and Black-footed albatrosses can preserve a body temperature of about 100.4°F (38°C) when the air is 27° to 30.6°F (15° to 17°C) cooler. For these birds of tropical islands, to stay cool while resting on sand heated by strong sunshine is no less important than to avoid chilling. The young dissipate heat not only by vigorous panting but also by sitting on their heels with their broadly webbed feet raised above the ground and shaded by their bodies (plate 73). For albatrosses, as well as for boobies and related birds, loss of heat from the webbed feet, not only by radiation but also by convection when they are exposed to a breeze, helps to keep the body cool in hot weather (Howell and Bartholomew 1961).

Semialtricial goatsuckers, such as the Common Nighthawk, are not able to regulate their temperatures as well as gulls, terns, and albatrosses of similar age. Howell (1959) found nighthawk chicks to be intermediate between these aquatic birds and altricial passerines in the development of homeothermy. Often hatched in shadeless spots, such as the flat roof of a high building, nighthawks rapidly

PLATE 73. Downy Laysan Albatross resting on its heels while it dissipates heat from the webs of its raised, shaded feet, on Midway Island (photo from T. R. Howell and G. A. Bartholomew).

become overheated in full sunshine and might perish if the parent left them unshaded for many minutes before they are about two weeks old.

Nidicolous altricial birds vary greatly in the ability to regulate temperature during the first few days after they hatch. Born with a thick coat of down and open eyes, a Slender-billed Shearwater a few hours old can preserve the adult body temperature of about 100.4°F (38°C) while its burrow is at 71.6°F (22°C). Although usually brooded by a parent for only the first two days, the chick may be left alone even earlier (Farner and Serventy 1959). In the harsh environment of the South Orkney Islands, Black-bellied Storm-Petrels were alone in

their crannies amid tumbled rocks from the day they hatched, without ill effects, and were evidently able to maintain a high body temperature. Such early cessation of brooding is exceptional among petrels and shearwaters (Beck and Brown 1971).

In sharp contrast to the downy hatchlings of gulls, terns, albatrosses, shearwaters, and petrels, boobies hatch naked with slight ability to regulate their temperatures. The air temperature of the tropical islands where Masked or White Boobies breed is usually mild, but the open ground on which the nests are situated becomes dangerously hot in sunshine. Although a newly hatched chick resorts to gular flutter to cool itself, it soon becomes overheated when exposed to the sun's rays; in shade, its temperature rapidly drops. Until, at the age of about two weeks, its rapidly increasing size and tardily acquired coat of fluffy white down give it efficient temperature control, the young booby depends upon its parents for brooding and shading. While still small, it rests upon the parent's webbed feet—the eggs, it will be recalled, are incubated beneath the booby's feet (Bartholomew 1966).

Tropicbirds, which often nest on the same islands with boobies, escape from their shells with downy, instead of naked, bodies and when newly hatched preserve their temperatures quite well when left unbrooded in the shade. In sunshine, they prevent their temperatures from reaching lethal heights by vigorous panting. Nevertheless, they are closely brooded most of the time (plate 74); when they grow too big to be kept beneath the parent's body, they are held beneath a wing (Howell and Bartholomew 1962a).

Whether they hatch completely naked or with sparse down, the familiar altricial birds of field and woodland are at first poikilothermal. They are wholly dependent upon their parents to maintain their body temperatures, which fall while the adults are absent, to rise again when brooding is resumed. If left alone for many minutes on a cool day, they become thoroughly chilled, and a short exposure to full sunshine may be fatal to them. However, they soon acquire the ability to hold their temperatures high. Budgerigars have effective temperature regulation when ten days old; Rock Doves, at eleven days; and Wrynecks, representing the woodpeckers,

PLATE 74. Red-tailed Tropicbird brooding its chick under one wing (photo from T. R. Howell and G. A. Bartholomew).

also at eleven days. In Barn and Cliff swallows, temperature control is apparently established at nine or ten days; in Northern House Wrens, at about eight days. In sparrows, including the Vesper, Field, and Chipping, which have shorter nestling periods, adequate homeothermy is attained still earlier, and they can hold their body temperatures above 98.6°F (37°C) in an environment at 68° to 77°F (20° to 25°C) when they are only six or seven days old (Dawson and Evans 1960, Stoner 1935, 1945). Even more rapid is the development of temperature control in nestling North American Ovenbirds, which three days after hatching could hold their temperatures almost constant around 100.4°F (38°C) during a half-hour exposure to air at 79°F (26°C). After their fourth day, they were no longer brooded in their covered nest on the ground in woodland shade (Hann 1937).

These altricial birds establish homeothermy at a later age, counting from hatching, than the birds that we earlier considered. However, if we count from the start of embryonic development, or of incubation, which is perhaps the more enlightening way, they do so much sooner. By this reckoning, a shearwater or storm-petrel that hatches with good temperature control is already six to eight weeks old; an albatross, over nine weeks old. In contrast to this, a Barn Swallow is able to regulate its temperature when only about twenty-five days old, a Northern House Wren at twenty-two days, and a Chipping Sparrow at eighteen or twenty days. Moreover, the transition from the poikilothermal state of the early embryo to the homeothermal condition of later life is much more rapid in the second group. Many precocial and semialtricial birds begin to acquire temperature control before they hatch, so that the developing embryo within an egg remains considerably above the ambient temperature when left unincubated. The Ring-necked Pheasant that begins to show temperature control two weeks before it hatches and is practically homeothermal three weeks after hatching requires about five weeks to complete the transition. The sparrow that shows slight control of its temperature three or four days after it hatches has good control four days later.

Although homeothermy has the great advantage that, with adequate nutrition, it permits sustained activity and development, poikilothermy is not without compensations. The "cold-blooded" reptile or amphibian can emerge unharmed from long periods of severe chilling that would be fatal to man and many birds. Similarly, before they establish temperature control, altricial nestlings can endure without ill effects intervals of reduced body temperature that might kill the chick of a pheasant or domestic hen. For certain birds, the capacity to reduce the body's temperature in times of stress promotes survival. In northern Europe, the success of the Common Swift's aerial flycatching varies greatly with the weather. In cold, wet spells, the yield of volitant insects is so poor that it is unable to nourish its nestlings. Instead of squandering their vital reserves in a vain endeavor to preserve high body temperature, the unfed young Swifts become poikilothermal, permitting their temperatures to drop to that of the air. They survive in a dormant state until, with better weather, the parents are again able to feed and brood them adequately and they resume growth. Because of the vicissitudes to which they are subject, their period in the nest may be as short as five weeks or as long as eight weeks, a range equalled by few other small birds (Lack 1956a).

Nestling hummingbirds are amazingly tough, able to withstand exposure to cold, rain, and strong sunshine that might be fatal to passerine nestlings several times their size. In the tropics at an altitude of about 2,500 feet (760 meters), where nights are often chilly, hummingbirds of several species cease to brood, even by night, when their young are from eight to ten days old, still largely naked with insignificant natal down and juvenal plumage still ensheathed. Even after the cessation of nocturnal brooding her nestlings until their eighteenth day. ing a daytime shower, but they are often left exposed to full tropical sunshine, which causes them to stretch up their necks and pant strongly. At high altitudes, nocturnal brooding may continue longer; in the Guatemalan mountains at 8,500 feet (2,600 meters), I found a White-eared Hummingbird brooding her nestlings until their eighteenth day. But at nearly 7,000 feet (2,100 meters) in Wyoming, Calliope Hummingbirds, the smallest birds north of Mexico, no longer covered their nestlings through the night after they were eleven or twelve days old.

These tiny Calliope nestlings, which even when mature would weigh only about 0.1 ounce (3 grams), had already developed considerable temperature control and, unbrooded in their downy open nest, remained mostly between 71.6° and 86°F (22° and 30°C) on nights when the thin mountain air fell to within a few degrees of freezing. When two weeks old, unbrooded nestlings of Anna's Hummingbird, a slightly larger species, remained about 27°F (15°C) warmer than the night air. This is the more surprising when one recalls that at similar nocturnal temperatures adult hummingbirds not attending nests reduce their metabolism and become torpid while they sleep (Calder 1971, Howell and Dawson 1954).

The much bigger Oilbird or Guácharo takes an exceptionally long time to establish temperature control. Raised on a narrow ledge in a dark cave in northern South America, the Oilbird nestling does not even approach homeothermy until it is about three weeks old, when it begins to acquire insulation by growing a belated coat of down and putting on fat. Its temperature does not rise to the adult level until several weeks later, and it remains in its nest until three or four months old (D. W. Snow 1961).

CHAPTER 20

The Male Parent's Discovery of the Nestlings

When the male and female share incubation, either may be present when the first egg hatches and bring the first food. But they are not always equally competent in attending the newly hatched young. Once I watched a male Mountain Trogon who behaved most queerly. On the second morning after the eggs hatched, he came with a small insect but, instead of offering it to a nestling, he sat holding it stupidly in his bill for nearly an hour while he brooded. When his mate came to replace him, he flew away with the insect. Four more times in the course of the morning he brought food, only to hold it while he brooded instead of feeding his offspring. Of the five insects that he took to the nest, a nestling received only one, and that largely by accident. As the trogon was about to leave with the food still in his bill, his departure was delayed by a squirrel that noisily rustled the dry ground litter in front of the nest cavity. While the parent hesitated to go in the presence of a potential nest robber, he seemed suddenly to remember that he held food, backed farther into the nest, and placed it in the upturned mouth of one of the nestling trogons. The female also brought five items of food, all of which she delivered promptly to the nestlings. If she had been as inept as her mate, the young might have starved. But the male trogon was capable of im-

provement; two days later, I found him feeding his offspring efficiently.

As already related (p. 215), while the first egg was hatching beneath him, a male Black-hooded Antshrike refused to relinquish the nest to his mate, who had come to relieve him. Evidently he failed to inform her of the momentous event that was taking place, for when she returned after half an hour she had nothing for the nestling. This time he made way for her, flew off through the thicket, and returned thirteen minutes later with the nestling's first meal.

Preparation for the Discovery

Although the male antshrike seemed not to communicate to his mate the important news that the eggs they had been patiently attending for the past fortnight were at last hatching, the regular alternation of the two parents on the nest ensured that both would soon learn of the nestlings' presence. But what happens when only the female incubates, as in many passerine birds? How does the male learn that it is time to bring food to the young? Does she inform him, or must he learn for himself? Here we touch upon a problem not only of considerable importance in the parental behavior of birds but also of perhaps even greater interest for

PLATE 75. Male Swainson's Warbler presents an insect to his incubating mate. Such feeding prepares the male for prompt attendance on the newly hatched nestlings (photo by S. A. Grimes).

the light it may throw on the parents' ability to communicate with each other.

To answer these questions, we must study the activities of the nonincubating male while his mate attends the eggs, for the habits he forms in this interval are obviously relevant to his eventual discovery of his offspring. Indeed, the chief importance of the male's association with the nest at this time seems to be that it prepares him for the work that he must later perform, and for this reason we have deferred until now the detailed consideration of his behavior during the incubation period.

While the female incubates, the nonincubating male may show his interest in the nest in one or more of the following ways: (*a*) by adding material to it, (*b*) by guarding it, (*c*) by bringing food, (*d*) by escorting the female when she returns to her eggs, and (*e*) by visits of inspection.

Of these five ways of maintaining contact with the nest during the incubation period, bringing material to it is the least frequent. Both the male and female Rose-throated Becards add to their great, globular, swinging nest until the nestlings hatch; but the male of at least one pair placed all his contributions on the outside, so that this activity did not bring him into close contact with the eggs, which were separated from him by a thick wall. Males who incubate, as in the ovenbird family and bushtits, more commonly bring material, but this does not concern us at present.

Guarding the nest itself appears to be less widespread than guarding the territory that contains it. A few male birds stand like sentries above or beside their nests while the females go off for food, then fly away when they return to resume incubation. The alternation of the two at the nest may be as regular as that which occurs when incubation is shared, with the difference that the male watches but does not warm the eggs. The habit of closely guarding the nest is, understandably, best developed in fairly big birds that are able to drive away at least the less formidable of would-be predators. I have noticed it chiefly among jays and in one of the largest of the American flycatchers, the Boat-billed. Nevertheless, nests are sometimes guarded by male birds so small and weak that I could not imagine what enemy they could hold aloof. In such birds,

standing sentry is likely to be an individual peculiarity rather than a habit of the species as a whole. At one nest you may find the male guarding during his mate's recesses; at another of the same species in the same locality, you may watch in vain for this behavior. This is true, for example, of the Orange-billed Nightingale-Thrush, Yellow-bellied Elaenia, and McCown's Longspur (Mickey 1943).

Bringing food to the nest during incubation is widespread in certain families, such as crows and jays, titmice, wood warblers (plate 75), and tanagers, whereas in others, such as American flycatchers and wrens, it is rare. Two kinds of food bringing may be distinguished. When the male supplies so much nourishment to his sitting partner that he substantially reduces the time she must devote to foraging and enables her to incubate more constantly, his activity may be called "sustaining food bringing." Such abundant feeding is regular in certain goldfinches, siskins, crossbills, and jays; among hornbills, the male supplies all his partner's needs and she never leaves the nest. We have already noticed such food bringing in our discussion of incubation and need say no more about it here.

The second kind, which I call "anticipatory food bringing," was long overlooked by ornithologists. While watching a nest of the Pink-headed Warbler in the Guatemalan highlands, I was greatly puzzled by the behavior of the male, who sang in neighboring pine trees while his mate incubated but never covered the eggs himself. After a while, he came with a caterpillar and passed it to her through the sideward-facing doorway of the nest roofed with pine needles. A few minutes later, she left the eggs, and during her absence he brought a small green insect, stood in the doorway, lowered his head into the nest, and murmured softly. Why did he not take this offering directly to the female, who was foraging nearby, as doubtless he knew as well as I did? After repeating his low notes in the doorway, he left the nest and hopped among the surrounding bushes, still holding the insect. Presently his mate returned and rested near him; but instead of giving her what he held, he went again to the nest, blocking her return while he stuck his head inside and uttered soft, low notes as before. As soon as he left the way clear, the female returned to her eggs, and then, after some hesitation, the male stood in the doorway and gave her the food that he had long held.

Later in the day, this male Pink-headed Warbler came with a whole mouthful of insects, lowered his head into the nest, and uttered a thin twitter. He retired a short distance, then returned to the doorway and twittered as before; he did this five times. All this while I could plainly hear the *chip chip* of his mate, hunting insects down the steep slope. How stupid, I thought, to try so persistently to feed the female at the nest, when she was obviously elsewhere. But presently I understood. The male warbler was not trying to feed his mate but to feed his nestlings. Since they were still tightly enclosed in the shells and could not take what he had brought for them, the female received it when she returned, but this was only incidental. At this nest I saw the male bring food four times while his mate was absent, but only once while she was within. Since she was present twice as much time as she was away, it was evident that he preferred to come when he could see the eggs.

In later years, I have seen male birds of a number of other kinds anticipate the hatching of their nestlings, although few tried as hard as the Pink-headed Warbler to make the eggs eat the food. These impatient fathers included several other kinds of wood warblers, tanagers, finches, the Tropical Pewee, and the Masked Tityra of the cotinga family. A male Variable Seedeater even offered to regurgitate food at a second-brood nest in which his mate had not yet laid an egg! Sometimes the female, when present, ate what her mate had brought; but at other times it was not even offered to her. Even if it was presented to her, she might refuse it, as happened at nests of the Buff-rumped Warbler and Crescent-chested Warbler. A male Crimson-backed Tanager, whose eggs were four days from hatching, brought food to the nest three times in ten hours. On two of these visits his mate was absent and he presented the food to the eggs with low, cheeping notes. On the third visit he found the female sitting and surrendered his caterpillar to her with apparent reluctance. She seemed nowise eager for it, passed it back to him once or possibly twice, but in the end was persuaded to swallow it.

More rarely than the male, the female presents food to her unhatched eggs, as I have seen at nests of the Fasciated Antshrike, Tropical Pewee, and Orange-billed Sparrow. The female's activities at the nest seem to be more closely controlled by the actual progress of the breeding operations than are those of her mate, and she less often does inappropriate things, such as offering food to eggs or withholding it from the nestlings she broods, as the Mountain Trogon did.

Sustaining feeding has present importance, for it increases the female's time on the eggs and in cold weather may save them from freezing. Anticipatory food bringing has future importance, for it helps the male to find the nestlings promptly. In addition, there is a third kind of food bringing, which may be called "occasional." Some male birds bring food to their mates only once or twice in a morning, in quantities insufficient to reduce significantly the time that the females devote to foraging. Yet they do not offer this food to the eggs and are not obviously anticipating the nestlings. Although behaviorally this occasional food bringing differs from anticipatory food bringing, which involves an earnest attempt to feed the eggs, functionally its importance is much the same, for it prepares the male for prompt attendance upon his nestlings. Occasional food bringing may be anticipatory food bringing that has become formalized and routine, in the individual or perhaps in the evolutionary history of the species—a development that would be favored by natural selection because it increases reproductive efficiency. Indeed, as Nolan (1958) suggested, it is not improbable that sustaining feeding and courtship feeding are derived, in the evolutionary sequence, from anticipatory food bringing, which in turn has evidently developed from the normal feeding of nestlings.

Anticipatory food bringing raises fascinating psychological questions. Does the male bird who tries to feed the eggs have a mental image of nestlings? Is this behavior restricted to males with previous breeding experience, or are the more earnest and sustained attempts to deliver food to eggs made only by males who remember nestlings of previous broods? Nolan concluded, from indirect evidence, that even male Prairie Warblers breeding for the first time engage in anticipatory food bringing, but further studies of this matter are needed.

Male birds who neither incubate nor guard their nests when the females are absent often accompany the latter on their outings. Sometimes the male calls the female from her eggs to go off and hunt food with him, and often he returns with her. Such behavior is widespread among American flycatchers and tanagers, and often it occurs in cotingas, wrens, mockingbirds, vireos, wood warblers, finches, American orioles, and other families. The male who escorts his mate to her eggs may go right up to the nest or turn back at a point some distance from it. The latter is the more prudent behavior, for the approach of two birds to the nest is more likely to draw attention to it than the approach of one. A male White-breasted Blue Mockingbird would leave his returning mate at a point six feet or more from her nest in a low thicket; he scarcely ever came closer. But when a male Yellow-green Vireo saw his mate returning to her nest, he would hasten to stand close beside it while she settled on her eggs in front of him. A male Streaked Flycatcher, whose mate was incubating in a nest box in a mango tree, sometimes escorted her to the tree, sometimes to the doorway in the side of the box, and more rarely he entered with her. When this occurred, a low, twittered conversation took place within. In most birds, such courteous attention to the female is most frequent early in the morning; later in the day the male is more neglectful of his partner.

In a few kinds of very small birds of which the male and female together build a closed nest with a round doorway in the side, the female's return to her eggs is the occasion of a spectacular ceremony. As she flies toward the nest from a point a short distance in front, the male flies with her, accompanying her so closely that he seems to be racing her to the doorway. But she always wins the race and enters, while at the last moment he veers aside and goes off to forage. I have often witnessed this performance in euphonias, which build globular nests in crannies in trees and fence posts, and tody-flycatchers, whose pensile structures swing from slender twigs in trees or shrubs. Larger birds that escort their mates to the nests in somewhat similar fashion include the Black-faced Grosbeak, whose

shallow open cup is hidden in a cranny among epiphytes, and the Bright-rumped Attila, whose open nest is situated in the deep embayment between the plank buttresses of a great tree.

With the exception of anticipatory food bringing when it occurs in the female's absence, none of the foregoing contacts with the nest provides the male with an infallible means of learning what it contains. Bringing material during the incubation period is rare unless the nest is large and elaborate, and then the male's contributions are likely to be added only to the outside, so that he does not come into contact with the eggs. When the male guards the nest, he may habitually stand at a point that does not afford a view of its contents. When he brings food, he may deliver it to the female while she covers and conceals whatever the nest holds. When escorting his mate, the male may turn back at a point some distance from the nest, or he may dart past it too rapidly to see what is inside. A more certain method of learning what is happening at the nest is the visit of inspection, when the male goes with empty bill to stand on the rim, or in the doorway of a closed nest, and bends forward to scrutinize the contents, sometimes remaining in this attitude for a minute or more. The male who habitually guards at a spot some distance from the nest may from time to time advance to its rim for a close examination. These visits of inspection are not frequent; few male birds make them more than three or four times in a day. Some never do.

But why, you may ask, must the male bird make a special effort to learn what is happening in the nest? Wouldn't the simplest way be for his mate to inform him, by a special call or gesture, that he has become a father and must now bestir himself to feed his young family? The well-developed vocal organs of birds are capable of a variety of utterances, some of which serve to keep mates in contact with each other, some to warn of danger, some to call for food, and so forth. Why should they not have an utterance to tell their partners of an event of prime importance in their mutual endeavor to rear a family?

Why, indeed, should they not have such a note? It may be difficult to believe, yet many birds lack it. I have, usually from concealment, carefully watched nests of about twenty species while the eggs were hatching, trying to learn how the female informed her partner of this momentous event, yet I could never convince myself that she had any direct means of telling him. Occasionally it seemed to me, as it has seemed to other watchers, that the female was trying by voice or other means to call her mate's attention to the newly hatched nestlings; but the male's failure to respond to such notes or gestures makes it almost certain that we misinterpreted them. Sometimes the notes by which the female seemed to be trying to call her mate's attention to the newborn nestlings were of kinds that she uttered before they hatched, or after the male had begun to feed them, from which it is evident that they were not specific for this occasion. It would be premature to conclude, from the relatively small number of birds that have been studied at this critical time, that in no species does the female inform her mate by voice or gesture that her eggs have hatched, but the sample has been sufficiently varied to make us sure that this ability is not widespread among small birds (Skutch 1953*d*).

Promptness of the Discovery

The promptness of the male's discovery of his offspring depends, then, largely on habits that he formed while incubation was in progress. Usually he discovers the nestlings by seeing them with his own eyes, which he is almost certain to do when he makes a visit of inspection to the nest—I say "almost certain" because such inspections may degenerate into a carelessly performed formality and he may overlook the nestlings, as happened to a certain male Yellow-green Vireo. A food-bringing visit, whether sustaining, anticipatory, or occasional, will reveal to the male that the eggs have hatched if the female is absent; even if she is covering the newly hatched nestlings, her increased eagerness for his offerings, especially if she was previously disdainful of them, may stimulate him promptly to bring more. Or she may rise up to pass his food to a nestling in his presence, or even leave so that he can deliver his billful directly to it.

At times, apparently, the male is prompted to bring food to the nest by seeing his mate gather it; in certain cases, this is the most probable explanation of the short interval between hatching and the first feeding by a male whose visits to the nest during incubation were widely spaced. If the nest is enclosed and the male does not enter it to inspect its contents, he may not become aware of the nestlings until he hears their voices issuing from it, in which case a day or more may elapse before he begins to feed them. Other possible means of learning that nestlings have hatched are by seeing the female remove an empty shell or a dropping; but probably only old, experienced, and unusually alert birds would draw the correct inference from such clues. We lack direct evidence that male birds do sometimes learn about the nestlings in this way.

Although there is often a considerable interval between the hatching of the first egg and the male parent's first view of a nestling, this view usually leads very promptly to food bringing. An incubating female Gray's Thrush, whose nest I was watching, became very restless soon after 6:00 A.M., constantly rising up to look beneath herself and sitting high in the nest, instead of well down in the bowl as formerly. At 6:15 A.M. she carried off part of an empty shell, thereby apprising me that an egg had just hatched. She promptly returned to resume sitting. When, at 6:24 A.M., her mate alighted a few feet from the nest, she appeared to take no notice of him, and he left at once, apparently no wiser. He did not again approach the nest until 8:59 A.M., when the female was absent. Resting on the rim, he spent two minutes intently regarding his newborn offspring, lowering his head into the bowl, and mincingly opening and closing his bill. Then he flew off, to return in three minutes with food for the nestlings. Nearly three hours had elapsed before he learned that the eggs had hatched, but he brought food only three minutes after making the discovery. At other nests the course of events has been similar.

Although the male Gray's Thrush, and many other birds that I have watched, discovered the nestlings at a stroke, at times the discovery is more gradually achieved as the culmination of growing curiosity. During the period of incubation, as day follows day with monotonous sameness, the male bird sometimes becomes less interested in the nest than he

TABLE 11
The Beginning of Feeding by Altricial Parents[a]

| Species | Approximate Interval between Hatching[b] and First Arrival with Food by: | |
	Female	Male
Rose-throated Becard	—	1.5 days or less
Vermilion-crowned Flycatcher 1	32 minutes	6 hours 34 minutes
Vermilion-crowned Flycatcher 2	almost immediately	6–10 days
Gray-capped Flycatcher 1	24 minutes	4 hours 32 minutes
Gray-capped Flycatcher 2	26 minutes	6–32 hours
Bran-colored Flycatcher	27 minutes or less	4–29 hours
Yellow-bellied Elaenia	46 minutes	49 minutes
Southern House Wren	98 minutes	25 minutes
Highland Wood-Wren	—	1.5 days or less
Gray's Thrush	19 minutes	2 hours 49 minutes
Orange-billed Nightingale-Thrush	7 minutes	9 minutes
Yellow-green Vireo	4 minutes	49 minutes
Buff-rumped Warbler 1	7 minutes	3 hours 22 minutes
Buff-rumped Warbler 2	more than 2.5 hours	56 minutes
Golden-masked Tanager	50 minutes	51 minutes
Silver-throated Tanager	8 minutes	2 hours 33 minutes
Scarlet-rumped Black Tanager 1	38 minutes	38 minutes
Scarlet-rumped Black Tanager 2	36 minutes	1 hour 32 minutes
Red-crowned Ant-Tanager	—	0.5–1.5 days
Buff-throated Saltator 1	—	2 hours or less
Buff-throated Saltator 2	61 minutes	2 hours 50 minutes
Streaked Saltator	30 minutes or less	40 minutes or less
Yellow-faced Grassquit	9 minutes	5 hours 22 minutes

[a] From Skutch 1953*d*.
[b] The interval was timed from the female's first departure from the nest after the first egg hatched, or from the removal of the first empty shell, whichever came first.

was when it was being built or newly made. But finally comes a day when the monotony is broken. The female comes and goes more frequently, bringing food to the newly hatched nestlings. She rests in unfamiliar attitudes while she offers it to them. Perhaps, coaxing them to take their meal, she voices notes that have not hitherto been heard. These changes arouse the male's interest; he rests closer to the nest than formerly; he looks more attentively. Perhaps he notices food projecting from his mate's bill as she approaches. He hovers near the nest, which may be roofed, so that only a careful inspection will reveal what it contains. His interest grows warmer and warmer, until finally he goes to look in. He sees the nestlings, spends a minute or two examining them attentively, then leaves and promptly finds food to bring to them. Such was the course of events at a nest of Gray-capped Flycatchers, whose

story I have told elsewhere (Skutch 1960). The male Gray-cap did not see his nestlings until some four and a half hours after the first of them hatched, but he fed them in less than two minutes after he saw them.

At twenty-three nests of eighteen species, representing seven passerine families, that I watched carefully in Central America, the interval between the hatching of the first egg (or the beginning of daytime activity if it hatched at night) and the first feeding by the male ranged from nine minutes to between six and ten days (table 11). Eight males brought food within an hour after the first egg hatched; eight in from one to six hours; six in from six hours to one and a half days; one between the sixth and tenth day after hatching. Even in the same species in the same locality, there were considerable individual differences, which seemed to depend in part upon the male's alertness and in part upon apparently trivial circumstances, such as the nearness of the male's preferred perch to the nest. Thus, for example, one male Vermilion-crowned Flycatcher first brought food six and a half hours after his mate's first egg hatched; another was still not feeding the nestlings six days after they hatched, but after four more days I found him doing so.

In the north, males may take somewhat longer to discover their nestlings, perhaps because they do not develop such close relations with their partners as do tropical birds that are mated throughout the year. At five nests of the Prairie Warbler, the male who fed most promptly did so three hours after the first nestling hatched. Another delayed five and a half hours and a third nearly ten hours, while two others failed to bring food during the five and nearly nine hours, respectively, that elapsed between hatching and the end of the day's activities (Nolan 1958). A male Black-throated Green Warbler sang profusely, apparently oblivious of his offspring until his attention was drawn to them by the bustle of their departure from the nest, when the oldest was nine days of age. Then, at last, he brought food for them, but he tried to treat them like newly hatched nestlings rather than fledglings. Other male Black-throated Green Warblers, however, have begun to feed their young considerably earlier (Nice and Nice 1932). When I placed a newly hatched Brown-headed Cowbird in a Gray Catbird's nest that contained three eggs, the male catbird brought food to it three-quarters of an hour later, after he had seen his mate feed the intruded nestling.

The zeal with which some male birds present food again and again to unhatched eggs suggests that they lack adequate occupation and are eager to begin feeding their offspring. The same impatience to undertake parental duties is revealed occasionally by the male who brings food to a neighbor's nestlings, usually those of another species, while his mate incubates. When at last he is able to attend his own nestlings, the male frequently, in an initial burst of activity, brings more food than the newly hatched birdlings can consume. Often he entrusts it to his brooding mate for delivery to them, but sometimes he feeds them directly, in her ab-

sence, or when she rises to uncover them. When the tiny nestlings are slow to swallow what is offered to them, the parents show the utmost solicitude, presenting the same morsel over and over, often mashing it between their mandibles between presentations, and continuing for minutes together to press the nourishment upon their little ones, while they coax with soft notes. Frequently they fail to make the nestlings eat, and in the end their mother devours what they cannot swallow.

The surplus that the male brings for the nestlings permits their mother to brood more constantly than she could do if she were obliged to find food for both herself and her young—sometimes she spends more time with the newly hatched young than she did with the eggs that she was incubating. In mild weather and with small broods, this increased attentiveness may be of slight importance; in the tropics, for example, many females rear alone as many nestlings as others do with a mate's help. But if the eggs hatch in cool or rainy weather, or if the nest is exposed to direct sunshine, the more constant brooding that the male's prompt and liberal feeding makes possible may save the lives or some or all of the nestlings; hence the importance of his early discovery of the nestlings, and of the habits formed during the period of incubation that promote it.

Sometimes, when we watch a pair of birds working together in closest harmony to build a nest and raise a brood, we imagine that such efficient cooperation would be impossible without an effective means of communication—the ability of each to tell the other what needs to be done at each step in their common enterprise. But the study of how the male finds the nestlings reveals a failure of communication at a critical point where the conveyance of factual information would greatly facilitate the undertaking. Although birds seem well able to communicate their moods and emotions, we have no reason to suppose that mates are better able to communicate facts to each other at other stages of the nesting than at this point. Accordingly, we must conclude that their admirable cooperation is due to the perfection of their innate endowment, to the fact that the instinctive activities of the two partners fit together like key and lock, rather than to their ability to exchange ideas. We may further conclude that birds, for all the satisfaction that constantly mated kinds appear to find in each other's company and the distress they show when separated, have not reached that higher stage of psychic development at which joy is enhanced by the sharing of good news. If they had attained this level of development, we would hardly expect a female bird to permit her closely associated mate to remain for hours, or even days, ignorant of the fact that their nestlings had hatched.

CHAPTER 21

Feeding the Young

Until their young hatch, altricial and precocial birds face the same task, that of keeping their eggs warm. After hatching, their occupations are quite different. Parents of precocial chicks lead their broods to productive foraging areas, where the young pick up their own food or, in the case of subprecocials, receive it from their parents' bills (plate 76). These parents seem not to work so hard as the parents of altricial nestlings, who must find all the food for their offspring as well as for themselves and carry it to their nests from a distance more or less great. For these parents, the nestling period is the most strenuous stage of the whole reproductive cycle, when they are most likely to lose weight. In chapter 18, I perhaps said enough about how precocial chicks are nourished. Ornithologists have given much more attention to the feeding of altricial nestlings, which remain in one spot and, accordingly, lend themselves better to quantitative studies. In this chapter we shall be concerned only with altricial and semialtricial birds.

The Nestlings' Food and Water

Most parent birds nourish their young on the same foods that they themselves eat: insect eaters feed their offspring insects; seed eaters give them seeds; fish eaters bring them fish; reptile eaters feed them reptiles, and so forth. It is frequently said that birds that are largely vegetarian nourish their nestlings on insects, which contain more protein for building their growing bodies. The difference in the diets of parents and offspring is proportional rather than absolute, since most seed-eating and fruit-eating birds vary their diet with a few insects, or even many at times when they are easy to catch, and nestlings of frugivorous birds often receive much fruit, especially as they grow older. After the first few days, manakins and waxwings feed their nestlings largely with berries, seed eaters with seeds. Crossbills and Linnets usually rear their young on vegetable food alone, and other cardueline finches may do so occasionally. The diet of young Red Crossbills may consist of only one type of seed (Newton 1967). Oilbirds and Bearded Bellbirds appear to give only fruits to nestlings of all ages (D. W. Snow 1961, B. K. Snow 1970).

The parents do not indiscriminately bring their young the same food that they choose for themselves; on the contrary, they try to select appropriate items, especially as to size. Very small articles are chosen for newly hatched nestlings, and the size of the offerings increases as the young grow. This is especially noticeable in fish-eating birds, such as kingfishers (plate 77) and puffins. Purple Martins

PLATE 76. A Whooping Crane broods and feeds its subprecocial chick (photo from William Conway, New York Zoological Society).

bring small crushed insects to very young nestlings, but older ones receive large dragonflies, grasshoppers, and beetles. When parents forage for nestlings, they seem to eat at once food articles that are too large or too small, preserving only those of appropriate size to take to the nest. Not only does this effect an economy of effort but it also conduces to the safety of the young, particularly in birds that bring only a single article at a time; for, the fewer visits a parent makes, the less likely it is to betray the nest's situation to lurking predators. Birds of the tropical forest, such as trogons and antbirds, often seem to bring the largest articles that they can find and that the nestlings can swallow, and thereby they reduce visits to their nests to a minimum. But, if for any reason they fall behind in supplying their young and the latter become very hungry, they may bring smaller articles at a more rapid rate.

When gathering food for their young, birds depend largely on the same skills that they use when foraging for themselves: aerial flycatching, underwater fishing, hovering before flowers, gathering fruits, drilling into wood, or whatever other specialty they possess. Many kinds of birds, including ovenbirds, antbirds, kingfishers (plate 77), and trogons, usually bring a single article at a time to the young, but others, including songbirds, come with several or many pieces at once. When numerous articles are carried in the mouth or bill, we sometimes wonder how the bird manages to gather the last of them without dropping the first, and our perplexity increases when the objects are slippery, elusive fishes or flying insects.

A bill is less versatile and flexible than a five-fingered hand, yet even with a hand we should have great difficulty seizing one fish while holding others.

A Common Puffin, however, may capture over twenty fishes before it takes its haul to the nest—a feat that puzzled Edmund Selous and led him to indulge in some rather fanciful speculations. Lockley (1953) pointed out that the roof of a puffin's mouth, "under the upper mandible, is furnished with a double row of backward projecting spines or serrations which must materially help to keep the fish in position. Moreover the curious thickened finger-like tongue must be useful, when the bill is opened to gather the last of the load of little fishes, to press the fish already caught against these spines." As was earlier mentioned in our discussion of the gathering of nest materials, when a pair of Hawfinches who were damaging peas in a garden were shot, each was found to be carrying a number of larvae clamped between the tongue and the lower mandible, an arrangement that should leave the upper mandible free for picking up more. Perhaps a similar method is used by the flycatcher who continues to capture insects in the air until it has a whole billful of them. Nevertheless, when the prey is as elusive as darting minnows or flying insects, the available explanations seem hardly adequate to account for the observed results. Certainly, in addition to whatever structural aids the bird possesses, it needs amazing skill.

When food is brought in the bill or throat, the preparation it receives differs greatly with the species, the age of the nestlings, and the nature of the food. For very tiny nestlings, the food may be well mashed in the parent's bill before delivery. For young of any age, living prey is often killed or disabled by beating it against a branch; by this method the wings are frequently knocked from large insects. Jays prepare food for their nestlings by holding it beneath a foot, if the item is large, and tearing off pieces with the bill. The fragments are then crammed into the parent's throat, but the last piece may be carried conspicuously between its

PLATE 77. Male Belted Kingfisher carrying a fish to his young (photo by S. A. Grimes).

mandibles. After catching an insect in the air, Acorn Woodpeckers often place it on a horizontal branch, or in a crevice in the side of an erect trunk, while they pull off the wings, and probably also soften it, before they take it to their nestlings.

When an ani brings to a communal nest an article too large for the young to swallow, it may wait for the arrival of another parent, who then seizes the free end of the insect or lizard. Between them they pull the victim in two, and each gives its part to a nestling. I have watched Groove-billed Anis struggle for many minutes with a refractory lizard. Standing on opposite sides of the nest, two parents tried valiantly to break it between them, but they succeeded only in pulling it from each other's bill. In addition to such ineffectual attempts to dismember the lizard, one parent or another would take it to a clear space on the ground and work strenuously for five or ten minutes to reduce its size, but in vain. At intervals, the still-intact lizard was presented to a nestling, who struggled fruitlessly to gulp it down. At last, a parent carried it off, and I saw it no more. When possible, nestlings of anis, roadrunners, and other lizard-eaters swallow the reptile headfirst, leaving the tail projecting from the mouth until, as the foreparts are digested, it can be gulped down.

Birds of many kinds make little effort to prepare food for their nestlings, giving them, for example, fruits that consist of a large, indigestible seed surrounded by a thin pulp, or such a seed enclosed in a saclike aril, when it would be relatively easy for them to remove the aril from the seed. After they digest the surrounding flesh, the young birds regurgitate the seed, as they do with the hard, chitinous parts of insects and other indigestible matter.

A few forehanded birds store some or most of the food for their young. In Scandinavia, Thick-billed Nutcrackers nourish their nestlings chiefly with hazelnuts that they buried in the ground in the preceding autumn and with remarkable memory can find even under the deepest snow (Swanberg 1956). In the western United States, Piñon Jays give cached piñon seeds to their nestlings and fledglings, although these form only a minor part of their diet (Balda and Bateman 1971). Lewis' Woodpeckers, which lay up large stores of acorns in the fall, show a shorter prevision while feeding their nest-

lings. When flying insects are very abundant, these woodpeckers capture more than enough for immediate consumption and insert the excess in cracks in their nest stubs, to be given to the young as needed (Bock 1970). Most extraordinary is the practice of the Crested Bellbird, which inhabits arid parts of Australia. It gathers living grass caterpillars, partly paralyzes them by squeezing them in its bill, then deposits them among the eggs in its open nest. The bellbird's forehandedness seems to be related to the circumstance that it breeds at a time when increasing drought will make caterpillars harder to find. Although this stored food ensures that the newly hatched nestlings will be promptly fed, it hardly suffices for their needs until they are fledged (Chisholm 1952*b*).

Some birds prepare in their own alimentary canals a special food for their nestlings. This has long been known of pigeons, which give their young a curdlike substance that is aptly called "pigeon's milk" (plate 84). Like milk, it is rich in proteins, fat, and ash, but it lacks sugar. Moreover, it is produced from fatty cells that become detached from the crop's epithelium, by a process somewhat similar to that which yields milk in the mammary glands of mammals. The analogy does not end here, for in both pigeons and mammals the secretion of the milk is stimulated by prolactin, a hormone formed in the anterior lobe of the pituitary gland. Unlike the milk of mammals, pigeon's milk is produced by both sexes of parents, which share rather equally the care of the young. Newly hatched pigeons and doves receive only this secretion from their parents' crops. As they grow older, an increasing proportion of the seeds and other solid foods taken by the parents is mixed with the milk regurgitated to the young. In the Wood Pigeon, the proportion of milk in the nestlings' diet decreases from 92 percent during the first three days to 49 percent between seven and nine days, to 33 percent between ten and fourteen days, and to around 20 percent during the remaining week of nest life (Murton 1965). Among parrots, which feed their mates and young by regurgitation, crop milk has been found in Budgerigars.

Albatrosses, petrels, and shearwaters secrete oil in the proventriculus for the nourishment of their

PLATE 78. Black-browed Albatross regurgitates to its chick on Campbell Island, New Zealand (photo by John Warham).

young (plate 78). Although the presence of this secretion was long known, its nutritive function was obscured by the fact that certain members of this group forcibly squirt ill-smelling oil from their mouths or nostrils at the intruder who molests them —a defensive reaction to which we shall return in chapter 32. However, albatrosses, which do not deliberately eject oil for defensive purposes, secrete it liberally, beginning shortly before their eggs hatch. Nonbreeding birds, and those in the early stages of incubation, when roughly handled, vomit a greenish fluid that contains only a trace of oil. This stomach oil is the only food fed to albatross chicks during their first few days, and even thereafter it enters largely into their diet. Since in albatrosses, as in

petrels and shearwaters, stomach oil serves primarily as food for the young, it is analogous to pigeon's milk (Rice and Kenyon 1962).

The food given to most young birds contains all the moisture that they need and their parents do not bring them water to drink. A notable exception is the sandgrouse, which live in arid parts of Africa and Asia. The chicks leave the nest as soon after hatching as they are dry and pick up their own food, consisting largely of seeds, which do not contain as much moisture as insects. To remedy this deficiency, the parents bring them water from a distance. This is normally done by the male, whose specially modified abdominal feathers have a much greater water-holding capacity than the plumage of

PLATE 79. Male Namaqua Sandgrouse soaking his specially modified breast feathers with water that he will carry to his distant chicks (photo by G. L. Maclean).

most birds—greater even than the abdominal feathers of the female sandgrouse. The sandgrouse father goes in the morning to drink and soak up water in his feathers at one of the few available sources in the dry country where he dwells (plate 79). Although some of this water evaporates on a long flight through desert air, the Namaqua Sandgrouse can carry enough to satisfy the needs of his chicks a distance of 15 to 20 miles (24 to 32 kilometers) from the nearest waterhole in the Kalahari Desert. When he reaches his destination, his chicks emerge from their hiding places beneath the sparse vegetation. They crowd around him and with their bills strip the water from his abdominal feathers while he stands upright among them with out-fluffed plumage (plate 80). They are dependent upon their

parent for water until, when nearly two months old, they can fly to the waterhole (Cade and Maclean 1967, Maclean 1968). Thus, although sandgrouse are fully precocial in respect to food, in respect to water they are semialtricial—reminding us once more that we can draw no sharp boundaries between the several categories of young birds. The water-carrying habit of sandgrouse is not altogether unique, for Little Ringed Plovers bring water on their breast feathers to chicks hatched on very hot days (Gatter 1971).

Methods of Feeding

Birds have two principal methods of feeding their young, directly from the bill or mouth and by regurgitation. The term "regurgitation" is often care-

PLATE 80. Namaqua Sandgrouse chicks drink water from their father's breast feathers (photo by G. L. Maclean).

lessly used and is perhaps impossible to define with precision. It is preferable to restrict this designation to the delivery of food that has been swallowed or secreted by the parent and is carried in the crop or stomach, whence it is brought up by considerable muscular effort when the young are fed. When food is brought to the nest in the bulging throat in addi-

PLATE 81. Red-bellied Woodpecker brings a billful of berries to nestlings inside the hole (photo by S. A. Grimes).

tion to that in the mouth and bill, as occurs in certain jays, or only in the swollen throat, as is the case in swifts, it does not seem proper to speak of regurgitation.

Nestlings of passerines and certain other small birds fed by regurgitation, including hummingbirds, often have a dilatation of their oesophagi that when filled with rapidly delivered food forms a prominent swelling on the front and one side of their necks. In redpolls and some other seed eaters, such oesophageal diverticula persist into adult life (Fisher and Dater 1961). Nestlings that are fed from the bill usually lack this swelling and have more symmetrical necks.

Although feeding by regurgitation seems to be followed without exception in certain families, such as albatrosses, petrels, pigeons, and hummingbirds, in many families both methods of feeding are found. Whether a bird brings food to the nest in its bill and mouth or deeper in its alimentary tract depends largely on the kind of food it selects and also on the distance that the food is carried. When the parent collects fairly large items not far from its nest, it would only be wasting its effort if it swallowed them, then laboriously brought them up a few seconds later. But obviously it would be impossible for a hummingbird to hold much nectar in its narrow bill, and the minute insects and spiders that supplement the nestlings' liquid diet are also most easily carried in the alimentary canal. Petrels and shearwaters, which gather small marine organisms at great distances—sometimes apparently hundreds of miles—from their nests, could hardly nourish their young if they did not swallow the food and regurgitate when they arrive at the nest.

Although most finches bring food in their bills and mouths, grassquits and seedeaters, which nourish their young chiefly with minute grass seeds, feed by regurgitation, as do seed-eating goldfinches, siskins, and crossbills. Mannikins, waxbills, grassfinches, and seed-eating weaverbirds likewise feed their young by regurgitation, but some of the insect-eating weaverbirds bring food in their bills. Among tanagers, the euphonias and chlorophonias, which subsist largely on mistletoe berries and gelatinous fruit pulp, regurgitate food to their nestlings, but most other species carry insects, berries, and pieces

PLATE 82. An American Flamingo regurgitates food for its newly hatched, downy chick on its moundlike nest of mud (photo from William Conway, New York Zoological Society).

PLATE 83. A Green Violet-ear Hummingbird feeds her nestlings by regurgitation. Their growing bodies have burst the green mossy nest asunder and they rest upon its ruins in the Guatemalan mountains.

of larger fruits in their bills and mouths. Among honeycreepers and their allies, the fruit-eating species carry food in their slender bills and their mouths, but the nectar-drinking flower-piercers and bananaquits feed by regurgitation and their nestlings have prominent neck pouches like those of the nectar-sipping hummingbirds. Woodpeckers that feed mainly on fruits and fairly large insects carry food to the nest in their bills (plate 81), but those that gather enormous numbers of ants and their larvae and pupae find it convenient to swallow this food and then regurgitate to their young. An exception is the minute Olivaceous Piculet, for which even an ant's pupa is a relatively large object. When a woodpecker that feeds its young from the bill wishes to give them liquids, such as honey found at a feeding station or sap from a boring in the bark of a tree, it may gather a billful of insects, seeds, or bark, soak it in the liquid, and carry this to the nest,

as has been observed in the Gila Woodpecker and Yellow-bellied Sapsucker (Antevs 1948, Blackford 1950).

When food is regurgitated, it must often have been acted upon by the parent's digestive enzymes while in its body, and the amount of such predigestion should vary with the interval that elapses between swallowing and delivering the food to a nestling. It is likewise possible that in some cases the food is enriched by secretions from the parent's alimentary tract, analogous to the milk of pigeons and the oil of albatrosses but in smaller amounts and less easy to detect. This is a matter of which we still know too little.

Food is transferred from parent to young in diverse fashions. Whether fed directly from the bill or by regurgitation, most passerine nestlings, especially those that grow up in open nests, lift up their gaping mouths, displaying the vivid linings bordered by broad, light-colored flanges, and wait passively for the parent to stick in the food. Nestlings in dark holes often take solid food from the parent's bill with a grasping movement. Before regurgitating, a hummingbird pushes her long, sharp bill so far down a nestling's throat that one witnessing the procedure for the first time fears that she is about to commit infanticide (plate 83). Usually she feeds her two nestlings several times, alternately, on each visit. When a swift returns to its nest, its throat distended with a mass of small insects caught high in the air and stuck together with saliva, it pushes much of its head into the enormous gape of an older nestling to deposit the food in the youngster's throat with its small bill. A parent pigeon takes a nestling's bill in its mouth (plate 84), usually one at a time when the young are newly hatched and sightless but often two simultaneously, on opposite sides, after they can see. The heads of the parent and young bob up and down together as the food is forcibly regurgitated from the parent's crop. A parent petrel or shearwater takes the bill of its single nestling into its mouth somewhat as a pigeon does; the parent Diving Petrel "then squirts forth a red cream, thick and ribbon-like, as if from a tube of tooth paste" (Richdale 1943).

When a parent gull arrives with food, a chick pecks at its bill until it regurgitates. In the Herring

PLATE 84. Wood Pigeon regurgitates "pigeon's milk" to its nestling in an English copse (photo by R. K. Murton).

Gull, the target at which the chick aims is the red spot on the lower mandible of the yellow bill. This stimulates the old gull to regurgitate a mass of partly digested food onto the ground, from which it picks up pieces and passes them to the chick, although at times the latter helps itself to the food before it. Unlike certain other gulls, Kittiwakes always regurgitate directly into the mouths of their nestlings and thereby avoid soiling their narrow nest ledges with remains of food, a consideration that hardly applies to ground-nesting gulls whose young wander from the nest (E. Cullen 1957). Among terns, the Wideawake pecks at its parent's

bill and is fed by regurgitation, but other species receive insects or small fishes that are brought visibly in the parent's bill (J. M. Cullen 1962). The pecking responses of young gulls and terns have fascinated experimenters, who have presented them with a variety of models, different in color and shape, and counted the number of pecks that each elicited (Tinbergen 1953, Hailman 1962).

Herons, egrets, and bitterns also regurgitate to their nestlings, at first a predigested liquid soup, which as they grow older is replaced by whole or partly digested fish, frogs, and other items. The growing heron or bittern seizes the base of the parent's bill with a scissorlike grip, often wrestling strenuously with it until the food comes up. Spoonbills and ibises open their long bills and permit their young to pick regurgitated food from their throats. This method of feeding, often in an exaggerated form, is widespread among pelicans, cormorants, boobies, and related birds: For naked, newly hatched nestlings they regurgitate liquid aliment. When the chicks grow stronger, the parent opens its bill widely, permitting them, one at a time, to push in their heads and rummage for food in the capacious parental gullet. It looks as though the parent is trying to swallow its offspring headfirst!

Great White Pelicans less than ten days old may peck feebly at the cherry red, naillike tip of the parent's bright orange-yellow bill, much as the Herring Gull chick pecks at the red spot on its parent's bill; usually, however, the old pelican must coax the small nestling to take nourishment. Young from ten to twenty-five days old sometimes spontaneously solicit food by pecking at the parent's breast or beak, but mostly they must be persuaded to take their meals. "The process is brutal, the young being seized by the neck and shaken, sometimes flung roughly about," until it pecks at the base of the parent's bill or at its breast and is fed. When somewhat older, the growing pelican pursues its parent with calls and vigorously beating wings, finally grasping the old bird's bill and dragging it roughly down. Then the parent usually lies on the ground and opens its great beak so that the youngster can stick in its head. Since it cannot see what it is doing inside the capacious gullet, it struggles blindly to reach food, often staggering about and dragging the adult with it. Although in the Pink-backed Pelican this sort of struggle often results in injuries to the young or their parents, Great White Pelicans are rarely hurt by these violent transactions (L. H. Brown and Urban 1969).

When a hawk or falcon brings food to its eyrie, it holds the prey beneath a foot while it tears off pieces and passes them to its young, from time to time swallowing a bit itself. A Double-toothed Kite that I watched dismembered even small insects in this manner. One insect, which might have been gulped down whole by a nestling of one of the larg-

er flycatchers, was broken into about thirty-three tiny fragments, each of which was daintily eaten by a nestling or its parent. Yet, strangely enough, lizards, much larger than any of these insects, were delivered whole to the young kites and gulped down headfirst, apparently because the parent found it too difficult to dismember them. Older raptorial nestlings often receive their food entire and themselves tear off the pieces that they swallow.

How Meals Are Apportioned among the Young

When a nest contains several young, all of which look very much alike to the human eye, one wonders how the parents distinguish them, in order to apportion the food fairly among their offspring and make sure that none is neglected. Usually when a passerine parent reaches the nest it delivers food so rapidly that there is hardly time for a critical weighing of claims. As a rule, the nearest or the highest gaping mouth receives the meal. But if, because it is satiated, the nestling does not promptly swallow what it received, the parent may remove the food to place it in another mouth, or perhaps the same one again, until it finally goes down with due speed. Ultimately, receiving a meal depends on the rapidity of the swallowing reaction, which in turn is evidently determined by the nestling's need of food. When I watched a nest of White-tipped Brown Jays at which seven grown birds attended three nestlings, food was very often removed from the throat of a sluggish, satiated nestling. The removal was a slower, more conspicuous act than placing it there; it looked as though the nestlings were supplying food to their attendants!

At a nest of the Southern House Wren in a gourd with a wide doorway through which I could watch the four feathered nestlings, the one that opened its orange-yellow mouth most promptly received all the insects that a parent brought on each visit, and food was never taken from it, as occurred with the jays. An instant later, all four of the nestlings might be clamoring for food, which meanwhile had been swallowed by the one who had been most alert, doubtless because it was the hungriest. In this case, prompt delivery by the parent ensured that the meal went where it was most needed. After eating, the nestling would often turn around to deliver a

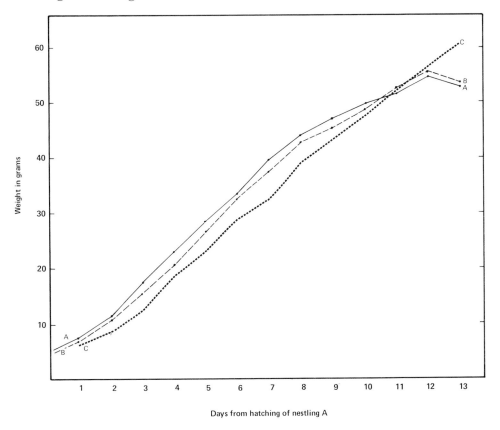

FIGURE 19. Growth of three nestlings of Gray's Thrush. The youngest (*C*) hatched a few hours after its siblings (*A* and *B*) and for the first ten days remained the lightest, but finally overtook them. Weighing was discontinued two days before the young left the nest; they would not have remained if disturbed.

dropping to the parent, and this often caused it to lose the preferred place nearest the doorway. There was a constant shifting about that prevented any one of the nestlings from receiving most of the food.

When the nest has an opening just wide enough to accommodate a single bird, as in woodpeckers and many other hole nesters, a feathered nestling often occupies the orifice and while there nearly always receives all the food that the parents bring. Sometimes it appears that the same head fills the doorway for hours together, most unfairly depriving its nest mates of their share of nourishment. But if the observer can recognize the nestlings individually, he may find that the head in the doorway changes rather frequently. At a nest of Golden-naped Woodpeckers with a female and two male nestlings, a male once received 7 consecutive meals and each of the others received a single uninter-

rupted series of 4 meals. Otherwise, in six hours of watching, I saw no young woodpecker get more than 3 consecutive meals. Since 122 meals were delivered in this interval, it was evident that none of the three nestlings succeeded in establishing a monopoly.

Newly hatched nestlings may be, as we have seen (p. 258), coaxed long and patiently to take their meals. Soon, however, they spontaneously raise their open mouths when the arriving parent shakes the nest; any gentle swaying, made for example by the wind or a human visitor, causes the same reaction. Likewise any movement above the nest, as of a hand, prompts them to gape for food, and they also respond to a variety of low sounds. But after their eyes are open and they learn to recognize their parents, they gape only to them. The appearance of a hand or some other strange object above the nest now causes them to crouch down, as though in fear, rather than to reach up for food, as they formerly did. If drowsy nestlings fail to gape when a parent arrives, they are often brought to attention by a special call.

Despite certain reactions of parents and young that, in passerines and some other birds, prevent one nestling from being overfed while its brood mates go hungry, when the attendants cannot supply enough food for all, the smallest and weakest nestling, which is often the youngest that gets off to a bad start, often receives less than the others (fig. 19). It may be so underfed that it leaves the nest underweight and frequently dies. In certain birds, especially raptors, if the youngest does not succumb from malnutrition, it may be killed or driven from the nest by its elder siblings, a matter to which we shall return in chapter 23.

Rates of Feeding

The rate at which nestlings are fed varies enormously from species to species. In general, parents that regurgitate feed far more seldom than those that bring food in the bill. This difference is well illustrated by the woodpecker family, in which parents who feed by regurgitation may bring food at the rate of only once per hour for each nestling, or even less frequently. Woodpeckers who carry food in their bills may bring six or seven meals per nestling per hour and at intervals, especially in the early morning, they feed much more rapidly. Another peculiarity of birds that feed by regurgitation is that, in many instances, the frequency of delivering meals increases little, if any, as the nestlings grow older, and it may even decrease. In five hours of the morning, two White-crested Coquette Hummingbirds received twelve meals when they were about three days old, nine when they were about ten days old, and twelve when they were about twenty days old. Each meal evidently became more copious as the nestlings grew. As pigeons and doves grow older, the parents regurgitate to them fewer times in the course of the day. Although continuous regurgitation may now last longer than when the nestlings were newly hatched, the total time devoted to feeding them diminishes. Similarly, in the Greater Flamingo, the regurgitated meals, at first

infrequent and brief, became still less frequent but much longer as the young grew, increasing from five to fifteen seconds when they were newly hatched to eleven to fifteen minutes when they were about to fly (L. H. Brown 1958).

In all these birds that are fed by regurgitation, the bigger the young, the more rapidly they exhaust the contents of the parent's crop. Petrels and shearwaters are fed by the parents at night, and often they pass one or more nights—up to ten in the case of the Sooty Shearwater—without a meal. Although infrequent, their meals are liberal; a Sooty Shearwater may, in exceptional cases, receive as much as 10.6 ounces (300 grams) of food in a single night, from parents who weigh about 28.2 ounces (800 grams) (Richdale 1945). Young King Penguins, whose regurgitated meals are also widely spaced, may receive more than 2.2 pounds (1 kilogram) at one feeding (Stonehouse 1960).

In contrast to the massive, infrequent meals of these and many other birds that feed by regurgitation, the rate of food bringing by those that carry food in the bill is often amazingly rapid. In the Netherlands, a family of nine Great Tits, seventeen days old, were fed 990 times by both parents in the course of a day; but even for titmice such rapid feeding is exceptional (Kluijver 1950). A pair of Red-throated Rock-Martins in Central Africa fed three well-grown nestlings at the rate of once per minute, maintaining this strenuous activity for at least six days (Moreau 1947). Five or six older nestlings of the Northern House Wren may receive 491 meals in a day from both parents, but this again is an exceptionally high rate for the species (Kendeigh 1952). The most rapid feeding that I have recorded in many years of watching nests was 65 meals in a single hour, by a pair of Gray-capped Flycatchers attending three nestlings eleven days old. Four to 12 meals per nestling per hour is the average rate of feeding by a wide variety of small birds. Others that bring food in the bill have much lower rates. One meal for each nestling in an hour is about the average rate of feeding of trogons, and some antbirds and ovenbirds feed only slightly more rapidly (tables 12 and 13).

If the male takes any active part in the nesting, other than guarding the nest, he nearly always feeds

TABLE 12

Rate of Feeding Nestlings with Food Brought in the Bill, Mouth, or Throat[a]

Species	Hourly Rate of Feeding Visits per Nestling during Week of Life			
	First	Second	Third	Fourth
Black-throated Trogon	0.5	1.0	—	—
Broad-billed Motmot	—	2.2	—	1.6
Blue-diademed Motmot	1.2	1.7	2.3	1.6
Rufous-tailed Jacamar	—	—	—	2.4
White-whiskered Softwing	0.5	0.4	0.4	—
Tawny-winged Dendrocincla	0.75	1.0	1.3	—
Rufous-fronted Thornbird	—	6.6	5.5	—
Great Antshrike	1.0	2.6	—	—
Black-hooded Antshrike	0.7	2.4	—	—
Chestnut-backed Antbird	2.1	4.3	—	—
Orange-collared Manakin	1.2	1.7	—	—
Masked Tityra	1.0	1.2	1.0	2.0
White-winged Becard	—	2.3	2.1	—
Rufous Piha	1.0	1.3	1.7	1.8
Eastern Kingbird	2.7	5.3	5.8	—
Yellow-bellied Elaenia	8.9	9.4	13.6	—
Wire-tailed Swallow	—	—	12.3	—
Black-eared Bushtit	3.3	4.7	9.0	—
Southern House Wren	—	6.6	5.5	—
Carolina Wren	1.1	2.6	—	—
Long-tailed Silky-Flycatcher	2.9	6.8	6.8	5.6
Shining Honeycreeper	3.2	7.2	—	—
Scarlet-rumped Black Tanager	6.5	6.1	—	—
Black-faced Grosbeak	1.4	2.5	—	—

[a] See note to table 13.

TABLE 13

Rate of Feeding Nestlings with Regurgitated Food

Species	Hourly Rate of Feeding Visits per Nestling during Week of Life		
	First	Second	Third
Yellow-shafted Flicker	—	—	1.1
Rufous-winged Woodpecker	0.25	0.5	0.5
Scaly-breasted Hummingbird	1.2	—	1.2
White-crested Coquette	1.2	0.9	1.2
Long-billed Starthroat	0.6	—	1.0
Tawny-bellied Euphonia	0.75	1.3	1.3
White-vented Euphonia	1.2	1.2	1.4
Variable Seedeater	4.5	7.6	—

NOTE: Each entry in tables 12 and 13 is based upon at least five hours of observation, and often much more. The records were all made by the author, except the following: Eastern Kingbird (Morehouse and Brewer 1968), Wire-tailed Swallow (Moreau 1939), Carolina Wren (Nice and Thomas 1948), Yellow-shafted Flicker (Sherman 1952).

the young. In many species, he feeds but neither builds nor incubates; in others, he builds and feeds but does not incubate. When the nestlings first hatch, their nonincubating father may bring most or even all of their food while their mother spends much of her time brooding them. Occasionally a male passerine is at first reluctant to approach the nest while the female is absent, and he may flit about a short way off with food in his bill until she returns to the nestlings, when he advances to deliver his offering, as I have seen in the Golden-crowned Spadebill, a diminutive flycatcher. In the White-whiskered Softwing, the male, who takes the leading role in incubation and attends the eggs through the night, is alone responsible for brooding the newly hatched young, while the female brings all their food. After some days, the male softwing helps to feed the nestlings, but the female seems at all stages to be the chief provider.

As the nestlings grow older and need less warming, their mother, in many passerine species, brings more meals to them, finally equalling, or even surpassing, her mate in this activity. Although fairly equal participation in feeding by the two parents, after brooding ceases, seems to be the rule among songbirds, the female Scarlet-rumped Black Tanager nearly always feeds more often than her mate. At some nests she brings twice as many meals. In a number of species, it has been found that each parent compensates for any excess or deficiency in the activity of the other, so that, if for a few days the male brings more food than he usually does, the female brings less and vice versa; or, if one parent dies or deserts, the survivor may feed as often, or nearly as often, as both had been doing together. In many birds that do not regurgitate, the rate of feeding increases for some days after hatching, then tends to become stationary, with minor daily fluctuations. If the young remain long in the nest, the rate reaches a peak, then declines after the nestlings have passed their period of most rapid growth. This decline in the rate of bringing food is especially likely to occur if the nestlings, having surpassed the adults in weight, become lighter in preparation for flight.

Many birds feed their nestlings most actively in the early morning, when they are hungriest after a night of fasting. After the young are satiated, the parents relax their efforts, doubtless foraging for themselves, then bring more food after their young have digested their first meals. So feeding continues throughout the day, with alternating intervals of greater and less activity. In most passerines and other birds that feed at a fairly rapid rate, there is probably no hour, and with large broods probably

no half hour, in which some food is not brought to the nest. Sometimes a peak of activity comes rather late in the day. Parents often continue to feed their nestlings in a light shower. A hard downpour depresses their rate of feeding, but, if it is long continued, the parents make an effort to nourish their young despite the deluge.

As a rule, the larger the brood, the more often the parents bring food to the nest. Employing relays of native observers whom he had trained in Central Africa, Moreau (1947) collected vast amounts of data on the rates of feeding of a variety of birds. These observations showed that in several species the number of meals served at the nest did not increase in proportion to the number of its occupants, so that a member of a small brood received more per day than a member of a larger brood. Thus a solitary nestling of the Wire-tailed Swallow was fed at an average rate of 16 times per hour, but the parents brought food only 13.2 times per hour for each member of a brood of two, and only 8.5 times per hour for each of a brood of three. This relation between brood size and the number of meals brought for each nestling was found by Moreau to hold for several other African swallows, the Palm-Swift, Little Swift, Paradise Flycatcher, and Mountain Wagtail. Elsewhere it has been demonstrated in Red-crowned and Red-throated ant-tanagers in British Honduras, Northern House Wren in the United States, Pied Flycatcher in Finland, Starlings in the Netherlands, and in Great Tits, Blue Tits, and Robins in England. House Sparrows increased the frequency of their feeding visits in proportion to the brood size in broods of one to three, but for larger broods they failed to increase their feeding rate. Yet the most common brood size was four (Seel 1969).

It does not necessarily follow from this that nestlings in large broods are undernourished. As Royama (1966) pointed out, each member of a large brood needs less food than each member of a small one because by huddling together the nestlings keep each other warm. From the thermal aspect, they may be considered to form a larger animal. As is well known, the bigger the animal, the larger the ratio of the heat-producing mass to the heat-dissipating surface, and the less the cost per gram of body weight of keeping warm in cool weather. The member of a large brood spends less energy to maintain its body temperature and can devote a larger portion of its nutriment to growth.

The lower the temperature, the greater the advantage of belonging to a large brood. At an ambient temperature of 64.4°F (18°C), three nestling Great Tits kept warm with much less metabolic cost than a single nestling; but further increases in the size of the brood reduced the energy expenditure only slightly. When the air temperature was reduced to 53.6°F (12°C), however, the amount of organic fuel that each nestling had to burn in order to keep warm fell steadily and substantially as the brood was increased to twelve. When too many nestlings were placed together, they tended to become overheated (J. A. L. Mertens 1969). We may conclude that it is more economical, in terms of parental effort per nestling, to raise large broods than small ones, within the limits placed by availability of food and the number of eggs and young that the parent can adequately cover, and that, the cooler the weather, the greater the economy. This, however, will not help us to understand why tropical birds raise smaller families, even in the cool highlands, than those at high latitudes.

There are, moreover, many exceptions to the widespread tendency for parent birds to bring relatively less food for each member of a large brood than for each member of a small one. Moreau himself found such an exception in the White-rumped Swift, in which the hourly rate of feeding was twice as great at nests with two young as at nests with one, so that the same number of meals were brought for each nestling whether it belonged to a large or a small brood. In contrast to the findings in England, Kluijver's massive data from the Netherlands failed to substantiate the decrease in the number of feeding visits per nestling Great Tit as the brood becomes larger; he believed that this occurs only when the parents cannot find enough food. Later, in an intensive study of Black-capped Chickadees, Kluijver (1961) found that, even in broods of three, each nestling is not fed more often than in a brood of five or six. The number of feedings per nestling depends more on the size of the items brought than on the number of young in the brood. Food bringing

is regulated by the nestlings' needs, as indicated by their begging cries. The parents do not ordinarily work to the limit of their capacity feeding their young. If kept away from their nest, such as by photography, they compensate by bringing food much more rapidly, continuing to do so up to a half hour longer in the evening and also starting earlier on the following morning.

Once when I had two nests of the Scarlet-rumped Black Tanager, each with two nestlings of about the same age, I transferred a young tanager from one nest to the other in such a way that on consecutive mornings I watched the same two parents attend two, three, and then one nestling. They adjusted their rate of bringing food to the size of their family with an accuracy that surprised me. In five hours, two nestlings received 75 meals, three nestlings received 106 meals, and the single nestling was fed only 36 times, so that the number of meals brought per hour for each nestling was, respectively, 7.5, 7.1, and 7.2.

I was impressed by how much more noise the occupants of this tanager's nest made at meal time when there were three than when there was only one. The lone nestling was soon silenced by the food placed in its mouth. But, when three were present, two unfed young remained to clamor for food each time a parent arrived, and this seemed to stimulate the adult to return more promptly with another meal. This matter was investigated in Finland by Haartman (1953), who devised a number of ingenious experiments with the Pied Flycatcher. In a specially made nest box, he reduced the number of nestlings from seven to two; after the appetite of these two had been satisfied, they were rapidly replaced by two others that had become very hungry; after these were fed, two more were substituted for them. The parents continued to bring food to the succession of two constantly hungry nestlings about as frequently as they had done for the nestful of seven. In another experiment, a double nest box was devised. In one side, which the parents could enter, a single nestling was left. In the adjoining compartment were six nestlings that the parents could hear but could not see or reach, so that they became very hungry and clamorous. The parents brought food to the single accessible nestling far more rapidly than they would have done if they could not have heard the others. In yet another experiment, mirrors were placed on two sides of the nest box to increase the apparent number of the nestlings, but this arrangement caused no increase in the rate of feeding.

These experiments clearly demonstrated that the rate at which the flycatchers brought food was determined, not by the number of nestlings, but by the pleas of the hungriest of them. In this hole-nesting species, the hunger cries of the nestlings were evidently the most effective stimulus to the parents; but, in birds with open nests, the speed and strength of the nestlings' gaping reaction may be equally or even more important. I have noticed with a variety of birds that when the nestlings receive their food eagerly and clamorously the parents promptly return with more, whereas if the meal is taken sluggishly the attendants remain absent a longer time.

Although clamoring by the nestlings increases the number of parental visits with food, it is certainly not an indispensable condition for bringing food. We have already learned that birds, especially males, sometimes offer food to eggs that cannot beg for it. Moreover, parent birds may continue for some hours, or even days, to bring food for nestlings that have died, or have been carried off by predators, sometimes along with their nest, so that the bereaved parents have not even a nest at which to offer their billfuls. I have witnessed such behavior in trogons, nunbirds, wrens, vireos, tanagers, and icterids. A pair of Golden-naped Woodpeckers continued for six days to bring food to their desolated nest. Others have recorded similar behavior in hummingbirds, kingfishers, and finches. Murres or guillemots (*Uria*) perform the food-offering ceremony at their nest site for two or three days after they have lost a pipped egg or a chick. In captivity, parent Wood Thrushes carried food for eight days after the death of newly hatched nestlings (Ivor 1952). This persistence in food bringing after the loss of young may be compared to the continuation of incubation far beyond the normal period when the eggs fail to hatch, which we earlier noticed. In either case, the persistence of the appropriate ac-

tivity provides a wide margin of safety for the eggs or young.

Increase of Parental Efficiency with Age

In recent years, we have had mounting evidence that in many species, especially of long-lived birds, experienced individuals are more efficient as parents than those nesting for the first time. Although this increased efficiency may be manifest at any stage of the nesting, it is frequently evident when feeding the young, so we may consider it here. We have already noticed that, in species which build elaborate nests, the first efforts of young birds often result in inferior structures, and that as Mallee-Fowls grow older they construct better and more efficient incubator mounds.

Some of our best evidence on this point comes from gulls and terns. Kittiwakes nesting for the first time reared fledglings from only 36 percent of their eggs; those breeding for the second time, from 56 percent; and those breeding for at least the third time, from 70 percent of their eggs. Moreover, the older birds laid larger sets of eggs. In several instances, Kittiwakes nesting for the first time failed to incubate their eggs properly, deserted before their eggs hatched, or failed to feed newly hatched chicks, which accordingly died. When inexperienced parents did rear young, they grew, on the average, about as rapidly as the nestlings of older parents if there was only one in a nest; but, with two in a nest, the offspring of experienced parents gained weight more rapidly than those of parents breeding for the first time (Coulson and White 1958). The performance of Kittiwakes is improved not only by practice in attending nests but also by sustained cooperation with a certain mate. Long-established pairs of Kittiwakes lay earlier and larger sets and hatch more of their eggs than do newly formed pairs, even if these new pairs consist of birds that have had nesting experience with different mates (Coulson 1965).

Arctic Terns on the Farne Islands, Northumberland, may live for more than twenty years, and they were not found nesting until they were three years old. Those breeding for the first time were mostly unsuccessful. Maximum success was achieved by terns over five years old, and at ten years and over they continued to breed with no decrease of fecundity (Horobin 1969).

Similar increase of breeding success with age has been found in other northern sea birds. Although Gannets nesting for at least the third time hatched 86 percent of their eggs, pairs nesting for the first time hatched only 62.5 percent, because they lost more eggs and more of the remainder failed to hatch, probably as a result of ineffective incubation and infertility (Nelson 1966). Young murres breeding for the first time seem seldom, if ever, to succeed in hatching their eggs. They tend to lay on the fringes of the colonies, in the most vulnerable locations, and they readily desert their eggs. Moreover, they lay so late in the short clement season of the high Arctic that they would hardly have enough time to raise their young, even if they did hatch (Tuck 1960).

In Latvia, Tufted Ducks start to breed when one year old, lay about ten days later than two-year-olds, and are less successful in rearing their broods. Strangely enough, the two-year-olds were more successful than those a year older (Mihelsons 1968).

Turning now to small land birds, we find that among Blackbirds at Oxford yearling pairs laid fewer eggs and nested less successfully than pairs of older birds, raising on the average only 3.4 young per pair while their elders raised 6.3 young (D. W. Snow 1958). Yearling female Skylarks in England hatched only 42 percent of their eggs and raised young until they could leave the nest from only 15 percent of their eggs, whereas older females hatched 97 percent of their eggs and fledged 51 percent of the total laid (Delius 1965). In California, older male White-crowned Sparrows held territories on the average more than twice as large as those of yearling males. All six pairs that included an experienced male succeeded in fledging young, but only three of nine pairs with a yearling male did so. Nesting success depended more upon the age of the male of a pair than upon that of his mate; yet in successful pairs the mean combined ages of both partners was 3 years, whereas that of unsuccessful pairs was only 2.3 years (Ralph and Pearson 1971). Northern House Wrens, on the contrary, seem not to improve as parents as they grow older. A large number of records showed that yearlings of both sexes fed their nestlings as often as older wrens (Kendeigh 1952). In small land birds, whose average life span is much less than that of many seafowl, it seems more necessary to attain high breeding efficiency at an early age.

When we considered the improvement in nest construction of certain birds as they grew older, we asked how much this was to be attributed to practice and how much to increasing maturity itself, so that the bird might improve even if denied an opportunity to practice. In an effort to answer a similar question in regard to later stages of the breeding cycle, Lehrman and Wortis (1967) experimented with captive Ring-necked Doves. They found that doves who have already nested once are more efficient in hatching eggs and rearing young than others of the *same* age nesting for the first time. This difference in efficiency largely disappears in the next nesting, showing that in these doves the major part of the experience needed for high success in breeding is gained in raising a single brood, which, of course, involves many repetitions of the principal activities. When an inexperienced female was paired with an experienced male, she delayed longer to lay her first egg than did an experienced female; but, otherwise, pairs in which one partner of either sex was experienced and the other inexperienced did as well as pairs in which both partners were experienced. One partner compensated for the other's deficiencies. Another way in which Lehrman and his associates demonstrated the importance of experience in parental behavior was by injecting doves with prolactin, which stimulates the production of crop milk. Of twelve experienced birds, ten were induced to feed nestlings by this treatment; but, of an equal number of doves with no parental experience, none did so. Evidently practice, apart from age and the maturation it brings, improves the performance of parental offices.

CHAPTER 22

Sanitation of the Nest

Plutarch, in his essay on the relative intelligence of land and water animals, remarked that the wryneck and swallow are so cleanly that one always removes and hides its excrement, whereas the other teaches its young to turn their tails out of their nest before they void their droppings. From this statement it is evident that already in the first century of the present era men were familiar with the two principal methods by which nests are kept clean.

Nest sanitation is most efficiently performed by the more highly evolved birds, especially the passerines. In most species of this order, the excrement of the nestlings is enclosed in a fairly tough, gelatinous sac, which facilitates its removal by the parents, who either swallow it or carry it away in their bills. At many nests, one may see both methods of disposal; but the former occurs more commonly when the nestlings are very young, the latter after they grow older. Frequently the nestlings defecate just after being fed, sometimes turning around to deliver a dropping into the parent's bill. If not swallowed, the white sac is nearly always carried to a distance before it is released; sometimes the parent has a special spot for depositing these objects. The female Superb Lyrebird carries the droppings of her single nestling to the nearest stream, where she submerges them. If there is no water within one hundred yards of her covered nest, she buries this waste. If the white sacs that young dippers shoot from the nest do not fall into the stream, the parent picks them up and takes them there. So careful are many songbirds of nest cleanliness that they even remove droppings that have accidentally fallen on surrounding foliage.

The habit of taking a dropping after delivering food is strong, and I have occasionally seen a parent solicit one from a fledgling that had recently left the nest and rested some yards away from it. A parent Yellow-bellied Elaenia frequently tried to catch a perching fledgling's dropping by diving in pursuit as it fell earthward; sometimes it quickly swallowed the food that it was bringing to the young in order to retrieve the excrement. The parent's eagerness for the young bird's droppings suggested that they had an agreeable taste. Sometimes hunger causes a parent to eat its nestlings' excrement, which may contain a certain amount of undigested nourishment available to the more efficient digestion of the adult. When a female Great Tit undertook the care of ten nestlings, without the help that the male tit usually gives, her plumage became badly worn, she lost her tail, and she swallowed the droppings instead of carrying them away, as Great Tits with older nestlings commonly do (Kluijver 1950).

A few birds, including the Blue-crested Plantain-

PLATE 85. Female Splendid Blue Wren carrying a nestling's dropping from the nest in Australia (photo by John Warham).

eater and White-cheeked Coly, gently massage their nestlings' anal regions with their bills in order to stimulate excretion (van Someren 1956). The Yellow Wagtail may prod with its bill, or tug at the down on the nestling's back, to produce the same effect. The parent may even seize the nestling's head and shake it when defecation is delayed. A female was so eager for a dropping that she flew violently at her mate who had taken a sac after both had waited for it to be voided (S. Smith 1950).

Among passerines, nest sanitation is most conspicuously deficient in some of the smallest species. The older nestlings of manakins defecate over the rim of their slight nests, and the foliage beneath them is often heavily soiled with excrement that is dark with fruit juices. As their nestlings grow older, parent goldfinches, siskins, crossbills, and Purple Finches fail to remove all the waste. The young, perhaps in an unsuccessful attempt to void over the edge of the nest, deposit many of their white droppings on its rim. A recently abandoned nest of these finches is heavily soiled with dried excreta. The diminutive, colorful mannikins, waxbills, and other grassfinches of Africa and Australia are in general neglectful of the hygiene of their covered nests. Although much feces accumulates in them, it dries promptly and does not foul the plumage of the occupants. Older nestlings of some grassfinches come to the entrance of their nest to evacuate (van Someren 1956, Immelmann 1962). Black-faced Weavers keep their nest clean for about a week after their young hatch but thereafter permit the droppings to accumulate in the bottom of the closed structure (Morel et al. 1957). The same is true of Yellow-faced Grassquits. In the euphonias, among the smallest of tanagers, the droppings are not enclosed in sacs, but the parents manage to keep their roofed nest clean, seeming to lick up the waste (Skutch 1972).

In at least some of the woodpeckers, the nestlings' excrement is voided in gelatinous sacs, similar to those found in passerines, and removed from the hole by the parents. Yellow-shafted Flickers are so impatient to receive the droppings of their offspring that they nip a nestling's heels, or more often the fleshy protuberance that will bear the tail, to stimulate excretion, sometimes inflicting a dozen bites that are by no means gentle (Sherman 1952). Young woodpeckers remain in the nest much longer than passerines of corresponding size, and, before they leave, the sacs cease to be formed. The removal of excrement now becomes more difficult for the parents, who may dig wood particles from the walls of the nest cavity in order to sop up the waste and make it easier to carry away. Some kinds of woodpeckers abandon the attempt to keep their nests clean. This occurs about the time the young woodpeckers can climb up and take their meals through the doorway, so that the parents do not need to enter in order to feed them. The bottom of the hole now becomes filthy, but the nestlings spend much, if not all, of their time clinging to the vertical walls above the waste matter.

Woodpeckers that use their nest holes as family dormitories, including the Golden-napes and piculets, keep them clean at all times, often removing a number of heaping mouthfuls of waste in rapid succession and carrying these to a certain spot before dropping them. In the Golden-naped Woodpecker, this house cleaning is done chiefly by the male. Barbets and toucans, some of which likewise use their nest holes as dormitories, are similarly careful of sanitation. Prong-billed Barbets remove waste from their cavities in trees along with considerable amounts of wood particles; these are so bright and fresh that they appear, as in woodpeckers, to have been newly bitten away from the walls or floors for this special purpose. The parent barbets clean house most actively in the evening, just before both retire to sleep with their nestlings. Although they are related to woodpeckers and barbets, I have never seen puffbirds or jacamars make any effort to clean their burrows.

In pigeons and doves, nest sanitation varies greatly from genus to genus. The White-fronted Dove and Ruddy Quail-Dove at all times keep their nests as clean as any songbird's. They swallow the droppings, and I have watched a quail-dove, in the intervals of regurgitating to very young nestlings, eat their excrement. I could not imagine how it avoided feeding the young doves their own droppings with the next installment of pigeon milk. The Scaled Pigeon and Red-billed Pigeon also keep their nests clean while the nestlings are younger; but, before the young fly, the outer parts of their nests become heavily soiled. In contrast to these more or less cleanly members of the family, many doves, including the Blue Ground-Dove, Ruddy Ground-Dove, and Inca Dove, allow their nests to become filthy. The incubating parent sometimes soils the nest with its excreta—a habit rare among birds—and the accumulation of waste grows as long as the nestlings are present. Indeed, the heavy deposit of excrement may be of some importance in the economy of these doves, for it cements together the sticks and straws of the often flimsy nest, weighs it down, and makes it more stable for later broods that may be reared in it, thereby decreasing losses of eggs and young.

When hummingbirds are very young, their mother takes their droppings in the tip of her long bill and flings them from the nest, or else carries them a short distance away. After the nestlings are older, they rise up and squirt their liquid excreta forcefully beyond the nest's rim, as their mother sometimes does while she incubates. In hermit hummingbirds, the nest is fastened by cobweb beneath the drooping tip of a palm frond or some other kind of leaf or beneath a strip of a frayed banana leaf. The nestlings always rest with their heads toward the leaf and their rear ends outward, just as their mother does while she incubates and broods, and this orientation favors the sanitation of the nest. Ani nestlings try to shoot their liquid feces beyond the nest much as hummingbirds do, but projecting ends of the long sticks of which it is built catch much of the ejecta, to give the nest a characteristic odor. Among altricial nonpasserines with open nests, shooting the excreta outward is the most common method of disposing of it. Spattered with excrement and littered with decomposing food that has been dropped and the putrefying remains of fallen nestlings, a crowded colony of herons or egrets is most offensive to human nostrils.

Nestling Black Terns attend most efficiently to the sanitation of their floating nest. When only a day old, they climb to the edge, turn around, and immerse their abdomens before they defecate. A day later, they swim a few inches away before voiding (Cuthbert 1954).

The incubating female hornbill carefully shoots her excrement through the narrow slit in the plug that closes her doorway, and this is the practice of the young hornbills, at least after they are strong enough to climb up to the aperture. The nestlings of swifts that breed in cavities of various sorts likewise drop their excrement through the opening, and their parents may remove at least some of the waste that the young do not succeed in casting forth. Many other birds that nest in holes and burrows pay no attention to sanitation, with the result that their chambers soon become foul with the nestlings' droppings, the indigestible seeds, fish bones, or hard parts of insects regurgitated by both adults and young, uneaten food, shed feather sheaths, and the crushed eggshells from which the nestlings emerged. Sometimes the nest cavity is pervaded by a strong odor of ammonia, generated by the decomposition of nitrogenous wastes. Such is the condition of the holes of trogons in termitaries and decaying trunks

and the burrows of kingfishers and motmots in the earth. Yet the fledglings that finally emerge from these filthy nests are often clad in lovely plumage that, surprisingly, has escaped defilement.

When nestlings die, they may be carried away by the parents if they are small; but larger corpses are often permitted to remain and putrefy, or in dry climates to mummify, beside the living young, creating most unsanitary conditions.

Despite the parents' failure to remove waste, some nests are kept fairly wholesome by the scavenging insect larvae that inhabit them. The Golden-shouldered Parrot of tropical Australia breeds in a hole that it carves in a termite mound, and this cavity is kept clean by the larvae of the moth *Neossiosynoecha scatophaga*, which live in silken recesses in its floor. From time to time they emerge to eat up every scrap of waste, not only from the bottom of the nest but also from the feet and feathers of the nestlings. The life cycle of these insects is closely adjusted to the progress of the parrots' nesting, and they seem in no way to harm the birds. The nest of another Australian parrot, the Eastern Rosella, which breeds in a hole in a tree or post, is kept clean by dipterous larvae, *Passeromyia longicornis*; although apparently only beneficial to the parrots, they have been found to suck the blood of certain other nestlings. These same larvae also act as scavengers in the nests of a kingfisher, the Kookaburra. The nests of certain Australian grass-finches that are careless of sanitation are kept clean by a beetle, *Platydema pascoei*, which eats droppings and possibly stray feathers but does not harm the birds. As many as 345 of these beetles have been taken from a single grassy nest (Chisholm 1952a).

Although hornbills eject their excrement through the narrow slit in the doorway, it seems impossible that by this means alone the nest could be kept sanitary during the months when it is occupied, first by the incubating female and later by her and the nestlings. Insects, largely larval, come to the birds' aid and devour fallen scraps of food, raveled feather sheaths, and similar debris. One nest of the big Silvery-cheeked Hornbill contained 438 insects, including 201 lepidopterous larvae, belonging to eight species. They kept the cavity in a tree "astonishingly inoffensive and nearly odourless" even after fifteen weeks of continuous occupancy (Moreau 1942c).

Birds of the most varied kinds make an effort to exclude intruding insects from their nests; I have watched them spend many minutes picking off ants and, less often, termites. Ants seem unable to reach the contents of an unbroken egg, but sometimes they cause the incubating parent to fidget until it cracks a shell, when they enlarge the gap and swarm in to devour the developing embryo. Despite valiant efforts, birds are unable to stem a heavy attack by myriads of ants, and many eggs and nestlings are destroyed by these aggressive insects.

Even the most careful of avian parents are unable to keep their nests sanitary according to the high standards of civilized people. A structure of straws or sticks or the rough interior of a hole in a tree or termitary cannot be kept as clean as a smooth enamelled surface or bedsheets that can be removed for washing. Hence, lice and many other kinds of vermin breed in the nests of certain birds. Usually they do no great harm to the occupants, but if they become very abundant they may retard the development of the nestlings or even kill them. It has been suggested that the advantage of a brief interval for the multiplication of nest parasites should be included among other advantages of short incubation and nestling periods. Birds do well to make a new nest for each brood, rather than reuse an old one that may harbor vermin. But they often diminish the value of their new construction by including materials pulled from old nests, thereby running the risk of transferring noxious inquilines.

CHAPTER 23

The Interrelations of Nestlings

Parents sometimes exhort their small children to agree together like little birds in their nest. If ornithologically sophisticated, a child might ask, "Which birds do you wish us to imitate?" Although some birds do provide an example of fraternal concord, others behave toward their siblings in a way that no parent who loves his children would wish them to copy. The parent who knows birds might reply, "Songbirds, of course, not birds of prey."

I have never seen nestlings of passerines, hummingbirds, pigeons, or anis quarrel with each other, no matter how tightly packed in the nest they were. Yet, despite the harmony among the members of a brood, the stronger nestlings make no sacrifices for their weaker brothers and sisters. As long as the parents can bring enough food for all, the reactions that we earlier considered ensure its equitable distribution among the nest mates. But, if food is scarce, the weaker, which usually means the younger, nestlings fail to receive their due share, with the result that they fall behind the more vigorous ones in weight and may even succumb. We should hardly expect the stronger nestlings to display abnegation for the benefit of their weaker nest mates; indeed, such altruism does not often appear even among human children until they are a few years old and have been prompted by their elders. Accordingly, any arrangement to ensure the equal division of food among all the nestlings, even in times of scarcity, must be innate, fostered by natural selection. For reasons that will presently appear, natural selection is not likely to favor such an arrangement.

Those who have watched the nests of woodpeckers through openings made in their walls have sometimes seen the feathered nestlings fight furiously for possession of the place in the doorway that ensures receipt of the next meal brought by the parents. There is evidence that in the Great Spotted Woodpecker of Europe these struggles may, on rare occasions, cause the death of a nestling. In the Yellow-shafted Flicker, fighting among nestlings seems to be confined to a few days in the fourth week of their lives (Sherman 1952). While watching the nests of a number of kinds of tropical American woodpeckers from a point in front of the doorway, I have never noticed any suggestion that the young occupants were quarreling for this coveted place.

It is chiefly among raptorial birds that the young attack their siblings with intent to kill. Fratricide, which seems usually to be followed by cannibalism, has been recorded for a large and growing number of eagles, hawks, and owls, including the Common Buzzard of Europe, Swainson's Hawk, Red-tailed Hawk, Golden Eagle, Bald Eagle, Kestrel, Goshawk,

Peregrine, Hen Harrier or Marsh Hawk, Short-eared Owl, Long-eared Owl, Great Horned Owl, Barn Owl, and others. Short-eared Owls, which nest on the ground in northern lands, deposit surplus food at points three or four inches from the nest's edge, apparently so that the nestlings, especially the older ones, may reach and eat it as soon as they are hungry, instead of devouring their younger nest mates, as they might otherwise do if the parents were delayed in feeding them. Nevertheless, cannibalism seems to be frequent in this owl and has actually been witnessed by an observer in a blind (Ingram 1959, 1962).

A circumstance that favors cannibalism in birds of prey is the disparity in ages of brood mates. Eggs are often laid at intervals of several days, and effective incubation may begin before the last is deposited, so that those laid earliest hatch long before the last—a situation just the opposite of what we find in pigeons and many other birds whose incubation habits promote nearly simultaneous hatching. The Snowy Owl of Arctic regions lays up to nine or ten eggs, which sometimes hatch over an interval of fifteen days, so that the oldest nestling is well grown before the youngest escapes from the shell. In the larger broods, several of the nestlings sometimes succumb (Sutton and Parmelee 1956).

The spread of hatching over several days is by no means confined to birds of prey, and an ecological explanation has been advanced for it (Lack 1954). Food for the young is not equally abundant in all years. The amount available depends on such factors as weather, which affects the insects on which swifts and swallows feed, or the abundance of animals whose populations periodically rise and fall, as is true of the lemmings, voles, mice, and other rodents on which many hawks and owls prey. The birds, of course, cannot foresee how much food will be available for their young weeks or months after they begin to lay, so that, if the female is adequately nourished while her eggs are forming, she tends to lay a number that corresponds to the number of young that might be properly fed in an average year. If food for the nestlings turns out to be scarce, the survival of the species is promoted if the parents rear a few sturdy offspring, rather than a larger number of weak, undernourished young that

are likely to perish soon after they leave the nest, or even before they are fledged. The reduction of the nestlings to the number that can be adequately nourished will occur more surely and swiftly if some, because of greater age, have a marked initial advantage over their siblings than if all are fairly equal at the outset; and the sooner the reduction is accomplished, the less waste of precious food. Even in songbirds and other families in which hatching is more or less simultaneous, this reduction occurs to a certain extent when food is scarce; and, since it is advantageous for the species, natural selection cannot operate against it. But it is chiefly in raptors that the elimination of the excess nestlings is hastened by fratricide.

Although this ecological explanation seems fundamentally sound, it is certainly not the whole story, as far as raptorial birds are concerned. In a year of scarcity, the pinch should be greatest after the nestlings are well grown and make the maximum demands for food. Yet their fratricidal tendencies are manifested while they are still very small and the parents have no difficulty in supplying their needs. That the fratricidal impulse of birds of prey is primarily an expression of the deep-rooted ferocity engendered by the predatory habit seems evident from certain astonishing observations made in Africa at a nest of Verreaux's Eagle by Rowe (1947). When the second eaglet hatched, the first was already three days old and about one-half larger. Two minutes after the mother stood up and revealed the newly hatched nestling, the older one squatted on its back and sat on its head while itself being fed. The younger one had to fight at least half a dozen times before it received its first meal. Many of its subsequent feedings were cut short by its truculent nest mate, and it rarely got a full meal unless the older one was sleeping. By repeated savage attacks, by deliberately squatting on and smothering it for intervals that sometimes exceeded one-half hour, and by reducing its meals, the older eaglet murdered the younger one when it was six days old. The observers did not learn what happened to the corpse. The record shows that the killer was amply fed by the parent eagles, who did not try to stop the brutal attacks.

Even the parent raptor may sometimes devour its

offspring. While I watched a nest of the Laughing Falcon in a hollow high in a great tree in the Costa Rican forest, a Tayra climbed up and killed the single nestling, but I drove the big, black weasel off before it could take its victim. After a while the nestling's mother, who was resting in front of the niche when the tragedy occurred, returned and ate her offspring. This surprised me the more because otherwise I have seen Laughing Falcons eat only snakes (Skutch 1971).

Because of their potential for adjusting the rate of recruitment of a species to the carrying power of its habitat, Wynne-Edwards (1962) regarded prolicide (killing of offspring by the parent), fratricide, and cannibalism as very advanced adaptations "seemingly almost confined to the highest animals, found in the arthropod and vertebrate classes." If this is the way evolution is advancing, and such behavior its highest product, I believe that we must agree with T. H. Huxley (1947) that it is incumbent upon the moral man to resist evolution or the cosmic process with all his strength. Perhaps, however, the adjustment of the birth rate, or the rate of conception, to the species' need of recruitment is a higher achievement of evolution, as it is a neater, more economical, and more humane method of population control. We shall return to this question in chapter 34.

In another bird of predatory habits, related to the jaegers and gulls rather than to the raptors, the situation is somewhat different. The two eggs of the South Polar Skua hatch with an interval of one to three days, rarely more. If the interval is short and the first chick weighs only 0.07 to 0.25 ounce (2 to 7 grams) more than the second when the latter hatches, both are likely to live until they fledge, unless they succumb to starvation, unfavorable weather, or predation by adults of their own kind. If the interval between hatching is longer and the initial weight difference between the nest mates is 0.28 to 0.63 ounce (8 to 18 grams), the younger has a very poor prospect of survival.

The chief cause of chick mortality in this skua is persecution of the younger chick by its older sibling, which begins when the former is a few days old, usually at a time when the elder is temporarily short of food. Stretching up to its full height with raised wing stumps and uttering a shrill challenge call, the older chick attacks the younger with vicious pecks on the head and back. Without attempting to defend itself, the younger one flees over the rough Antarctic ground. If later it returns to the nest, the attack will before long be renewed. At first the parents try to stop the persecution by brooding the chicks or interposing their bodies between them, but such is the pugnacity of the elder that they soon abandon the effort to keep peace. Obviously perturbed by the behavior of their offspring, they continuously repeat the alarm call and perhaps engage in displacement scrape making.

Often the evicted younger chick settles on a different part of the parental territory, where it is attended by one parent while the other parent takes care of the first chick. Occasionally a young skua driven from its own nest wanders into the territory of a neighboring pair and is accepted by them. This has dire consequences, for if the adopted chick is older than the foster parents' own chicks, it fights with them and ultimately causes their deaths. One pair of skuas, still incubating, neglected and later ate their eggs after they adopted a displaced younger chick (E. C. Young 1963, Spellerberg 1971).

Kittiwakes also lay two eggs, which usually hatch a day or two apart. Although the two young of these small gulls may rest peacefully together for hours, when a parent arrives with food and the younger nestling tries to take its share, the older one pecks it away. Because it rests on a tiny shelf high on a seaside cliff, the persecuted chick cannot flee, but it allays hostility by an appeasement display, which consists in turning its head aside, hiding its bill in its down, and presenting the black band across its nape. When its elder sibling is satiated or sleeping, it gets its share of food. Ground-nesting gulls have no need of an appeasement display, for brood mates soon scatter over their parents' territory and do not fight for food or any other reason. Lacking a means to placate hostility, chicks of the Herring Gull and Black-headed Gull which, in an experiment, hatched in a Kittiwakes' nest were more fiercely attacked by their foster brothers than one Kittiwake chick would have attacked another (E. Cullen 1957).

PLATE 86. Young Brown Boobies from neighboring nests. The bird on the right still wears its nestling down, through which dark wing feathers are pushing; that on the left has acquired its dark juvenal plumage and is almost ready to fly. They pecked and bit each other, just as they did to the photographer.

Like South Polar Skuas and Kittiwakes, Brown Boobies and White Boobies regularly lay two eggs, yet only with extreme rarity and in years when food is exceptionally abundant do they rear two young (Simmons 1965). Incubation begins with the laying of the first egg, with the result that it usually hatches about five days before the second egg, and the younger chick nearly always dies or disappears soon after its birth. What happens to it? The parents do not discriminate between them but preen both by turns. Indeed, they readily accept the offspring of other parents, even when they are several weeks older or younger than their own. It appears that the

older chick is responsible for the death of its sibling. When Dorward (1962) placed a newly hatched booby in the same nest with one six days old, he saw the larger chick push the smaller one with its head each time the brooding parent stood up. Sometimes it gripped the neck or wing of the smaller one in its beak and pushed with its feet. Yet it seemed to lack strength to force the younger one out of the nest. When a small chick was found outside, the parent never tried to retrieve it, as it does with an egg that rolls out of the nest. All the evidence points to the conclusion that, when only a few days old, the stronger booby tries to establish supremacy and expel its sibling, who dies unless it is adopted by a neighboring pair of boobies, as rarely happens.

As one would expect, nestlings intolerant of their own siblings will hardly welcome visitors from neighboring nests. Years ago, I visited a breeding colony of Brown Boobies on a tiny island in the Caribbean Sea. Whenever I approached a large but still flightless young booby, it protested with loud, hoarse cries, as with spread wings it backed to the farther side of the bare circle on the rough coral limestone that represented its nest. As I drew nearer, its cry grew louder until, with bill opened to its fullest extent, it emitted a continuous *ah-h-h-h.* Flapping its widely spread wings, it tried to bite my hand, my shoe, or anything else that I placed within reach. Backing away from me, it sometimes came too near another young booby, who greeted it with pecks and bites, which it repaid in kind. A young booby treated an intruder of its own species exactly as it treated me (plate 86). Gannets lay only a single egg and their young are no less intolerant of a nest mate than Brown Boobies and White Boobies are. However, by placing together two chicks of different parentage and of the same age within a day, therefore evenly matched in weight and strength, Nelson (1964) succeeded in making a number of "twins" grow up together.

Thus far we have considered the relations between nest mates of the same species. Frequently parasitic birds are hatched in a nest with young of the host species, and occasionally a nesting tangle results in the presence of young of two different, nonparasitic species in the same nest. In these accidental nesting associations, both kinds of young are rarely reared unless they are fairly equal in size and age and require the same kind of food. When a parasitic nestling is present, the course of events differs greatly according to the species of the intruder. Nestlings of American cowbirds seem never deliberately to attack their nest mates. But often they are larger, develop faster, and take more food than the offspring of their foster parents. Thus the smaller of the parasitized species may fail to rear their own young along with a cowbird. The larger host species, however, frequently raise nestlings of their own in a nest that harbors a cowbird. The Song Sparrow has been known to rear as many as five of its own young and one Brown-headed Cowbird in the same nest (Nice 1937, Friedmann 1963). Likewise in the parasitic weaverbirds or viduines, the foreign nestlings grow up along with the young of their hosts (Friedmann 1960).

Turning now to the parasitic cuckoos, of which there are many species in the Old World but few in the New, we find a very different situation. Only in the three species that impose upon members of the crow family do the parasitic young commonly grow up with the nestlings of their host. This rarely, if ever, occurs in nests parasitized by any of the numerous species of *Cuculus* or of *Chrysococcyx,* the glossy cuckoos. Not only do these cuckoos, like the cowbirds, remove one or more eggs from each parasitized nest, but also their offspring when very young eliminate all the fosterer's progeny. This behavior has been longest known and most carefully studied in the common Cuckoo of Europe. A few hours after it hatches, the blind, naked cuckoo moves restlessly until it maneuvers a nest mate or an unhatched egg onto its slightly concave, sensitive back, where it is supported between the cuckoo's upstretched wings. Then, with an effort amazing in such a tender nestling, it climbs backward up the side of the nest, using its head and neck as a prop, and heaves the nest mate or the egg out of the nest, exercising all its skill to avoid tumbling out with its victim. The impulse to throw things from the nest is not satisfied until the young cuckoo is alone, so that two cuckoos cannot be raised in the same nest. Since the foster parents are usually much smaller than the fosterling, it needs all the food they can bring to it. They seem invariably to neglect their proper offspring, which the cuckoo has thrown from the nest, even when it lies dying in plain view. Few birds pick up fallen nestlings and replace them in the nest, and those that fall or are thrown out before they are feathered are in any case doomed (Friedmann 1968).

The recently hatched Lesser Honey-guide of Africa is just as intolerant of company as the young parasitic cuckoo, but, since it is usually reared in the nest of a barbet, woodpecker, or some other hole-nesting host, it cannot eject its competitors as easily as the cuckoo throws them from an open nest. But the bill of the newly hatched honey-guide is equipped with two fairly long, sharp hooks, one at the tip of the upper mandible and one at the tip of the lower. With these formidable weapons, the blind, naked nestling ferociously attacks its nest mates, gripping, biting, and drawing blood, continuing this savage treatment until the young barbet or woodpecker dies from its wounds and the parasitic nestling remains alone in the hole. The hooks at the tips of the honey-guide's mandibles are developments of the egg teeth, which in the related woodpeckers are present on both mandibles (Friedmann 1955).

We began this chapter by considering nestlings that lie peaceably in their nest, neither harming nor helping each other, like those of most passerines, doves, and other small birds. From these we passed to some examples of young birds that worry, drive away, or murder their nest mates, reminding us of Saint Augustine's views on the sinfulness of the newly born. Let us end the chapter on a more pleasant note by glancing at a few nestlings that are mutually helpful, performing small services for each other. Although the eggs of herons and egrets hatch at intervals, with the result that the members of a brood may differ much in size, on the whole they seem to agree well together. From time to time they squabble a little, but the young alternately preen their siblings' plumage and their own. A young heron from a neighboring nest who tries to intrude into the family circle is, however, expelled, sometimes after a strenuous and prolonged struggle (Bent 1926, Cottrille and Cottrille 1958). Among subprecocial chicks, rails preen each other, in some species beginning when about two weeks old. A ten-day-old American Coot preened its companions (Nice 1962). Chicks of the Little Penguin also preen each other (Warham 1958*b*).

Rarely nestlings offer food to each other. Young European Cormorants do this and likewise help to keep their nest in order by arranging its materials. In the Crowned Hornbill, as in certain other members of the family, the female, who has been enclosed in her nest hole since she started to lay, breaks out before her young are ready to fly. The latter then proceed to plaster up the doorway again, working from within, and leaving a gap just wide enough for their parents to pass food in to them. Three young Crowned Hornbills removed from their nest and placed in a box with a small opening cooperated in reducing its size with mud supplied to them, along with particles of food and their own droppings. They attended efficiently to the sanitation of their box, and one offered the others food which, if not accepted, was finally added to the plaster (Moreau and Moreau 1940).

In this chapter we have been concerned only with the relations of young birds with other members of the same brood. Far more often than feeding or preening their nest mates, they help their parents to attend their younger brothers and sisters of a later brood, as will be told in chapter 29.

CHAPTER 24

Leaving the Nest

One who follows the activities of a pair of nesting birds until their young attain independence sees much of interest and beauty. But perhaps the most exciting event in the history of the family is the departure of the young birds from the nest. This is, indeed, the climax of the whole fascinating drama. It is the moment of triumph for the parents, who by the exercise of diligence and caution have successfully brought their eggs and little ones through many perils to which neighboring nests have succumbed. To reach maturity the young must still escape many dangers, but they are no longer so likely to be lost all together as while they lay in a helpless cluster in the nest. To classify the types of hatchlings, as was done in chapter 18, it was necessary to say something about the time of their departure from the nest. Here we shall look more closely at this critical point in a young bird's life.

The Departure of Some Precocial Chicks

In a Costa Rican forest, I studied a Marbled Wood-Quail's nest, which was situated on the ground and roofed with dead leaves and twigs. The female alone incubated the four white eggs, taking each day a single outing that began around sunrise and lasted from about one and a half to three hours. Her mate came each morning to call her from the nest, and at the end of her excursion he escorted her back to it, but he always prudently stopped short some yards away, thereby diminishing the probability of revealing its exact location to the many hostile eyes that lurked in the rain forest. The four eggs hatched in the interval between the hen's return from her recess at 8:30 one morning and the following dawn. On this day, the mother delayed in her nest well past her usual hour for departure, while her mate, who had come for her at sunrise, moved impatiently through the undergrowth and called with growing insistence.

At last, the hen answered him softly, increasing his suppressed excitement. Then, with more low murmuring, she rose and pushed halfway through her doorway. Her chicks, whose soft, variegated down was already dry, needed no urging to leave the nest. One pushed out beneath her while she hesitated in the entrance. When at last she stepped forth, four chubby birdlings tumbled out around her. As she slowly advanced to join her partner, they followed with short runs punctuated by brief pauses. Her way led across a long, low, decaying log, which presented a formidable barrier to chicks who had escaped from their shells less than twenty-two hours earlier. Three soon managed to scramble over it and their mother led them off, while her mate returned to encourage the weakest one until it, too, found a spot where, unaided, it could surmount the obstacle. Then all vanished into the dark undergrowth of the forest.

In a neighboring valley, I studied a "nest" of the Little Tinamou, in which two beautiful, glossy, purplish drab eggs lay on some bits of decaying weed stems amidst the dense weeds of a neglected plantation of coffee and bananas. The grayish brown tinamou covered them so steadfastly that I could almost touch him. On the day when one of the eggs was pipped, I began to watch from a blind. The parent sat almost motionless in the rain from 2:00 P.M. until nightfall. When I returned at dawn, he was still at his post. Soon I heard the soft peeps of a chick, and presently I caught fleeting glimpses of its little bill, or the top of its head, pushed up through its father's out-fluffed plumage. In the middle of the morning, the parent rose up, arranged his feathers, then walked slowly toward the neighboring thicket, repeating a low, soft whistle as he went. The single tiny chick, clad in long, dense, silky down, chestnut and buff in color, had tremendous difficulty following on wobbly legs over the rough ground. Sometimes it fell into a little depression, from which it extricated itself with great effort; almost insuperable barriers, such as fallen banana leaves, blocked its way. But lured onward by the parent's repeated whistles, it pushed dauntlessly forward, and in slightly over half an hour it managed to traverse the eight difficult feet that separated its birthplace from the thicket where its father waited. A heroic journey for a chick less than one day old!

The majority of nidifugous chicks are hatched in nests on or near the ground or close above water, so that they may easily walk or swim away. But in a number of species, including some of the guans of tropical America and certain ducks, geese, and mergansers, they are born well above ground, in a cavity in a tree or in an open nest in a tree or on a cliff. How do these chicks, still without flight feathers on their stubby wings, manage to reach the earth? One might suppose that the parent would carry them down, perhaps in its bill, one at a time. Although sometimes they do this, it is most exceptional. Nearly always the chicks simply jump from the nest; they are light and their abundant down

cushions the impact, so that their apparently reckless plunge is rarely fatal.

The female Wood Duck permits her ducklings to remain in the high nest box or hollow in a tree until the day after they all hatch, when their down has thoroughly dried. While in the nest cavity, they are never fed. On the day of their departure, she waits until the rising sun has dried the dew from the herbage; on a wet day, she may delay for hours. Finally, from the doorway of her nest she surveys the neighborhood to make sure that no enemy is in sight. When assured that all is safe, she begins to repeat a soft call, difficult to describe, either from the doorway, a nearby limb, or the ground below. The ducklings answer with peeps and soon appear in the opening. Spreading their tiny wings, they

PLATE 87. Wood Duck duckling jumping from a high nest box (photo by W. J. Breckenridge).

PLATE 88. Mountain Trogons, twelve days old and almost ready to leave their hole in a low stub in the Guatemalan highlands.

plunge boldly downward (plate 87). They may hit a hard spot with a thump and bounce up a few inches, but nearly always they are on their feet within seconds of landing. Sometimes the falling duckling strikes an obstacle and is momentarily stunned; but in watching the departure of more than fifteen broods from high nests, F. Leopold (1951) never saw one so disabled that it could not soon follow the duck to the water, as earlier described (p. 232).

Even more spectacular was the departure of a brood of two Torrent Ducks from a crevice in a vertical cliff, sixty feet above a turbulent Andean stream. First the female fluttered from the nest down to the water, where she swam around in an eddy, calling loudly and continuously. Within half

a minute the first duckling leapt, bounced off a salient angle of the cliff and fell upon bare, jagged rocks below. The second followed, bouncing from the rocks and plunging into a thorny thicket. Yet both ducklings promptly joined their mother and swam skillfully in the rushing torrent (Moffett 1970).

Although Canada Geese usually breed on the ground, at times they choose a ledge on a cliff, or lay their eggs in the old nest of an eagle or a hawk high in a tree. When six goslings leapt from a height of fifty feet, five landed without mishap. The sixth struck a projecting branch and fell as though dead, but after a few minutes it recovered enough to join its parents in the nearby river (Hornocker 1969). Among other waterfowl that have been seen

PLATE 89. Nestling Blue-diademed Motmot, almost ready to fly from its burrow.

leaping from high nests are the Common Goldeneye and Common Shelduck, also the Mallard when occasionally it chooses a high site.

The Departure of Altricial Land Birds

Altricial or nidicolous birds include all the passerines, hummingbirds, hawks, and nearly all birds that breed in holes and burrows, such as woodpeckers, trogons, toucans, kingfishers, and jacamars (plates 88, 89, and 90). If undisturbed, the young of these diurnal birds most frequently leave their nests in the morning, when they are most energetic and active. Yet they may do so at any hour of the day. Once I watched a seventeen-day-old Yellow-bellied Elaenia forsake its nest late on an afternoon of long-continued, drenching rain. It did so by climbing up among the close-set leaves of the sup-

porting bough until it was well above its nest, and there it fell asleep in the rainy dusk. Apparently, it departed at this unusual hour because its downy nest had become sodden and uncomfortable.

Birds that grow up in open nests have ample opportunity to stretch and exercise their wings before they take their first flight (plates 91, 92, and 93). One may see young hummingbirds standing up in the nest and beating their well-feathered wings into a haze, while they cling with their feet to avoid being carried off. It is important for them to have a clear space around the nest where they can take these exercises without striking anything. If an obstruction is present, they may so injure a wing by beating against it that they cannot fly. I once saw this happen to a young Bronzy Hermit whose nest beneath a banana leaf was enclosed like a tent by the drying leaf blades. Many nestlings rest on the nest's rim, or even venture a few inches beyond it over the surrounding boughs, then return to the familiar home. Rarely they make a longer excursion, only to rejoin their siblings who have stayed behind. But my impression is that the majority of true nidicolous birds never sever contact with the nest until the final leave taking, then they do not return to it, save in those relatively few species that use the nest as a dormitory, as do certain woodpeckers, swallows, wrens, and other birds (see chapter 30).

Not only are young birds reared in open nests able to exercise their wings before they fly, but also their nests, if not on the ground, are usually surrounded by branches, so that their first flight need not be long. Those that grow up in holes and burrows lack space to exercise or even fully expand their feathered wings, and moreover the doorways often face the empty air high above the earth or else a wide expanse of open water. Thus a weak, inexpert fluttering from the nest would be disastrous, and the first flight must necessarily be strong and long. To witness the emergence of one of these birds is highly dramatic, especially when one reflects that its safety depends upon the efficient performance of an activity that it has never had an opportunity to practice.

One morning, soon after sunrise, I saw a young Ringed Kingfisher come to the mouth of a long burrow in the bank of a Guatemalan river. For a while

PLATE 90. Green Kingfisher fledglings, ready to fly from their burrow in the bank of a Guatemalan river.

he rested there, gazing across the broad stream that sparkled in the sun's earliest rays. At intervals he bobbed his head up and down and uttered a loud rattle. Then suddenly he launched forth, turned downstream, and rose to perch in a willow tree. I estimated that his first flight carried him two hundred feet, with a rise of about forty—not a bad performance for a bird who had passed all his days in the dark earth, unable to spread his wings after they were fully grown.

This kingfisher took the decisive step spontaneously, while his parents were out of sight. This is nearly always the way with altricial birds: Unless they are alarmed, as perhaps by the too-near approach of a human observer, they leave their nests in response to their own growing strength and need of activity; they require no more parental urging to come forth from their nests than at an earlier stage

they needed to break out of their eggshells. But as the time for departure draws near, they are in a state of unstable equilibrium. The slightest stimulus may set off their flights—just as the gentlest touch will cause the sudden expansion of some explosive seed pod, like that of the jewel-weed, which without interference would have burst open a short while later. Often this stimulus is the departure of a parent who has just fed a nestling. Although the parent has made no apparent effort to lure the young bird from the nest, the latter may follow as the adult leaves, perhaps at the prompting of an awakening social impulse or following reaction. I watched six Black-faced Antthrushes, belonging to five successive broods, leave the same hollow stub in the tropical forest when a parent arrived with the first morning meal; this occurred before sunrise on the day when the fledglings were eighteen days old,

PLATE 91. Crimson-collared Tanagers, about twelve days old, ready to leave their open nest in the Caribbean lowlands of Honduras.

rarely a day more or less. As far as I could tell, the parent behaved just as it had done on preceding mornings.

In other instances, some slight accidental disturbance provides the stimulus for departure. While I watched two young Violet-headed Hummingbirds in their downy nest above a mountain torrent, a small, harmless flycatcher flitted past and caused one of them to fly out. Half an hour later its nest mate did likewise, quite spontaneously. When a man or some other animal approaches a nest while the parents are nearby, the excitement of the adult birds, their calls of alarm, may prompt the young to jump out and scramble to safety several days before they would otherwise have left.

There are many accounts of parent birds coaxing or luring their reluctant or timid young from the nest, often by withholding food from them. Since I have from concealment, or from a good distance, watched the departure of many young birds, ranging from kingfishers and trogons to tanagers and finches, without ever seeing a successful attempt of the parents to persuade fledglings to leave, I am skeptical about the interpretation of some of these episodes. It appears that in many of these cases the observer's presence excited the parents, which would greatly alter the course of events.

How difficult the interpretation of the birds' behavior can be was impressed upon me, some years ago, while I watched a burrow of the Rufous-tailed Jacamar. When the first nestling came to the mouth of the tunnel in a bank, its mother, approaching with a dragonfly in her long, slender bill, twice hovered in the air in front of it, then flew off without delivering the food. This may have been an attempt to lure the young bird into the open, but more probably the parent did not deliver the dragonfly, simply because the fledgling was now filling the mouth of the burrow and left her no place to alight. The next meal that the parent brought was given to the stay-at-home rather than to the fledgling who meanwhile had emerged. This seemed to indicate that the adults did not care whether their young remained in the nest or came out.

Similarly, I have seen a Southern House Wren, bringing food, push past a fledgling at the entrance of the nest box on the point of departure and take it to another inside, probably because the one halfway through the doorway was in a position where it could not conveniently be fed. Immediately after three house wrens of another brood made their first exit from the gourd in which they hatched, the parents started to go in and out, in and out, just as they do when they lead fledglings back into the nest at nightfall. It appeared that they were coaxing the newly emerged young to reenter the gourd, but soon they changed their tactics and led them off to forage. A male Carolina Wren apparently tried to coax his young out of the nest; but, after some of them emerged, their mother seemed to try to lead them back into it (Nice and Thomas 1948).

After the first of three young Black-crowned Tityras left their nest in an old woodpecker hole high in a branchless trunk on a mountainside, the parents seemed eager to bring the other two into the open. But all their coaxing in front of the hole was of no avail. The three young left on three consecutive mornings, all at very nearly the same hour. This regularity in the time of their departures was to me convincing evidence that they took the bold step in response to inner prompting rather than to parental urging. As each of these fledgling tityras flew forth, one or both parents followed it closely in shielding flight, a special mode of protection to which we shall presently return. After two young Masked Tityras were feathered, their mother seemed eager to have them leave the nest hole. But one lingered for three days and the other for five days, after I first heard the calls that were apparently intended to bring them out. Until they were ready to go, all her urging was ineffectual.

In Iowa, Sherman (1952) watched many broods, representing more than a dozen species, leave their nests, yet she saw "nothing that in the remotest degree resembled coaxing." With sharp wit, she ridiculed the notion that parent birds try to entice their young from the nest. How, she asked, do they know that a fledgling has reached the age of nest leaving unless they can keep records, count, and reckon? When the young of the same brood differ in age, this reckoning must be somewhat complicated. Moreover, they would have to make allowance for the varying speed of development of different broods and individuals. Her experience with birds of the temperate zone, as mine with tropical birds, convinced her that in nearly all cases young birds when undisturbed leave their nests when they feel the physiological impulse to do so.

The Heinroths (1959), who had extensive experience with young birds, many of which they reared by hand, threw light on the supposed withholding of food by the parents in order to force their offspring to quit the nest. The young bird, they believed, loses its appetite shortly before it leaves. In some species it takes no food at all at this time and becomes restless until it has flown, whereupon its appetite returns. Even if the young make begging movements at this time, one cannot induce them to take any food. They found this true not only of

PLATE 92. Rufous-tailed Hummingbirds, almost ready to leave the cup that their growing bodies have flared outward.

songbirds but also of eagles, owls, and many other species. Possibly this loss of appetite just before leaving the nest is adaptive. The less the fledgling eats, the lighter it will be and the better it should fly on its first aerial journey, which in some instances must be long to bring it to safety.

Although enticing altricial young to leave the nest by displaying or withholding food is exceptional, and many reports of such parental behavior seem to include an erroneous interpretation, it evidently occurs in certain cases. Young Horned Larks, still flightless, usually leave their nest on the ground by following a parent who has just fed them, but in one instance a female enticed a belated nestling from the nest with a morsel of food (Bent 1942). Likewise, Elgon Olive Thrushes withheld food from their young to make them leave (van Someren 1956). In a few species this may be the customary procedure. Meinertzhagen (1954) told vividly how parent Ospreys at a Swedish lake displayed food to their nestlings from a distance, until the latter were

PLATE 93. Groove-billed Ani, about six days old, with feathers just emerging from the ends of the long sheaths. Already it can perch and will jump or climb from its open nest if alarmed.

forced by hunger to fly from the nest and receive it.

M. K. Rowan (1955) reported that it was the regular practice of the Red-winged Starlings that nested on her porch in South Africa to withhold food from their young when the time for their departure was at hand. When one brood, which had clung to the nest somewhat beyond the usual time for leaving, failed to respond to this lure, their father lost patience and proceeded forcibly to evict the young. Despite his mate's effort to shield her offspring from his rough treatment, he lifted two of them bodily out of the nest and dropped them on the floor of the veranda. Then he pecked and pushed them over the edge of the high porch into the sheltering vegetation of the garden. In somewhat similar fashion, a parent Pygmy Nuthatch seized a nestling's shoulder

with its bill and, after a vigorous tussle, pulled it through the doorway of the nest hole and let it flutter to the ground (R. A. Norris 1958). I have never myself witnessed a comparable instance of forcible expulsion from the nest. In my experience, parent birds lead or lure to a safe place fledglings who have already ventured forth far more often than they try to make them quit the nest.

Even when they do nothing to hasten the exit of their nestlings, parents often seem to sense, doubtless from the young birds' increasing restlessness, that something out of the ordinary is about to happen. They may reveal their excitement by an unusual amount of singing, as I have noticed in the Southern House Wren. At times the parents appear to be emotionally stirred by their fledglings' departure.

For four mornings I watched for the exit of a young Vermilion-crowned Flycatcher who delayed longer than usual in its covered nest. Early on the fourth morning it flew out while the parents were absent. When one of them returned with food and found its young in the open, it sang a long-continued, rapid sequence of low, soft notes that I had not heard on the first three mornings. This chant was very like that with which the yellow-breasted Vermilion-crown selects her nest site and begins to build her nest. It was a fitting celebration for the successful conclusion of her nesting.

Years later, I had a similar experience while studying the little Pale-headed Jacamar in Venezuela. In the middle of the morning, a fledgling flew from the burrow in a bank and rose into a treetop. Here it was promptly joined by both parents, who immediately began to sing. Again and again they repeated their song, consisting of a crescendo of sharp *weet*'s and twitters, running off into high, thin trills. They seemed to be congratulating their fledgling, or themselves, on the successful conclusion of its nestlinghood. From time to time, the young jacamar joined in with its weaker voice, flagging its tail as they did. When, later that same morning, another of the four young spontaneously left the burrow, the song that celebrated its departure was less prolonged. In the same locality, a pair of Rufous-fronted Thornbirds sang much when their two fledglings flew from the high, swinging nest of interlaced twigs (Skutch 1968, 1969a).

Shielding Flight

As an altricial fledgling flies from the nest, especially one that is high and exposed, a parent often follows it closely, usually flying above or a little behind the young bird until it alights. I have seen such close escort of the fledgling on its earliest flight by birds so various as Montezuma Oropéndolas, White-tipped Brown Jays, Rough-winged Swallows, White-capped Dippers, Black-crowned Tityras, and several kinds of American flycatchers. Danforth (1930) saw a parent Sparrow Hawk closely follow one of the young hawks as it flew from the old flicker's hole where it was hatched.

The value of such shielding flight seems obvious.

Should a predatory bird pursue parent and young flying so close together, the adult would probably veer aside at the critical moment. The raptor would then probably follow the uppermost of the two, on which its eyes were fixed, giving the weakly flying fledgling a few moments' grace in which to reach shelter. Among birds that are more or less gregarious in the nesting season, including oropéndolas, brown jays, and swallows, the fledgling's first flight may be a spectacular event, for the young bird, untried on the wing, is often closely followed not only by its parents but also by such neighbors or helpers as happen to be near when it leaves home.

A pair of Tropical Kingbirds reared two fledglings in a slight, open nest built low above the water, among the tangled stems and roots of the epiphytic growths covering an old, partly submerged tree standing near the shore of Gatún Lake in the Panama Canal Zone. I happened to be watching from a cayuca when the young kingbirds left. Soon after sunrise, the first suddenly flew from the nest. Directing its course to the wooded shore, about 80 feet (25 meters) distant, it flew quite well, high above the water. As soon as the parents noticed what was happening, they hurried in pursuit of the fledgling, and while it was still many feet from the shore one overtook it and flew directly above it, apparently in contact with it. The parent was not in a position to support the fledgling; it may even have forced it lower. Together parent and young reached the shore, where the latter alighted in a bush in plain view. Immediately both parents flew at it and knocked it from its exposed perch into the midst of the foliage, where it was well concealed. The second act was commentary upon the first. They appeared to be trying to prevent the young bird from making itself conspicuous. The fluttering flight of a fledgling looks very different from the controlled flight of an adult, even when their speed is the same. A fledgling kingbird, weakly flying, would be tempting prey for a hawk, which the adults, far from fearing, would chase and harass.

A flock of black Yellow-rumped Caciques had attached their long, woven pouches to the branches of an epiphytic shrub growing at the top of the same decaying trunk where the kingbirds nested.

About two hours after the departure of the first young kingbird, the second left the nest. As it flew out over the water, three caciques hurried after it and one or two of them bumped against it. When one of the parents darted up to the rescue, the caciques turned back to their nests. Although this fledgling seemed just as capable of flying to the shore as its nest mate was, the interference by the caciques made it fall into the water a few yards from shore. Before I could untie my canoe and go to rescue it, the young flycatcher had flapped its way to the land, where it crawled out on the sloping bank. Here the parents flew down to it and tried to coax it farther from the water.

Why did the caciques pursue the fledgling flycatcher? It could hardly have been with hostile intent; they had had ample opportunity to attack it in the nest, in its parents' absence, had they been so inclined. Birds of many kinds perched among their swinging pouches without provoking hostility, unless they were parasitic Giant Cowbirds. It may have been the strangeness of the flying kingbird that caused the caciques to rush in pursuit of it. Its slow, fluttering flight was very different from the swift, direct flights of the adult birds that had been coming and going from the nesting colony. More probably, the caciques were actuated by the same parental instinct to shield a fledgling on its earliest flight that caused the parent kingbirds to hurry after the first fledgling; the unfortunate consequence to the young kingbird was the result of the disparity between its own size and that of its would-be protectors.

Behavior like that of the kingbirds who knocked their fledgling from an exposed perch has been seen in other birds. According to L. Howard (1952), parent Great Tits "sometimes stand on the back of their young to push them from an exposed perch if a Hawk flies overhead, also, when the young are full-grown and demanding food too roughly, the parent will occasionally subdue them by standing on their backs. . . . If the warning note to take cover from above is unheeded, the parent Great Tit sometimes pushes the fledgling from its exposed perch by kicking from above with its feet, scold-notes accompanying the push." When Mountfort (1957) approached a young Hawfinch sitting rather pre-

cariously on a wire, the agitated parents buffeted it off its perch and forced it to fly.

The Departure of Some Marine Birds

Natural history contains many facts so unexpected, so intrinsically improbable, that we could hardly believe them were they not repeatedly attested by competent observers of proved integrity. Among these natural happenings more marvelous than fairy tales are the journeys from nest to water of certain sea birds. The two species of murres or guillemots (*Uria*) are hatched in crowded colonies on cliffs, usually high above a turbulent sea. When between two and three weeks old or rarely more, and weighing about a third as much as the adults, the young become restless. Emerging in the evening from beneath their brooding parents, they walk around their ledge, preen, flap their incompletely feathered wings, and bob up and down. A parent also displays with its chick, bowing toward the chick and toward the sea, while it utters a long, high-pitched growl. Parent and chick bill and preen each other's head and neck. These mutual displays attune parent and chick for the coming event. When the activity reaches a climax, the chick suddenly leaps from the ledge. Although it cannot fly, its rapidly beating wings and webbed feet reduce the velocity of its descent and even enable it to glide forward. From an inland cliff 500 feet (150 meters) high, chicks of the Thick-billed Murre sailed across ½ mile (800 meters) of intervening low ground to reach the water. Even if they strike jagged rocks, the young murres bounce off and are rarely injured.

The descending chick is almost always closely accompanied, usually followed, by a parent. In the Thick-billed Murre, a number of other individuals join the parent and chick, spreading out in V-formation and frequently proceeding with a beautiful, slow, butterflylike flight. The chick strikes the water with a splash and immediately begins to squeal shrilly, while the parent and other adults gather around it, crying *arrr-rr-r, arrr-rr-r* loudly and hoarsely. Even if they reached the water some distance apart, parent and chick, guided by each other's cries, soon come together. At first the chick is worried by the excited crowd of other murres that

PLATE 94. Adult Common Murres block the way when chicks try to go to sea before they are able to survive the shock of cold water (photo by L. M. Tuck).

surround it closely, chasing and pecking it, diving to come up beneath it. These attacks are more playful than vicious, and the hangers-on are soon left behind as the parent leads its chick out to sea. The departure normally takes place in the evening twilight or, in the high Arctic, when the continuous daylight is dimmest and ravenous gulls are least menacing.

This is the usual course if all goes well. However, sometimes a murre chick tries to leave prematurely,

before it can maintain its body temperature. Adults may try to prevent its departure by blocking its way to the water, for if it plunges into the cold northern ocean it is doomed (plate 94). Sometimes parent and chick fail to meet on the water and the adult swims away, leaving the young murre to float around aimlessly until it perishes or a gull captures it. Some observers have claimed that the murre chick is always escorted to sea by a parent of either sex, while others have believed that an unrelated adult may at times take charge of it. There is evidence that the latter sometimes occurs, at least in the Thick-billed Murre. One wonders how parent and offspring find and recognize each other among the hundreds of birds that often mill around at the foot of the cliff when they descend, but the chick learns to recognize its parent's voice while it is hatching. Probably the parent continues for some time to feed the young bird that it accompanies to sea; information on this point is difficult to procure. The young Razorbill goes to sea in much the same way as the related murres (Perry 1946, Pennycuick 1956, Tuck 1960, J. Greenwood 1964, Tschanz 1968).

Other sea birds are abandoned by their parents while still in their nests. This is true of the Common Puffin, which after about six weeks of abundant feeding on fish has become very fat. Then it remains alone in its burrow for a week or ten days longer, living on its reserves and reducing its weight. Finally, under cover of darkness to avoid the murderous gulls, it makes its way alone to the edge of the seaside cliffs, flutters down into the surf, and turns its course toward the open sea (Lockley 1953). Likewise, Manx Shearwaters, after becoming well feathered, spend a week or more alone, hiding in their burrows by day, by night emerging for air and exercise. Finally, on their own initiative, the abandoned young set out for the sea, often hundreds of yards away from their natal burrows in the middle of a small island. If a steady breeze is blowing or they find a suitable elevation, they may take flight while still far from a seaside cliff; but in calm weather neither they nor their parents can take off from level ground, and they must shuffle overland to the shore (Lockley 1942). Apparently no shearwater, petrel, or albatross receives any parental care after it goes to sea.

The same is true of many other marine birds. After losing weight during their last days in the nest, Yellow-billed Tropicbirds, sixty-five to eighty-five days old, headed straight out to sea, flying unsteadily and barely maintaining altitude (Stonehouse 1962). Contrary to earlier published statements, the young Gannet is fed by its parents up to its last day or two in the nest, and sometimes within an hour of leaving. Then, weighing about a third more than the adults, it flies down alone to the water, sometimes alighting with a mighty splash about half a mile from the high cliff from which it started. Once upon the water, it cannot rise until it reduces its weight. For some weeks it floats on the waves, living on its great reserves of fat until it is light enough to fly and plunge to catch its own fish (Gurney 1913, Nelson 1964, 1966). In receiving no food from its parents after its first flight, the Gannet differs from the related boobies, whose parents may attend them for a long while after they fledge.

Penguins also appear to be wholly independent of their parents after they enter the water. Sladen (1955) watched young Adelie Penguins swim off without their parents, although they went in the same direction. Soon after their young go to sea, the adults begin the annual molt, during which they stand and fast, unable to feed their young.

There is an interesting parallel between the babyhood of these marine birds and that of certain mammals also born on islets or seaside ledges. During their first fortnight or so, infant Gray Seals suck their fasting mothers and grow fat on abundant milk far richer than that of most terrestrial mammals. Then their mothers swim off and desert them. After fasting for a week or more on their sea-washed ledges or barren islets and shedding their white natal pelage, the young seals venture alone into the broad ocean to find their own living. In a few weeks, they may travel hundreds of miles from their birthplaces (Lockley 1954).

Carrying the Chicks

Nearly always young birds, when not dragged away by some predator, leave the nest by their own exertions, but very rarely their parents carry them away. This seems most often to happen when the adults are alarmed for the safety of their little ones. One of the most detailed accounts of such behavior was written by E. L. Turner (1924). While with readied camera she watched from a blind before a nest of the Water Rail, the parent returned and carried off four newly hatched or hatching chicks, one after another, grasping each in its bill by the head, neck, or shoulder. One of these chicks was just escaping from the shell when the parent whisked it away, shell and all. Then the rail, with great difficulty, carried off an addled egg that remained in the nest.

A Virginia Rail whose eggs were hatching behaved quite differently. As a photographer approached, it slipped from the nest and was followed by two downy chicks a few hours old. While the photographer was focusing his camera on the nest, a parent rail appeared with one of the chicks in its bill and dropped it into the nest in front of the lens, then settled down to brood there. Another Virginia Rail picked up and carried away a chick that fell into a footprint and could not escape (Bent 1926). Clapper Rails have also been seen carrying their young. A Common Gallinule bore two chicks, one by one, from their nest two yards up in a bush down to the water, where the other parent was swimming with three more chicks (Nice 1962). Somewhat similar behavior has been reported of an African cuckoo. When a coucal's nest was threatened by a brush fire, the parent carried away its nestlings, one by one, between its thighs (Kilham 1957).

As Alvarez del Toro (1970, 1971) approached a nest of the Sungrebe beside a quiet stream in the lowlands of the Mexican state of Chiapas, the male parent jumped into the water and swam away. The second of the two eggs had been laid only ten days earlier. Examination of the empty shells showed that they had hatched, yet the nestlings had vanished! Later investigation revealed that the naked, almost embryonic nestlings had been carried away by their father, one under each wing, in a pouch formed by a fold of skin, where feathers of the ventral plumage helped to support them. Afterward, the parent brought them back to the nest. He could carry his nestlings while flying as well as while swimming.

Although parent birds have been seen to replace their young in the nest less often than they have been seen to remove them for greater safety, the former is occasionally done by raptors. When a nestling Montagu's Harrier was placed thirty inches from the nest, its mother picked it up in her beak and carried it back (Frisch 1966).

Although passerine birds have not, to my knowledge, been seen either carrying their young from the nest or replacing them when they fall out, they may on rare occasions transport them when they are in trouble. When a fledgling Rough-winged Swallow fell from its nest into a stream and lay quietly with outstretched wings, an adult, presumably a parent, promptly came to its aid. It dropped lightly upon the young bird's back and with rapidly beating wings propelled it across the water to the bank five feet away (Drinkwater 1963). The nearest parallel to this behavior that I have found is provided by the Franklin's Gull, which breeds in vast colonies in shallow, reedy lakes in the interior of the North American continent. When some chicks entered the water and were drifted away by the wind, the parents, greatly agitated, seized them by their necks, rose with them three or four feet into the air, and tossed them toward safety. This was repeated again and again, by a succession of adults, until the chicks were at last flung into some nest, not necessarily their own, exhausted and bleeding from the treatment they had received (Bent 1921).

Transporting the young has been recorded much more often of precocial and subprecocial birds than of altricials and may be regarded as one aspect of the greater mobility of their chicks. Probably the bird that has most often been reported carrying its young is the European Woodcock, which not only removes them from danger but also carries them over obstacles, such as fences and rivers, and transports them to feeding grounds not otherwise accessible. The most usual method of carrying the chicks is between the legs of the flying woodcock, or between its legs and breast. Less often they are grasped by the parent's bill. Occasionally the chick, held between the legs, is partly supported by the parent's bill or forward-turned tail. Rarely the young woodcocks are given a ride on the parent's back. In one way or another, they may be carried in flight for considerable distances until they are almost as big as the parent. Rarely two chicks are carried simultaneously, but usually they are transported one by one. A number of observers have watched a parent return repeatedly until the three or four young were removed to a safer place. The generically distinct American Woodcock has a similar habit.

Less often other shore birds, including the Lapwing, Old World Golden Plover, Redshank, and

PLATE 95. African Jaçana carrying young under its wings. Only the chicks' long, dangling toes are visible (photo by J. B. D. Hopcraft).

Common Snipe, have been seen carrying their chicks between their legs or in their feet (Alexander 1946). Although carrying the young, or anything else, between the legs is certainly rare among birds, it has, curiously enough, been reported as an occasional happening in the goatsucker family (Robin 1969; see also p. 148). The African Jaçana and some related species carry their chicks away from danger by holding them between the wings and the body. The parent walks over the lily pads with one or two chicks on each side, their long legs and toes hanging exposed below the adult's body (plate 95). Apparently it cannot fly with its chicks (Hopcraft 1968).

Other birds transport their young on their backs, which seems the easiest way. Parent loons may partly submerge themselves to make it easier for their chicks to climb aboard, then rise high in the water and swim away with them. Grebes not only swim over the surface bearing their young among the feathers of their back but also may even carry them when wholly submerged, a habit rarely reported of other birds. Among waterfowl, swans have been most often seen, and photographed, giving their downy cygnets a ride on their backs, one or more at a time. The parents do not help the young aboard, but they spontaneously climb up between the folded wing and the tail, while the par-

PLATE 96. Trumpeter Swan carrying cygnet on its back (photo by D. A. Hammer).

ent, of either sex, floats on the water. This habit has been recorded of swans that breed in middle latitudes, including the Mute Swan of Europe, Black Swan of Australia, and Black-necked Swan of southern South America, and at least once (Hammer 1970) of the Trumpeter Swan of North America (plate 96). Apparently, swans that nest in the far north do not carry their young. Ducks seem to carry their young on their backs less often than swans; but the Australian Musk Duck is said to regularly transport her one or two ducklings in this manner, sometimes even diving while they cling to the feathers of her back. A few other ducks, two sheldgeese, and two mergansers have been reported to swim with their young aboard. We even have a few accounts of ducks flying with young on their backs. One of the more recent tells how a female, either a Mallard or Black Duck, transported four ducklings, one by one, from a small pond to a lake 250 yards (230 meters) away (Shchepanek 1968).

Ducks rarely carry anything in their broad bills. They build their nests with what they find close around their sites, and the young collect their own food. Yet reliable observers have seen them, on rare occasions, remove empty shells from their nests, and even carry away broken or addled eggs in their bills. Accordingly, it is certainly not impossible for them to transport their downy young in the same manner. Although, as we have seen, newly hatched Wood Ducks nearly always descend from their nests by leaping, they may exceptionally be carried down in their mother's bill, and even less frequently by riding on her back. Similar methods of bringing their young from a high nest to the ground have been recorded of the Common Shelduck, Barnacle Goose, and a few other waterfowl; even more surprisingly, the Javan Whistling Duck has repeatedly been seen flying with its young held in its feet (Johnsgard and Kear 1968).

Although the transportation of young on the back is understandably most frequent in aquatic birds, it has been reported of partridges and grouse (Alexander 1946). Despite many recorded instances, the carrying of young, like the carrying of eggs, is by no means common among birds. In many years of bird watching, I have never seen any wild bird transport its young. Not all of the foregoing cases of carrying young refer to their removal from the nest. But if a parent is capable of transporting its offspring, it is likely to carry its newly hatched chicks from a threatened nest. This was true of the Water Rail, the coucal, and some of the numerous instances of European Woodcocks carrying young that were collected by W. B. Alexander.

Nestling and Fledging Periods

"Nestling period" and "fledging period" are sometimes used interchangeably, but this is hardly correct. The nestling period is the interval between hatching and the young bird's departure from the nest; the fledging period is the interval between hatching and the attainment of the ability to fly. Even if undisturbed, some altricial birds, especially those that nest on the ground, hop from the nest before they can fly; whereas others cling to the nest for several days after they have some ability to fly, as we can prove by chasing them out prematurely. Accordingly, the nestling period may be either shorter or longer than the fledging period.

Comprehensive studies of altricial birds nearly always give the nestling period rather than the fledging period. To determine the former accurately, one must be careful not to alarm the young and cause their premature departure by approaching their nests too closely. For subprecocials and precocials, whose "nestling period" is hardly more than the time required for their down to dry, our chief interest is to know the fledging period, or age of attaining the power of flight. Unless the young birds are kept in an enclosure, this is usually difficult to learn with accuracy, for while still unfledged the chicks may wander widely and be difficult to follow and to recognize.

When we consider nestling periods, the semialtricials pose a special problem. After they begin to wander from the nest at the age of a day or two, they are no longer nestlings in the strictest sense. But, instead of following their foraging parents, as precocial and subprecocial chicks do, they are fed on or near the nest with food that their parents bring from a distance, just as in true altricial nestlings. Hatched on the ground in skimpy nests or none at all, semialtricials find it easy to make short excursions beyond their natal spots, and correlated with this they frequently develop homeothermy at an earlier age, but otherwise they are attended much like altricial nestlings. Accordingly, it does not seem to be stretching the point to consider that semialtricials have a nestling period that terminates when they begin to accompany their foraging parents or, in certain marine birds, when they become independent of them. If we take this broader view, the longest known nestling period is that of the Wandering Albatross, which extends over 280 days, and the next longest is that of the Royal Albatross, which lasts about 240 days (see table 7, p. 201).

When we gave attention to incubation periods, we found that both the size of the egg and the safety of the nest influenced their lengths, and that nest safety seemed to have greater weight. These same two factors strongly influence the lengths of nestling periods, but their relative importance is somewhat altered. Increasing size tends to lengthen nestling periods more than it does incubation periods. One reason for this seems to be that many birds defend their young more zealously than they defend their eggs, and, the larger the bird, the more effectively it can protect its offspring. Again we see the tremendous influence of the hazards to which eggs and nestlings are exposed on the rapidity of their development. Since nestlings grow up in the nest where they were hatched, they are exposed to about the same risks as were the eggs from which they came, now increased by the greater activity of the food-bringing parents. Hence, birds that have a short incubation period usually have a short nestling period, and those that have a long incubation period have a long nestling period. Albatrosses, which breed on islands where their eggs and young are relatively safe, have both the longest incubation periods and the longest nestling periods.

In many small birds, the incubation and nestling periods are of nearly the same length, but some curious exceptions are known. The antbirds of the tropical forests have, for birds their size, rather long incubation periods, usually from fourteen to sixteen days, and much shorter nestling periods, in many species from nine to twelve days. Manakins, which breed in the same woodlands, hatch their eggs in from eighteen to twenty days, but their young remain in the slight nests for only thirteen to fifteen days. In hole-nesting birds, whose eggs are relatively secure, nestling periods are often considerably longer than incubation periods. This tendency is extremely pronounced in woodpeckers and toucans, whose nestling periods are from about two to three times as long as their incubation periods— a situation rare among small birds, for which an explanation was earlier offered (p. 206).

Even in a single species, the length of the nestling period varies more or less. The weightiest single influence on this intraspecific variation is the amount of food that the young receive. This in turn depends on weather, seasonal fluctuations in abundance of food, number of nestlings in the brood, and the order in which they hatched. Swifts forage much more successfully on dry days than on rainy ones, when it is difficult for them to catch the flying insects on which they subsist. In England, where the weather varies greatly from summer to summer, the different amounts of food available to the Common Swift are reflected in the great range in the length of its nestling period—from thirty-seven to fifty-six days—the shortest periods being found in fine weather, the longest in cold, wet summers (Lack 1956a).

Hummingbirds nesting at high altitudes in the tropics also encounter spells of bad weather during which it is more difficult for the mother to find food,

and consequently the development of the young is retarded. The nestling period of the Green Violet-ear in the Mexican highlands ranges from nineteen to twenty-eight days and is longest when cold, rainy weather inhibits growth of the young (Wagner 1945). In the harsh environment of the high Peruvian plateau, the Andean Hillstar remains in the nest from thirty to thirty-eight days, by far the longest nestling period recorded for any hummingbird, although the species is of average size (Dorst 1962). In Ohio, Song Sparrows remain in the nest ten or rarely eleven days in May, but later in the season, when days are warm and food is apparently more abundant, they often leave when only nine or even eight days old (Nice 1937). In Costa Rica, two young Yellow-bellied Elaenias remained for twenty and twenty-one days, respectively, in an unusually late nest; however, earlier in the season, when food is evidently more abundant and most of these elaenias breed, the nestling period is only seventeen or eighteen days.

In the Oilbird the nestling period of young reared in the same cave ranged from 88 to 125 days. These nocturnal birds lay their eggs at intervals of from about four to nine days, and the last egg of a set of three may be deposited from six to eleven days after the first. The interval between the hatching of successive eggs is about the same as that between their deposition, so that the oldest member of a brood has grown considerably before the youngest hatches. The younger nestlings are accordingly at a considerable disadvantage at meal time. Being less adequately nourished than their elder siblings, they develop more slowly. In a brood of four, the youngest may have a nestling period two weeks longer than that of the oldest, and in a brood of three the difference may be ten or eleven days (D. W. Snow 1961).

In certain small birds, however, the youngest members of the brood tend to leave at an earlier age than their nest mates, because nest leaving is contagious and all sally forth at about the same time. In the Great Tit and Blue Tit, the nestling period of the youngest members of broods averages about 1⅓ days less than that of older members (Gibb 1950).

As we earlier learned, the parents in many instances bring food more frequently for each member of a small brood than for each member of a larger brood. Although a nestling in a large brood may need less food because its companions help it to keep warm, it may nevertheless be less adequately nourished than one in a small brood. In consequence, it sometimes happens that nestlings in small broods develop more rapidly and leave the nest sooner. Northern House Wrens in broods of two or three were fed at an average rate of about seventy-three times per capita per day and left the nest at an average age of 15.3 days. When the brood consisted of four, five, or six nestlings, they were fed at the rate of about forty-nine times per day for each of them and they departed the nest box at an average age of 16.6 days, or 1.3 days more than in small broods (Kendeigh 1952).

In the Red-throated Rock-Martin of Africa, nestlings in broods of two flew at an average age of 22.5 days, whereas trios left the nest when 24 days old. Although the number of parental visits with food was in direct proportion to the number of nestlings present, the more frequent meals brought to the larger broods may have been smaller. In other species studied by Moreau (1947), the nestling periods for small and large broods were not significantly different, even when the number of feeding visits per nestling fell as the family increased in size. In Costa Rica, the average nestling period of sixteen Gray-capped Flycatchers in broods of two was not significantly different from that of eighteen individuals in broods of three, being slightly over twenty days in each case. One solitary nestling stayed in its nest for twenty-one days. Whether nestlings in larger broods are retarded may depend largely on the abundance of food available to their parents. On Kent Island in the Bay of Fundy, broods of three Tree Swallows left the nest slightly younger than larger broods; but in Connecticut, where conditions were more favorable for the swallows, this relation between brood size and nestling period was not found (Paynter 1954).

When broods of the same species differ in size, two factors may influence the length of the nestling period in contrary directions. The more adequate nourishment that members of small families often receive should hasten their development, relative to that of larger broods. On the other hand, because they enjoy more room, these nestlings may linger in the nest longer than those who are cramped and jostled in a more crowded nest.

CHAPTER 25

The Education of Young Birds-I

The young bird's passage from the nest into the outer world is a transition hardly less abrupt than its earlier escape from the eggshell. Formerly it rested in one spot where it was warmed and protected, where everything it needed was brought to it; now it moves through a perilous world, where dangers beset it on every side, where it must learn to gather food, avoid enemies, and find a safe spot for the night. Fortunately, for most young birds, the transition from the sheltered life of the nest to the far more difficult life of the outer world is softened by the continuing care of the parents, from whom the fledglings or chicks receive not only food, or assistance in finding it, but also protection and an education that is of the greatest value to them. Only a few exceptional birds, including megapodes, certain swifts, and a number of shearwaters and other seafowl, leave the nest without a parent to guide and protect them. We know little of how they meet the problems that now confront them, whether they promptly associate with older birds whose example is helpful to them or face the world alone.

Possibly some readers will object to the use of the word *education* for the process that equips the young bird to take care of itself. Birds, they may say, have been endowed with "instincts" adequate for their needs, and instincts are not learned. But modern research has shown that the boundary between instinctive and learned behavior is not as sharp as it was once held to be. A process hardly to be distinguished from learning is needed to perfect the innate modes of behavior and adapt them to the complexities of the external world. As we saw in the preceding chapter, birds that grow up in a space too narrow to spread and flap their feathered wings fly surprisingly well when they first emerge into the outer air, but they lack control and have difficulty in alighting; they need much practice to perfect their flight. All of our human activities, whether piloting an airplane or writing a book, are based upon capacities that no teacher could give us if they were not innate in us. Thus we must recognize a learned component in much of the behavior that we call innate or instinctive, and an innate component in that which is learned or acquired. The activities that we classify as instinctive and those that we designate as learned differ in the relative amounts of these two components that are needed to perfect them (Hebb 1953).

The education of birds is, no doubt, largely the leading forth and perfecting of aptitudes already latent in them; but this is, as etymology attests, the original meaning of the word *educate*—a meaning that teachers have too often forgotten, to the pupil's detriment. In this process, the young are greatly helped by their elders, from whom they learn where to find food, what to avoid, and in many species also the song of their kind. But first of all they must learn to recognize the parents themselves, so as not to become separated from them.

Imprinting, or Attachment of the Young

When I was a small boy, I had five downy Indian Runner ducklings a few days old, who followed me, single file, wherever I went. None of my walks about the farm was too long for them, provided I went slowly enough for their little legs. Unfortunately, I was still much too young to make and record observations on this remarkable association between ducklings and boy, and I do not recall how it began. But, in a general way, we can reconstruct what happened. The ducklings had become "imprinted" by me, and imprinting is a subject that in recent decades has received much attention from students of animal behavior.

Nidifugous birds, who leave the nest soon after they hatch, face a special problem not confronted by nidicolous nestlings. In the short interval between the escape from their shells and the departure from the nest, they must learn what to follow, and they must learn it so well that they are not diverted from their parent by the unfamiliar objects of all sizes and shapes among which they will soon be moving. One might suppose that nature would have endowed them with an innate image of the parent or an innate propensity to follow an object of just the size and shape, making just the sounds, characteristic of an adult of their species of whichever sex takes charge of the young. Or, to use the language of the ethologists, the "releaser" of their innate propensity to follow should be an object bearing the salient features of the parent. Strangely enough, many nidifugous birds are not so equipped at birth. Instead, they have been endowed with a remarkable capacity to learn at a very early age, in a very short while, in a quite special way, and with surprisingly lasting results. This special mode of learning is called "imprinting."

It has been said of the newly hatched Greylag Goose that it regards as its parent the first living thing it sees. If one wishes an incubator-hatched gosling to become attached to a goose mothering other goslings, he must be careful not to permit the

hatchling to see the person who transfers it, for in this case it is likely to refuse to remain with the goose family but persistently follow humans. Imprinting has been studied chiefly in ducklings, goslings, and domestic chicks, which by simple procedures can be made to follow a great variety of objects, ranging from their own mother to the unrelated hen who hatched them, a bird of some other kind, a wooden model of a drake, a small boy, a man, a football, a box containing a ticking alarm clock that is pulled slowly along a rope from which it is suspended, or a slowly moving shadow low on a neighboring wall. The object by which the hatchling becomes imprinted must show at least one of the characteristics of a living animal; it must either move or emit a short, low, repetitious note; if it does both, the imprinting will be more rapid and effective. The requirement that the object that causes imprinting give some sign of animation avoids the danger of the young bird's becoming attached to some immobile thing beside its nest, such as a rock or a plant, which can afford neither guidance nor protection—a fixation that might be disastrous to the newly hatched bird.

Following the method developed by Eric Fabricius in Finland, Nice (1953) imprinted ducklings of a number of kinds by placing them in front of her in a nest improvised in a basin. Covering them with her hand, she repeated *kom kom kom* many times over, then slowly withdrew her hand. At first they remained motionless, but after a few minutes they started to climb out of the nest, coming straight toward her. When she moved slowly away, they followed. Soon they were taking long walks with her. F. H. Hess believed that, the greater the effort made by the young bird to follow the imprinting object, the more strongly attached it becomes, but the evidence on this point is conflicting (F. V. Smith 1969).

For imprinting by models, the chick or duckling may be placed in a long, narrow runway, along which the model is moved as the little bird advances toward it. Ramsay and Hess (1954) used an elaborate model, a papier mâché Mallard drake, mounted on off-center wheels that caused it to waddle as it moved, equipped with a loudspeaker that repeated *GOCK, gock, gock, gock, gock*, and also with a heater. A ten-minute run with this model, during which the duckling accompanied it for 150 to 250 feet (45 to 75 meters), was sufficient for imprinting. However, as already suggested, ducklings and chicks can be imprinted by much simpler objects.

As a rule, imprinting by the parent or the object that substitutes for it must be done while the precocial bird is still very young. Nice imprinted Shovelers, Mallards, Gadwalls, and Wood Ducks that hatched under her eyes and came to her as soon as they could crawl; some of these ducklings were only one and a half to three hours old. Ramsay and Hess found that Mallards were most effectively imprinted when between thirteen and sixteen hours of age; by their usual method, only one could be imprinted after twenty-four hours and none after twenty-eight hours. But, if permitted to accompany other ducklings that were following a model, a duckling might be partly imprinted up to the age of 38 hours.

The upper age limit for imprinting seems to be set in part by the waning of internal readiness and in part by the rise of fear reactions, which cause the young bird to avoid any strange object larger than itself. This is well illustrated by the observations of Alley and Boyd (1950) on the European Coot. When a man approaches a nest with recently hatched young, the parents give alarm notes, which cause all the chicks more than an hour or two of age to flee, while even those still too weak to run try desperately to crawl down the side of the nest farthest from the intruder. But the young coots have no innate fear of man, and if carried away from the nest before the parents can condition them against him by their alarm notes, they can be imprinted to accept him as their foster parent. This can be achieved only if they are taken when less than eight hours old, after which avoiding reactions prevent acceptance of the human fosterer.

Chicks of the domestic fowl differ from ducklings in showing an innate response to at least one of the characteristics of their natural parent, the clucking of a broody hen. Incubator-hatched chicks that have never been near a hen are drawn by her clucks. Even chicks who have been kept blindfolded since they hatched grope their way toward the voice of a hen they have never seen, as Spalding

(1873) demonstrated long ago. Indeed, a chick that has just pipped its shell and is giving distress calls within the egg may become quiet if it hears a hen's soothing notes (Collias 1952).

Most ducklings show no such ability to discriminate between the calls of their natural parent and other soft, reiterated sounds. When the eggs of Mallards were exposed to a rhythmically repeated call for the twenty-four–hour period that ended when they hatched, the ducklings gave no evidence of having been affected by the sounds that reached them through their shells. Whereas Mallards and other surface-nesting ducks are imprinted primarily by a visual object, which they later associate with a sound that the object frequently emits, hole-nesting ducks, such as the Wood Duck, may be imprinted by a sound from an unseen source, which they try to reach. Later experiences in following a seen object do not alter their preference for this sound. It is important for ducklings that are hatched in dimly lighted holes to become promptly imprinted by their parent's voice, for they jump out of the cavity in response to her notes before they can see her well. In another hole-nesting duck, the Muscovy, Ramsay (1951) found that ducklings who a few hours after birth were removed from the duck who hatched them could already distinguish her calls from those of other Muscovy Ducks. Their memory of their parent's voice persisted through four days of separation but was lost after two more days.

Although early observers claimed that imprinting is permanent and irreversible, but this is not always true. After a bird has, at the appropriate early age, been imprinted by a certain object, it can, in some species, be trained to follow other things that differ greatly from the original one. Thorpe (1956) and his associates found this to be so of European Coots and Common Gallinules; particularly responsive individuals were made to follow three different models alternately by being given one run with each per day. The models were as diverse as a wooden gallinule, a white box, a black box, a man, and a canvas blind six feet high by two feet square. This capacity to shift the following-reaction from one object to another of very different aspect persists from the third to the sixtieth day of the chick's life. However, gallinules who were not accustomed to the experimental set-up when only a few days old became wild and fled from any model that was later presented to them. Coots from twenty to sixty days old will follow a familiar object again and again, sometimes for over one hundred runs at one-minute intervals, showing no waning of the response, although they are obviously tired.

The remarkable part of this performance was that the bird received no reward in the form of food, brooding, or some other pleasant experience; on the contrary, after each run it was chased, caught, and picked up with its legs dangling—treatment that it did not relish. This calls attention to one of the unique aspects of imprinting: it is a mode of learning for which rewards are not necessary, unless the activity of following the parent or model can be regarded as self-rewarding. In ordinary learning, like learning to run a maze or solve a problem, effort cannot be sustained unless success is rewarded with food or some other pleasant experience.

The transference of imprinting was also observed in a gosling Lesser White-fronted Goose who had become separated from its family in Arctic Norway when between one and two weeks old. By this age it had evidently been imprinted by its parents and when first acquired by Steven (1955) was wild and tried to avoid its captors. But after a few days, it no longer attempted to flee from them; after six days in captivity it had become thoroughly attached to its human companions, following them whenever possible and resting in contact with their feet.

Although coots and gallinules that followed strange models seemed to show no after effects in adult life, for most other birds imprinting has momentous consequences. A thoroughly imprinted young bird cannot be lured from the object by which it was imprinted, or from a foster parent that differs from its natural parent even as much as a hen does from a duck or a Turkey. It cannot even be lured by a parent of its own species with a brood of young birds its own age. When Cushing and Ramsay (1949) formed heterogeneous families by hatching the eggs of a variety of domesticated birds under the same hen, they found that, after a few days, they could mix these families together in a pen, yet every young bird would go back to its own

family group led by the hen who hatched it. These motley families that held together consisted of a Barred Rock hen with Muscovy ducklings, Pekin ducklings, and Barred Rock chicks; a Bantam hen with Bantam chicks, Barred Rock chicks, and Mallard ducklings; a Rhode Island Red hen with young Turkeys; and numerous other combinations.

The effects of imprinting sometimes persist into adult life, when the bird, disregarding others of its own kind, chooses as its social companions, even as its "mate," birds of a different family or order, at times even the human whom it came to regard as its "mother" in early life. The disturbing effects of becoming imprinted by a member of a different species or by a lifeless object may often be avoided if the young bird is placed with others of its kind and age at the conclusion of the experiment.

The Mutual Recognition of Parents and Young

The young bird does not by imprinting learn to know the animal or object by which it was imprinted as an individual but rather as a type. Imprinting is, in nature, learning to recognize a species but not an individual; the reason for this is apparent when we recall that it influences not only the bird's attachment to a parent but also, in many instances, its subsequent choice of a mate. However, many precocial and semialtricial birds soon distinguish their own parent from other individuals of the same species. The process of recognition is sometimes hastened by the pecks or other rebuffs that the young bird receives if it tries to join the mother of another brood. In the European Coot, the chick's ability to distinguish strange coots from its own parents comes in two stages. From eight to eleven days old, the young coot learns to avoid strange adults who approach it in a menacing attitude when it invades their territory. At the age of three weeks, the young coot can distinguish strangers from its own parents even when they do not threaten it; it now recognizes its parents as individuals.

When only two days old, Mallard ducklings can distinguish the voice of their foster parent, a hen, from that of two or three others, as Ramsay (1951) demonstrated by enclosing the hens in boxes and releasing the ducklings at a point equidistant from them. With scarcely any mistakes, each duckling went to its own invisible mother, guided by her calls. Two-day-old chicks of the Ring-necked Pheasant and wild Turkey showed almost equal ability to distinguish their foster mother from other calling hens. When Collias (1952) mixed domestic chicks of several broods in the dark, they showed a tendency to sort themselves out and go each to its own mother, at least when the hens were of different breeds. In another experiment, chicks from mothers of three different colors were placed in an enclosure with three strange broody hens of the same colors. The chicks taken from the black mother followed the strange black hen, those from the red mother followed the strange red hen, and those from the white mother went to the strange white hen. These experiments demonstrated that older chicks have some ability to recognize their parents by either voice or appearance.

While the young precocial bird is learning to distinguish its parent from similar adults, the parent comes to know its chicks individually, and by this dual process the family becomes a closely knit group, often closed against strangers. The parent often takes longer to recognize its offspring individually than the latter take to recognize the parent, a consequence of the fact that the young have only one or two attending parents, whereas the latter may have many young, also probably because in birds, as in men, adults have more individuality than babies. Moreover, the appearance of the growing young changes as the days pass, while that of the parent remains much the same.

Before it recognizes its young individually, the parent bird of numerous species will accept and attend others of approximately the same age and general appearance as its own brood, although those conspicuously larger or smaller, or of a different color, may be repulsed with pecks if they try to join the family. After the young are recognized individually, this acceptance of strangers ceases in all but the most tolerant species, such as some wild ducks. Usually a bird mothering young of a different species, as a hen with ducklings or a duck with chicks, will peck a young bird of her own kind that is suddenly added to her brood. But the experience of raising a brood of a certain species does not affect the acceptance of a later brood of another kind that the female may hatch; a hen who has reared a brood of chicks may later raise ducklings.

Parent European Coots with young less than two weeks old will tolerate, feed, and brood strange young that resemble their own, and they may even adopt them permanently. After the young coots are much over a fortnight old, their parents, who now evidently recognize each offspring as an individual, vigorously attack intruding coots of whatever age (Alley and Boyd 1950). In the American Coot, the situation is somewhat different. The young of a single brood, who may differ in age by as much as a week, go through a series of rather abrupt changes in plumage while still attended by their parents. The adults appear to recognize their offspring by the appearance of the majority, and, when the oldest start to turn lighter in color, they are occasionally attacked by parents confused by this transformation. Later, when most of their young have reached this stage, the parents accept them as their own. Now the youngest chicks, still wearing their black natal down, are the odd members of the brood and are attacked by the parents, sometimes quite severely. Sometimes the perplexed parents alternately feed and persecute an odd-colored member of their family (Gullion 1954).

In semialtricial birds that breed in crowded communities, the mutual recognition of parents and young assumes special significance. Successful reproduction is promoted by each parent's feeding its own young; if feeding were random and adults gave food to the first chick that begged for it, many young might starve. To ensure the orderly feeding of chicks who at an early age wander from the nest and may mingle with many of their contemporaries, parent and young must recognize each other individually. On their crowded sea-cliff ledges, Common Murres learn to distinguish their parents' voices while they are hatching (Tschanz 1968). Chicks of Black-billed Gulls, which breed in compact colonies on dry stream beds in New Zealand, could distinguish the voices of their parents from those of other adults when two to four days old, as Evans (1970) demonstrated by playing back the gulls' recorded voices through a loudspeaker. In Herring Gulls, which have larger nesting territories, parents do not recognize their chicks individually until they are about

PLATE 97. Snares Island Crested Penguins with downy chicks in crèches (photo by John Warham).

five days old. Before their young reach this age, they will accept and feed strange chicks of similar age, but beyond this point they peck them (Tinbergen 1951). Sooty Terns, which nest on the ground like Herring Gulls, learn to recognize their own chicks in about four days; but Noddy Terns, whose nests in trees and shrubs are not so likely to be invaded by wandering chicks, seem not to recognize their unfledged young as individuals (Watson and Lashley 1915).

Laysan and Black-footed albatrosses take about ten days to become familiar with their young as individuals. Parents with chicks younger than this readily accept other small chicks, but, by the time their young are six or seven weeks old, most parents know them as individuals and repulse strange

chicks. Since young albatrosses of all ages return to be fed at their nest or near it, the parents do not need to distinguish their own offspring in order to avoid feeding strange chicks (Rice and Kenyon 1962). Parent King Penguins with small chicks will brood and feed chicks of other parents, including those very much bigger than their own. After about a week, the parents learn to recognize their own chick by its feeding calls, and the chick can distinguish the call of its parents. After this stage is reached, a parent who has accepted a strange chick for brooding will attack it if it breaks silence (Stonehouse 1960). After the young of penguins, flamingos, pelicans and certain other birds leave their nests to gather in herds or crèches, sometimes containing hundreds or even thousands of individuals,

PLATE 98. Australian Pelicans with downy chicks in crèches. Parents recognize and feed their own young amid the throng (photo by John Warham).

mutual recognition of parents and young, often by voice, ensures that each parent feeds its own offspring (plates 97 and 98). This system prevents the more vigorous or aggressive young from depriving their smaller or weaker associates of their meals.

In strictly altricial or nidicolous birds, the situation is quite different from that in precocial and semialtricial species. In the first place, it would be impossible for most newly hatched altricial birds to recognize their parents by sight, for their eyes are tightly closed and in most species they remain so for some days. In the second place, there is no need for the nestlings to recognize their parents as individuals or even as species, for they do not follow them but remain inertly in one spot where food is brought to them. Very young altricial nestlings fail to dis-

tinguish their parents from other birds or to form an exclusive attachment to some particular object presented to them instead of their parents; in various species, they react to anything that gently shakes the nest, makes a low sound, or moves above it by lifting up their gaping mouths for food.

After their eyes open and they can make visual discriminations, the nestlings learn to distinguish their parents from contrasting forms, gaping toward the former but crouching down into the nest if a strange object approaches. In the Common Grackles studied by Schaller and Emlen (1961), these diverse responses to the sight of the parents and to that of other objects first occur at the age of eight or nine days, some days after they first crouch when strongly shaken. In a variety of other small passerines, crouching at the sight of strange objects begins at five to nine days of age, but in the Blue Jay this reaction does not appear until fourteen days and in the still larger Crow not until eighteen days. Nestling Silver-throated Tanagers first greeted my hand by crouching, instead of gaping, when they were eleven days old; but they shrank into the bottom of their open cup somewhat earlier if they could see my face as well as the hand that I held above them. It is doubtful whether all these passerine nestlings even at this stage recognize their parents as individuals or can distinguish them from birds of other species of about the same size. But nestlings of large, colonial-nesting birds, which stay in the nest much longer than most passerines, do finally distinguish their parents from other adults in the colony. Until over two weeks old, the young Fulmar spits oil at any bird that approaches it incautiously, making no distinction between its parents and strangers. In its third week, however, the chick recognizes its parents and refrains from spattering them (Duffey 1951).

Not only do altricial nestlings fail for many days to recognize their parents, but also the latter do not even recognize their offspring as individuals. The parents' attachment is to the nest. So long as they do not mistake its location, there is little danger that they will attend nestlings of some other bird to the neglect of their own. Even in crowded colonies, birds have a remarkable ability to pick out their own nest from the many similar ones that surround it. Usually they make the appropriate response to the contents of their nest without showing any concern for the individual identities therein. Thus Emlen (1941) demonstrated that a female Tricolored Blackbird who has just finished laying her eggs will feed strange young introduced into her nest, and the male blackbird will do so even during the laying period. When nearly fledged nestlings were substituted for very young ones, or recently hatched nestlings for older ones, the parents continued to feed the new occupants of their nest as though no alteration had been made, thereby considerably shortening, or prolonging, the interval during which they cared for young in the nest. Black-crowned Night-Herons will also accept and attend nestlings of any age that are placed in their nest, even when they have been substituted for eggs (R. P. Allen and Mangels 1940). Brown Booby parents will adopt strange chicks that differ in age from their own by three or four weeks (Dorward 1962). Although, like many gulls and terns, they are colonial ground nesters, they are much more altricial, and this seems to account for their different treatment of alien young.

The altricial bird's concentration upon the contents of its own nest—to the neglect of what is beyond it even if it be its own nestling—is on the whole beneficial to the species. It leads to the chilling and starvation of unfeathered young that have fallen from the nest or been cast out by a parasitic cuckoo, but these fallen babies could hardly be saved in any case.

After the young are feathered and begin to give location calls, the behavior of the altricial parents changes abruptly. Now their concern is with their fledglings, wherever they may be. Thus female Red-winged Blackbirds, who earlier had accepted strange nestlings placed in their nests, begin to recognize their young individually and to follow them to new locations when they are ten days old and about ready to leave. Recognition is evidently by voice (Peek et al. 1972). Nevertheless, individual recognition of altricial young by their parents seems to be of minor importance when the territory is fairly large and the family remains within it. If dependent young wander beyond the parents' territory, as often happens, to recognize the young individually may be of consequence. After D. W. Snow (1958) color banded young Blackbirds in their nests, he noticed that, when two families became mixed, each parent invariably fed its own fledglings. But in similar circumstances parent Pygmy Nuthatches fed each other's young (R. A. Norris 1958). This, of course, is not proof that the parents did not distinguish their own young from their neighbor's, for passerine birds not infrequently feed young of species so different from their own that they could hardly have mistaken them for their own progeny (chapter 28).

The Bird's Recognition of Individuals

The ability of birds to distinguish each other individually stirs the admiration of bird watchers who, even with the help of powerful field glasses, can rarely recognize individuals without marking them with colored bands or in some other way. It has been demonstrated that birds as diverse as Song Sparrows and Herring Gulls recognize as individuals not only their mates but also some of their neighbors. Indeed, according to Tinbergen (1953), among the gulls, neighbors appear to acquire a reputation, so that the alarm calls of some, known to be nervous and panicky, tend to be ignored, whereas the similar calls of more reliable neighbors are taken seriously. Some flocks of birds develop a social hierarchy, often manifested in the peck order. The top bird, A, pecks all others but is pecked by none of them. The next bird, B, pecks all except A; C pecks all except A and B, and so forth. Such a system seems impossible unless each member of the flock recognizes every other individually. The extent of the personal acquaintanceship of such birds is suggested by Sabine's (1959) studies of Oregon Juncoes, in which she discovered the personal standing of each member of a flock of forty-two.

Birds recognize each other by a number of characteristics, of which facial expression and voice appear to be the most important, just as among ourselves. A male swan in the Berlin Zoo once attacked his own mate while she fed with her head submerged, but as soon as she looked up and he could see her face, his hostile attitude changed. A Herring Gull sleeping in the daytime on its territory or while incubating is usually not wakened by the calls of other gulls passing a short distance away, but it awakes instantly when it hears the voice of its approaching mate. Since the latter utters no special call, it is evidently the quality of its voice that is distinctive. A Herring Gull can recognize its flying mate at a distance of thirty yards (Tinbergen 1953). Domestic hens recognize their flock mates largely by the appearance of their heads and necks. Altering the comb or dyeing the head or neck produces a much greater effect than a change in the outline or coloration of the body. A hen whose comb has been artificially altered may no longer be recognized by her companions but pecked as a stranger. Nevertheless, no single feature is the sole means of recognition, which depends upon the hen's deportment as well as her appearance (Guhl and Ortman 1953).

Since the appearance of a bird's plumage changes considerably while it is molting, and greatly when it undergoes seasonal changes in color, the advantage of depending largely on facial expression for recognition is obvious. The bird watcher may sometimes, if he takes great pains, recognize as individuals birds that at the first glance appear quite alike, and often he finds the feathers of the head most helpful. But he can scarcely ever distinguish his birds so readily, or so many simultaneously, as the birds themselves appear to do.

Just as birds recognize each other individually, and chiefly by facial expression, so they learn to recognize humans who are closely associated with them. There is a well-known story of a pheasant who courted the German ornithologist Oskar Heinroth and fought Frau Heinroth as a rival. In an attempt to deceive the bird, the Heinroths exchanged garments. The pheasant started to attack the dress, but he paused, looked up at Dr. Heinroth's face, then flew at Frau Heinroth in trousers. When she and her sister exchanged clothes, the pheasant still knew his supposed rival. A male European Robin that L. Howard (1952) tamed while a fledgling was driven off by his father and settled down at a point about a mile from her cottage. Nevertheless, the young bird continued to perch upon her hand and to take food from her, although he would do so with nobody else. The robin recognized his benefactress despite her changes of attire, and even after an interval of eight months during which he did not see her.

The case was somewhat different with a Redwinged Blackbird who had become highly antagonistic to Sherman (1952) because of her frequent visits to the nests of his mates. Although he battered Miss Sherman whenever she approached, he showed no hostility toward her sister. When the two women entered his territory after they had exchanged clothes, the Red-wing, after some hesitation, struck the sister, but without his customary vigor, and he refused to touch his old "enemy." This bird apparently identified Miss Sherman by her total aspect and was confused when parts of the familiar pattern appeared on two separate persons.

I had a somewhat similar experience with an exceptionally nervous and querulous female Gray's

Thrush who in two consecutive years nested near our house. During the first year, I gained her dislike by occasional visits to her nest and by gathering fruits from surrounding trees. She never attacked but protested with her plaintive *quee-oo* whenever she saw me, although mostly she ignored other people. A year later, when her nest was in a more secluded spot, she remembered me and displayed her special antipathy to me even before I discovered where she had built. Whenever she saw me approaching, she would fly far to meet me and would follow me, incessantly complaining, over an area of five or six acres. She would raise her mournful cries even when, through an open window, she glimpsed me inside the house, although most birds pay no attention to anything indoors.

To test the thrush's ability to recognize me, I walked about the garden disguised in a dressing gown and a hat that I never wore on the farm. These appearances might be greeted by a complaint or two, but never by the close following and interminably reiterated protests that I usually evoked. After I had satisfied myself that the thrush failed to recognize me in my disguise, I decided to see how she would react to a sudden change in my attire. After walking a short way in my usual clothes with the thrush following and complaining loudly, I suddenly, in her presence, put on the dressing gown and felt hat that I had carried inconspicuously under an arm. After repeating *quee-oo* a few times more, she fell silent and ceased to pursue me. Then I removed my extraordinary garb, which set off a new series of complaints. It was clear that the thrush distinguished me by my total appearance rather than by my face alone, but sometimes she seemed to suspect my identity beneath a strange attire.

This permanently resident thrush may have kept me in view through the long interval between nesting seasons; but the Red-winged Blackbird recognized his "enemy," Miss Sherman, after a winter's absence, and the European Robin recognized his friend, Miss Howard, after eight months. I have found no direct evidence that birds remember each other after so long a separation, although it is likely that they remember others of their own kind at least as well as they remember people. Domestic

hens recognize their old companions after a separation lasting from a week to a month, the persistence of their memories depending on the length of their previous association. A hen or a rooster that has been away longer than this is treated as a stranger. Birds have excellent memories for places, as is evident from the fact that migratory species return to their last year's territories after an absence of seven or eight months, during which they may have visited another continent, where also they settle in the same localities in successive years. When two migratory birds remate after a winter's separation, it may be that they are brought together again simply by their attachment to the familiar territory, but it is also probable that they remember each other personally. The subject merits further study.

Learning to Eat

To pick up and swallow food, birds need little instruction. To forage competently in a natural environment, selecting from the immense variety of swallowable objects those that are wholesome and nutritious and avoiding those that are harmful, is an accomplishment that is only gradually perfected. The development of competent foraging behavior involves several stages. First, the bird must be equipped with the movements, often complex, with which it seizes or captures its food. Second, it must learn what to eat. Third, it must learn to avoid objects that are apparently edible. Finally, in many habitats, the bird that takes full advantage of the opportunities for nourishing itself must learn not to be deceived by objects that appear dangerous, or that resemble inedible objects, although actually they are wholesome food.

The movements birds use to seize and swallow food, despite the high degree of sensory and muscular coordination involved, are largely innate, at least in some species. This was proved for the domestic chick by the careful experiments that Spalding (1873) made a century ago. A chick does not eat the moment it emerges from the eggshell; it must dry and gain strength before it can stand, hold up its head, and peck. This requires about four or five hours, during which the newly hatched chick might, conceivably, learn much of importance, especially how to coordinate its visual and other impressions with its muscular responses. To eliminate the possibility of such learning, Spalding placed little hoods over the heads of a number of chicks just as they were hatching, before they had opened their eyes. He kept them blindfolded for one to three days, after which he unhooded them, one by one, on the center of a table covered with a large sheet of white paper on which he had placed a few small dead and living insects. The chicks seemed somewhat stunned by their first exposure to light and they remained motionless for several minutes. But, often at the end of two minutes, they followed the movements of crawling insects with their eyes, turning their heads with all the precision of a grown hen. "In from two to fifteen minutes they pecked at some speck or insect, showing not merely an instinctive perception of distance, but an original ability to judge and measure distance, with something like infallible accuracy. They did not attempt to seize things beyond their reach, as babies are said to grasp at the moon; and they may be said to have invariably hit the objects at which they struck. . . . I have seen a chicken seize and swallow an insect at the first attempt; most frequently, however, they struck five or six times, lifting once or twice before they succeeded in swallowing their first food."

The foraging of a chicken involves scratching with the feet as well as picking up the food thereby uncovered, and this, too, is innate. Chicks that Spalding kept quite isolated from their kind, so that they could not learn by imitation, began to scratch when two to six days old. Spalding's conclusion that the chick's pecking is innate has been confirmed by later investigators, who have pointed out that the total reaction, which includes swallowing, improves in accuracy with practice (Wood-Gush 1955).

I have seen newly hatched songbirds raise their gaping mouths for food as soon as they escaped from their shells. This receptive attitude requires much less coordination than the active pecking of a day-old chick, and comparable behavior does not appear in altricial nestlings until some days later, when they begin to peck at specks or small objects that attract their attention while they rest in the nest. Such exploratory pecking has been observed in the Black-billed Cuckoo at the age of six days, in

the Ovenbird at eight days, and in the Song Sparrow at twelve days (Nice 1943). Birds with longer nestling periods seem to begin somewhat later to peck at things. Before they leave the nest at the age of eighteen days, Southern House Wrens peck at small creatures crawling over the wall of their box or gourd, or they come to the doorway and try, usually without success, to catch insects that fly past. As soon as they emerge from their hole in a tree, young Golden-naped Woodpeckers climb over the trunk and branches, pecking frequently, much as their parents do. Although they are now about five weeks old, these young woodpeckers appear to be no more advanced in their foraging behavior than a day-old chick. They seem rarely to find anything edible.

The young of precocial and altricial birds are alike in that they begin to peck before they can distinguish edible from inedible objects. The domestic chick at first pecks indiscriminately at any spot or small object that contrasts with the surface on which it rests, including its own toes. It may even try to eat its droppings, but after one or two attempts it avoids what is obviously disagreeable to it. Thus, by a process of trial and error, it learns to pick up and swallow what is edible and ignore other small objects. If for some days a chick is given no food but only spots on the floor to pick at, the repetition of unrewarded efforts extinguishes the pecking response and it becomes incapable of feeding itself. To keep such a chick alive, it must be laboriously taught to feed on mash, and it is most difficult for it to learn to eat anything else.

I have watched fledgling songbirds and hummingbirds try to eat buds, flowers, leaves, fragments of bark, or other inappropriate things, which they finally dropped. Like the domestic chick, they seem to learn by trial and error what is good for them. Since they begin to peck at things even before they leave the nest, passerines, woodpeckers, and other altricial birds may, more or less accidentally, find a little of their own food on the day of their departure; but a number of days must pass, and they must become more expert, before they can nourish themselves. Meanwhile, they are closely associated with the parents who feed them. The example of their elders—where they are led and what they are given—must greatly influence the development of their own feeding habits during their days of complete and then partial dependence.

What part does imitation play in the development of the foraging habits of birds? Spalding observed that a young Turkey has an efficient method of catching resting flies by stealing toward them with measured steps; when sufficiently near, it slowly and stealthily advances its head until its bill is only an inch or two from the prey, which the bird finally seizes with a sudden dart. This mode of stalking was innate, for his young Turkeys had no opportunity to learn from adults. Noticing that his domestic chicks were no less fond of flies than the Turkey, he placed one of them with a Turkey so that it might learn, by imitation, this superior method of flycatching. But although the two remained close companions throughout the summer, the chicken never acquired the efficient trick of its larger comrade. It seemed "wholly blind to the useful art that was for months practiced before its eyes." However, we should bear in mind that domestic animals, especially chickens, tend to be less intelligent than the wild stock from which they sprang. Intelligence is inimical to the docility required of domestic animals, which for generations have been selected for qualities that are not enhanced by mental alertness.

That imitation plays a role in the development of the feeding habits of birds seems also to be ruled out, as Thorpe (1956) remarked, by the common observation that in a mixed flock each species sticks rather closely to its own way of foraging without copying the methods of its associates. In the British Isles, Sanderlings, Dunlins, and Ruddy Turnstones are constantly seen feeding together on the shore, but the first two have not been known to engage in the stone turning that the Ruddy habitually practices when searching for food. In contrast to the rather stereotyped modes of foraging of shore birds, some other birds exhibit extraordinary versatility in finding food. Great-tailed Grackles eat fruits, catch insects, turn over stones for what may be lurking beneath them, catch small fish by shallow dives, pluck vermin from cattle, plunder bird nests, and so forth. We can hardly exclude the possibility that the grackles acquired some of their diverse foraging techniques by imitating other birds with more specialized modes of feeding.

Students of animal behavior have been reluctant to concede that birds, or indeed any animals below the higher primates, engage in imitation, other than the vocal imitation that is widespread in birds. Imitation, they say, involves self-consciousness, and there is no evidence that birds are self-conscious. Apparently they fail to distinguish between two quite different kinds of imitation or copying, exemplified respectively by the actor and the learner of some practical art. Successful acting, whether on the stage or in the parlor, no doubt involves seeing ourselves as others see us and as we see those whom we mimic. But when we try to learn some skill, such as tying a knot or using a tool, by watching another person, consciousness of self is often lost in our intense concentration on the operation, which we then try to perform as our teacher did. This second kind of imitation, which involves no self-consciousness, may not be beyond animals that are incapable of the actor's objectification of self.

This question of imitation or copying becomes important when a new feeding habit diffuses rapidly through a population of birds. Some years ago, it was noticed that titmice were opening milk bottles that delivery boys had left on porches and doorsteps in England and drinking the cream. The habit was acquired by several kinds of tits as well as by Jackdaws and Great Spotted Woodpeckers, and it spread rapidly over the country. No one supposed that every tit who opened milk bottles acquired the trick independently. Those who were skeptical of birds' ability to learn by watching their companions, even in the case of titmice whose intelligence may be somewhat above the average for small birds, attributed the origin and spread of this epidemic of petty larceny to trial-and-error learning and "local enhancement."

Tits hammer on hollow-sounding objects, when opening nuts for example, and they pull loose bark from trees to expose insects lurking beneath—a habit that seems to be the origin of their strong propensity to tear paper or cardboard whenever they find it. Apparently, the tits who first discovered how to open milk bottles hammered experimentally on the hollow caps of thin, soft metal until they had made a hole, then tore at the tops of waxed cardboard. The unexpected discovery of nourishing cream beneath the cardboard encouraged them to repeat this procedure at other milk bottles until they had improved their technique. After the more enterprising tits had become habitual milk drinkers, the attention of other tits was drawn to the spots where they were busily engaged and found food—the phenomenon of local enhancement, which is widespread among animals who congregate at spots where they see others eating. However, the novices, it is held, did not learn how to open the bottles by watching experienced birds but rather by discovering the trick themselves by the same process of trial and error that the pioneers had followed.

House Finches in California provided a rare opportunity to study the whole process of learning a new method of feeding. After they had exhausted the supply of figs on a neighboring tree, they turned their attention to a standard hummingbird nectar bottle, which delivers a sugary solution through a narrow, downwardly directed tube, beneath which hummingbirds hover while they insert their bills upward into the orifice. Some finches learned unobserved to hover, while others clung to the tube, head downward, to reach its mouth. After the tube was greased to discourage such clinging, all who desired a sweet drink were obliged to hover, and those that attempted to do so were carefully watched. Their efforts were at first indirect and clumsy. But, within two weeks, adults acquired the art of hovering motionless on rapidly beating wings, like hummingbirds, while they drank. For a variety of reasons, immatures took longer, about six weeks, to learn this mode of foraging unusual in the finch family. After they became proficient, the male House Finches could hover in front of the feeder for as long as five seconds, but females for only about one second. One can hardly doubt that the hummingbirds provided the local enhancement and directed the finches' attention to the feeder, but whether the latter tried to copy the method of the hummingbirds remains an open question (P. M. Taylor 1972).

The problem of whether small birds can learn how to procure hidden food by watching other birds was attacked experimentally by Alcock (1969),

working with Fork-tailed Flycatchers, White-throated Sparrows, and Black-capped Chickadees. First, he trained individuals of these species to find mealworms by removing a light wooden cover from a food tray. Then he permitted the experimental birds to watch the trained birds uncover and eat the food, either in company with them or from a distance. When an experimental bird found and ate a worm in company with the trained companion, sometimes by taking it from the latter, the learner was frequently able, on a subsequent trial, to find the covered mealworms while alone. But when, after an interval of watching without being able to reach the food, an experimental bird was allowed access to the covered tray in isolation, this bird approached, touched, and pecked at the tray significantly more often than control birds who had not watched a trained companion, but it always failed to uncover the prize.

It is evident from these experiments that, by watching the trained companion that it could not reach, the experimental bird learned that food was available in a certain place, but it had not paid enough attention to the companion's procedure to repeat it. But when more closely associated with the trained companion, able to participate in the rewards of its skill, the experimental bird did learn to uncover the food by itself. This is not greatly different from the way that we ourselves often learn, but of course feats more difficult than removing a cover from food: by watching somebody else perform an operation, we learn that by doing something in a certain place a desired result may be obtained; but without closer participation in the operation, we may be unable to repeat it alone.

Especially among raptors does parental example seem to play an important role in the development of feeding habits. Fledgling hawks and falcons, which while nestlings saw only prey that their parents had killed before bringing to the nest, appear to lack innate recognition of living animals as potential food. But they are curious about novel objects and have a strong propensity to seize and manipulate them, especially those that move. By nipping and tearing an animal that they have caught, they may discover that it can be killed and eaten, and such an experience would increase their readiness to capture and dismember a similar animal. Observations on free birds, no less than the experience of falconers, show that young raptors are slow to capture prey for themselves in the absence of guidance by their parents or human trainers. Since, after leaving the nest, many birds of prey associate for a long while with their parents and continue to be fed by them until they become proficient in hunting, they have ample opportunity to learn from them. It appears that in these circumstances the food preferences of the parents may be transmitted to the progeny not by genetic inheritance but by example or tradition. This is probable when a local population of a raptorial bird consistently follows a method of feeding quite different from that of the species as a whole. An example of this is a population of Peregrine Falcons in New Mexico that for years captured bats in a cave, although most Peregrines prey on birds. Cushing (1944) discussed the evolutionary significance of the nonheritable food habits of raptors.

Certain finches also have local food preferences that seem to be perpetuated by tradition rather than by genetic inheritance. European Goldfinches at Oxford, England, ate many seeds of *Senecio vulgaris*, but in Holstein, Germany, where this composite was also abundant, they neglected it. In Holstein, Linnets consumed many cultivated cereals, while at Oxford they apparently took none, although much grain was grown there. The Linnets at Oxford preferred the seeds of pigweed, which those in Holstein did not eat (Newton 1967).

Birds that seize their food on the wing or dive into the water for it depend upon special skills, which they practice while still supported by their parents. Fledgling swallows and martins often rest upon some exposed perch, such as a dead branch or an overhead wire, while their parents bring food to them. Sometimes, as a parent approaches with a billful of insects, the juvenile flies out, meets the attendant high in the air, and receives the food from it. This method of taking its meals gives the young swallow training in controlled flight, which will be helpful when it undertakes to capture volitant insects for itself. When a White-fronted Nunbird catches an insect or some other small creature for its fledged young, it perches in one spot, holding the item conspicuously, until the juvenile flies up, sometimes from a distance of one hundred feet, and, without alighting, snatches it deftly from the parent's bill. This gives the young valuable practice in the nunbird's habitual method of foraging, which, as in other puffbirds, consists in flying up and plucking an insect from foliage or bark while on the wing.

Kingfishers appear to require much practice to perfect their art, for fish are swift and, in consequence of the refraction of light at the water's surface, are rarely seen in their true position by the bird watching from above. Even adult kingfishers sometimes plunge repeatedly without catching anything. A parent Belted Kingfisher provided practice for her young newly emerged from their burrow by catching a fish while they perched nearby, beating it into quiescence against a branch, then dropping it into the water for them to retrieve. After some hesitation, one of the hungry juveniles would dive for it. At first the inexperienced kingfisher missed more often than it caught the moribund fish, but it persisted until it secured its meal. With this practice, the juveniles' skill improved so rapidly that within ten days they were catching uninjured fish for themselves (Bent 1940).

Inca Terns catch fish in the Humboldt Current off the coast of Peru by plunging to the surface without submerging completely and by dipping in their bills while skimming over the water. Ashmole and Tovar (1968) watched juveniles, who had recently learned to fly, practice these methods by plucking small pieces of seaweed or other flotsam from the ocean's surface. After carrying the object for a short distance in flight, the tern would drop it, then frequently retrieve it. It might drop and catch the same object as many as ten times in succession. Often several young terns engaged simultaneously in this play, competing to catch the same object or picking different ones from the water. These young terns stayed close to their island base and were still fed by their parents. The young of frigatebirds, which also capture their food by difficult aerial maneuvers, gain proficiency in flight by attacking incoming boobies and by a sort of game that consists in catching feathers or strands of seaweed from each other high in the air (Stonehouse and Stonehouse 1963).

Learning What Not to Eat

To learn what to eat and how to procure it is only part of the young bird's education in dietary habits. It must also learn what not to eat. The insectivorous bird soon discovers that not all insects are palatable. Rothschild and Lane (1960) believed that some warningly colored insects are not merely distasteful but contain chemical substances that cause physical pain, distress, or fright in the bird who incautiously seizes them. These inedible insects are just the ones most likely to attract the young bird's attention because many of them have bright colors, which the bird does not instinctively recognize as a warning. The first experience with one of these protected insects, such as a wasp or a distasteful butterfly, often makes the young bird display vehemently and utter notes of alarm. If a similar insect is later placed in a cage with the same bird, it may chatter in alarm and try frantically to escape. Evidently its first experience is so distressing that it makes a deep impression on it— an impression the strength of which is also indicated by the time that it is remembered. Small insectivorous birds of several kinds retained for three to fourteen months the memory of a single experience with a warningly colored insect (Thorpe 1956).

The cries and attitudes of alarm of an experienced bird that sees a warningly colored insect may cause its less experienced companions to avoid the insect. In this manner parents may, perhaps unintentionally, save their offspring the necessity of learning the hard way. Thorpe mentions experiments in which Swynnerton studied the effect of warningly colored insects on two species of swallows. When one swallow was being tested with a new kind of insect, its companions anxiously watched its response, as though with the intention of profiting by its reaction and thereby avoiding even one unpleasant experience with the distasteful insect.

As is well known, insects that are unpalatable to birds and other insectivorous animals are mimicked by other kinds of insects that lack this means of protection. Birds who have had a single distressing experience with a certain kind of insect tend to avoid the edible species that resemble it, but the persistence of this avoidance varies with the closeness of the resemblance and a bird's power of discrimination. Recent experiments reported by Rothschild and Lane, Tinbergen (1958), and others leave no doubt that this so-called Batesian mimicry gives important protection to numerous kinds of insects. Indeed, the warningly colored creatures, which often display some shade of red, appear to have taught some birds to be wary of insects that bear this color, even if they do not closely resemble a protected species.

Birds, like cautious people, distrust novelty in food. Hand-raised Blue Jays, Common Grackles, and Red-winged Blackbirds feared and avoided butterflies that were conspicuously different from insects that they were accustomed to eating. If they attacked such butterflies, they did so inefficiently and dropped them after the first seizure. Apparently, conspicuous coloration itself, apart from unpalatability, protects insects from hungry birds (Coppinger 1970).

Many birds include berries and other fruits in their diets, and we must ask whether they do not have to learn which kinds are edible and which are noxious, just as in the case of insects. Some of us as children were taught by prudent parents to avoid as poisonous every berry that we found in the woods and fields, except the few kinds that we were told we could eat. However, a ripe fleshy fruit is not likely to harm animals, for its whole biological purpose is to be eaten by animals, which in return for this nourishment disseminate the enclosed seeds, either discarding them or passing them unharmed through their alimentary canals. If a fruit is poisonous or even disagreeable to any particular animal, this is probably a chemical accident rather than an adaptation, as it is in the case of protected insects. A familiar example of such an accident is the reaction of humans to the poison ivy widespread in North America; all parts of this plant cause severe inflammation of the skin and mucous membranes of many people, but birds eat its berries with impunity, especially during winter when other food is scarce. Although it is seldom advantageous to plants to produce poisonous fruits, to have poisonous or otherwise inedible seeds is an advantage; the fruit should be eaten, the seeds spread abroad intact.

What excites our wonder is the smallness of the recompense that birds demand for transporting the seeds of plants. The plants often set aside a minimum of carbohydrates for the berries or fruit pulp, reserving the more valuable and nutritious proteins for the seeds. Many fruits and arils (the fleshy outer covering of certain seeds) that are eagerly sought by birds contain little nutriment and are decidedly distasteful to ourselves. If birds demanded a richer reward for their disseminating services, plants might have evolved a greater variety of highly nutritious fruits, and as a consequence of this our agriculture and dietary habits would be far different from what they are, to our immense advantage.

To become expert foragers, birds must learn not only to avoid certain things, especially aposematic or warningly colored insects that they may spontaneously take to be edible, but also to eat nutritious objects that they may spontaneously avoid. Among the latter are a number of insects that rely on their dull, concealing coloration to escape detection but if found by a keen-eyed bird display a bright patch or large eye-spots with alarming suddenness. An example is the Eyed Hawk Moth of Europe, which by day rests immobile on a tree where its spread forewings match the bark. If, despite its excellent camouflage, the moth is discovered and touched by a foraging bird, it opens its forewings more widely, exposing a large, bright eye-spot on each hindwing, then slowly flaps its wings up and down. Some birds, alarmed by the sudden appearance of these staring eyes, drop the moth as though it were red hot. Then they may follow it to the ground and hop around it in great excitement, not daring to touch it again, as happened with a Yellow Bunting watched by Tinbergen. When captive Chaffinches were presented with an Eyed Hawk Moth, they gave the insect an exploratory peck, then jumped back as though stung when it exposed its "eyes." Although some of the birds thereafter avoided the moths, even when they were not displaying, others returned to eat the corpulent insects, whose display is all bluff.

Thus birds, innately endowed with movements for seizing and swallowing food, must learn to use this hereditary equipment in a complex, baffling world, which presents many surprises, many deceptions, many perils. And often, no doubt, they make a fatal mistake before they have learned what to eat and what to avoid. Such errors, no less than the greater vulnerability to predators and other dangers, result in higher mortality among birds that have recently left their parents than among experienced adults.

Learning to Drink and Bathe

Birds need not only food but also water, yet they have no innate ability to recognize this colorless liquid. Even when thirsty, they do not spontaneously drink when they first see or hear water or feel it on their wings or tail. Although an extensive water surface may fail to draw their attention, a glittering drop is sufficiently small and distinct to excite an exploratory peck, as many another small object does. It is probable that young birds often learn to drink by tasting dewdrops on grass or raindrops hanging from foliage. Other birds may discover water by pecking at its glittering surface, at a small floating object, or at a speck on the side or bottom of a shallow vessel. By whatever accident a bird first gets water into its mouth, its presence there sets off an innate swallowing reaction, which in many birds includes lifting up the head so that the fluid runs back into the throat.

Some birds require several such accidental experiences before they learn to drink. One of Wallace Craig's young doves quickly learned to associate the action of drinking with various stimuli, such as the shape and appearance of the dish of water, the person who brought it, the sound of pouring it, and so forth; but the dove took longer to recognize water as such. Even when thirsty, it would not try to drink if given water in an unfamiliar dish. While learning, the dove often went through the movements of drinking as soon as it saw the familiar dish being brought but before it had arrived. Or it might stand in the dish and lower its bill outside, appearing surprised when its attempt to drink failed. Evidently, birds need a longer time to become familiar with a transparent, elusive substance like water than with solid food (Thorpe 1956).

The bathing movements of birds, like their drinking movements, are innate, but the young bird must learn by experience to perform them in the proper medium. When hand-reared birds are first shown water, they may go through the motions of bathing outside the dish. If the water is stirred, it attracts them more strongly and they hop in for their bath. Even nestlings have been seen to perform bathing movements when water approached their nest. Virginia Rails first bathed when little over a day old, American Coots at three to five days, and Sora Rails at six to eight days. Even when encouraged to do so, a Killdeer made no effort to bathe until it was ten days old; not until it reached the age of two weeks did it perform all the bathing movements of an adult (Nice 1962).

The ways of bathing of birds are fascinatingly diverse. Perhaps the most usual is the splash bath, in the performance of which the bird enters shallow water and splashes it over itself by vigorous movements of its wings, from time to time dipping in its head. More dainty is the dip bath of a hummingbird in a still reach of a woodland stream. With wings beating too rapidly to be seen, the graceful bird drops down and partly immerses itself in the limpid water, remaining only an instant, rising a short distance and then dipping again, sometimes repeating this several times. Then it perches on a neighboring twig to shake and preen its plumage. Sometimes hummingbirds bathe by pressing against a saturated tuft of moss on a twig, or by gliding with depressed breast over the horizontal surface of a large leaf, such as that of a banana plant, in the early morning when it is laden with dew or drops from the night's rain. Other birds, from tiny honeycreepers to large jays, perform their ablutions by fluttering about in a spray of small leaves that are heavily laden with rain or dew, until they are thoroughly wet, then shaking and preening. Slessers (1970) has described in great detail how bathing birds fluff out their feathers to expose their skin to the water and make a variety of vigorous movements to wet themselves thoroughly.

Some birds, including house wrens and sparrows of various kinds, "bathe" in dust as well as water. Still other birds, such as domestic chickens and other gallinaceous species, bustards, sandgrouse, larks, and other birds of arid regions, only dust. This does not seem a good way to get clean; but after vigorously shaking out the dust that it has worked well into its plumage and preening thoroughly, the bird becomes surprisingly fresh and neat. Moreover, the dust probably helps to control insect parasites, perhaps by suffocating them. Farmers in tropical countries have discovered that they can protect their beans from weevils by keeping them imbedded in the dry dust and fine debris among which they settle during the process of threshing. Some precocial chicks dust at a surprisingly early age. When a Red-legged Partridge only three hours old stumblingly followed its parent's calls back to the nest from which it had been removed, it happened to pass over some fine, dry earth. It immediately stopped and commenced to dust itself, making all the characteristic movements for working the earth into its natal down, but in a rather fumbling manner (Goodwin 1953).

Preening does not present the young bird with the same problem as does bathing, for it has no difficulty finding its own sprouting feathers. In several altricial birds, rudimentary preening movements have been observed as early as the fourth or fifth day of life. Before they leave the nest, young birds spend much time preening, thereby hastening the shedding of the horny feather sheaths and the expansion of the enclosed plumage. Sometimes crowded nestlings preen each other's plumage.

CHAPTER 26

The Education of Young Birds-II

Learning to Recognize and Avoid Enemies

From the moment they hatch until they die, birds of all kinds pass through three stages in relation to the many enemies that menace them. First is the confidence of the hatchling, which neither recognizes nor attempts to avoid enemies. As the young bird's senses and mobility improve, this pristine innocence is succeeded by general, undiscriminating wariness, which causes it to avoid all unfamiliar objects above a certain size, especially those that move, regardless of whether they are dangerous or harmless. Finally, with growing experience, the bird distinguishes enemies from neutral or indifferent objects and avoids only the former.

This general course of the development of the avoiding reactions is complicated by two factors. The first is the influence of parents and other companions, who by warning notes or infectious behavior may cause the young bird to avoid things to which, alone, it would be indifferent. At a later stage, such example may also help to reconcile the bird to things that it spontaneously avoids. Second, there is the possibility that, in addition to its general wariness, the bird may instinctively recognize its hereditary enemies and respond appropriately to their presence. In the megapodes, which go off alone as soon as they emerge from their natural in-cubators, the chicks' reactions to the creatures that surround them are apparently uninfluenced by parental example. The study of these birds should throw much light on the recognition of enemies and related problems of avian psychology. For convenience, we shall speak of "fear" whenever a bird flees or avoids anything, although its emotions in these, as in other situations, are unhappily hidden from us.

When studying imprinting in the preceding chapter, we learned that many newly hatched birds respond in much the same way to individuals of their own species and to other moving objects, and that this pristine fearlessness permits them to become attached to, and accept as their parents, a wide variety of things, animate and inanimate, that they can follow. In the European Coot, this stage of initial indifference lasts about eight hours; in ducklings and domestic chicks, slightly longer. In this stage, domestic chicks fear only a loud noise and loss of balance. But, even when newly hatched and unable to distinguish friend from foe, precocial birds respond instinctively to the warning notes of their parents; in obedience they become silent and may try to scramble from the nest to hide, even before they can stand erect and walk.

When a brood of precocial chicks, following their parents over the ground in search of food, hear the alarm note, they instantly squat down and lie motionless and silent wherever they happen to be. They may take advantage of whatever stones or vegetation are at hand to screen them; but their safety depends chiefly upon their small size, concealing coloration, and immobility. The parent may also squat and try to avoid detection, or it may, by giving a distraction display or otherwise making itself conspicuous, attempt to lure the enemy from its hidden chicks. The young remain where they first squatted until, when danger has passed, the parent returns and draws them together with the rallying call. In Seton-Thompson's (1898) story of "Redruff," the male Ruffed Grouse returned to call together and take charge of the brood of chicks whose mother had been killed. What happens when no parent survives to reanimate and lead away the chicks that lie "frozen" on the ground? It is sometimes said, I know not with what truth, that they remain until death overtakes them on the spot where they squatted in obedience to the alarm note. In any case, without a parent they are not likely to survive.

The habit of squatting motionless when alarmed is of the utmost importance to the chicks of terrestrial precocial birds, which in their downy, flightless condition have little chance of escaping enemies if they make themselves conspicuous by trying to run away. One reason why the domesticated Turkey cannot return to the wild life of the woods is that the chicks have lost the habit of "freezing," hence are highly vulnerable to predators. In hybrids between domestic and wild Turkeys this behavior is incompletely developed; some chicks fail to squat in the face of danger and others do so imperfectly, often rising up and revealing themselves at the critical moment, as A. S. Leopold (1944) observed. Perhaps this is also one of the reasons why the domestic chicken has been unable to establish itself in the feral state in the New World, even in tropical lands where conditions seem not to differ greatly from those in its ancestral home in southern Asia and where its keepers often permit it to find much of its own food, to roost in trees, and to lay its eggs in nests well concealed amid vegetation; but the chicks still do not lie immobile in response to the parent's alarm note.

In altricial birds the stage of pristine fearlessness

is longer than in precocial birds, a consequence of the less advanced state in which they hatch. At first, they are too undeveloped to respond in any significant way to a direct menace or to the calls of apprehensive parents. Soon, however, violent shaking of the nest makes them crouch down, although gentle movement causes them to lift their open mouths for food. Somewhat later, they stop calling for nourishment and shrink down in the nest in obedience to the parents' warning call. Feathered nestlings that crouch in the nest in the face of an intrusion, such as a bird watcher's visit of inspection, may suddenly jump out if at the same time they hear the notes of an excited parent. Young that jump from low nests often hide amid the ground cover, where, like precocial chicks, they are most difficult to find. Silent immobility amid foliage contributes much to the safety of many kinds of altricial fledglings that leave the nest before they can fly strongly.

The stage of undiscriminating wariness begins in some precocial birds at the age of two days or less, when even in the absence of a parent's warning notes they avoid objects that a few hours earlier they would have accepted as foster parents. Altricial birds enter this stage while still nestlings, crouching when a strange object approaches the nest instead of greeting it with eagerly gaping mouths. After leaving the nest, the young show alarm or fear in the presence of a variety of things, especially those that are large, unfamiliar, or in motion. To go through life trying to escape every strange or moving object that attracted its attention would greatly handicap any bird, and to avoid this undesirable condition its innate wariness must be modified, becoming associated chiefly with what is dangerous.

Nice and Ter Pelkwyk (1941) found hand-reared Song Sparrows especially favorable subjects for the study of innate fear reactions. When alarmed they call *tchunk*, with elevated crests, tails raised and flipped, wings twitching, and restless changes of position. In "fear" they call *tick*, crouching with compressed plumage and extended necks. In stronger fear or "fright" they call *tik-tik-tik*, pant with open mouths, fly to hide, or, when confined, flutter in an attempt to escape. Moreover, the number of *tchunk*'s or *tick*'s uttered per minute provides an index of the sparrow's degree of alarm or fear.

The hand-reared Song Sparrows displayed alarm when shown cats and kittens, a large mounted hornbill, a mounted Ruffed Grouse, a stuffed teddy bear about a foot long, and a pewter pitcher. Fear or fright was caused by a large tame rabbit, a dog, and a Common Starling flying rapidly past the window. A model of an accipitrine hawk and a cardboard rectangle caused fear when they were moved quickly in front of the window, but neither did so when passed slowly. Small snakes and white rats, however, elicited only curiosity when placed beside or even within the sparrow's cage. It is evident that the intensity of the young bird's response bore little relation to the dangerousness of the object presented to it. Rabbits, Ruffed Grouse, and Starlings are not enemies of Song Sparrows, to say nothing of teddy bears and pewter pitchers; but cats, which aroused only mild alarm, are perils to be avoided, and snakes doubtless take a heavy toll of the sparrow's eggs and nestlings. Other hand-reared birds, such as Common Mockingbirds and Blue Jays, are far less timid, boldly attacking things that alarmed the Song Sparrows.

The extremely diverse character of the objects that excite alarm or fear in some young birds makes it difficult to decide whether they recognize, innately and prior to experience or conditioning by older birds, any particular animal or class of animals as enemies. One of the best-attested cases of the innate recognition of an enemy is the response of certain young precocial birds to flying birds of prey. When a flat model of a long-tailed, short-necked hawk (a typical *Accipiter*) is attached to a wire and pulled rapidly over young geese, ducks, or gallinaceous birds, they become alarmed and try to escape. Curiously enough, if this same model sails above them with the "tail" foremost, so that it resembles not a hawk but a long-necked, short-tailed goose or crane, it fails to alarm the same young birds. Since in this model the forward and rear edges of the wings have the same shape, the birds' reactions evidently depend upon their interpretation of the long extension as a tail or as a neck according to whether it follows or precedes the wings. Not outline alone, but outline combined with direction of movement, forms the stimulus to which the birds react (Tinbergen 1951).

Some birds have an innate fear of owls. The sight of a living owl excites in a young passerine, such as a Song Sparrow or Common Mockingbird, a degree of terror that cannot be attributed to the owl's size or to its nearness; likewise, a stuffed owl, or even a model of an owl, alarms them greatly. So strong was the impression that a mounted Barred Owl made upon Mrs. Nice's hand-reared Song Sparrow that he was greatly perturbed when he came in sight of the spot where he had last seen the owl four months earlier. Tests carried out by Nice and Ter Pelkwyk with captive birds, and by Hartley (1950) in the field, revealed that to be effective a model must have an owllike outline, appear solid, be colored in browns and grays or in contrasting shades of these colors, and bear a streaked, barred, or spotted pattern. A flat model is recognized as an owl only if shaded to give the illusion of solidity. A cardboard model with painted head and eyes but unpainted body excited Song Sparrows almost as much as did a fully painted model with partly spread wings.

To recognize a stationary owl, a bird requires an image more detailed and complex than the flat, colorless form that suffices for recognition of a soaring bird of prey. The reason for this difference seems obvious. To avoid flying predators, a bird must act quickly, even at the risk of making a mistake. But if an animal is at rest, or moving rather slowly over the ground, the bird can safely take more time to make sure that it is dangerous. And this seems to be the reason why, with the partial exception of snakes, birds seem not to innately recognize wingless creatures as their enemies. In many continental areas, the predatory mammals, for example, are numerous and varied, and in general outline some of them closely resemble harmless kinds. It seems hardly possible that a bird could be innately endowed with one or a few simple, generalized patterns that would suffice for the recognition of so many potential enemies. For the avoidance of the majority of its predators, the bird depends upon its inborn wariness as modified by the example of its companions and its own experience.

The case is different with snakes, all of which have much the same form, despite great variation in size, and most of which prey upon eggs and young birds. Although Song Sparrows, Blue Jays, and some other birds do not fear the first snakes that they see, Curved-billed Thrashers instinctively recognize them as enemies, as Rand (1941) demonstrated. Since many people have a deep-rooted dread of serpents, it is of interest to consider to what extent fear of snakes may be innate among the warm-blooded vertebrates as a whole. According to Thorpe (1956), the Wood Rat recognizes them innately by sight. On the other hand, young African Giant Rats, born in captivity in Paris, were wholly indifferent to a fairly large python and even approached to sniff its nose. However, the parent Giant Rats, who evidently had been familiar with pythons in their native Africa, attacked the snake violently (Bourlière 1954). In regard to monkeys, the evidence is conflicting, some students stating that their well-known dread of serpents is innate and others that it is learned from their companions. In young Chimpanzees and human children, fear of snakes appears to be acquired rather than inborn.

At least some kinds of owls have an innate aversion to crows. Hand-reared Screech Owls, taken from the nest before their eyes opened, showed extreme anxiety when they first saw a stuffed museum specimen of a crow; the intensity of this reaction was hardly diminished by the almost daily presentation of the specimen over a period of three months. Even when only the front half of the crow's head was shown to the owls, they were alarmed (Kelso 1940). Crows have an equal antipathy to owls. This mutual aversion was known to Aristotle, who in the ninth book of his *History of Animals* attributed it to the fact that by day, when the owl is dim-sighted, the crow preys upon its eggs; whereas at night, the owl preys upon the crow's, "each having the whip-hand of the other, turn and turn about, night and day." Nevertheless, when Ramsay (1950) removed Common Crows from the nest, after raising them beneath a mounted specimen of a Barred Owl, and placed them at the age of about five weeks with a living Barred Owl that was eight weeks old, neither species revealed alarm. Soon they were roosting side by side, and one of the crows begged food from their strange companion. The owl, of course, had no need to prey on the crows, for it was fed by its captor. And this brings to mind another remark of Aristotle, that, if there were no lack of food, animals would not be at enmity with each other, or with man, but all would dwell together in harmony.

Sometimes, while wandering through woodland, one's attention is drawn by chattering voices to a crowd of small birds gathered around an owl that perches somnolently in a tree. The protesting birds may be all of one kind, but often several kinds are present, each repeating its characteristic notes of alarm or complaint, while it flits restlessly from twig to twig. From time to time, one of the small birds darts threateningly toward the motionless owl, but rarely does it strike the nocturnal bird. This hostile demonstration may continue for many minutes, until the owl flies off, or, more often, the complaining birds drift away to forage or attend their nests. Such a display is called "mobbing," which was defined by Hartley as "a demonstration made by a bird against a potential or supposed enemy belonging to another and more powerful species; it is initiated by a member of the weaker species, and is not a reaction to an attack upon the person, mate, nest, eggs or young of the bird which begins it."

The mobbing of owls has been observed in many parts of the world, including the Eastern and Western hemispheres, the tropics, and temperate zones. It was familiar to the ancient Greeks, who evidently believed that the small birds gathered to admire this bird, which they made the emblem of Athena, goddess of wisdom. Aesop composed a fable about this curious avian habit. In a famous discourse delivered at Olympia in A.D. 97, Dio Chrysostom, a wandering Stoic philosopher, likened the audience that had gathered about him to a crowd of small birds surrounding an owl, "some alighting near and others circling about her."

Although in northern lands birds seem to mob owls more frequently than anything else, they are by no means the only creatures that are so treated. Sometimes a resting or even a flying hawk is the center of excitement. In the tropics, a motley crowd of small birds often gathers about a snake, especially one that has climbed into a bush or tree. Flitting back and forth, the birds protest "in fifty different sharps and flats," while the bolder of them now and then nip the snake's hinder parts, causing it to make a sudden movement that stirs up a fresh spasm of excitement among the surrounding birds. In California, about twenty birds of four species mobbed the shed skin of a large rattlesnake that lay on the ground, partly exposed and partly extending into a hole (Banks 1957). Even a large, bedraggled moth hanging amid foliage may incite behavior hardly to be distinguished from mobbing, if some particularly nervous bird, such as the Scarlet-rumped Black Tanager, is present to serve as the nucleus of a protesting avian crowd. Indeed, I have seen these tanagers "mob" a pair of new shoes that, while armed men were pillaging the country during a revolution, had been hidden amid shrubbery.

What is the significance of mobbing? Can we offer an explanation more convincing than that of the ancient Greeks, who thought that the birds were admiring the owl? In considering this question, it is necessary to bear in mind that the birds seem seldom to discomfit the creature they mob, except perhaps in the case of a small snake, or if the mobbers are as big as jays or crows. Altmann (1956) saw a male Red-winged Blackbird repeatedly strike the back of a stuffed Screech Owl's head, but such boldness is exceptional among mobbing birds. Rarely, one of them is attacked and killed by the animal they worry.

It is advantageous to birds, as to an army, to know exactly where the enemy is. Mobbing certainly advises all the feathered creatures in the vicinity that a dangerous character is lurking nigh. "Birds," wrote Hudson (1920), "ever fly reluctantly from danger." More than that, they frequently approach danger and try to keep it in view. To advance toward their enemies is more characteristic of birds than to flee from their enemies. Hudson described how Ypecaha Rails hastily converge on the spot where one of them raises the alarm call. When he shot a rail, the survivors, after taking refuge among the surrounding reeds, promptly turned around to watch and follow him, sounding their powerful alarms the whole time. Similarly, I once saw a Gray-necked Wood-Rail follow a large opossum that was running clumsily after the rail's mate.

L. Miller (1952) could draw a variety of birds by whistled imitations of the notes of various owls.

Large birds, such as jays, came to the calls of big owls but not to those of small ones; small birds were attracted by the notes of small owls but ignored those of big ones; the birds were interested only in the kinds of owls that preyed on them.

In flying swiftly away from a man, rather than gathering around and keeping him in view, birds behave quite differently than they do in the presence of most of their foes. The reason for this difference seems to be that man is the only enemy who can kill at a distance, so that in regions where he has long persecuted birds their survival has depended on developing a special response to his presence. In the case of a nonhuman enemy, however, birds avoid the danger of a surprise attack by discovering where it is and keeping it in sight, often by mobbing it. Moreover, mobbing must play an important part in the education of young birds, who in these gatherings learn to recognize certain animals as enemies. Such education must be particularly valuable in the case of predators that they do not recognize innately.

The reactions of hummingbirds to predators are most curious and difficult to explain. I rarely see owls in my forest, but on the single occasion when I watched one being mobbed, hummingbirds were conspicuous among the mobbers, and they ventured much closer to the big Spectacled Owl than did the trogons and honeycreepers who were likewise protesting. In California, Altmann watched Anna's Hummingbirds fly two or three inches from stuffed owls and make jabbing movements toward their eyes. When a hawk sails across the sky at no great height, a hummingbird or two often flies with shrill notes close behind or even beside it. Yet I have never known an adult hummingbird to be captured by any predatory bird, and few reports of this have been published (Mayr 1966, Skutch 1973).

Habituation: Learning What Not to Fear

The education of young birds includes learning not only what to avoid but also what not to fear. We have seen that, after their pristine fearlessness, birds enter a stage when they are alarmed by almost any unfamiliar object, especially those that are large and in motion. A bird that fled from every animal, every swaying bough, and every loud sound would waste much time and energy, forage inefficiently, and perhaps even suffer nervous disorders. To avoid dangerous things effectively, it must distinguish them from harmless ones and respond differently. The process by which a bird or other animal loses its distrust of harmless sights, sounds, or other stimuli to which it is repeatedly exposed is called "habituation." When, from time to time, my feeding shelf rots and I replace it with a new board, I receive an interesting demonstration of this process. When the birds return and find a banana lying on a bright new board in the spot that the dingy one occupied a short while before, they are distrustful, utter notes of alarm, and look down from a higher branch without descending to eat. Presently a bolder one drops down beside the coveted fruit, only to shoot up again as though it had alighted on a hot stove. After a while, one of them gathers courage to take a few nibbles at the fruit. In a day or two, the birds become habituated to the new board and are eating there as freely as ever. Likewise, when a blind is set in front of a nest, its acceptance by the parent birds usually involves habituation.

For many free birds, habituation is a remarkably effective process that achieves some amazingly fine adjustments. They cease to be alarmed by such rapidly moving, noisy, and doubtless originally terrifying human inventions as trains, motorcars, and airplanes, and they may raise their families close beside a busy railroad or highway. They become indifferent to the harmless herbivorous and frugivorous animals in their environment and even learn to exploit them in the quest of food, like anis and Cattle Egrets that follow grazing cattle to seize the insects stirred up from the grass by them, or cowbirds and grackles that pluck ticks from their bodies. More than this, it has been claimed that birds learn to distinguish the moods of their habitual enemies, becoming more wary of them when they are hungry or hunting than when they are satiated or resting. Experimenting with a tame Red-tailed Hawk, F. Hamerstrom (1957) found that it was mobbed more often and by larger gatherings of birds when it was hungry, or "sharp-set," than when it was well fed. When hungry it rested with body upright, neck stretched up, and plumage compressed—an attitude of alertness that contrasted with the more horizon-

tal posture, contracted neck, and relaxed body of its restful, well-fed state. The subject, however, requires further study, as Grier (1971) cast doubt upon the ability of birds to distinguish between a hungry and a satiated hawk. Yet it has been reported that antelopes regard resting lions, and ducks treat playing otters, with an indifference that contrasts sharply with their wariness when these carnivorous animals are hunting.

The capacity of birds and mammals to recognize the intentions of predators hardly requires finer powers of discrimination than the annoying ability of an intelligent horse at pasture to tell, without seeing a rope or other obvious sign, whether his master is approaching to put him to work or to caress him. He becomes hard to catch only if he suspects that his services are required. Free birds develop finer powers of discrimination than those that have long been subject to man. I have often heard domestic chickens sound the alarm for an aerial predator when a pigeon, parrot, or carrion-eating vulture flies overhead—a mistake that I have not known a wild native bird to make.

Man greatly affects the abundance and prosperity of birds of all kinds. He often makes conditions more favorable for some of them, either incidentally, such as through clearing land and agriculture, or deliberately, such as by putting up birdhouses and providing food; or he may decimate or even exterminate them by destroying their habitat or, more directly, killing them. Similarly, birds differ vastly in their reactions to man, some seeking his company, alighting on his shoulder or taking food from his hand, whereas others flee the moment he comes into view and scarcely permit him to glimpse them. In general, the birds of continental areas are shy, but those of oceanic islands that were long without human inhabitants are tame—to their own great detriment when these remote sanctuaries are invaded by callous men. I have also found birds remarkably tame on high tropical mountains where few people intrude.

Even a single species may vary greatly in tameness in different parts of its range. Thus, as J. S. Huxley (1947) pointed out, Moorhens or Common Gallinules are tame in England but wild in Switzerland and North America; Robins are tamer in England than on the continent of Europe. Such contrasts in behavior may also be found in the same region, Wood Pigeons and Black-headed Gulls being very tame in London where they are fed and protected but, especially the pigeons, shy and wary outside the city.

What is the reason for these striking contrasts in tameness? With regard to man, is fear or tameness in birds innate or is it acquired from the example of parents and companions? No one appears to have demonstrated that any species of bird innately recognizes man as an enemy, so he is not in the same category as soaring hawks and perching owls, which many birds innately recognize as foes. As Goodwin (1948*b*) pointed out, such inborn fear persists much more stubbornly in domesticated birds than does the general wariness or potentiality for fear that, when directed toward man by social example or individual experience, seems to be the foundation of birds' avoidance of him. The domestic Ring-necked Dove, for example, is remarkably placid and not easily frightened by strange people, sudden movements, loud noises, and the like. But, at the sight of a soaring hawk, it becomes alarmed like any wild pigeon, even though for countless generations it has been reared in situations where predatory birds could hardly attack it.

That genetic or heritable differences in the wariness of birds exist, especially with reference to terrestrial animals, is evident from A. S. Leopold's (1944) study of Turkeys. He found that when the wild northern race is crossed with the domestic breed, derived from Mexican ancestors, the hybrids are intermediate between the parents in a number of characters, including tameness. If these hybrids roam free in the woods, the wilder individuals survive better and in consequence the population becomes wilder; in captivity the tamer, more placid ones thrive better, and the special traits of the wild ancestor tend to disappear.

Not only races or populations of birds, but also members of the same brood, differ in their genetically determined wariness, just as they differ in other heritable characters. When Ivor (1944*a*) raised two young Wood Thrushes from the same nest, the male became one of the tamest birds he ever had in his aviary; but, even after four years,

this bird's sister remained wary and would not permit Ivor to approach her closely. That this difference in tameness was not somehow associated with sex was evident from the fact that another hand-reared female Wood Thrush became quite fearless. Similar differences in the tameness of siblings appeared in Ivor's Common Bluebirds. That fear of man may be a trait that is inborn, rather than acquired from parental example, is shown by his observations on Blue Jays. When a pair that was quite tame reared a brood in his aviary, the young jays remained persistently wary. Evidently the parents were not homozygous for the genes that determine tameness or wariness.

One reason why birds are so often distrustful of man seems to be that his conduct is so unpredictable. Other animals, domestic and wild, follow hereditary patterns of behavior. They act today very much as they did yesterday, and the other creatures associated with them become familiar with their habits and know what to expect of them. But men are constantly doing something different, acting in strange ways, producing sounds never heard before, or constructing novel things. Birds and other animals are bewildered by this unexpected diversity of man's activities and remain wary of him.

Taking into account all the evidence, which is too extensive to be presented here in detail, I believe we may conclude that there are specific, racial, and individual differences in the innate or heritable wariness of birds. This wariness, which is manifested as a tendency to flee strange objects and to resist the friendly advances of man, is often slight or almost wholly lacking in birds of oceanic islands and other localities with few or no terrestrial predators; it is usually strong in species that for many generations have been exposed to the teeming predators of continental areas. Birds deficient in innate wariness may, even in the face of cruel persecution by man, fail to develop caution rapidly enough to save themselves from extinction by him. At the other extreme, birds with a strong vein of innate wariness may resist all efforts to be tamed. Many birds, perhaps the majority, fall between these two extremes. Where persistently persecuted, especially for several generations, they become exceedingly wild and unapproachable by man. If gently and kindly treated, perhaps for only a short while, they become tame and confident, at times to the point of alighting on a human body and taking food from the hand.

In the process of becoming tamer or growing more wild, social companions exert a strong influence: A wild bird loses some of its fear of man if it joins a flock of tame ones, and, conversely, a tame bird who joins wild companions may become more wary. Thus the tameness or wildness of a bird is influenced by the example of its companions and its personal experience, that is, by education. And in this process the example of the parents while the bird is young may be a decisive influence.

Learning to Sing

Without language or some substitute, such as the sign language of the deaf, no human society could hold together. Because of the greater adequacy of their innate endowment, birds depend less on communicating vocally with their fellows than we do. Yet I believe that, if they were suddenly to be stricken dumb, many species of birds would become extinct. Without voice, the family could not hold together; fledglings could not advise their parents where they are and that they are hungry; parents could not warn their young when danger approaches. Adult males could not proclaim possession of a territory or attract a partner. Mated birds could not keep in contact while they forage out of each other's sight in dense vegetation. For some of these purposes, especially courtship, some birds use nonvocal sounds, frequently made by modified feathers of their wings or tails; but such birds are quite exceptional.

The utterances of birds are commonly classified as songs and call notes. The latter consist of the usually briefer, less musical vocalizations, often monosyllables, that birds employ to reveal where they are, warn of danger, threaten enemies, and so forth. What the ornithologist means by "song" is difficult to define in a manner that will include all examples of it and exclude everything that is not song. To say that bird song is melodious or musical will cover most cases; it is just because the avian utterances that we have in mind are so often delightful that we call them "songs" rather than by some technical term. But some bird songs are insectlike buzzes, un-

melodious squeaks, or raucous shouts. Songs usually consist of several syllables and are longer than call notes, yet there are monosyllabic songs.

If we try to define song not by intrinsic characteristics but by function, we are in hardly better plight. Song is most frequently used to proclaim possession of territory and (or) attract a mate, and any utterance so employed, no matter how harsh to the human ear, should, I believe, be classified as song. But birds often sing from pure ebullience or joyousness, when we can hardly ascribe a utilitarian function to their singing, and females sitting in the nest sometimes hum little ditties expressive of contentment. It is so difficult to frame an inclusive definition of song that I am tempted to give an exclusive definition: "A song is any utterance of a bird that is not a call note." But this, too, fails to relieve us of all perplexities, for some birds use a long, melodious phrase for the same purpose that other birds use simple calls. To decide which of the sounds made by birds are songs one must use an educated judgment, and experienced ornithologists will sometimes disagree.

Our present interest is in how birds acquire their various utterances. Are they innate or learned? The call notes of birds are largely innate, for they are given in typical form by individuals reared in isolation, with no opportunity to hear others of their kind. Among the exceptions to this rule are certain notes of the Western Meadowlark, Raven, House Sparrow, and Blackbird, which are either not given at all, or not uttered in typical form, by individuals reared in isolation from others of their species. Lanyon (1960), who cites these examples, adds the caution that perhaps the abnormal conditions of captivity prevented the development of these call notes.

When we turn from call notes to songs, it is convenient to discuss passerines and nonpasserines separately. The former, with a more elaborately constructed syrinx or vocal apparatus, produce, in many cases, songs of a musical complexity unattainable by the latter. The simpler syrinx of the nonpasserines, usually without intrinsic (wholly internal) muscles, can produce only simpler sequences of notes. Yet the songs of some of these birds, including certain tinamous, motmots, and trogons, are notable for their fullness and mellowness of tone, which compensates, or more than compensates, for their lack of diversity. Indeed, the very simplicity of some of their utterances, combined with their purity of tone, gives an impression of sincerity and depth of feeling that in certain moods moves us more strongly than the more brilliant, and seemingly more studied, performances of many of the passerines. Among the nonpasserines that I have heard, the most elaborate songs, notable for their length and animation, have been given by jacamars.

So far as we now know, the songs of nonpasserine birds, like their call notes, are largely inborn and may be perfected even by individuals who have had no opportunity to hear others of their kind. Some students of avian sounds believe that no division should be made between the call notes and songs of nonpasserines. It is, indeed, even more difficult to separate them than in the case of the songbirds; yet in some families, like the trogons and jacamars, the songs are longer, more complex, and more melodious than the calls. The song of the domestic chicken —Chanticleer's crow—seems distinct enough from all its other notes. Vocal imitation, which enters into the learning of song by many passerines, appears to be absent from all the nonpasserines except the parrots, and even among them it appears to be confined to captive birds. It is remarkable that a bird with the parrot's flair for imitation should be content to utter the raucous notes that are all we hear from many species in their native haunts.

A large division of the passerines, the suboscine Tyranni, have simpler vocal organs and simpler, often less melodious utterances than the true songbirds. Mostly tropical, and less easy to rear in captivity than many songbirds, the development of their vocalizations has been little studied. The acquisition of full song by a songbird is a gradual process that passes through several stages. Soon after leaving the nest, and rarely while still within it, the young songbird begins to utter low, sweet, warbling notes, which often ramble on and on, without definite phrasing. At times, recognizable call notes are introduced into this recital, and these alone may reveal its authorship, for species that as adults have quite different songs now sing in much the same manner. This juvenile song, or "subsong" as it is often called, gradually passes into the phase

known as "rehearsed" song. The young bird now includes in its rambling music phrases or motifs that resemble those of the definitive song, although they lack the refinement they will later acquire. Since call notes and warbles still enter largely into the medley, it becomes even more varied than at first.

As the motifs of the adult or "primary" song are perfected, the juvenile warbles and call notes gradually drop out of the sequence, and, when they have (in many species) been wholly lost, the adult mode of singing is attained. Thorpe (1958) compared the development of song by a bird to the learning of speech by man. Just as the sounds comprised by any single language represent a selection from the much larger number of sounds uttered by a babbling infant, so the notes of a bird's adult song are only a portion of those that enter into its juvenile song. The young bird, like the human child, acquires the language of its kind by retaining those of its varied sounds that its elders repeat and dropping the rest.

The passage from juvenile to adult song is not pure gain, and some ornithologists have regretted that birds do not persist all their lives in their artless juvenile chanting. The adult song is usually louder, more stereotyped, more distinctive of the species or even of the individual, but often it lacks the sweetness and charming spontaneity of the young bird's first essays, and often, too, it is less generously poured forth. The Highland Wood-Wren, who as a songster never grows up but retains a rambling, juvenile type of song, sings far more continuously and profusely than its close relative, the Lowland Wood-Wren, whose adult song consists of brief, exquisite, cameolike phrases. In another wren, the Chinchirigüí, the contrast between the diffuseness of the juvenile song and the stereotyped brevity of the adult performance is most striking. However, in all birds the adult type of song has practical advantages that have favored its evolution: it is distinctive of the species as juvenile songs rarely are, and its greater carrying power makes it more effective for proclaiming territory and attracting a mate.

The age at which songbirds begin to sing varies greatly with species and individual. Nice (1943) listed sixteen species in which song was first noticed at the age of thirteen to twenty-four days and fifteen species that began to sing between four and eight weeks of age. Seven hand-raised Song Sparrows began their juvenile warbles at ages ranging from thirteen to thirty-one days. Gray-cheeked Thrushes started singing when fifteen to twenty-five days old. Blackcaps, Song Thrushes, and European Dippers have been heard singing while still in the nest; but, since their nestling periods are longer than that of the Song Sparrow, they may not have begun to practice their songs any earlier than the thirteen-day-old sparrow. Singing in the nest by young songbirds is rare, and I have never known any species to do so; but I have heard suboscine passerines give very creditable, although weak, versions of their parents' characteristic songs. Among them were sixteen-day-old White-winged Becards, who answered their father's dulcet notes with similar ones; nestling Bright-rumped Attilas, who could repeat much of the adults' varied repertoire; and nestling Streaked-headed Woodcreepers, whose clear trills closely resembled those of the parents.

Recent investigators of the development of song, especially in Europe, have concentrated on the problem of how much the young songster depends on accomplished individuals for his motifs. To study this subject it is necessary to rear birds in different degrees of isolation. Taken from the nest, or even hatched artificially, they are placed in soundproof rooms, where they grow up alone, or are permitted to hear only birds of the same or of different species, as the investigator wishes. Already it has become evident that songbirds differ immensely in their need of, and receptivity to, models for their songs. At one extreme are those that can give the typical song of their species without ever hearing it from another individual; at the other extreme are birds wholly dependent on experienced individuals for the motifs of their adult songs. Among the birds in which almost the whole of the primary song is inborn Thorpe (1956) lists the Corn Bunting, Reed Bunting, Tree Pipit, Grasshopper Warbler, Wood Warbler, and others. If reared in isolation from older individuals of their species, but exposed to the songs of different species, these birds, apparently uninfluenced by the foreign motifs, develop the songs of their ancestors. Probably the same is true

of suboscines, such as American flycatchers and becards, but this awaits investigation.

Greater dependence on instructors is found in such birds as the Chaffinch, Blackbird, Whitethroat, and Canary. If reared from the egg or nestling stage in soundproof rooms, these birds, passing through the developmental sequence that we have already considered, acquire songs that bear a general resemblance to the typical adult songs of their respective species. Although the basic pattern of their songs is innate, the finer details and much of the pitch and rhythm are acquired by learning. Birds of these species that grow up in complete isolation develop turns of phrasing peculiar to themselves. If several young male Chaffinches are kept together with no opportunity to hear older individuals of their kind, they stimulate and imitate each other, so that all develop similar patterns, and they acquire a repertoire larger and richer than they would have if they lacked all opportunity to hear others of their kind. Although members of the same isolated group develop similar songs, the songs of different groups may differ greatly. These studies of hand-reared birds help us to understand geographical variations in song. As has frequently been observed, the birds of a single species in one neighborhood may have rather similar songs that differ recognizably from the song pattern prevalent in another community.

Most interesting is the situation in the Bullfinch, which develops an imperfect song if deprived of the opportunity of hearing the male who rears it. Normally a young Bullfinch appears to learn his song from his own father. A Bullfinch reared by Canaries acquired the song of his foster parent. Even when exposed to the songs of mature Bullfinches, such a Bullfinch will sing a Canary-like song, so strong is his tendency to imitate the male bird who feeds him in the nest.

Birds whose inborn endowment of song patterns is slight and who accordingly depend heavily on imitation for their motifs appear to fall into two groups. Some, when deprived from an early age of the opportunity to hear songsters of their own kind but exposed to the songs of other species, develop a repertoire that consists of original motifs plus imitations of the alien songs. According to Lanyon

(1960), Blackbirds and Garden Warblers belong to this group. Other birds, reared in the same conditions, have failed to sing any songs typical of their own kind but acquire a repertoire that consists wholly of phrases heard from other species. This situation seems to prevail in the Baltimore Oriole, Bobolink, Red-winged Blackbird, meadowlarks, and other species of icterids. Yet these birds, like most others that are not true mimics, have an innate preference for the songs of their own species; in the free state each young male, who doubtless hears a variety of other birds while he is acquiring his repertoire, nevertheless, as a rule, confines himself to the songs of his own kind. Thorpe noticed that the Chaffinch, while in the "rehearsed" phase of song development, may repeat phrases heard from other species, yet he does not retain them in his adult song.

Male hummingbirds of certain species are among the most persistent songsters in the whole avian class. Most of those that I have heard tirelessly repeated squeaky or metallic, unmelodious notes; but a few, of which the Wine-throated Hummingbird of the Guatemalan highlands is an outstanding example, have such long-continued, sweetly varied songs that, if only their voices had a little more volume, they might rank among the best of feathered musicians. Numerous kinds of hummingbirds perform tirelessly in singing assemblies, in which a number of males are stationed within hearing of each other. Members of the same assembly tend to have similar songs, which often differ strikingly from the songs of other assemblies in the same locality. Since the hummingbirds in these assemblies are not reproductively isolated, the situation of uniformity within an assembly coupled with diversity between assemblies can hardly be explained by genetic differences. Evidently these hummingbirds learn their songs, probably from older members of the assembly that they join as they mature. Rarely a hummingbird learns the song of some other species, as happened with a Blue-chested Hummingbird who, year after year, repeated the song of the Rufous-tailed Hummingbird, a congeneric species abundant in the same garden. Yet this peculiar Blue-chested Hummingbird was stationed within hearing of the very different songs of several males of his own kind—a glaring exception to the rule that members of the same assembly repeat similar songs (Skutch 1972).

In our earlier discussion of the development of feeding habits, we noticed that students of animal behavior are reluctant to concede that birds imitate their companions, because imitation involves self-consciousness or the objectification of self, and we lack evidence that birds are capable of this. Birds have so little notion of their own appearance that if, perchance, a male sees his reflection in a mirror, window pane, or polished metal surface, he will often stubbornly fight his own image, mistaking it for a rival who has invaded his territory. The case of vocal mimicry is, however, in a class apart from other kinds of imitation, because, although free birds may never see themselves as a whole, they presumably hear the notes they utter just as we hear our own voices. Hence they can objectify their utterances as they cannot objectify their actions, and they can compare their rendition of a sound with its original, repeating it, if need be, until their reproduction is satisfactory. Vocal imitation may be, as Hartshorne (1958) suggested, the most unequivocal manifestation of avian intelligence.

Although imitation, as we have learned, enters largely into the development of the full repertoire of many songbirds, especially those with elaborate songs, the majority of them copy by preference the notes of other individuals of their own kind. The case is quite different with the true mimics, outstanding among which are the Common Mockingbird and Black Thrush in America; the lyrebirds, scrub-birds, and bower-birds of Australia; the mynahs of Asia; the cossyphas or robin-chats of Africa; the Common Starling, Red-backed Shrike, Marsh Warbler, and Reed Warbler of Europe; and of course the nearly cosmopolitan parrots, the only nonpasserines in this assemblage. Most of these birds, even with full opportunity to learn from others of their own kind, freely imitate notes of other species, seeming to delight in their mimicry. Nevertheless, since a bird like the Common Mockingbird has a very large repertoire, which includes notes that resemble those of other birds in the vicinity, the question arises whether each young mockingbird does not learn his motifs from older mockingbirds alone. The resemblance of some of his utterances to notes of neighboring species may be fortuitous, and some observers have denied that *Mimus* is indeed a mimic. But there is no obvious reason why a bird with a tendency to repeat a certain sound, let us say the song of a Whip-poor-will, should not learn this sound from a Whip-poor-will as readily as from another mockingbird.

Laskey's (1944) study of the acquisition of a repertoire by a hand-raised Common Mockingbird showed conclusively that the bird did imitate, not only the vocalizations of other kinds of birds but also mechanical sounds, like the squeak of a washing machine. Often he answered birds of other species with their own notes, and twice, when a Yellow-shafted Flicker appeared outside the window, he greeted it with its characteristic *wicka*, although the flicker was silent. In one lively performance that continued for sixteen minutes, the mockingbird sang forty-two different songs of twenty-four species of birds, interspersed among his own songs. Other observers have recognized the songs and calls of about thirty-five species in the performance of a single Common Mockingbird. Most songbirds have a much more limited repertoire.

Vocal imitation is often promoted by an emotional bond between learner and teacher. This may explain why young male Bullfinches, said to be "highly affectionate," copy the songs of their father or foster father in preference to the notes of any other songster. The best mimics among the parrots are likewise reputed to be exceptionally affectionate birds. Armstrong (1963) pointed out that "there is significance in the fact that a bird being taught by a person is receiving such individual attention as it rarely receives even as a chick or from its mate."

The process of learning to sing, which in songbirds may start in the nest but usually does not begin until after fledging, is as a rule completed by the following breeding season. By this time, the bird has acquired his full repertoire, which thereafter he seems to retain for life with little alteration or addition, except in the true mimics and apparently also the more versatile songsters with large reper-

toires, such as Great-tailed Grackles, Yellow-rumped Caciques, and certain thrushes. When a Chaffinch has learned a song and sung it at full intensity for only a few days, it becomes a permanent possession that he preserves from year to year with scarcely perceptible changes, as Thorpe has shown. Each year, as the season of song approaches, many older birds recapitulate the learning process, starting off with subsong and then passing through rehearsed song to the primary song, but running through the whole sequence much more rapidly than they did as juveniles just learning to sing.

Just as a young precocial bird learns one particular thing (the appearance of its parent or some substitute) at a particular period of its life and may thereafter lose the aptitude for such learning, so in the typical young songbird the capacity for another kind of learning (that of its song repertoire) is restricted to a particular stage of its development. In this, as in certain other respects, the learning of song resembles imprinting and might be considered as a variety of it. Like the imprinting of the parental image, a brief exposure to an appropriate auditory stimulus in the sensitive period may make a surprisingly lasting impression. The bird may remember for months a phrase that it has heard but not yet repeated, as in the case of Lanyon's meadowlarks that were exposed to certain songs while in the subsong phase in June, but were not heard to deliver them until they began practicing rehearsed song in early September. A Chaffinch who in earliest youth hears the song of his kind, but is isolated from other Chaffinches from September onward, retains latent impressions that will profoundly influence his singing in the following spring. A Nightingale reproduced the song of a Blackcap that it had heard for only one week, six or eight months earlier (Thorpe 1956). Some birds register a motif very swiftly. A Blackbird of the Old World, who for less than one minute heard the song of a Wood Thrush of the New World, incorporated it into his repertoire (Lanyon 1960).

CHAPTER 27

The Duration of Parental Care

For nearly all birds, parental care covers two distinct stages in the life of the young, while they are in the nest and after they leave it. In the first stage, the young receive the nourishment and warmth necessary for life and growth but little training that will help them to confront the perplexities and perils of the wider world that they will soon enter; this education, which we considered in the preceding chapters, comes largely in the second stage, when the young birds accompany their foraging parents.

The relative lengths of these two stages vary enormously: In highly precocial birds, the first stage may be very short and the second stage prolonged; in a few altricial birds, the situation is reversed, with a long nestling period followed by a short interval of parent-young association after leaving the nest, or none at all. The superprecocial megapodes may work hard to regulate the temperature of their eggs, but they wholly ignore the chicks that emerge from them; parasitic Black-headed Ducks receive a minimum of care from their foster parents. The great majority of young birds, however, are carefully attended for at least a few weeks, either within or after leaving the nest, and commonly in both stages. We have already considered the length of the nestling period. Now, to know the whole interval over which parental care extends, we must learn how long it continues after the young begin to move about. Rarely can this interval be determined with the same precision as the nestling period, for the young are no longer in one spot where they can be visited at daily or shorter intervals, but they wander about and are often very elusive.

Marine Birds

Marine birds present great contrasts in the lengths of the two stages of dependency. As we noticed in chapter 24, some, after long intervals in their nests, go off to sea alone and apparently receive no further attention from their parents. Indeed, in some of these birds, the duration of parental care is less than the nestling or fledging period, for the young remain unfed, living on their accumulated fat and reducing their weight, which may greatly exceed that of the adults, for some days before they fly or flutter into the water.

Manx Shearwaters, which have received much attention from British ornithologists, remain fasting, unvisited by their parents, for 2 to 15 days, often losing more than 3.5 ounces (100 grams) of weight, before they take to the water. The terminal fast of sixty-eight chicks averaged 8.5 days. Accordingly, to learn the duration of parental attention, we must subtract this interval from the nestling period, which ranges from 62 to 76 days and averages 70 days.

Evidently the average duration of parental care by Manx Shearwaters is 61.5 days (Harris 1966).

Richdale (1954*a*) found that Sooty Shearwater chicks in New Zealand might be fed on the last of their 86 to 106 (average, 97) days in the burrow, or they might remain unfed for as many as 27 days. For fifty-eight normal chicks, the mean interval of desertion was 12 days. Some of the young shearwaters left their burrows weighing so little that their prospects of surviving were poor. Richdale concluded that the chicks were abandoned because their parents migrated, and that such desertion is peculiar to migratory species of shearwaters, which obey the impulse to travel irrespective of the age of their young. Banded-rumped or Madeiran Storm-Petrels appear likewise to abandon their young in about the eighth week of their nestling period of 59 to 72 days (Allan 1962).

Most chicks of the Fairy Prion of southern waters remain unfed for one to four nights before their departure; a few fast for five or six nights; and about one in six receives a meal during its last night ashore. The nestling period of this petrel ranges from forty-four to fifty-five days (Richdale 1944*a*). A few White-faced Storm-Petrels are fed on the last of their fifty-five to sixty-three nights in the burrow, but the majority fast for one, two, or three nights (Richdale 1943–1944). Young Greater Shearwaters remain on their island birthplaces in the South Atlantic for about a month after the adults depart. During their long fast, the hungry young fill their stomachs with grass, earth, feathers, and other indigestible materials, as do Common Puffins and other young seabirds similarly abandoned by their parents (M. K. Rowan 1952, Lockley 1953).

Although we have abundant evidence of the habitual desertion of unfledged chicks by a number of marine birds, especially among shearwaters and petrels, the extent of this custom was exaggerated by early writers. We should remember that, even during the period of most rapid growth, some of these chicks pass several days, and frequently a week or more, without feeding visits from parents, who may forage many miles away. It must frequently happen that one of these long foodless intervals precedes the juvenile's departure, and a parent, returning at last with a meal for it, finds the nest empty. Richdale discovered that adult Sooty Shearwaters and White-faced Storm-Petrels, presumably parents with food, visited their burrows after the fasting chicks had departed. The same has occurred at nests of albatrosses, which, contrary to earlier reports, have no terminal starvation period (Richdale 1954*b*, Rice and Kenyon 1962). It is exceptional for a chick of the Diving Petrel to be left unfed on the night before its departure (Richdale 1943). But we lack evidence that any tube-nosed swimmer is attended by a parent after it has left the nest or its immediate vicinity. Accordingly, the period of parental care in this group ranges from about 50 days in its smaller members, such as the Fairy Prion, up to 280 days, which is the interval that the young Wandering Albatross remains ashore.

Among the auks and their relatives, the puffins appear to be exceptional in undergoing a period of starvation before they go to sea; but the habits of many members of this family, especially the quaint little auklets and murrelets of the North Pacific Ocean, are little known. Subtracting six to nine days of terminal fasting from the Common Puffin's nestling period of forty-seven to fifty-one days gives us about forty-two days as the duration of parental care (Lockley 1953). Cassin's Auklets remained in their burrows for forty-one to fifty days. Only the minority that stayed longer than forty-five days lost weight, evidently because they were no longer fed, so that the duration of parental attention in this alcid of the Pacific appears to be about the same as that of the Common Puffin of the North Atlantic (Thoresen 1964). Chapter 24 told how, when from two to three weeks old and still flightless, Razorbills and murres go to sea escorted by a parent. We do not know how much attention the old bird gives to the juvenile after its departure; but hand-raised Common Murres could catch fish alone when twenty-five days old (Oberholzer and Tschanz 1969).

Raised in a cranny among rocks, rather than on an exposed ledge, the two young Black Guillemots leave their safer home when about forty days old and immediately enter the ocean, where they are said to be attended by their parents for a short while (Winn 1950, Storer 1952). In contrast to this protracted growth in a sheltered site, downy, black-

and-white Ancient Murrelets leave their burrow or crevice on a rocky island in the northern Pacific when only a day or two old. This exodus occurs around midnight, when many parents call chirpingly from the sea near the shore, and hundreds of the tiny chicks pour down the rough slopes to the water's edge, where they fearlessly plunge into the pounding surf. Likewise, Xantus' Murrelets, when downy chicks only two or three days old, are conducted by their parents to the sea from their birthplaces on a rocky islet off the coast of Baja California or southern California. Here, again, we do not know how much care these amazingly hardy young murrelets receive after entering the ocean; but half-grown Ancient Murrelets were found in the Pacific, up to four hundred or five hundred miles from land, with parents whom they followed closely on the longest dives (Bent 1919).

Among the boobies and gannets, which are sometimes classified in the same genus, the latter are exceptional in going to sea, and apparently severing forever the bond with their parents, before they can really fly. In the North Atlantic Gannet, the duration of parental care is the same as the nestling period, about ninety days, with a range of eighty-four to ninety-seven days, as, despite certain published statements, it undergoes no starvation period (Nelson 1966). In the boobies of tropical waters, the situation is very different. The young of the widespread Brown Booby remain at or near their nests until they take their first flights at ages varying from 86 to 103 days, or even longer in unfavorable years. After fledging, they continue for a long while to return to the sites of their birth and receive food from their parents. Simmons (1967) actually witnessed such feeding thirty-seven weeks after fledging; but some juveniles persistently begged for about a year after they began to fly, and they may well have been fed up to this time. Accordingly, Brown Boobies sometimes continue to attend their young for a full year, and possibly as much as fifteen months, after they hatch. After a nestling period that ranges from thirteen to nineteen weeks, Red-footed Boobies return for a long while to receive meals at their arboreal nests; a juvenile nearly a year old was fed by an adult attending a later nest (Verner 1961). Abbot's Booby, which nests high in

forest trees on solitary Christmas Island in the Indian Ocean, feeds its young at the nest site for at least ninety days after it begins to fly, and probably much longer (Nelson 1971). Similarly prolonged parental care has been recorded in Blue-footed Boobies and White Boobies.

One great difference between gannets and boobies is that the former, nesting in cool northern seas rich in oxygen and marine life, can nourish their young much more profusely than boobies that fish in warm tropical waters with less dissolved oxygen and accordingly less abundant food. Nine or ten weeks after it hatches, the young North Atlantic Gannet may weigh one-third to one-half more than the adult. It loses some of this weight before it goes to sea two or three weeks later, but its reserves of fat are still large enough to sustain it until it can fish for itself. Nestling boobies grow much more slowly, fly from the nest when still lighter than the adults, without large reserves of fat, but they return over a long period for further nourishment from their parents.

Like boobies, frigatebirds, which often nest on the same tropical islets, have a very long period of dependence on their parents. The Ascension Island Frigatebird begins to fly in its sixth or seventh month but is at least partly dependent on its parents for three or four additional months. The flying juveniles return at intervals to the nest site to be fed, and one continued to receive food until at least ten months old (Stonehouse and Stonehouse 1963). Great Frigatebirds are fed even longer, sometimes for fourteen months after they fly, or until they are at least nineteen months old (Schreiber and Ashmole 1970). Another full-webbed swimmer, the Shag of Europe, leaves the nest when forty-eight to fifty-eight days old and may continue to receive food from its parents until at least ninety-six days old (B. K. Snow 1960).

The only other marine birds known to attend their offspring for about a year after hatching are King Penguins, which continue to feed their single young for ten to thirteen months. This long interval includes the harsh winter of high southern latitudes, when meals are very widely spaced; much of the young penguin's nourishment must be used to maintain its body temperature, and development languishes. The larger Emperor Penguins, hatched from eggs that are incubated during the bitter Antarctic night so that they can grow up in the milder and brighter half of the year, are dependent on their parents for only about six months (Stonehouse 1960, 1953). Smaller penguins go to sea at a much earlier age: the Rockhopper at about seventy days (Warham 1963); the Adelie at from forty-one to fifty-six (mean fifty-one) days (R. H. Taylor 1962); the burrow-nesting Little Penguin at sixty days or slightly less (Warham 1958*b*).

As they grow older, penguin chicks of several species leave their nests to gather in flocks, or crèches, which may contain many individuals. Earlier visitors to penguin colonies concluded that, in these massed assemblages, a parent returning from the sea with food might feed any juvenile that begged for it; but recent studies of marked birds have demonstrated that, with very few exceptions, parent Emperor, King, Adelie, and Rockhopper penguins feed only their own offspring. The supposed guardians of the crèches are not a few parents keeping watch over the young while others gather food for all of them, but nonbreeding birds, or parents who have lost their chicks, standing idly by (Sladen 1953, Prévost 1955, Pettingill 1960). We lack evidence that any penguins are fed by their parents after they swim away from the nesting colony, so that the duration of parental care is the same as the penguins' childhood ashore.

Terns likewise continue to feed their offspring for a long while. Sooty Terns are attended by their parents at least until they fly away, at an age that varies from eight to nine weeks or more, according to how well they have been nourished (Ashmole 1963*a*). Diminutive Fairy Terns at Ascension Island were still fed by their parents when ninety to ninety-five days old (Dorward 1963). In at least some species of terns, parental care does not cease when the young leave the nesting colony. In December and January, young Royal Terns were being fed on the coast of Peru, presumably by the parents whom they had accompanied on a long migration, for the bands on some of them showed that they had been hatched on the other side of the equator, thousands of miles away. Such feeding continued until the juvenile terns were about seven months

old. The continuance of parental care by terns on migration or in their winter homes is perhaps widespread, for it has been recorded also in the Elegant Tern, Sandwich Tern, and Common Tern (Ashmole and Tovar 1968). Gulls of various kinds attend their young at or near their nest sites for about thirty to sixty days, and Herring Gulls continue to feed them on their subsequent wanderings. Swallow-tailed Gulls of the Galápagos Islands are wholly dependent on their parents for food until they are four or five months old (Snow and Snow 1967).

Waterfowl

The duration of the association between parent and young in waterfowl varies enormously. With the exception of the aberrant Magpie-Goose of Australia, none feeds its young from the bill; but guidance and protection by the adults are valuable to the offspring. Among geese and swans, families hold together for a long while. Cygnets of Bewick's Swan accompany their parents on twice-yearly migrations of 2,600 miles (4,200 kilometers) and may stay with them for another year or two, as they do not breed until four or five years old (Scott 1970). Parent geese migrate with their latest brood of young. These goose families often unite in larger winter flocks, as occurs in the Canada Goose; but the surviving young return with their parents to the breeding area, after which they separate (Elder and Elder 1949). In this species, clannishness and continuing attachment to the locality of birth have led to the formation of numerous geographical races, some of which differ strikingly in size. Although cranes are not waterfowl, they resemble geese in their family cohesiveness. Young Whooping Cranes migrate southward with their parents, and the whole family stays together on a defended winter territory; but the parents become antagonistic to their offspring before they move northward the following spring.

Duck families are less cohesive. The drake may desert his mate while she incubates or remain to help her lead the ducklings, as in the Shelduck, Australian Pink-eared Duck, and others. In the Canvasback, the mother alone takes charge of the brood, but she abandons them before, at the age of nine to eleven weeks, they can fly. The cause of such parental desertion in this species, as in other kinds of ducks, is the onset of the postnuptial molt. During this period the adult sheds all its flight plumes simultaneously and remains flightless until they are renewed and it can migrate southward at summer's end. Since the mother's, rather than the duckling's, need determines the time of desertion, this occurs at widely different stages in the development of the young. Broods that hatch early may have the benefit of maternal guidance until they are nearly ready to fly, but late broods may be abandoned when only two or three weeks old.

The parentless young gather in flocks composed of ducklings of different ages from different broods. Still needing a leader, the younger ducklings follow the older ones instead of a parent. Much less wary than young under parental tutelage, these motherless ducklings are more vulnerable to predation (Hochbaum 1960). Redheads, which can fly when ten to twelve weeks old, usually separated from their mother at eight or nine weeks, but it was not clear whether she deserted her brood or they drifted away from her (Low 1945). The abandonment of still-flightless ducklings is widespread among diving ducks, but surface-feeding ducks, such as the Pintail and Blue-winged Teal, attend their broods more faithfully.

Ducklings abandoned while still flightless often gather into crèches, composed of few or many broods accompanied by a few adults, who may be either parents or ducks who have failed to raise families. In Manitoba, Hochbaum noticed a White-winged Scoter followed by eighty-four young, all under two weeks old, and Lesser Scaup ducklings in flocks of twenty or more, sometimes with two or more females leading the combined families. When Shelducks desert their broods to migrate to the distant shallow seas where they molt, the resulting flotillas may become spectacularly large, sometimes consisting of over one hundred ducklings with a few faithful guardians (Hori 1964). The crèching system may be less advantageous to the ducklings than to their parents. Sometimes emaciated after three or four weeks of faithful incubation, parents are free to devote their whole time to foraging and recuperation, while neighbors, possibly adults who failed to hatch their eggs, guide their young families. This is especially true of the Common Eider, who often finishes incubation in such poor condition that she may die (Milne 1969). By combining their broods and banding together, often with the help of nonbreeding yearlings, eiders protect their downy young from voracious Great Black-backed Gulls, their worst enemies (H. F. Lewis 1961).

Terrestrial Precocial Birds

The chickens that we raise are of many colors, from white to black, sometimes in beautiful patterns. They forage at large over the dooryard and pastures and into the edge of the tropical forest, and roost in a tree with a band of metal around the trunk to prevent predatory animals from climbing. Probably they are, in habits and mentality, a little closer to their wild Asiatic ancestors than are the poultryman's named, uniform breeds. The chicks, always hatched under a hen rather than in an incubator, forage more widely as they grow older, but they follow her by day and sleep with her at night in a box or, at a later stage, roosting beside and beneath her in the tree, until she finally shakes them off. Sometimes the hen sings, pecks away her chicks, and resumes laying when they are only two months old. If she does not lay so soon, she continues to lead the brood until they are about three months old; one hen attended a dozen chicks for nearly four months. The hen, rather than her progeny, is usually responsible for the dissolution of the family (Wood-Gush 1955).

Among quails family bonds seem to be more enduring. Both northern and tropical species of bobwhites (*Colinus*) remain in closely integrated groups, evidently composed of parents and their young, until the latter can no longer be distinguished from the former. The same is true of the wood-quails (*Odontophorus* spp.) of tropical America, but their flocks are smaller, as is to be expected of forest birds. For long intervals, I have watched at close range coveys of Marbled Wood-Quails and Spotted-bellied Bobwhites without noticing the slightest friction or any domination of one member by another.

Young Ostriches usually remain closely associated with their guardians for a year. The brood of young, even when numbering up to twenty-one, is never led by more than a single male and a single female; but one of these may be a step-parent, while the real parent engages in courtship with another adult. The growing Ostriches pick up their own food, but the parents guard them from danger, sometimes performing elaborate distraction displays. While the father tries to confuse the enemy, the cryptically colored young crouch motionless, just as smaller precocial chicks do, or sometimes the female leads them unobtrusively away, while the male continues to divert the intruder's attention (Sauer and Sauer 1966).

Among shore birds, a male Killdeer brooded his chicks until they were twenty-two days old. Young Killdeers may remain with their parents up to the age of six weeks, while the adults are raising a second brood (Nice 1962). In the far north, where the short favorable season makes rapid development a necessity, White-rumped Sandpipers fly well at seventeen days of age and soon thereafter leave their mother (Parmelee et al. 1968). In China, young Pheasant-tailed Jaçanas separated from their father during their third week (Hoffmann 1949).

Subprecocial and Semialtricial Birds

Among grebes, as among rails and their allies, all of which feed their mobile young from their bills, family bonds are strong. A pair of Least Grebes, who in the course of a year laid eight sets with a total of thirty-five eggs in a Cuban pond, occasionally fed the young of their second preceding brood while attending the first preceding brood and incubating another set of eggs. After these hatched, the downy young naturally received most of the parents' attention; but from time to time they gave food to the young of the preceding brood, already able to find much for themselves (A. O. Gross 1949a).

King Rails may remain with their brood for more than a month after they hatch, and Clapper Rails for five or six weeks (Meanley 1956, Nice 1962). American Coots begin to feed themselves when sixteen to twenty days old and at thirty days are no longer dependent on their parents for food, but they still keep company with the parents and may sleep on the nest with them until they are at least forty-six days old (Gullion 1954). European Coots dive

for their food when thirty-two days old but continue to receive some from their parents up to the age of fifty-four days. At ten weeks they leave the parental territory (Cramp 1947). Young Common Gallinules or Moorhens remain with their parents and often help them to raise a later brood, as has been seen in England, the United States, and Central America (Skutch 1961b). Grey (1927) wrote a charming account of how parent Moorhens gave pieces of bread to full-grown young hatched in May, who then passed this food to downy chicks born in July.

After leaving the nest at an early age, flamingo chicks, as already told (p. 316), gather in huge crèches or herds, where their parents feed them by regurgitation. L. H. Brown's (1958) observations on the Greater Flamingo, at Lake Elmenteita in Kenya, make it almost certain that parents and young recognize each other, and each of the former feeds its own offspring. By such individual attention, a meal was brought even to a young flamingo with a broken wing, who in a general scramble for food would doubtless have gone hungry. These herds, containing thousands or, in the Lesser Flamingo, hundreds of thousands of flightless young, are accompanied by a few adults, who lead them from danger. At times, the herds, pushed blindly onward by those behind, drive off adults still incubating or brooding chicks that hatched late, or run over parents who remain stanchly sitting on their nests. Surprisingly, few eggs are broken or small chicks crushed by the passage of many older ones over them. Parent Lesser Flamingos continue to feed their young until they fly when seventy to ninety days old (L. H. Brown and Root 1971).

Because of the mobility of young goatsuckers and the crepuscular or nocturnal habits of their parents, it is most difficult to learn how long the adults continue to feed their offspring, and I have repeatedly failed to obtain this information with the Pauraque so common in tropical America. A Lesser Nighthawk raised on a roof, who became very tame, was fed by its mother for more than a month after it hatched (Bent 1940).

Altricial Inland Birds

We have already noticed some very long periods of parental attendance in altricial marine birds. According to present information, the only altricial land birds with comparably prolonged parental care are certain large raptors. The young Crowned Eagle of Africa returns to the nest to receive food from a parent, chiefly the female, for as long as 11.5 months after its first flight, or until it is about 15 months old. Long before this, it can capture some prey for itself (L. H. Brown 1966).

Hatched naked and helpless, pelicans remain on or near their nests for about the first month of their lives. Then, having grown a downy coat, the ground-nesting species leave their natal spots and join their contemporaries in groups, or "pods," which may become quite large. Nevertheless, each parent recognizes its own young amid the crowd, evidently by sight rather than by voice, and it feeds only them. Great White Pelicans of Africa are fed by their parents until they fly at the age of about seventy days, after which parental attendance rapidly declines. Tree-nesting Pink-backed Pelicans are fed a few weeks longer (L. H. Brown and Urban 1969, Burke and Brown 1970).

Accurate records of the duration of parental attendance on fledged pigeons are, surprisingly, difficult to find. In one of the species that has been most carefully studied, the Wood Pigeon, parental feeding "certainly lasts for up to a week" after the young leave the nest at the age of twenty-two days (Murton 1965). In some of the larger members of the family, feeding may continue much longer. A captive young Victoria Crowned Pigeon was first seen to feed itself when slightly over thirteen weeks old, but it was still fed by its parents for several days more (Fleay 1961).

Although many marine birds cease to attend their young when, or even before, they leave the nest, land birds rarely permit their offspring to confront a perilous world without parental guidance. Among altricial land birds, such abrupt cessation of parental care appears to be confined to swifts. When, at the age of thirty-five to fifty-six days, European Common Swifts fly from the nest, their parents give them no further attention. The adults now begin to migrate southward, evidently leaving the young to find all their own food and their way alone to the winter home in Africa. Even the nonmigratory White-rumped Swift of tropical Africa appears to

lose contact with its parents when it leaves the nest, for Moreau (1942b) and his assistants watched fledglings about six weeks old emerge from the nest and fly beyond sight, often in the old birds' absence. Fledgling Chimney Swifts, however, may return to roost for some nights in their natal chimney with their parents; it seems not to be known whether they are fed after their first flight. Since the aerial insect catching of all swifts requires much skill, one might suppose that they would need a parent's help until they develop proficiency.

Hummingbirds have long been classified in the same order (Apodiformes) with swifts, which like them have superb powers of flight. But many differences distinguish the somberly clad swifts from the glittering hummingbirds, and it is gratifying to notice that some of the latest classifications place them in separate orders. Young hummingbirds are inconspicuous creatures to detect amid foliage, and until recently estimates of the duration of parental care were far too short. Unlike many other fledglings, hummingbirds do not follow their foraging parent and receive food where it is found. Accordingly, the mother, who alone attends them in nearly all species, must know where she can meet them. The young wait in a definite spot, usually not far from the nest, to which the parent returns after she has filled her crop with nectar, small insects, and spiders. If one can find this spot, he may watch the parent regurgitate to them day after day and learn the length of parental attendance. One Scaly-breasted Hummingbird was fed until at least fifty-two days old, and one from a different brood for sixty-five days; a Rufous-tailed Hummingbird, for fifty-eight days after it hatched; a Band-tailed Barbthroat, for fifty-six days; a Long-billed Starthroat, for forty-eight days; a White-eared Hummingbird, for at least forty days, by a mother who was incubating another set of eggs. All these young hummingbirds left the nest when twenty-four to twenty-six days old and, with the exception of the White-ear, were fed for at least twenty-three to forty-one additional days before they became independent (Skutch 1972).

Because of a dearth of information on the duration of parental care in free birds, we must pass over the parrots, rollers, hornbills, barbets, toucans, and other families of altricial nonpasserines. We know that some of these birds remain associated with their parents for a long while after they fledge, often sleeping in the same hole with them and doubtless receiving much valuable guidance; but we do not know how long they are fed. In Borneo, Scarlet-rumped Trogons still received much food from their parents seventeen weeks after they fledged, and several species of malcoha cuckoos and forest kingfishers for at least ten weeks (Fogden 1972).

For the nearly cosmopolitan woodpeckers, observations are more abundant; they show that in warm regions parents continue to attend their young much longer than in cold northern lands. In northern Europe, Great Spotted Woodpeckers leave the nest when about three weeks old and receive food for an additional week or two, or until four or five weeks old. Green Woodpeckers remain in the nest about four weeks and are fed for about two more weeks, or until about six weeks old (Blume 1961). In Canada, Downy Woodpeckers become independent at the age of about forty-one days, the bigger Hairy Woodpeckers at forty-one to forty-four days, and Yellow-bellied Sapsuckers at thirty-six to thirty-eight days (Lawrence 1966). The female sapsucker deserts her family four or five days after the young fly and settles in a neighboring area, where she tolerates neither her young nor her mate, who continues to feed their offspring for a while longer (Foster and Tate 1966).

In warmer regions, the family bonds of woodpeckers are stronger. In one of the less social species, the Red-crowned Woodpecker, I have known a female to feed her young for seventeen days, a male for thirty-six days, after they left the nest hole when about thirty-two days old. A male Golden-naped Woodpecker occasionally fed a son, ninety-four days old, for two months after the latter left the nest and long after he could forage for himself (Skutch 1969b). Most indulgent of all are Red-cockaded Woodpeckers in Florida, who feed their young up to the age of five or six months (Ligon 1970). How long parental attention continues depends not only upon the young bird's need but also upon the sociability of the species. Golden-naped

and Red-cockaded woodpeckers are much more social than Hairy and Downy woodpeckers.

Information on the duration of parental care in passerines is more abundant and, as is to be expected, reveals wide variations in this great order. Among the few nonoscines for which I have found such information are the Bicolored Antbird and Eastern Kingbird. In the Panamanian forest, young Bicolored Antbirds, which leave the nest at the age of about fourteen days, are fed by their parents until eight or ten weeks old (Willis 1967). In a family of kingbirds, the last feeding was seen thirty-five days after the young left the nest at the age of fourteen to seventeen days. Because of the mobility of juveniles, it is difficult to learn how often their parents feed them. Nevertheless, Morehouse and Brewer (1968) found that the rate of delivering food continued to increase not only during the whole nestling period but also until the young kingbirds' second week out of the nest, after which it declined until feeding ceased at about the juveniles' fiftieth day. Banded Broadbills in Borneo have been seen feeding young twenty weeks after they left the nest (Fogden 1972).

In the crow and jay family, parental attendance continues longer than in most other passerines. Blue Jays in Tennessee were fed until four months old, and Thick-billed Nutcrackers in Scandinavia until 105 days and possibly nine days more (Laskey 1958, Swanberg 1956). In a family of very small birds, Pygmy Nuthatches leave the nest from twenty to twenty-two days after they hatch and are still partly dependent on their parents for food when forty-five to fifty days old. For Brown-headed Nuthatches, the corresponding figures are eighteen or nineteen days and forty-four or forty-five days (R. A. Norris 1958).

Nice (1943) gathered evidence that many small passerines become independent when twenty-five to thirty-five days old, two or three weeks after they leave the nest. However, some instructive variations may be noticed. Second or last broods tend to receive more prolonged care than first broods, which are often neglected as the parents turn their attention to another set of eggs. When this occurs, the male may continue to feed the early brood while his mate builds a new nest or resumes incubation in the old one. Male Blackbirds continue to feed the young of the first brood for fifteen to twenty-five days, usually about twenty days, after they depart the nest at the age of thirteen or fourteen days. When the female lays again, she gradually loses interest in her fledged young, although an occasional individual feeds them as much as sixteen days after they leave the nest, while she is in the midst of incubating another set of eggs. At the end of the breeding season, when the female has stopped laying, she usually feeds her latest brood longer than the male, up to fifteen or even twenty-four days after they leave the nest; the male, whose attendance is now rather erratic, may wholly neglect them or continue to feed them as long as their mother does. In any case, the young Blackbirds become independent between their thirtieth and fortieth day of life (D. W. Snow 1958).

In the Great Tit, young of the first brood are fed only for six or eight days, rarely as much as nineteen days, after they leave the nest, but young of the second brood continue to receive food for twelve or fourteen days. Black-capped Chickadees in Massachusetts were fed longer, for ten to twenty-five days after a nestling period of sixteen to twenty days (Kluyver 1951, 1961).

At high latitudes, where the breeding season is short but long days permit prolonged feeding, young songbirds, like young woodpeckers, tend to become independent earlier than in tropical and subtropical regions. This difference has been noticed even in the same species in the North Temperate Zone. Some migratory White-crowned Sparrows in northern Washington became independent when twenty-five days old, and none was seen to receive food from a parent after the age of twenty-seven days. In central California, however, resident White-crowned Sparrows of a slightly different race were fed by their mother up to thirty-two days of age and by their father until thirty-five days old. In both regions, most young left the nest when ten days old (Blanchard 1941).

In Costa Rica, one young Black-striped Sparrow (a bird about the size of the White-crowned) was fed by its parents until it was at least fifty-eight days

old, and another, a member of a second brood, at least to the age of sixty-three days. Buff-throated Saltators, slightly larger members of the finch family, were fed for fifty-one days in one case and fifty-four in another. Black-striped Sparrows usually leave the nest at eleven or twelve days of age and Buff-throated Saltators at thirteen to fifteen days. In the same locality, a Buff-rumped Warbler, who left the nest when thirteen days old, was fed for the next fifty-one days, or until at least sixty-four days of age. None of the many migratory wood warblers that nest in the North Temperate Zone is known to receive parental attention for a period remotely approaching this. Protracted feeding has also been recorded for another species of tropical and subtropical distribution, the Natal Robin, which leaves the nest at seventeen days of age, first attempts to feed itself in its fourth week of life, becomes practically self-supporting after the sixth week, yet begs freely up to the eighth week, when parental care begins to decline (Farkas 1969). A related species, the Cape Robin, widely distributed in tropical Africa and resident also at the Cape of Good Hope, may feed its young for as much as fifty days after leaving the nest at the age of sixteen or seventeen days (M. K. Rowan 1969). In the dipterocarp forest of Sarawak, where parental care by a variety of birds continues for an exceptionally long time, a number of bulbuls, tree babblers, flycatchers, and drongos still fed young that had left the nest ten to twenty-three weeks earlier (Fogden 1972).

Division of the Brood

As already related (p. 289), in the South Polar Skua the older nestling attacks its younger sibling and drives it from the nest, and, in an effort to raise both, the parents may attend them in different parts of their territory, each caring for a different chick. This is an abnormal example of a habit widespread among birds, the division of the brood between the two parents, each of them confining its attention to certain fledglings. However, except in skuas, this is not done to save the young from fratricidal fury.

In a wide variety of monogamous birds that rear more than one young, the family splits up after leaving the nest, the male taking charge of some of the offspring and the female of the rest. This procedure has been noticed in the Whimbrel and Old World Golden Plover by Williamson (1948), Common Snipe by Tuck (1972), Red-bellied and Hairy woodpeckers by Kilham (1961, 1968*a*), Great Spotted Woodpecker by Blume (1961), Bicolored and Spotted antbirds by Willis (1967, 1972), Northern House Wren by Kendeigh (1941), Blackbird by D. W. Snow (1958), Ovenbird by Hann (1937), Snow Bunting by Tinbergen (1939), and Smith's Longspur by Jehl (1968*a*). These few examples, of the many that might be cited, are enough to suggest how widespread the practice is among the families of birds. In the Blackbird, the division of the family may be carried to such lengths that one parent will no longer respond to the begging of a fledgling who is being attended by the other parent, and the young, in turn, solicit food only half-heartedly, or not at all, from the parent who is not caring for them. Such strict division of labor is more likely to occur toward the end of the breeding season than with earlier broods, and it is probably exceptional. When a hawk killed a male Spotted Antbird, his mate fed the fledgling that he had been attending, in addition to the one already in her care (Willis 1972). Some parents feeding fledglings will at times give a billful to young of a different family, as has been recorded, for example, in the Pygmy Nuthatch (R. A. Norris 1958).

This splitting up of the brood corresponds to a tendency of certain young birds, who hitherto have huddled close together in the nest, to avoid each other after leaving it, as has been noticed in Snow Buntings and several kinds of thrushes by Tinbergen. The obvious advantage of this separation of the young is that it reduces the chances that all will fall victims to the same predator, and it may promote efficiency in parental foraging and feeding. However, it is by no means universal. The fledglings

of certain American flycatchers, far from avoiding their siblings, often perch close together while waiting for both parents to feed them, and at night they roost in contact with each other. While awaiting their meals, fledgling white-eyes sit bunched together, with the two outer ones leaning against the central one. Between feedings, they often preen each other's heads (Skead and Ranger 1958).

I have sometimes seen how this division of the brood may arise. At nests of the Broad-billed Motmot, Black-hooded Antshrike, and Variable Seedeater, the parent who was present when the first fledgling left went off with this fledgling and did not return to feed the young in the nest, who were attended only by the other parent as long as they remained there. The seedeater family was, however, reunited after the last young took wing; the motmots and antshrikes lived amid such heavy tropical vegetation that I lost sight of them soon after they left the nest.

As soon as they become self-supporting, some young birds are driven away by their parents, as is especially likely to happen to those of an early brood that is followed by a later brood. In highly gregarious birds, like grackles and starlings, parents and young together leave the vicinity of the nest and join the large flocks that now form. In species that are settled on their territories throughout the year, independent young may be permitted to live on the parental domain for a variable period, only to be driven away, or to leave spontaneously, before the next breeding season. In many other species, especially among the nonmigratory birds of milder climates, parents and young remain closely associated much longer; the young doubtless profit by this continuing contact with their more experienced elders, and they may help the parents in the following breeding season, as will be told in the next chapter.

CHAPTER 28

Helpers-I

One of the most exciting results of the intensive, widespread bird studies of recent decades is the growing awareness of helpers at the nest and their role in population regulation in a number of species resident in milder climates. Although most birds attend only their own mates, nests, and offspring, in the preceding chapters reference was occasionally made to the participation of more than two individuals in the activities at a nest. More rarely, a bird feeds an adult to whom it is not mated, perhaps an individual of a different species. A bird who assists in the nesting of an individual other than its mate, or feeds or otherwise attends a bird of whatever age who is neither its mate nor offspring, is designated a "helper." Helpers may be of almost any age; they may be breeding or nonbreeding individuals; they may aid other birds of the most diverse relationships to themselves, including those of different species; and they may assist in various ways. Accordingly, to understand this fascinating mode of cooperation among birds, we must pay attention to three points: (*a*) the status or condition of the helper, whether young or old, a parent or a nonbreeder; (*b*) its relationship to the bird or birds whom it assists; and (*c*) the activities in which it engages.

Status and Activities of Helpers

In age or degree of development, helpers range from nestlings to mature breeding birds, and perhaps even those no longer able to reproduce. Rarely a nestling, or a chick still dependent on its parents, makes gestures of helpfulness, such as arranging nest materials or passing a morsel to a sibling, reminding us of a little child who tries to help an older person at a task beyond its strength, perhaps only getting in the way. In either case, the assistance rendered is of little value but the inclination to give it is precious.

More commonly, young who have recently ceased to be dependent on their parents for food help, more or less substantially, to feed their younger brothers and sisters of the following brood, thereby earning the designation of "immature helpers." Among permanently resident birds whose families are not disrupted by wandering or migration, the young may remain with their parents from one breeding season to the next, and as yearlings, or even older birds, aid them in reproductive activities; or they may attach themselves to, and assist, some other established pair. Since some birds do not mature sexually until several years after they cease to grow, these yearling or older helpers may or may not be physiologically able to breed; the point is difficult to determine in living birds. If they are definitely known to be sexually immature in a breeding season after that in which they hatched, they may be called "innubile helpers" (the adjective is the negative of the Latin *nubilis*—marriageable). Sexually mature birds that serve as helpers instead of breeding are conveniently known as "nonbreeding adult helpers." Finally, we have birds who, while actually engaged in rearing their own family, sporadically or consistently attend the young of other pairs. According to whether this assistance is or is not reciprocated, these breeding birds may be designated as "mutual helpers" or "unilateral helpers." To summarize, we have the following classes of helpers:

1. Nestling helpers
2. Immature helpers
3. Innubile helpers
4. Nonbreeding adult helpers
5. Breeding helpers
   a. Mutual helpers
   b. Unilateral helpers

As to their relationship to the birds that they assist, helpers are divided into "intraspecific helpers," who aid other individuals of their own kind, and "interspecific helpers," who attend birds of other species. Known examples of intraspecific helpers include all of the above-mentioned developmental classes. Among free birds, interspecific helpers are nearly always breeding adults, but in captivity immature and other nonbreeding birds often become helpers. Helpfulness is usually displayed in nesting activities and rearing the young; but occasionally one adult bird feeds or preens another, other than its mate, when neither is engaged in breeding. This, too, is most likely to be witnessed among captive birds.

Birds most frequently help each other by sounding the alarm when danger threatens and by repelling animals of all kinds from the vicinity of their nests. The alarm notes of one bird alert other individuals of the same and different species. Even mammals learn the meaning of these cries and by heeding such timely warning may save their lives (Riney 1951). Birds of the most diverse kinds may join in mobbing a snake, owl, hawk, or some other predator, thereby giving the whole avian community the valuable information that danger lurks in a certain spot. By driving predators from their own

nests, birds may save nearby nests of their own or other species. Certain birds gain protection by nesting in a colony of terns or gulls, which vigorously expel dangerous intruders (p. 95). Or birds with nests close together may try, sometimes with success, to lure away their enemies by simultaneous distraction displays (p. 413). Most of these aids to survival appear to be incidental rather than intentional; a bird may be unaware of the presence of another who is alerted, and possibly saved, by the warning cry that it broadcasts. No more need be said about this very widespread mode of mutual helpfulness.

The most frequent mode of deliberate, individually directed helpfulness among birds is feeding. Since the perpetuation of most avian species depends upon giving food to other individuals, particularly the young, the impulse to feed has become so persistent and strong that it is readily directed beyond its primary context. It is one of the first forms of parental behavior to appear in young birds and occasionally even occurs in nestlings. In breeding birds, it often arises prematurely, as anticipatory food bringing, when the parents, especially the male, offer food to unhatched eggs (p. 253). It persists for hours or days in parents who have lost their young and find no recipient for the food that they bring (p. 278). From the context of parent and young it has passed to that of parent and parent; as nuptial feeding, it helps a male to win and retain a mate, and it may substantially increase the hours that the female can devote to incubation. It crops up in the most unexpected situations, as when a captive Raven passed food through the bars of its cage to a free Black Vulture; a Jackdaw pushed food into the mouth or ear of the scientist whom he regarded as his mate; and a male Cardinal for days put food into the mouths of seven goldfish that came to the pool's edge to receive it (M. Davis 1952, Lorenz 1952, Lemmons 1956). Finally, the urge to feed others is one of the last modes of parental behavior to disappear when a species becomes parasitic. Glossy cuckoos have repeatedly been seen to feed fledglings and even nestlings of their own kind that were being reared by foster parents, and similar behavior has less often been noticed among cowbirds. Glossy cuckoos also retain nuptial feeding (Bent 1958, Friedmann 1968).

Less frequent modes of helping are nest building, incubation, brooding the young, cleaning the nest, and allopreening. Since building, incubation, and brooding are successive stages in a nesting cycle, the helpers who perform these activities generally go through the complete cycle and are most often mutual helpers, as in anis, although sometimes they are unilateral helpers, as in certain bee-eaters. Rarely juveniles help to build, rather ineffectually, as I have seen in the Golden-masked Tanager, or quite capably, as in the Golden-naped Woodpecker. Even nestlings may occasionally help to build or maintain a nest (p. 292). Immature birds who feed nestlings sometimes brood them, especially in aviaries (p. 354). Precocious incubation seems to be rare; but a month-old Domestic Pigeon, after some initial difficulty, incubated its mother's later set of eggs for about two hours each day (Goodwin 1947).

Birds that preen individuals other than their mates or dependent young seem to qualify as helpers by our definition. Among the numerous examples of allopreening given in chapter 2 are several cases of adults performing this service for other adults of their species to whom they were not mated, and more rarely for individuals of different species. Immature birds and even nestlings or small chicks sometimes preen their siblings, unrelated young of their species, and, in captivity, young of different species. Adults of several kinds respond to the solicitation display of cowbirds by preening them.

Birds who preen individuals that are neither their mates nor dependent young, often outside the breeding season, provide us with some of the few available examples of helpers not associated with reproduction. Rarely one adult feeds another to whom it is not mated, even an adult of a different species. Some reported instances of such feeding are derived from old, probably uncritical observations of which details are not available, or they are based upon circumstantial evidence. Sometimes a bird with a badly misshapen bill, or otherwise so handicapped that it appears incapable of foraging for itself, remains alive and well, suggesting that it

is being supported by one or more companions. Although this is certainly not impossible, especially among birds known to feed their mates, the inference is hazardous. Once I watched a Lineated Woodpecker manage to feed itself with a bill so grotesquely deformed that at first sight I doubted its ability to do so, and others have made similar observations. But an adult male Black-headed Grosbeak with a badly deformed bill was seen to be fed by a female, probably his mate, by whose care he had apparently been kept in good condition for an extended period (Fox 1952).

An ailing Black-faced Wood-Swallow was fed regularly all day long by several others, upon whose good offices it appeared to be wholly dependent (Immelmann 1966). Wood-swallows are such highly social birds, helping each other in many ways, that such charitable ministrations are quite in keeping with their habits. More surprising are the reports of an old, blind American White Pelican, an adult Brown Booby with only one wing, and an adult Magnificent Frigatebird in the same plight, all evidently kept alive by obliging companions (Baird and Stansbury 1852, Murphy 1936). In captivity, a male European Robin fed a rival with whom he had been fighting, after the latter broke his leg (Lack 1953). By assuming the attitudes of a begging fledgling, an incapacitated adult might well induce others to feed it. Although I cannot vouch for all the reported instances of free birds supporting disabled companions, I cannot summarily dismiss any of them as false. When a Raven can feed a Black Vulture, or a Cardinal assiduously nourish goldfishes, it would be rash to set any arbitrary limits to the range of activities of avian helpers.

Immature Interspecific Helpers

Although free immature birds not infrequently attend younger individuals of their own species, the only instances of juveniles helping other species that I know occurred in aviaries. With an abundance of readily available food and no need to avoid enemies, young captives sometimes display precocious parental behavior that in free birds of the same age is suppressed by the difficulty of self-maintenance. If no still-younger individuals of their own kind are within reach, they direct their cherishing impulses upon whatever callow young of other species share their compartment.

In an aviary in East Africa, a young Black-shouldered Kite fed and brooded nestlings of her own species, rearing them from the age of one day until they became independent. She also adopted a day-old buzzard, whom she continued to feed for nearly two months, until it was three times her own size. She "brooded" a red notebook and other lifeless things of the same color (van Someren 1956). In Canada, a six-week-old Common Bluebird fed young Wood Thrushes, Veeries, Bobolinks, Cardinals, orioles, and a cowbird, a total of fifteen individuals that were being reared in the same aviary. When slightly older, this same female bluebird helped to feed and brood a nestful of bluebirds, sometimes sitting beside their mother (Ivor 1944*b*). Free young bluebirds often help their parents to nourish second broods. A captive Chipping Sparrow thirty-nine days old fed a still younger Red-winged Blackbird (Nice 1943). These instances will suffice to illustrate an activity that is widespread among young birds confined in aviaries.

In the same aviary where the juvenile bluebird was so helpful, an old, unmated Wood Thrush also fed the fifteen young of various species already mentioned. Probably many similar instances could be found by one who had access to the records of aviculturists. But free birds seem rarely to attend young of other species, except when they are breeding and their parental impulses are strongest. Practically all of the many records of free interspecific helpers that I have found refer to nesting birds, who may assist other birds either unilaterally or reciprocally.

Breeding Unilateral Interspecific Helpers

As we earlier learned (p. 253), male birds whose mates are incubating are often so eager to begin feeding their nestlings that they offer food to the unhatched eggs. The feeding of a neighbor's offspring may provide an outlet for repressed energy; and, in all territorial birds, the nearest nests are more likely to belong to some other species than to other individuals of the same species. In these cir-

cumstances, an adult male Common Bluebird fed nestling Northern House Wrens, upsetting their parents. After his mate hatched out the young bluebirds, he transferred his attention to his own offspring (Forbush 1929). A Winter or European Wren fed nestling Great Tits while his mate incubated, continuing to do so for at least four days (Armstrong 1955). Similarly, a male Carolina Wren whose mate was incubating in a nest box fed not only her but also Great Crested Flycatchers in a neighboring box (Laskey 1948). A Northern House Wren fed Red-shafted Flickers in a hole fifteen inches below that in which his mate was incubating. After his own young hatched, he fed both them and the young flickers (Royall and Pillmore 1968). It is significant that wrens are among the most restless and energetic of birds, often building an excess of nests.

A male Scarlet Tanager took food to young Chipping Sparrows until his own nestlings hatched (Hales 1896). When a pair of Oregon Juncos and a pair of Bewick's Wrens nested on opposite sides of the interior of a garage, the juncos often chased the adult wrens as the latter came with food for the young wrens. Nevertheless, while his mate incubated, the male junco fed the nestling wrens and also cleaned their nest. The parent wrens did not try to drive away the juncos (L. Williams 1942).

More rarely, both members of a breeding pair feed the young of some other species. A pair of Song Sparrows helped a neighboring pair of American Robins feed their nestlings and clean the nest, continuing until their own eggs hatched (Brackbill 1952). A pair of Common Bluebirds fed a brood of Common Mockingbirds, both before and after they left the nest (Carr and Goin 1965). The appeal of young birds clamoring for food may at times cause birds to desert their eggs in order to attend them, as happened when a pair of Blue Tits built a nest in a box on top of which a pair of European Robins already had a nest. The female robin laid five eggs and the tit laid three. After the robin's eggs hatched, the tits covered their own eggs with feathers and fed the young robins. At first, the two pairs fought a little, but soon all four birds settled down to attend the nestlings in concord. After the robins were fledged, the female tit laid another set of seven eggs over her original three and raised a brood (L. Williams 1942).

Sometimes a parent bird gives food to young not its own because its intention to feed its own offspring is temporarily thwarted. While young American Redstarts were being photographed in the hands of children, their father fed them, but their more timid mother delivered her food to young American Robins in a nest twenty-five feet away. At other times, a parent bringing a meal to its offspring may pass other young birds, whose gaping mouths or begging cries appeal irresistibly to parental instincts. As a parent Gray Wagtail flew over a brood of young thrushes, they opened their mouths, whereupon the wagtail faltered in flight, turned, alighted, and gave its whole billful to them (Armstrong 1947). When a pair of British Nuthatches nested three feet away from a pair of Common Starlings, one of the former often carried food into the nest box of the latter, and it also removed droppings of the nestling starlings (Powell 1946). Similarly, starlings feeding their own nestlings often entered a Yellow-shafted Flickers' hole in a neighboring tree, apparently fed the young flickers, and removed their droppings (Prescott 1971).

Sometimes a mated bird who apparently has not even laid is diverted from breeding by the irresistible appeal of nestlings. Some years ago I found, in a tree in front of our house, a nest of Golden-masked Tanagers with two nestlings that were attended by a female Tropical Gnatcatcher in addition to their parents. The helper fed and brooded the nestlings and also cleaned their nest. As days passed, she became increasingly hostile to the parent tanagers, devoting more time to futile efforts to keep these slightly larger birds away and less to attending the nestlings. The parents usually ignored her, unless she became very annoying, when they chased her mildly. Her attendance at this nest continued for at least twelve days and was terminated only by the nestlings' departure. While the female gnatcatcher was so engaged, her mate, without her help, built a nest three yards away in the same tree, but she apparently never laid in it. He, in turn, took no interest in his neighbors, the tanagers.

In other cases, the helper has lost its own offspring or has reared its fledglings to independence, without exhausting its impulse to feed or otherwise attend young birds. After a Screech Owl lost her eggs, she turned her attention to a family of flickers in another hole in the same tree. She continued for five days to brood the young woodpeckers, and even brought them a small bird, while the parent flickers continued to feed their unharmed young (Bent 1938). A Mourning Dove brooded and fed nestling White-winged Doves a few days old, who had been neglected by their own parents for most of the day. When, at last, the female White-winged Dove returned in the late afternoon, she fought and drove away the fostering Mourning Dove. The latter, nevertheless, continued to minister to the young Whitewings until they fledged. Apparently her own eggs had failed to hatch. In aviaries, Mourning Doves and several other kinds of doves are quick to adopt, and assist in the care of, young doves of any species (Neff 1945a).

A female Eastern Phoebe whose first brood was becoming independent brought food to nestling Tree Swallows, continuing this for about a week, while the parent swallows tried to drive her away (Deck 1945). After rearing two of her own young, a female Blackbird continued for two or three weeks to offer food to any bird that came near, and an adult European Robin was among those who accepted. European Robins who had lost their own young fed a brood of nestling Song Thrushes (Lack 1953). A male Cardinal, whose nest had been destroyed, fed four fledgling American Robins a few days younger than his own lost nestlings. For a week, he was almost as active in bringing food to the immature robins as were their own parents. Perfect harmony prevailed among the three attendants. After the replacement brood of the Cardinals hatched, the male apparently brought food to both families simultaneously. His mate took no interest in the robins (Logan 1951). A male Brown Towhee, whose own first brood had just become independent, joined a pair of Cardinals in feeding three fledglings of the latter. These three adults worked together in complete harmony for about three weeks. A month after helping to feed the Cardinals,

PLATE 99. A pair of Eastern Phoebes at their nest. A female whose own young were becoming independent fed nestling Tree Swallows (photo by S. A. Grimes).

the towhee and his mate raised a second brood (Antevs 1947). Lark Sparrows fed young Common Mockingbirds whose nest was close to their own (G. M. Sutton MS).

Sometimes parent birds become helpers by accident. A pair of Mountain Chickadees were found feeding nestling Williamson's Sapsuckers that were also being attended by their own parents. This situation apparently resulted from the collapse of the thin partition that separated the holes of these two species in a decaying pine trunk and the subsequent loss of the young chickadees (Russell 1947).

Frequently, when a bird is found feeding young of another species, it is no longer possible to learn the circumstances that caused it to do so. One April, we noticed a male Blue Honeycreeper in full nuptial attire giving food to a fledgling Scarlet-rumped Black Tanager on the feeding shelf beside our house. For the next three days, the brilliant honeycreeper continued to attend the tanager, who was at least twice his size, giving it chiefly bits of banana or plantain from the board, but sometimes insects caught amid the foliage. Again and again, this strange pair returned to the feeder, and the honeycreeper stuffed the tanager with fruit, once passing it six billfuls of banana in rapid succession. The honeycreeper insisted on pushing his long, sharp bill into the throat of the short-billed tanager, who seemed not to relish this method of being fed. When satiated, the young tanager would turn its head away, whereupon the honeycreeper would flit over its back from side to side, presenting the morsel alternately on the right and on the left, until the fledgling flew away with its attendant following. Often the young bird pursued the honeycreeper through the neighboring trees, begging. But when the attendant started off on a high flight, the tanager, a member of a species that does not travel so high and far, did not follow. The tanager was beginning to feed itself, and it also received at least occasional meals from a male of its kind, probably its parent. It seemed, however, to prefer the attentions of the more complaisant honeycreeper. No female tanager was seen to feed the fledgling. In another year, a female honeycreeper fed a fledgling Yellow-green Vireo.

Among the many recorded instances of interspecific helpers that were apparently in breeding condition but whose histories were imperfectly known were the following combinations: A Gray Catbird fed and mothered a brood of orphaned Cardinals; another Gray Catbird fed a half-grown flicker that had been dislodged from its nest and separated from its parents in a severe storm; and still another fed nestling Northern House Wrens in their box (Bent 1948, Nolan and Schneider 1962). An Olive-backed Thrush helped parent American Robins to feed their nestlings (Bent 1949). A Black-and-white Warbler repeatedly fed nestling Worm-eating Warblers, despite the attacks of the parents when he approached their nest. Once they tore food from his bill and themselves gave it to their young. This helper gave a distraction display when the observers visited the Worm-eating Warblers' nest (Rea 1945). Another Black-and-white Warbler attended for several days a fledgling Ovenbird that had apparently become separated from its parents (Kendeigh 1945). A Worm-eating Warbler fed four nestling Ovenbirds and cleaned their nest (Maciula 1960). In a family not noted for its sociability, a Green Violet-ear fed a young White-eared Hummingbird (Wagner 1959).

In two widely separated localities, a female House Sparrow was noticed feeding three fledgling Eastern Kingbirds, whose parents were not seen. The fosterer's head was often caught by the closing of the wide mouth into which she placed the food, and she had to struggle to release herself (Fitch 1949, G. D. Hamilton 1952). Another House Sparrow cooperated with a pair of Red-eyed Vireos in feeding and defending their nestlings, and still another fed Tufted Titmouse nestlings in intervals of attending her own in a neighboring box (Bent 1958, Prescott 1967). A female Common Grackle fed and protected nestling Chipping Sparrows—behavior unexpected in a bird that sometimes preys on smaller birds (Bent 1958). A male Chaffinch, attracted by the persistent calling of newly fledged Hawfinches, fed them six times (Mountfort 1957). A female Rufous-sided Towhee fed two young Common Mockingbirds, continuing for hours together. The mockingbirds eagerly accepted insects but rejected the seeds that the towhee offered them (Westwood 1946). Another towhee cared for a young Brown-headed Cowbird that had been hatched by an Orchard Oriole (Neff 1945*a*).

The relations between the helpers and the parents of the young that they attend are various. Sometimes the helper is hostile toward the parents, as was the male Oregon Junco who fed nestling Bewick's Wrens and the Tropical Gnatcatcher who attended a nest of Golden-masked Tanagers. More often, the parents are troubled by the presence of their uninvited assistant, and they may try to drive it away. Examples of this are the Mourning Dove at the White-winged Doves' nest, the Eastern Phoebe at the Tree Swallows' nest, the Black-and-white Warbler at the Worm-eating Warblers' nest, and the Common Bluebird at the Northern House Wrens' nest. In other instances, parents and helpers work together in concord, as happened when a Worm-eating Warbler fed nestling Ovenbirds, when a male Cardinal attended fledgling American Robins, and when a Brown Towhee attended fledgling Cardinals. Rarely the parents take food directly from the helper and either pass it to their nestlings or eat it themselves. Black-headed Grosbeaks accepted food from a Northern House Wren, Coal Tits from a European Wren, and Yellow Warblers from Song Sparrows.

The same contrasts in the attitude of the parents toward their assistants, and of the assistants toward the parents, are found in intraspecific associations, and they suggest the two distinct routes by which birds are led to attend nests of other individuals. In some species, prolonged close association between parents and their offspring brings the offspring into intimate contact with the parents' subsequent broods, in the same breeding season or, in the case of birds that mature slowly, in some later year. Many immature and innubile intraspecific helpers, and even some fully mature intraspecific helpers, are led in this way to assist at others' nests. In these cases, it is difficult to decide whether the nonbreeding birds are stimulated to engage in parental activities by seeing the parents do so or by the direct appeal of eggs or nestlings to their latent parental impulses.

The other route is that followed by most interspecific helpers and even by some intraspecific helpers. These birds are not closely associated with the parents until accidental contact with their nest or young, at a time when they are particularly susceptible, releases parental activity, in the course of which they come into close contact with the parents themselves. The attitude toward each other of the birds thus suddenly brought together may be friendly, hostile, or neutral. In the first group of helpers, the social bond is primary and participation in parental offices by the nonbreeders arises secondarily from it. In the second group of helpers, the appeal to parental impulses is primary and mutual accord sometimes develops from it, as when parents and helpers cooperate amicably in the care of nestlings, the former sometimes even accepting food directly from their assistants. In other cases, however, participation in a common endeavor fails to overcome the antagonism between the parents and the intruding collaborators.

Mutual Interspecific Helpers

Some of the most curious cases of helpfulness arise when two birds of different species build their nests close together, or even lay in the same nest. Only eighteen inches apart in the same pine tree, a Rufous-sided Towhee and a Field Sparrow had nests, both with nestlings of about the same age. The male towhee frequently fed the young sparrows and removed their droppings, and a parent sparrow likewise brought food to the nestling towhees (Hoyt 1948). When an American Robin and a Gray Catbird each built a nest in the same clump of lilacs, both took turns at incubating the catbird's eggs. After the young hatched, they were brooded by both the robin and the catbird (Bent 1948).

More often, the mutual helpers have laid eggs in the same nest, in which the parents of both species incubate alternately or even together, sitting side by side, or perhaps one upon the other. If successful in hatching out the mixed family, they may cooperate in brooding and feeding the nestlings. Both species of North American cuckoos, the Yellow-billed and Black-billed, often lay eggs in each other's nests, as likewise in the nests of a number of other small birds, possibly because their own poorly built structures have capsized while their eggs were ready for deposition. On one occasion, a Yellow-billed Cuckoo laid two eggs in a nest of the American Robin, in which the robin also laid an egg. Then a Mourning Dove added two eggs to the mixed set and incubated along with the cuckoo. Both birds were found sitting side by side on the eggs of three kinds. The further history of this interesting partnership is lacking (Bent 1940).

American Robins sometimes share a nest with another bird, of the same or a different species. The most curious of a number of recorded instances is that of a robin and a Mourning Dove, each of whom laid two eggs in the same robin's nest. They took turns incubating, then fed and brooded the nestlings until they were eight days old. On the following day, the four nestlings died (Raney 1939). A House Finches' nest in Colorado contained four half-grown robins, two newly hatched finches, and four finch eggs. It was attended by two adult robins and two female finches, apparently mated to the same male. All five of them fed the young regularly. The large robin nestlings smothered their small nest mates. After the robins fledged, the three finches continued to feed them. At another nest in the same locality, adult robins and adult finches fed young robins, but there was no evidence that the finches had laid eggs in this nest (Bent 1949). European Robins enter into similar associations. One was found sitting with a Willow Warbler on six eggs in a nest built by the warbler. Another pair of robins raised a combined brood with a pair of Pied Wagtails (Lack 1953). A pair of Cardinals and a pair of Song Sparrows nested simultaneously in a nest built by the former and lined by the latter. Both females laid eggs and both incubated and brooded, the Cardinal sometimes sitting upon the smaller sparrow. Three Cardinals were hatched and reared to nest leaving. No antagonism between the Cardinals and Song Sparrows was noticed (Brackbill 1952).

Unless the nestlings of the two cooperating species hatch at about the same time, are of approximately equal size, and require similar food, it is unlikely that both kinds will be successfully reared. The young of the mixed brood of Mourning Doves and American Robins survived until they were eight days old, and I suspect that the very different feeding responses of doves and thrushes ensured that each nestling received nourishment only from its own parents. In the joint nesting of Cardinals and Song Sparrows, only the young of the bigger Cardinal were fledged. When House Finches and American Robins nested together, the nestling robins smothered their smaller nest mates. But when European Robins shared a nest with Pied Wagtails, the nestlings of both species appear to have been successfully reared. No mutual partnership of this sort has proved sufficiently productive to give rise to habitual symbiotic nesting by two species, corresponding to the intraspecific communal nesting of anis.

The fostering of the young of parasitic cuckoos, honey-guides, cowbirds, and weaverbirds by a wide variety of other species is a wholly unilateral relationship. As in the numerous instances of birds hatching the eggs that other nonparasitic species have occasionally deposited in their nests, such unintentional service to other birds does not qualify the fosterer to be called a helper. Helpers do more than incubate the eggs that some other bird has dropped into their nest, which perhaps they do not distinguish from their own, and rear the resulting progeny. In one way or another, helpers deviate from the typical breeding pattern to serve, or to enter into close association with, individuals other than their own mate and offspring.

In this chapter, I have given only a selection of the cases of interspecific helpers that have come to my attention. Doubtless I have missed the greater part of the instances reported in the most diverse publications, and it is certain that only a minute fraction of such occurrences among free birds are ever noticed by people. Nevertheless, the known combinations of species that enter into these relationships are diverse enough to suggest that every species of altricial birds has occasionally helped another altricial species of somewhat similar size with which it has been associated over a wide area for many years. Although passerines have usually been discovered helping other passerines, the helpers and the helped sometimes belong to different orders:

chickadees, wrens, and starlings have fed wood-peckers of several kinds; doves have nested with cuckoos and robins. Although a very great disparity in size would doubtless preclude this relationship, the Northern House Wren was only about one-third the length of the parent Red-shafted Flickers whose nestlings he fed, and the female Common Grackle was twice the length of the Chipping Sparrow whose young she attended. A bird that builds in the open may feed nestlings in a box or hole, as when a Gray Catbird attended Northern House Wrens; and hole nesters may help open nesters, as when a house wren fed Black-headed Grosbeaks and a Common Starling gave food to American Robins (Herbert 1971).

CHAPTER 29

Helpers-II

To learn how many birds of the same kind are attending a nest usually requires sustained observation, preferably of banded or otherwise individually recognizable birds; but the sight of birds of different species serving at a single nest, or of an adult of one kind feeding a fledgling of another kind, arrests the attention of the casual observer, who may tell of the incident in a popular magazine of natural history or in a daily newspaper, or perhaps enlist a competent ornithologist to study the occurrence. Accordingly, the reports of interspecific helpers tend to exaggerate their frequency relative to intraspecific helpers. The latter are certainly more common, at least in the tropics, where in some forty years of bird study I have discovered intraspecific helpers at or near thirty-eight nests of twenty species (not including the many communal nests of anis that I have seen), but I have noticed only three interspecific helpers, belonging to two species.

Interspecific helpers are always of such sporadic, unpredictable occurrence that their impact upon the regular course of reproduction is negligible. Intraspecific helpers, however, are of frequent or regular occurrence in a growing list of species, especially at lower latitudes, whose breeding arrangements they strongly influence. Interspecific helpers hold our attention because of the light they throw on bird behavior; intraspecific helpers have additional interest to the student of the perplexing problems of population regulation.

Nestling and Immature Intraspecific Helpers

As mentioned earlier, occasionally even nestlings engage in parental activities, including feeding each other and working at their nests. From an early age, the solitary nestling Short-tailed Shearwater plucks grasses and debris from the walls of its chamber at the end of a burrow in the ground and tucks them around itself to form a neat cup in which it sits. It continues this behavior throughout its long stay in the nest, so that it accumulates much material, especially after it ventures forward into the entrance tunnel and carries things back to the chamber when it returns (Warham 1960). From the day they hatch, albatross chicks may save themselves from being buried under wind-blown sand by kicking it out of their nest hollows on the open beach (Rice and Kenyon 1962). Nestling European Cormorants help to work loose materials into their nests and they sometimes feed each other. Passing food to nest mates has also been observed in the American Flamingo (Armstrong 1947) and in captive Crowned Hornbills.

After the emergence of the female Crowned Hornbill from the nest cavity in which she has been immured from the start of laying until the nestlings are half-grown, the young birds, working from inside the hole, plaster up the doorway again, leaving a gap just wide enough for their parents to pass food to them. Captive nestlings placed in a box with a small opening proceeded to reduce its size with mud supplied to them, particles of food, and their own droppings. They attended efficiently to the sanitation of their box. After the emergence of a female Red-and-white-billed Hornbill from her nest hole, her two nestlings, twenty-one to twenty-five days old, replaced the plaster seal with material that they found inside the cavity (Moreau 1936a, Moreau and Moreau 1940). Of all nestling birds, these hornbills seem to make the most important contribution to their own maintenance. Aside from them, and perhaps the albatross chicks, it is questionable whether any of the aforesaid precocious nestlings aid their parents enough to be called "helpers," but their early manifestation of parental modes of behavior is highly significant.

I have found scarcely any records of parental activity by free-living fledglings who have just left the nest. At this stage of their lives, they seem to need all the nourishment that their parents can provide, and their chief activity is following their parents and clamoring to be fed. Later, when the young birds become somewhat proficient in foraging for themselves, they sometimes become attentive to other individuals that are still younger. This habit is well developed in the Common Gallinule or Moorhen. One July, on a weedy pond in the highlands of Costa Rica, I watched a family consisting of two adults, two full-grown young birds in grayish plumage, and four downy chicks, who kept up a constant peeping. The full-grown young birds seemed to give as much attention to the chicks as the parents did, and once one of the former appeared to pass food to a chick. The feeding of younger brothers and sisters by immature gallinules has also been observed in the United States and England. Here Grey (1927) watched parents give pieces of bread to full-grown young hatched in May, who then passed this food to the downy chicks born in July. When a parent gave bread directly to a chick, one of the older young took it from the downy one's bill, then promptly replaced it there. They insisted on following the ritual! In the rail family, to which gallinules belong, chicks or juveniles of several species have been observed to engage in at least rudimentary nest building and to preen their siblings (Nice 1962).

I have found no instance of an immature free-living pigeon engaging in parental activities, but in captivity a fledgling Domestic Pigeon about twenty-five days old regurgitated food to a younger companion. When about a month old, another pigeon took sticks to the nest where its mateless mother incubated and presented them to her as an adult male does to his mate. This young pigeon also incubated its mother's eggs, as mentioned in the preceding chapter.

Among cuckoos, young anis sometimes help to feed and defend their parents' next brood. A Smooth-billed Ani brought food when forty-eight days old, and a Groove-billed Ani did so at seventy-two days. A hand-reared Smooth-billed Ani carried and arranged sticks and straws at the age of about six weeks. When the inclination to help is present in a species or family of birds, it often manifests itself at various stages of life and in different ways. Later we shall notice another class of helpers among anis (D. E. Davis 1940a, Skutch 1959b, Merritt 1951).

When the large family of barbets, widely distributed through the tropics of both hemispheres, becomes better known, many examples of helpers will doubtless be found, for along with species that are quite solitary are others that are social. In the Frankfort Zoo, nestlings of the Double-toothed Barbet were fed by young of the preceding brood and also by a second female. After the fledglings left the nest, the whole group continued to feed them. They were led back to sleep in the nest hole (Schifter 1972). At a nest of Pied Barbets in Africa, two extra individuals of unknown relationship helped the parents to feed the young (van Someren 1956). Four adult White-eared Barbets fed four nestlings in the same cavity. This may have been a case of communal nesting, as the species lays only two or three eggs (Moreau and Moreau 1937).

In the related woodpeckers, juvenile helpers are also occasionally found. On one of the rare occasions when I have known the Golden-naped Woodpecker to raise a second brood, three young females of the first brood lodged with their parents in the nest hole while the later set of eggs was being incubated and the nestlings were growing up. Apparently, the young females did not share in incubation; but in their fourth month they brought their younger siblings a little food. Soon after this second brood flew, the branch in which they were reared fell and the parents proceeded to carve a new hole to serve as a dormitory for the family of seven. At least one female of the first brood assisted in this task, as did the young male of the second brood when he was about fifty-seven days old, three weeks out of the nest, and still receiving occasional meals from his parents (Skutch 1969*b*).

Swallows nesting in compact or loose colonies often help each other in various ways. In both the European and American races of the Barn Swallow, juveniles of the first brood have repeatedly been seen helping to feed nestlings of the second brood. Sometimes they even contribute to the nest for this brood, bringing straws and other materials only a week after their first flight. Juveniles of early broods are sometimes among the numerous House Martins who help their neighbors to build and to feed nestlings (Bent 1942, Witherby et al. 1938, Lind 1964). When fifty-four days old, a hand-reared female Purple Martin tried to brood nestlings and was soon bringing insects to them (Richmond 1953).

Precocious helpfulness has also been noticed among wrens, a family in which we earlier noticed numerous instances of adult interspecific helpers. In two different years, a juvenile Cactus Wren, slightly over two months old, fed fledglings of a later brood (Anderson and Anderson 1962). Young Southern House Wrens seem to help their parents rear later broods only when they are exceptionally precocious. In a family of these wrens that I watched carefully for two years, the parents seemed to be more vigorous, or more proficient, in the second year, when they started to nest unusually early. After their emergence from the nest, the fledglings of some broods were led by their parents to sleep in the gourd in an orange tree where they had grown up. Although young house wrens who lodge in the nest space of their parents are usually driven away about the time the following brood hatches, two members of the

first brood of the second year were sufficiently strong and persistent to resist eviction. Consequently, they passed the nights near the newly hatched young of their parents' second brood, and soon they were feeding the single nestling that remained from four eggs. They brought so much food that the parents now began to build a nest for a third brood—the only instance of overlapping broods that I have known in this wren.

After a while, the very precocious female helper, only seventy-three days old, tried to keep her mother out of the gourd, to which the latter still brought food in the intervals of building. This gave rise to the most stubborn conflict that I have witnessed among birds. After a day of recurrent fights, the pugnacious young female was defeated and driven away. Her more pacific brother continued to feed the nestling, who turned out to be another male. When fifty-four days old, this wren of the second brood began to bring insects to the third brood. In other years, I have watched in vain for young house wrens to help their parents with later broods (Skutch 1953*c*).

A captive Long-billed Marsh Wren, about a month old, took food from its human fosterer and passed it to still-younger marsh wrens that were being hand-reared with it. A few days later, it began to pick up worms and catch flying moths, some of which it gave to the other fledglings. In the evening, when the room was slowly darkened, it called with a subdued rapid twittering until the four younger wrens joined it in the empty sparrow's nest where they slept (Kale 1962).

Among mockingbirds, thrashers, and catbirds, juvenile helpers seem to be rare. On Indefatigable Island, Hatch (1966) saw a recently fledged Galápagos Mockingbird beg from an adult with negative results, sing briefly, and later feed the next brood of its parents in the nest.

In the thrush family, juvenile helpers have been repeatedly noticed among bluebirds. Five young Common Bluebirds of the first brood, all less than two months old, diligently attended four nestlings of the second brood, beginning when the latter were three days old. They brought food and cleaned the nest (Laskey 1939). Three young Mountain Bluebirds of the first brood fed their siblings of the second brood, and young Western Bluebirds are similarly helpful (Mills 1931, Finley 1907). Likewise, Wheatears of an early brood sometimes feed their siblings of a later brood (Nicholson 1930, Wynne-Edwards 1952).

An example of the many-sided helpfulness of wood-swallows, to which we shall presently return, is that of juveniles, who feed not only later broods of their own parents but also young of other families (Immelmann 1966). Likewise, recently fledged juveniles are among the helpers who attend callow young of the Superb Blue Wren, Variegated Wren, and other wren-warblers or fairy wrens of Australia. One Superb Blue Wren began this activity when fifty-eight days old (Cayley 1949, Rowley 1965). Among white-eyes, juveniles about two months old fed fledglings of their own kind placed in the same cage with them. When older, these white-eyes gave food to still-helpless House Finches and House Sparrows (Eddinger 1967, 1970). White-eyes are highly social birds, perching in contact both while they sleep and by day and often preening each other, an activity by no means confined to mates.

Another family in which helpers are frequent is the tanagers. At four nests of Golden-masked Tanagers in Panama and Costa Rica, I have watched young in juvenal or transitional plumage help mated pairs to feed their nestlings. At each nest, only a single juvenile was in attendance. One of these young helpers, about forty-six days old, at first brought food about as often as either of the parents; but, after the novelty of this adult occupation wore off, it came much less frequently. Moreover, it brought less food at a visit than the adults did. If the attendance of the parents had been as desultory as that of their young helper, the nestlings would have fared badly. As was earlier mentioned, juvenile Golden-masked Tanagers sometimes help a little while their parents build a later nest. Likewise, immature Red-throated Ant-Tanagers sometimes assist in building and more often they feed nestlings, probably later broods of their parents (Willis 1961).

Although the widespread, abundant finches have probably received more study than any other family of birds (except possibly those with species of economic importance) intraspecific helpers have not often been found among them, probably because of the strong territoriality of many species. An instance of a juvenile helper has been reported in the Cardinal, of which an individual about seventy-eight days old, probably a female, brought food to a late nest of her own species. As frequently happens with young helpers who have not yet outgrown their infantine ways, she begged for food from the nestlings' parents and was occasionally fed by their father. The female parent tried to drive her away; yet on one occasion, when they rested together on the nest's rim, she took food from the young assistant and passed it to a nestling (Brackbill 1944).

Doubtless young birds that have recently become independent of parental care would more often help their parents to raise later broods in the same season if the adults did not so frequently drive them away from their subsequent nest. Although I looked year after year for immature helpers in the Southern House Wren, I found them only in the year when the young developed so rapidly that they could resist expulsion from the parental territory. The Cactus Wrens that helped to feed younger siblings when slightly over two months old were also exceptional, as most juveniles are expelled from their parents' domain before this age. Probably juveniles rarely increase nesting efficiency. As purveyors of food to the nestlings they are often unreliable; they sometimes annoy the parents; and they increase activity that may attract the attention of predators.

Older Nonbreeding Intraspecific Helpers—Passerines

From the standpoint of ecology and the regulation of population, the most important class of helpers consists of birds that defer breeding for a year or more after they are fully grown and meanwhile remain closely associated with their elders, helping them to rear their broods much more efficiently than juveniles only a few months old could do. Some of these helpers are innubile or sexually immature; others are mature or might become so if they asserted their independence and won a territory and a mate; but, because of the difficulty of determining this point without dissection, we shall consider both of these classes together. We are only beginning to learn how widespread these older helpers are among the permanently resident birds of milder climates, but it is already becoming apparent that they are more frequent in some families than in others. Although among migratory or wandering birds older nonbreeding intraspecific helpers are less common, they have been found in a few species.

Let us begin our survey with the jays, apparently the first family in which the prevalence and importance of yearling helpers was recognized, and in which they are now known to occur in a number of species. The White-tipped Brown Jay is a large, noisy, gregarious corvid common along the Caribbean side of Middle America from southern Mexico to western Panama. The bills, legs, and toes of young birds are yellow and blacken with increasing age. The bills, especially, tend to become black in an irregular pattern that varies from individual to individual and, therefore, facilitates recognition, so that with careful watching one can learn how many attend a nest. Every nest that I watched in widely separated localities was attended by one or more of these helpers. One nest had five, who with the parents made seven birds caring for three nestlings. Since these jays are not known to raise more than a single brood of three in one season, some of these five helpers were evidently not offspring of the pair that they assisted—unless, perchance, some had remained with their parents for several years. The probability that some of the nests with helpers belonged to jays breeding for the first time also points to the conclusion that yearlings may assist pairs other than their parents. These helpers occasionally deliver a stick while the building female sits in her nest calling for them; they may feed her while she incubates, as does her mate; but their chief contribution is feeding the nestlings, which they do freely. They also guard the nest, vociferously protesting intrusions, often more vehemently than the parents (Skutch 1935, 1960, Selander 1959).

The White-throated Magpie-Jay, a big, blue-and-white corvid with a loosely tufted crest of recurved feathers that inhabits drier parts of Middle America, also has helpers. Observations on nests with young are lacking, but an incubating female was so well fed by three or more attendants that she did not forage for herself but devoted her brief absences from the eggs wholly to resting and preening (Skutch 1953a). The rare, elegant Tufted Jay of northwestern Mexico travels in flocks of from four to twelve or rarely more individuals, each of which includes a single breeding pair, nonbreeding two-year-olds, and yearlings distinguished by their lower crests, blue rather than white malar patches, and lack of a white spot above the eye. Each flock attended a single nest, to which its several members, including the yearlings, contributed more or less building material. Only the single mated female laid and incubated the three to five eggs in a set; but other members of the flock brought food to her while she sat, and later all joined in feeding the nestlings, as in the Brown Jay (Crossin 1967). The Bushy-crested Jay of northern Central America is a double-brooded species with both juvenile and older helpers. The older helpers simultaneously feed juveniles of the first brood and nestlings of the second brood, while the yellow-billed juveniles, themselves being fed, give food to nestlings. One family consisted of about eight adults and three juveniles, all feeding four nestlings (J. W. Hardy in litt.)

Another highly social species is the Mexican or Gray-breasted Jay of the southwestern United States and adjacent Mexico. Throughout the year, it lives in flocks of up to twenty or rarely more individuals, consisting of both old birds with wholly black bills and younger ones with particolored bills, like Brown Jays'. These flocks may include two or more breeding pairs, who build their nests in the flock territory, sometimes two nests in the same tree. The whole flock takes an interest in nest construction, and several of its members may help at one nest, either bringing or arranging materials. Often, however, one member of a flock steals material from another's nest for incorporation in its own—a common habit of birds that nest in colonies. Later, several members of the flock attend the young. Fledglings may receive three-quarters of their food from flock members other than their parents, and four were seen to feed the same fledgling within ten minutes (Gross 1949b, Hardy 1961, J. L. Brown 1970). In the related Scrub Jay, a nest with two young in Florida was attended by three adults, at least two of whom brooded; in Mexico up to eight individuals cared for three young (Grimes 1940, Wagner 1966).

In the highly gregarious Piñon Jay of the southwestern United States, sociality takes a somewhat different form. These jays nest colonially in traditional localities. Gray nonbreeding yearlings do not, as in Brown Jays and Tufted Jays, assist at the nests, but the blue breeding adults help each other to feed older nestlings and fledglings. A banded mother of four fed nestlings in a neighboring nest, and as many as seven adults were seen around some nests. Although fledglings were attended chiefly by their parents, they received some food from other adults, including parents feeding their own young and those that had lost their young. Thus Piñon Jays seem to be mutual helpers, although the exchange of courtesies by two recognizable pairs was not recorded (Balda and Bateman 1971).

All our observations on helpers at the nest among jays come from the more southerly parts of the North American continent, from Arizona and Florida to Costa Rica. Doubtless helpers remain to be discovered among closely related jays in South America; but they appear to occur, if at all, only sporadically among the jays, crows, and other corvids of higher northern latitudes, which have received most attention from bird watchers.

Another group of birds in which yearling or older nonbreeding helpers are widespread is *Campylorhynchus*, a genus of large wrens sometimes known as cactus wrens. They are so named because some species dwell among the cacti of arid regions in tropical and subtropical America, although others inhabit clearings in rain forest. The Banded-backed Wrens of southern Mexico and Central America live throughout the year in family groups of up to a dozen individuals, which at all seasons lodge together in bulky dormitory nests each with a side entrance. The same or similar nests are used for breeding. Only one female was seen to incubate the five white eggs, but nestlings are fed by nonbreeding helpers. At one nest, a helper, who brought more food to the young than their mother did, led them to sleep in a dormitory after their first days in the open. The Gray-barred Wren, Spotted Wren, and Giant Wren, all of Mexico, appear to have similar

cooperative habits but have been less carefully watched (Skutch 1960, Selander 1964a). In the multiple-brooded Cactus Wren of the southwestern United States, so well known through the researches of the Andersons (1962), only juvenile helpers were found.

Among the elegant little wren-warblers of Australia, helpers are widespread. These highly social birds perch by day and roost by night pressed together in rows, and all the flock members preen mutually. In addition to the juvenile helpers that we have already noticed, yearling and even older individuals assist mated pairs. The most carefully studied species is the Superb Blue Wren, of which Rowley (1957) learned the composition of eighteen color-banded groups. Twelve consisted of only the single breeding pair, four had a second male, one had both a second male and a second female, and one had two extra males and one extra female, making a total of five. All these birds were entering their second year, if not older, and all were in full breeding plumage. At each of the six nests with extra attendants, the helpers joined the parents in feeding the young, both before and after they fledged.

In this multiple-brooded species, the female parent, occupied with a new nest, may cease to feed her offspring after they have been out of the nest about ten days, leaving their care wholly to the father and his assistants. These attendants usually keep the fledglings at a distance from the nest, thereby reducing the risk of betraying its location. If a parent dies, helpers may rear the brood. Thirty-two adult birds in sixteen simple pairs reared twenty-four young to independence, while thirty-seven attendants in twelve breeding groups reared thirty-four young. Accordingly, the simple pairs produced only 0.75 independent young per attendant, whereas the families with helpers produced 0.92 young. In each category were thirty-two nests, which in the case of simple pairs yielded 0.75 self-supporting young per nest, while those attended by larger groups yielded 1.06 per nest. The helpers increased breeding efficiency (Rowley 1965).

Among the colorful tanagers of tropical America, older as well as juvenile intraspecific helpers are not infrequent. Rarely, a third Golden-masked Tanager or a third Speckled Tanager in full breeding plumage attends a nest with only the normal complement of two eggs. At one of the three known nests of the Plain-colored Tanager, four grown birds, all alike in plumage, fed two nestlings, often coming in a little flock. The only nest of the Dusky-faced Tanager of which we have information contained two nestlings that were fed by at least three grown birds, and possibly by all the members of the group of about seven that often passed by (Skutch 1961*b*, 1972). In Trinidad, groups of four or even five Turquoise Tanagers in adult plumage often attend young of a single brood (D. W. Snow and Collins 1962).

After a single male and female had built the only known nest of the Black-faced Grosbeak of Caribbean Central America, and the latter had begun to incubate the three eggs, they were joined by a third adult indistinguishable from them (the sexes being alike), who thereafter was a constant attendant and took a large share in feeding the nestlings. In this grosbeak, as in the Dusky-faced Tanager, the regular occurrence of helpers would be consistent with their highly social habits, even in the breeding season. The adult helpers at the nests of the Golden-masked and Speckled tanagers may have been males who could not find mates; those at the Plain-colored Tanagers' nest may have been a breeding pair who had lost their young; but the helpers of the Dusky-faced Tanagers and Black-faced Grosbeaks were more probably nonbreeding yearlings. These were all birds of tropical forests. Far away on the Arctic tundra, a pair of the weakly territorial Smith's Longspurs who helped to feed nestlings of another pair had apparently lost their own young in a storm (Skutch 1972, Jehl 1968*a*).

After they leave the nest, fledglings of the Large Gray Babbler of India join a band and are apparently fed and protected by the whole group. Although as a rule only the parents attend young still in the nest, a late brood of two nestlings was fed by five grown birds (Hutson 1947). In a number of other babbling-thrushes, a small flock attends the young, but relationship of the several members apparently has not been elucidated. Babblers rest by day and roost by night touching each other in rows, and they preen each other (Andrews and Naik 1965).

Older Nonbreeding Intraspecific Helpers—Nonpasserines

Helpers have been recorded less often among nonpasserine than among passerine birds, but possibly only because the former have been studied less. Among Chimney Swifts in the United States, extra birds frequently aid parents in incubation, brooding, and feeding the nestlings. Some nests have two helpers simultaneously, in addition to the parents. These assistants are of both sexes, but more often males than females. Some are yearlings, others older birds apparently in their last year of life, still others of intermediate age, who become parents in later years. The helpers sleep clinging to the wall of the chimney, tower, or air shaft near the nest, in company with the parent or parents not incubating or brooding. Sometimes they roost with the fledglings (Dexter 1952, Sherman 1952). Swifts and swallows are exceptional among long-distance migrants in frequently having helpers other than juveniles.

Among colonial-nesting bee-eaters of several species, helpers appear to be frequent. The Red-throated Bee-eater is gregarious throughout the year and breeds in yard-long burrows that are usually grouped in colonies of twenty to thirty active nests. Of ninety-six nests, sixty-nine, or about two-thirds, were occupied by mated pairs without assistants. The remaining third were attended by larger groups: twenty-three trios, three quartets, and one quintet consisting of a breeding pair with three helpers. Pairs normally hold together for life, and their young often remain with them into the following year. Most helpers are males, which are more numerous than females in the ratio of three to two, but about a third are females. Yearling helpers predominate, but one was two years old. In one instance, the same individual helped the same mated pair in two consecutive years. These bee-eaters start to breed when one or two years old, more often the former. The assistants help to dig the nesting burrows, share incubation of the set of one to four eggs, and feed the young before and after they leave the nest. When not sleeping in their burrows, bee-eaters of several kinds roost in trees pressed together in compact rows, like wren-warblers.

Sometimes male helpers of the Red-throated Bee-eater courtship feed the breeding female and may even copulate with her. Nevertheless, the helpers must be regarded as essentially innubile birds, for, if their impulse to breed were stronger, all the females and at least some of the males might find mates and establish families of their own. Female helpers apparently never laid an egg in the burrow that they attended. Nests with helpers produced an average of 2.7 young each, whereas those attended by the parents alone produced only 2.3. But pairs produced 1.1 fledged young per attendant; multiple units, 0.9 young—a difference too slight to be significant.

A nest of the White-collared Bee-eater at Lake Chad was attended by five helpers, who in the course of a day brought food to the nestlings 87 times, while the two parents did so 128 times. Two other nests of this species each had two male helpers. In a colony of about 25,000 Rosy Bee-eaters nesting in a sandbar in the middle of the Niger River, adult males were greatly in excess, and up to six grown birds fed a single brood. Helpers have also been noticed in two other species of bee-eaters (Fry 1972, 1967).

Although helpers appear to be lacking among the few New World kingfishers, they do occur in some of the more numerous Old World members of the family. In a population of piscivorous Pied King-fishers in Uganda that contained about twice as many males as females, up to four males sometimes aided a mated pair, feeding the female and her brood. The males also fed each other, thereby strengthening the bonds between the several members of the cooperating group (Douthwaite *in* Fry 1972). Among the terrestrial, insectivorous Kookaburra kingfishers in New South Wales, half the breeding pairs are solitary and half have one or more helpers, who are their offspring of the preceding year. They aid in territorial defense, perform about one-third of the incubation, and bring about two-thirds of the nestlings' meals. These helpers

are of both sexes and comprise about two-thirds of the total population, in which males and females are present in approximately equal numbers (Parry *in* Fry 1972). The related todies of the West Indies also have helpers at the nest (K. Kepler *in litt.*).

At three nests of White-fronted Nunbirds in the Caribbean lowlands of Costa Rica, we found four, four, and three grown birds bringing food to the nestlings. At the only nest of this sociable, insectivorous puffbird where the number of young could be learned without digging up and destroying the long burrow in the forest floor, there were three, fed by four adults. The helpers could not be distinguished from the parents by appearance or behavior, but they were apparently nonbreeding yearlings (Skutch 1972). In another family of piciform birds, the toucans, I once found five Collared Araçaris, evidently a mated pair with three helpers, feeding and sleeping with at least three nestlings, in an inaccessible hole 100 feet (30 meters) up in a great tree in a Panamanian forest (Skutch 1958*a*).

We have already noticed the occasional occurrence of juvenile helpers in the Golden-naped Woodpecker. In the Red-cockaded Woodpecker of the southeastern United States, family bonds are even more enduring. The young are sometimes fed by their parents until five or six months old and may remain with them into the following breeding season; but after they fledge each individual sleeps in a separate hole, instead of sharing a family dormitory, as do the Golden-napes. Helpers were present at two of the six nests that were frequently watched. At one of these nests, the helper was a yearling offspring of the parents. At the other nest, the helpers were two males and one female of unknown relationship to the parents. These assistants did not incubate, but they fed and brooded the nestlings, cleaned the nest, and were as active as the parents in its defense. One helper begged for, and received, food from the male parent. Nestlings attended by parents with helpers gained weight more rapidly than nestlings fed by their parents alone. At two nests with helpers, 2.0 young fledged per nest, whereas the average number of young that emerged from seven nests without helpers was only 1.4 (Ligon 1970).

Among Acorn Woodpeckers in California, more

than a single pair of birds may carve the nest hole and assist in incubating the eggs. At least five, and possibly more, brought food to a single brood of nestlings. In Costa Rica, four male Acorn Woodpeckers and one female took turns incubating in the same hole, changing over very frequently. At another nest, two males and one female incubated; at yet another nest, two males and two females fed feathered nestlings. Unfortunately, at none of these high, inaccessible nests could the number of eggs or young be learned, and without this information the interpretation of these situations is difficult. But I surmise that in each case a mated pair was assisted by yearling helpers, as were the Red-cockaded Woodpeckers. Although only a single parent slept in the nest with the eggs, I have found as many as five individuals lodging together in a high hole (Leach 1925, Skutch 1969*b*).

Excess Males as Helpers

In some of the foregoing accounts of intraspecific helpers, we noticed that, although the yearling or older assistants were of both sexes, they were more often males than females. This situation suggests that despite certain indications of breeding behavior, such as were observed among Red-throated Bee-eaters, many of the helpers are not physiologically and psychologically ready to reproduce. If they were, we would expect the female helpers to take partners among the more numerous males and start their own nests, leaving only the excess males to assist mated pairs. In certain other species, however, all the helpers are males, strongly suggesting that, especially in the case of very small birds that we expect to mature rapidly, they are sexually mature individuals who attach themselves to mated pairs only because they cannot find mates.

The causes of the preponderance of males in many species of birds are obscure. After fertilization, eggs should be male and female in approximately equal numbers. Females who perform all, or most, of nest building and incubation are probably under greater strain and exposed to greater hazards than the males of their species, yet the latter are more numerous in certain species in which the shares of the sexes in nesting activities are approximately equal. Perhaps the frequency of male helpers is a cause, rather than the effect, of their numerical preponderance in the population as a whole. By remaining longer with their parents in familiar surroundings, the males avoid some of the hazards that confront females who at an early age leave the parental domain to seek mates.

Among Arctic Terns, some of the younger males, unable to breed because of a shortage of females, attached themselves to nesting pairs and helped feed their chicks (J. M. Cullen 1957). A similar situation exists in a bird of quite different ecology, the Pygmy Nuthatch of the western United States. Because males are much more numerous than females, many cannot find partners, and these bachelors serve mated pairs. Of thirty-six nests in California, eight were attended by three individuals. The helper was invariably a male, usually a yearling but sometimes older, who was not actually mated to the female. He assisted in nest construction, feeding the female while she incubated or brooded, feeding the nestlings, and cleaning the nest. He continued to feed the young birds after their emergence from the nest, and at all stages of the breeding cycle he slept in the nest hole with the other members of the family. The group continued to roost together over the winter.

When two families foraged together, an adult of one sometimes fed fledglings of the other. Up to six individuals excavated a nest cavity that was later occupied by a single pair. At one nest, nine nestlings, not an abnormally large brood, were fed by four adults. Likewise, in the Brown-headed Nuthatch of the southeastern United States, males are in excess and when unable to find partners may assist mated pairs in all nest activities; but they were not found sleeping in the nest cavity with the parents and their eggs or nestlings. In one nest, seven young were attended by four adults, who were evidently two pairs (R. A. Norris 1958).

In the Black-eared Bushtit of the Guatemalan mountains, adult males have black faces and dark eyes, while adult females have gray faces and light eyes. In the breeding season, black-faced birds are much more numerous than gray-faced birds, and they attach themselves to mated pairs that have already finished their exquisite, lichen-encrusted, pensile pouches. At times, while incubation is in progress, these supernumerary birds bring downy material to the nest, but more often they help to attend the nestlings, feeding and even brooding them. Three nests had, respectively, one, one, and three of these black-faced helpers, most of whom passed the chilly nights of the high mountains in the cozy pouch along with the two parents and the four black-faced nestlings that each cradled. The fact that all twelve nestlings that I saw were black-faced makes the interpretation of my observations somewhat uncertain. It is now known (as it was not when I studied these birds) that in certain races of bushtits juveniles of both sexes resemble the adult male rather than the adult female—a situation extremely rare among those passerine birds in which the sexes are differently colored. One must conclude either that all five of the helpers at the nests I watched were males or that the females of these diminutive birds do not assume the adult plumage until their second year, which seems improbable (Skutch 1960). At nests of the more northerly Common Bushtit, a third attendant sometimes helps to incubate and to feed and brood the young (Addicott 1938).

Among Blue-winged Warblers, and hybrids between this species and the Golden-winged Warbler, males also appear to be present in the breeding population in greater numbers than females, and they help to feed the nestlings of established pairs, thereby sometimes incurring the enmity of the nestlings' father. After the young of one brood fledge, diligent bachelor Blue-wings may perform the same service for another brood belonging to a different pair (Short 1964). With this exception, intraspecific helpers appear to be rare among wood warblers; but a second male, who seemed abnormal, fed an incubating female Kentucky Warbler, although he was occasionally chased by another male who sang better and appeared to be her mate (De Garis 1936).

The only nonparasitic cowbird, the Bay-winged of Argentina, rears its young in nests built by other birds, such as the Leñatero or Firewood-gatherer and the Great Kiskadee. Here the nestlings are often fed and protected by helpers, who are usually, if not always, males, the more numerous sex. After the

young fly, they are usually attended by helpers as well as their parents. After feeding the fledglings of one breeding pair, the helpers may turn their attention to those of a later brood of another pair. After his offspring become independent, a male sometimes feeds young not his own (Fraga 1972).

In the Tasmanian Native Hen, a large, flightless gallinule, the excess males behave somewhat differently, forming polyandrous groups rather than simply serving as bachelor helpers. About half the females have only a single mate; the rest have two or three "husbands," who may stay with them as long as they live and help to raise the young. The polyandrous families breed more successfully than the simple pairs, raising about 61 percent of their young, while the simple pairs raised only 45 percent (Ridpath 1964, Lack 1968). Polyandrous males do not properly fall under the heading of "helpers," but, because of a certain similarity in the arrangements of these gallinules to those that we have been considering, they seem worth mentioning here.

Breeding Helpers

Among breeding helpers we may include not only birds actually engaged in rearing their own progeny but also those that, after losing eggs or young, turn their still-unexhausted parental impulses to the service of more fortunate neighboring pairs, becoming unilateral helpers. Occasionally, a parent bringing nourishment to its own young passes close to other young and responds to their appeal for food in a casual helpful act that may never be repeated. When highly social birds nest close together, they may more or less frequently feed each other's young, becoming mutual helpers. Such mutual helpfulness is carried to the extreme in communal nesters, which lay their eggs in a common heap and rear their progeny as a single family, making no distinction between mine and thine.

When Wood Pigeons coming to feed their nestlings find a begging juvenile from another family perching nearby, they may give some food to it after they have fed their own young. One flying juvenile was found at the nest of its benefactors on four consecutive days. Not all parents are so complaisant; some attack and drive away the intruding young (Murton and Isaacson 1962). Both nestling and fledgling White-eared Hummingbirds are sometimes fed by strange females, but, if the mother finds these helpers, she chases them away (Wagner 1959). We hardly expect such helpfulness among individualistic hummingbirds. However, when sixteen-day-old nestlings of the Planalto Hermit in an aviary lost their mother, another female took charge of the orphans as though they were her own (Ruschi 1949).

At the snug, feather-lined purses of Long-tailed Tits, helpers are frequent. Some nests with more than two attendants hold unusually large sets of eggs that were probably laid by two females; but nests with normal sets often have one, two, or even more extra attendants, who in some cases appear to be breeding birds that volunteered assistance at a neighboring nest after having lost their own. Sometimes these helpers are present during incubation, but more often they bring food to the numerous young. The absence of territorial defense in this titmouse makes it easy for any unemployed bird to attach itself to the nest of a breeding pair (Lack and Lack 1958). A pair of Brown-headed Nuthatches who lost their nest helped to feed the nestlings of a neighboring pair (R. A. Norris 1958). Perhaps more often than they attend a neighbor's young of their own kind, bereaved parents among highly territorial species turn their attention to the offspring of some other species nesting nearby, as related in the preceding chapter.

Just as parents that have lost their own young may serve the young of a neighbor of the same species, so young separated from their parents may be adopted by a neighboring pair, to be raised along with their own offspring, as has been recorded of the European Coot, Great Tit, European Blackbird, American Robin, European Robin, and other birds (Alley and Boyd 1950, Skutch 1961*b*). Among precocial birds, lost or orphaned young are often guided and perhaps brooded, although not fed, by the parent of another family. Male Baird's Sandpipers and Stilt Sandpipers sometimes lead young other than their own, and both sexes of the Snowy or Kentish Plover may do so (Parmelee et al. 1968, Walters 1959). One of the two guardians of a flock

of Ostrich chicks may be a foster parent rather than a real parent (Sauer and Sauer 1966). Large flotillas of ducklings of various parentage are sometimes guided by a single duck (p. 345).

Among birds that nest in crowded colonies, we find extremes of attachment to neighbors' chicks, along with extremes of hostility. To the latter we shall presently return. Among the cherishing colonial breeders are the murres or guillemots, both the Common and the Thick-billed species. Although each female lays a single egg, from two to four chicks of five days or older may be sheltered under an adult's wings. Bereaved adults, and even parents with young, feed chicks not their own, although the parents of these chicks may rebuff such gratuitous attentions. As was mentioned earlier (p. 302), on leaving its natal ledge the young murre may be escorted out to sea by some adult other than its parent (Perry 1946, Tuck 1960).

Among King Penguins, which nest on islands in cold southern waters, helpful neighbors often save the lives even of chicks with attentive parents. On a chill, windy day, a young chick's chances of survival decrease rapidly after five minutes of exposure. Such a chick, temporarily dropped by its parent, is often picked up by a chickless adult, who rolls it onto its feet and broods it for a short while. If several such adults are present, they may pass the downy chick from one to another, each warming it briefly, and so keeping it alive until one of its parents claims it. None broods it for more than two or three minutes. Large chicks may foster tiny chicks with similar brevity. "The effectiveness of this behaviour rests in its apparent inefficiency; a temporary guardian which held a foundling chick for a number of hours would reduce the possibilities of its ever being recovered by the parent, and would find itself in charge of a chick which it could not maintain. Similar behaviour is seen in the Emperor Penguin." Three adult King Penguins may share the incubation of the single egg and feed the single chick (Stonehouse 1960).

A probable example of mutual helpers is the Piñon Jay, in which, as earlier mentioned (p. 366), parents feed young not their own on a rather large scale. In the White-breasted Wood-Swallow of Australia, the fledglings of several families rest in the same tree, where all the parents feed them in common, the young of other pairs along with their own. The many-sided helpfulness of this and related species is impressive. Juveniles feed not only nestlings of their parents' next brood but also young of other families. Up to five individuals may attend one nest. Moreover, adults feed not only their own mates but also other adults. Pairs of Black-faced Wood-Swallows feed by turns the nestlings in neighboring nests, six sometimes bringing meals to a single brood. These birds also nourish ailing companions (p. 353). Three nestlings of the Papuan Wood-Swallow were fed by four or five adults. The highly social wood-swallows feed and bathe together, preen each other, and sleep in tight clusters (Immelmann 1966, Gilliard 1958).

The helmet-shrikes of Africa are also extremely social birds with interesting modes of cooperation of which, unhappily, we lack detailed studies. Several nests may be built close together, and sometimes two females lay and incubate in the same one. A small flock may feed and defend the young (Gilliard 1958, Ames 1963).

The Bell Miner, one of the honey-eaters so numerous in Australia, nests in colonies that are divided into a number of groups, each containing up to a dozen individuals. A single female builds the nest and incubates the one to three eggs, but fledglings are fed not only by members of their own group but also by those of neighboring groups (Swainson 1970). In another Australian honey-eater, the Noisy Miner, several adults feed the young, both while in the nest and after leaving it. One nest had at least eight visitors in addition to the parents, who brought most of the nestlings' food (Dow 1970). Two pairs of adult White-naped Honey-eaters fed one brood of young, and, in another instance, two obvious sets of eggs were found in one nest (Mathews 1924).

Among mutual helpers we may include the builders of apartment nests, the Monk Parakeets, Palm-Chats, and Sociable Weavers, whose huge, many-chambered nests were briefly described in chapter 7. It would not be surprising if these very social birds gave more or less care to their neighbors' eggs and young; but their high, well-enclosed nests and

the confusion created by their multitudes make the study of intimate details of their domestic life peculiarly difficult, and we lack information. Another species in which pairs help each other to prepare their nests is the Sand Martin or Bank Swallow. In England, although apparently not in North America, the excavation of their burrows is a communal activity, the climax of an elaborate aerial display, in which many martins participate. The number of holes is at first approximately the same as the number of individuals present. At one display, from three to six birds may be active at a single burrow, while, at the next display, excavation is carried on at quite different holes. The same swallow may move from hole to hole, digging at several in succession. As far as we know, after the burrows are finished each pair of Sand Martins confines its attention to its own nest and young (Hickling 1959). Similarly, colonial House Martins help each other to build their nests of clay. After four to fourteen individuals have completed the walls of one structure, they proceed to prepare a nest for another pair. The members of a colony often feed each other's young and help lure them from the nest (Lind 1964). When parent Purple Martins were carried away for homing experiments, neighbors of both sexes, but more often females, fed their neglected nestlings (Southern 1968).

The climax of mutual helpfulness is found in the communal nesters, of which the tropical American anis, lanky black cuckoos with high-arched bills, are the best-known examples. About half the nests of Groove-billed Anis that I have found in Central America belonged to single pairs, the remainder to two, three, or, rarely, more pairs. The cooperating pairs join in building their common nest, a bulky structure of coarse sticks, lined with green leaves that are brought daily until the young hatch (plate 100). Each female lays about four eggs; I have found up to twelve in a nest, while a compound set of fifteen was reliably reported. All the cooperating birds of both sexes take turns incubating, only one at a time, as the ani in charge leaves when another comes to replace it (plate 101). A single male incubates through the night. The nestlings, hatched quite naked and developing with surprising rapidity, are fed, brooded, and vigorously defended by all the parents, who evidently cannot distinguish their own from those of their coworkers. The Smooth-billed Ani closely resembles the Groove-bill in its domestic arrangements, but the cooperating group may be larger, containing up to five females. These two species further show their high sociality by perching by day and sleeping at night pressed together in compact rows, and by mutual preening. The Greater Ani and Guira Cuckoo likewise nest communally, at least at times, but they have been less carefully studied (Skutch 1959*b*, 1966*b*; D. E. Davis 1940*a, b*, 1942).

Another communal nester is the White-winged Chough, a black, crow-sized mudnest builder of eastern Australia. It lives in closely knit groups of two to twenty, usually about eight, individuals, consisting of parents with their young of both sexes, who remain with them for several years. All mem-

PLATE 101. Male Groove-billed Ani, marked about the head with white paint for identification, incubating four eggs.

PLATE 100. Communal nest of the Groove-billed Ani, with eight eggs laid by two females.

bers of the group help to build the massive nest of mud in a tree, but yearlings do so inefficiently. The original female parent and her older daughters lay their eggs together, usually two for each female; thus, the nest may contain from two to nine eggs, according to the size of the family. All members of the group feed the young, of which rarely more than four are raised, probably because the nest is too small to hold more, and some are trampled at the frequent change-overs by brooding individuals. After the young of the first brood leave the nest, they are fed only by the nonbreeding helpers, leaving the parents free to start another brood. Two broods are reared in a season by pairs with helpers but only one by pairs without assistants. The

Apostlebirds of the same family and region, so called because they are often found in families of about twelve, have similar communal nesting habits (Thomson 1964, Lack 1968).

Hoatzins, large, rather awkward birds that dwell along the banks of lowland rivers in tropical South America, also appear to nest communally. From two to six birds join in building a crude nest of sticks on a branch overhanging the water; they then cooperate in incubating the two to five eggs for twenty-eight days, in rearing the young, and in defending them with admirable zeal (Grimmer 1962). Another large bird that often nests communally is the Magpie-Goose of Australia. Many males, probably more than half, mate with two females, both of whom lay in the same nest. Sometimes a trio of females nest together (Frith and Davies 1961). Cooperative nesting, of a quite different sort, is also practiced by certain megapodes, particularly the Common Scrub-Fowl. Several pairs often make and attend the great mound of decaying vegetation in which the eggs are incubated by heat of fermentation (Frith 1956).

In addition to habitual mutual helpers, birds of many kinds may become mutual helpers more or less accidentally. Indeed, as we earlier noticed, birds of different species or even families sometimes cooperate in rearing a mixed brood. Among intraspecific mutual helpers that departed from the usual habit of their kind were two Wood Ducks that shared the same box and warmed each other's eggs, two Blue Tanagers mated to the same male who laid their eggs in the same nest and incubated by turns, two female American Redstarts who attended a nest with a double set of eggs, and two female Reed Buntings who shared a nest (Bellrose 1943, Skutch 1972, Bent 1953, B. D. Bell and Hornby 1969). Although Cardinals are zealously territorial birds, in at least three recorded instances two females laid and incubated in the same nest, sitting side by side but often facing in opposite directions. In two of these joint nestings, some of the eggs hatched, and both females fed the young (Hawksley and McCormack 1951, O. O. Rice 1969). It is significant that anis, with a long racial experience in communal nesting, regularly incubate even large sets of eggs by sitting one at a time, while birds that

only rarely share a nest with another often sit side by side, sometimes with a partner of a different species. This seems to be a less successful method.

The Advantages of Helping

From the foregoing survey, it is evident that intraspecific helpers are of frequent occurrence in species of the most diverse taxonomic position, habitats, feeding methods, and nesting habits. The only thing they seem to have in common is a greater measure of sociality than those many strictly territorial birds that, in the breeding season, resist the intrusion of any self-supporting individual of their kind other than their mates. As we have noticed in passing, many of these helpful species are "contact birds" that do not insist upon maintaining an "individual distance," but rest by day pressed side by side and at night roost in a compact row on a branch, or else sleep packed together in a nest or hole. Often they show their mutual friendliness by allopreening that is by no means confined to mates, or by feeding companions other than their nuptial partners.

A substantial number of helpers, including certain swifts, bee-eaters, swallows, and wood-swallows, are aerial insect catchers, which, as Lack (1968) showed, often nest in colonies. Numerous others, including bushtits, nuthatches, wrens, and wren-warblers, glean insects from foliage or bark. Still others, among which are certain tanagers and grosbeaks, subsist on a mixed diet of insects and fruits; while the fare of the numerous jays with helpers is still more varied. Among marsh and water birds that often have helpers are the Magpie-Goose, Common Gallinule, Pied Kingfisher, and Kookaburra. Marine birds with helpers include murres, certain terns, and King Penguins.

In some tropical localities, helpers are surprisingly numerous. In a small area of high rain forest and adjoining clearings in the wet Caribbean lowlands of Costa Rica, we found, in a single season, helpers at nests of three species not previously known to have them: the White-fronted Nunbird, Black-faced Grosbeak, and Dusky-faced Tanager. In six other species that occurred there, I had earlier discovered helpers in other parts of their range. These included a toucan, an ani, a jay, a wren, and two tanagers. To these may be added a cotinga abundant in this forest, the Purple-throated Fruitcrow, which has since been found with helpers in Guyana (D. W. Snow 1971).

Whenever the social habits of a species are such as to encourage helping, we may expect to find several classes of helpers. Among anis and wood-swallows, we have noticed both mutual helpers and juvenile helpers. Among swallows, wren-warblers, tanagers, and penguins, both adult and immature birds feed or foster young not their own. White-winged Choughs have yearling and mutual helpers. In nuthatches with yearling male helpers, two pairs sometimes nest together as mutual helpers. However, immature helpers seem rarely to occur in species with nonbreeding yearling helpers, because these species are typically single-brooded, as in bee-eaters, Chimney Swift, White-fronted Nunbird, Red-cockaded Woodpecker, Brown Jay, Banded-backed Wren, Black-eared Bushtit, Pygmy Nuthatch, Brown-headed Nuthatch, and others.

The reason for the frequent association of yearling (or older) nonbreeding helpers with single-broodedness seems clear: a species that needs to raise two or more broods in a year in order to maintain its population could ill afford the loss of reproductive potential that results when breeding is delayed for nearly two years. Among the few species known to have both juvenile and yearling helpers are the Bushy-crested Jay in Nicaragua and the Superb Blue Wren in Australia. No explanation of this situation in the jay is presently available; but the occurrence side by side of both juvenile and yearling helpers in the blue wren is apparently associated with the peculiar climate of the island-continent, where in seasons of favorable rainfall many species must propagate rapidly to compensate for long periods of drought.

The widespread occurrence of intraspecific helpers, and perhaps even more that of interspecific helpers who lavish attention on the progeny of an alien species, points to the conclusion that birds have a very strong impulse to cherish young, which often manifests itself even before they are capable of reproduction. Since they live in a harshly competitive world, this impulse will doubtless be favored

by natural selection if it promotes the survival of the species, but suppressed if it imperils adequate reproduction. What conditions favor the helping habit? What are its advantages and disadvantages?

Looking to their parents for food, guidance, and protection, young birds of many kinds prefer to remain with them as long as they are permitted to do so. They evidently shrink from facing a perilous world alone, and often they must be forcibly driven away. If they can stay with their parents until the latter nest again, close association with the nest, eggs, and young may arouse dormant parental impulses and cause them to help. As was earlier told, before the next brood of Southern House Wrens hatches, the juveniles of the preceding brood are nearly always evicted from the nesting space where they have lodged; but when, on rare occasions, they can resist expulsion, they help to feed their younger siblings. Accordingly, the sedentary habit, which permits the persistence of family groups, favors helpfulness; migration or irregular wandering, which usually separates the female from her mate and parents from young, destroys the situations in which helpers most readily arise. The intraspecific helpers most often found among migratory birds are juveniles who attend later broods in the same season, as occurs in swallows and bluebirds, or else they are parents who have lost their broods. Helpers of all kinds are most frequent among the sedentary birds of milder climates, and it is chiefly in such climates that they become an important feature of the breeding system.

One count against helpers is that, by increasing the amount of activity at a nest, they increase the probability that it will be noticed by predators. This disadvantage of having helpers hardly applies to colonial birds, whose clustered nests are, in any case, so public that they must be safeguarded by inaccessibility. At isolated nests, the greater probability of detection by predators may be more or less offset by the likelihood that, with more pairs of watchful eyes and keen ears, the approaching enemy will be detected sooner, and perhaps circumvented. Four or five defenders of a nest might drive away a hungry snake, hawk, or small mammal that would be undeterred by a single pair. Yet even nests defended by this number of bold jays are sometimes pillaged.

Although the advantages that parents derive from their assistants are sometimes dubious, we can hardly doubt that juveniles and yearlings benefit immensely by helping their parents. They remain on the familiar parental domain, under parental tutelage, instead of wandering unguided through a perilous world until they can win a territory of their own. They learn more from their parents before they are cast on their own resources. Even before they have nests of their own, they gain that experience in nest attendance which, as we have noticed (p. 278), increases breeding success. Moreover, by keeping their offspring with them until they are quite mature, the parents increase their own reproductive success. Although, because it is usually the only practicable method, we commonly gauge nesting success by the percentage of eggs that produce fledged young, a more realistic measure would be based upon the number of young that reach reproductive maturity. This is what counts in the perpetuation of a lineage and gives any stock an evolutionary advantage. The increased life expectancy that helpers gain by continuing to associate with their elders may, in some cases, be a greater contribution to the reproductive success of their parents than the food that the helpers bring to their younger siblings.

From the standpoint of population control, the most important category of helpers consists of those that defer breeding for a year or more while they assist at the nests of older birds, at least when these helpers include a substantial number of females and not only excess males who, in any case, could not find nuptial partners. Although, as we have seen in the cases of the Red-throated Bee-eater and Superb Blue Wren, helpers may directly increase nesting success, they seem rarely to augment the population as rapidly as they might do if they started to breed at an earlier age. This is certainly true when the number of attendants at a nest of a single-brooded species, including the parents and helpers of both sexes, exceeds the number of nestlings that they feed, as often happens among jays.

In a successful species that rather fully exploits its range or ecological niche, the deferred breeding of helpers, as of marine birds, aids in keeping the population in balance with available resources. If, as is highly probable, young birds that remain long with their parents have a greater life expectancy than those that are cast off when they can barely support themselves, species with yearling helpers can preserve a satisfactory abundance by hatching fewer eggs than do less cooperative birds. Their population is maintained with a smaller annual turnover of individuals. Not those birds that raise the greatest number of young, but those that maintain their population with the smallest reproductive effort, are the truly efficient species. They enjoy the longest, and doubtless most satisfying, lives.

On the other hand, the system of helpers may be disadvantageous to a species threatened with extinction. The stubborn persistence of delayed breeding, which may once have been favorable, by a "relict" species with a shrinking range, such as the Tufted Jay, may be one of the causes of its decline. It needs a higher reproductive potential that probably could be gained by earlier breeding in single pairs. With a similar breeding system, however, the aggressive, adaptable Brown Jay has been expanding its area as the rain forest gives way to cleared lands.

As a method of regulating population, the system of helpers is evidently an alternative, or in some cases a supplement, to territoriality; it may prove to be a more sensitive method. Although some species with helpers, such as Smooth-billed Anis, Kookaburras, and Superb Blue Wrens, defend group territories, the presence of intraspecific helpers always signifies some relaxation of strict territoriality, which repulses from the defended area all independent individuals except the mate. As in other birds that defer breeding until over a year old, species with helpers may start to reproduce at varying ages, as in the Red-throated Bee-eater and Superb Blue Wren. At a low population density, they may start to breed earlier than at a high density. Thus the helpers form a reserve of experienced individuals ready to start efficient reproduction if the population is somehow reduced below the usual level, but perhaps refraining from reproduction if the population is already dense and unoccupied territories difficult to find. We should consider the possibility that innubile helpers remain sexually retarded by their close association with older, dominant individuals, often their parents, and that they would have become sexually mature if this influence had been absent. Proof of this appears to be lacking; but Rowley (1965) showed that dominant males of the Superb Blue Wren molt into nuptial plumage a month or more earlier than their subordinates.

Whatever the methods by which breeding adults may maintain dominance over their helpers, it is certainly very subtle. One of the outstanding, and most pleasing, features of birds that frequently have helpers is the harmony that prevails within the group, the almost complete lack of agonistic behavior. I have spent many hours watching nests of anis, Brown Jays, Banded-backed Wrens, Black-eared Bushtits, White-fronted Nunbirds, Collared Araçaris, Golden-masked Tanagers, Black-faced Grosbeaks, and other birds, at which from three to seven individuals were in attendance, without, as far as I can recall, ever seeing one member of the group peck, threaten, or otherwise behave aggressively toward another. All were perfectly friendly.

The Antithesis of Helpers

In this and the preceding chapters, I have emphasized the widespread tendency of birds to minister to young not their own, often those of different species. Some years ago, I enjoyed an amusing demonstration of the strength of this tendency. Two female Buff-throated Saltators, territorial finches, had somehow built their open cups in coffee bushes only eight feet apart. The dominant female tried hard to keep her more timid neighbor away. Yet, instead of destroying her rival's nestling, which would effectively have caused the departure of its parent, she from time to time fed it, and once, apparently absent-mindedly, she even brooded it briefly!

It would be wrong to leave the reader with the impression that this cherishing tendency is universal among birds, and accordingly it seems proper to close this chapter by calling attention to the exactly contrary behavior of certain species. We might expect to find tenderness toward neighbors' young most highly developed among birds that nest in compact colonies, but, surprisingly, this is just where

it is often most glaringly lacking, especially in certain marine birds. After feeding its own chick, a parent Laysan or Black-footed Albatross often rushes to the nearest unattended nest and attacks the young occupant, sometimes causing its death (Rice and Kenyon 1962). On Ascension Island, small chicks of the Frigatebird were invariably eaten by neighbors or interlopers of their own kind if left exposed for even a few seconds (Stonehouse and Stonehouse 1963).

North Atlantic Gannets attack and sometimes kill chicks in neighboring nests, even when they do not trespass (Nelson 1964). A wandering chick of the Wideawake or Sooty Tern that comes within range of an incubating or brooding neighbor, or any adult other than its parent, is likely to be pecked so mercilessly and repeatedly that it may succumb (Ashmole 1963a). Although, as we have seen, parent Common Murres frequently feed and brood their neighbors' chicks, they sometimes peck the chicks savagely (Perry 1946). Even more surprisingly, chicks of both the American and European Avocet that become separated from their parents are often attacked by other adults, who strike these unfortunates with their feet or pick them up in their bills and shake them violently (Gibson 1971). Yet parent avocets sometimes brood recently hatched chicks of their neighbors (Selous 1927). Anything that stirs up a colony of sea birds, such as the passage of a man through it, causes more chicks to wander from their nests and meet an untimely end.

Some of this harsh treatment of inoffensive young is evidently adaptive. In colonies where hundreds or many thousands of sea birds nest close together on the ground, it is all too easy for a semialtricial chick to wander from its own nest and end its journey in another nest. Some disciplining may be necessary to prevent this occurrence, which can have unfortunate consequences. Many marine birds, which bring food from afar, can supply only a single chick; in times of scarcity, they may fail even to do this adequately. If a second chick manages to push beneath a parent brooding its own chick and is adopted, either the interloper or the original occupant of the nest is likely to starve. Sometimes it is the latter. When a chick of the South Polar Skua is driven from its birthplace by its belligerent elder sibling (p. 289) and wanders into the territory of another pair of skuas, it may be accepted by them, whether they have eggs or young. This charitable act may result in the abandonment of the fosterers' own eggs or the death of their chicks. Similarly, a pair of Sabine's Gulls deserted their eggs after adopting a chick that came to them from a nearby nest.

In such circumstances, the parent who repels, rather than accepts, an alien chick is likely to leave more progeny of its own, with the consequence that the hostile attitude toward neighbors' offspring will be promoted by natural selection. Nevertheless, some of the examples just given make it clear that certain of these colonial birds are often gratuitously brutal. One of the reasons why passerine birds have covered the earth is the perfection of their parental instincts, which ensure not only the best care of their own offspring but also tenderness toward their neighbors' young.

CHAPTER 30

The Nest as a Dormitory

Hitherto we have considered nests merely as receptacles for eggs and young birds. The open structures built by the majority of birds do not correspond to men's houses, because they lack a roof, a deficiency that is made good by the brooding parent itself. Not the nest alone, but the nest plus the parent who forms the roof make an adequate shelter for tender young. As a rule, the adult itself gains nothing by sleeping in such a nest, and after the nestlings no longer need protection, it goes to roost elsewhere.

Many birds, however, build a covered nest, or else they lay their eggs in a hole in a tree, a burrow in the earth, a cranny in a cliff, or in some other situation that affords shelter for the parents no less than for the young. In such an enclosure, which may or may not be lined for the reception of the eggs, the parent's presence may be necessary to warm the tender nestlings but not to shed the rain from them. The parent itself is shielded from rain and wind, and it may also be safer from prowling nocturnal enemies than if it slept in the open. Appreciating these advantages, the parent may continue to pass the night with the nestlings after they no longer require warming. It may return to sleep in the nest after their departure, and sometimes it leads them back to this shelter. It may even begin to lodge in its nest before it has laid a single egg there. Thus the

nest may come to serve as a dormitory. In the present chapter, we shall attempt to trace the steps by which the habit of sleeping in dormitories probably arose, to survey the various sleeping arrangements that dormitory users have evolved, and to assess the advantages and disadvantages of the dormitory habit.

Leading their fledglings to sleep in a snug nest is one of the last refinements of parental care in birds. The continued use of the same dormitory is one of the bonds that cement family ties and lead to modes of cooperation that we surveyed in the preceding chapter. No study of the relations between parents and young would be complete without paying attention to this phase of avian life.

Ways of Sleeping

The sleeping habits of birds is a subject that has hardly received the attention it merits when we recall that birds spend almost half their lives in sleep, or at least resting quietly in their roosts. This is most obviously true of the purely diurnal birds resident in equatorial regions, where day and night are of approximately equal length. At high latitudes in summer, the daily period of activity of birds may be considerably longer than that of sleep. But if these birds remain in the same locality throughout the year, they will in winter have longer nights than days. The long winter nights compensate for the short summer nights, so that in this case, too, the birds spend about half their lives in their sleeping places. Migrants like the Arctic Tern, American Golden Plover, and Bobolink, which breed at high or middle latitudes in one hemisphere and then pass the nonbreeding season well beyond the equator in the other hemisphere, enjoy long summer days through much of the year and may devote a larger share of their lives to activity. But even in the continuous daylight of the Arctic midsummer, birds take a few hours of repose in each twenty-four–hour period. All animals appear to require sleep, not excepting fishes, insects, and other creatures that lack lids for closing their eyes.

Considering how much time they devote to sleep, it is surprising how few birds take the trouble to prepare comfortable shelters for themselves, which, with their skill in nest building, they could do if so inclined. In wooded regions, most birds roost amid foliage. Many sleep singly, perhaps with the mate not far away, if, like numerous tropical species, they remain paired throughout the year. Among the American flycatchers that live about my house, the Tropical Kingbird, Boat-billed Flycatcher, Gray-capped Flycatcher, and Yellow-bellied Elaenia follow this method. I have sometimes found a male and female of these species roosting a yard or two apart, but never in contact. Young, unmated flycatchers sometimes press close together in a row, perhaps with one of their parents, and make a charming sight as they settle down to rest on some slender twig in the evening. I often find Blue Tanagers sleeping amidst the dense foliage of an orange tree, the male and female usually not far apart, and this, as far as I can learn, is general in the tanager family. Smaller tanagers sometimes roost on a large, horizontal thorn of an orange tree. The local finches and most other birds are more secretive, and I have rarely discovered just how they sleep.

In some families, mates sleep on perches in contact with each other. This method of roosting is followed by the Wren-Tit in California, and by a number of African species of white-eyes or zosterops. Even when roosting in flocks, mated Golden-fronted White-eyes rest side by side; when they have fledglings, the whole family presses together on the perch, with a parent at either end of the row (Erickson 1938, Skead and Ranger 1958, van Someren 1956). Similarly, mated parrots, such as the big Yellow-headed Parrot and smaller Orange-fronted Parakeet, sleep touching each other, but at some distance from the other pairs that roost in the same tree. One night I found two wintering Tennessee Warblers pressed close together in a cashew tree, although they were probably not mated at this season. Each rested on one foot, with the other foot drawn up into its plumage. The two diminutive birds supported each other.

As we noticed in chapter 29, some highly social birds, including anis, bee-eaters, wren-warblers, and wood-swallows, roost in compact rows. Others cluster still more tightly together, in a ball rather than a row. In tropical Africa, bunches of six or more White-cheeked Colies cling to the topmost shoot of the tree or shrub that is their nightly roost (van Someren 1956). Ten or more White-breasted Wood-Swallows collect in a ball-like mass at the forest's edge. If disturbed, they explode in all directions, like a feathery bomb, probably frightening whatever nocturnal prowler aroused them (Gilliard 1956). In England, nine to twelve Long-tailed Tits roost pressed together in a row, or, in colder weather, they seek the shelter of a thicket, where they cuddle together in a tight ball, their tails sticking out on all sides (Lack and Lack 1958).

Gregarious birds often congregate in great numbers at their roosts. Those of starlings, crows, and grackles are familiar to many people in northern lands; but that American Robins also gather in multitudinous companies for the night is not so well known. In the tropics, parrots often roost gregariously, although not, as far as I have seen, in such huge flocks as many northern and migratory birds form. Often shade trees in a park or residential district of a city are chosen for roosting by parrots, who in the evening gather from the surrounding countryside, where one would suppose that they would sleep more tranquilly than amid urban lights and noises. Apparently, however, close to human dwellings they find protection from some of the nocturnal animals that persecute them in wilder areas. They draw much attention to themselves as they go noisily to rest, with many changes of position and much bickering before they settle down for the night. The small hanging parrots of the Oriental tropics sleep like bats, in pendulumlike clusters suspended from leafy boughs of trees. Still other parrots roost in hollow trunks and caverns. As we emerged at sunset from a cave in the limestone hills of eastern Peru, where innumerable Oilbirds nested on ledges and slept through the day, green parakeets with red foreheads were entering the wide portal, from which the Oilbirds in their hundreds would soon be streaming.

A tree with compact foliage, or a dense clump of vegetation growing in an open place, is favored for roosting by many birds. The roosts that I have seen in the tropics were more remarkable for the variety than the number of birds that resorted to them. One of the most interesting was in a patch of introduced elephant grass, more than head high, in a cultivated valley in Honduras. Here slept many small seed-eaters of four species, along with Black-cowled Orioles, wintering Baltimore Orioles, and, in autumn, a flock of migrating Eastern Kingbirds that delayed here for over a month, then resumed their long journey to South America. A tall, dense clump of timber bamboos beside the dwelling house on a great coffee plantation in Guatemala provided a safe sleeping place for resident Gray's Thrushes, Yellow-winged Tanagers, and Vermilion-crowned Fly-catchers, together with wintering Baltimore Orioles, Orchard Orioles, Rose-breasted Grosbeaks, and Tennessee Warblers. In other regions, more popu-lous roosts have been found. In the Transvaal, an estimated million Barn Swallows, on their way northward, passed the night in a single reed bed, along with thousands of birds of other kinds (Rude-beck 1955). At least several million Red-winged Blackbirds, Tricolored Blackbirds, Brewer's Black-birds, Brown-headed Cowbirds, and Common Star-lings roosted in one dense aggregation in the Colusa Marsh in California in the autumn (Orians 1961).

Many ambulatory birds that spend the day on the ground roost in trees or shrubs. All day the Great Tinamou wanders over the floor of the tropical for-est, seeking food, but as dusk descends it flies up to a horizontal bough at a good height, where it roosts. Likewise, Spotted Wood-Quails ascend high into trees to sleep. But Bobwhites sleep on the ground

in compact clusters, each bird facing outward, so that if attacked in the night they may shoot out in all directions without impeding each other's escape. In winter, Rock Ptarmigan, Prairie Chickens, and other terrestrial birds burrow singly into soft snow, where each sleeps at the end of a short tunnel. Thereby they find protection from the colder air above them. If sleet falls and glazes over the surface of the snow, they may be sealed in their tunnels and starve. Among more aerial birds, Snow Buntings burrow into snowdrifts for shelter, and Siberian Tit-mice dig through snow to roost in disused rodent burrows, each individual retaining the same hole through the winter (Zonov 1967).

Marine birds may sleep on either land or water and apparently also in the air. As the sun sinks low, long, sinuous strings of Guanay Cormorants stream in to the barren islets off the coast of Peru. Here in crowded thousands they pass the night on the guano-covered ground where they nested, returning next morning to fish in the cool coastal water. But petrels, albatrosses, auks, and other birds that spend the nonbreeding season hundreds of miles from the nearest land can hardly seek solid ground for their repose. Those who cross the stormy North Atlantic in midwinter may watch graceful Kittiwakes, who all day have followed the heaving vessel looking for edible refuse from the galley, glide in the dusk down to the billowy surface of the ocean, where they fold their pointed wings about them, face into the wind, and prepare to pass the long, cold night. Other gulls return to sleep on land. Herring Gulls and Great Black-backed Gulls have been found passing winter nights on the spots where they nested the preceding summer (Perry 1946). Throughout the year, each Black Noddy Tern roosts on the same narrow ledge of a seaside cliff where it nested. The Wideawake Tern comes to land only to breed, and it cannot stay long in the water without becoming waterlogged, which points to the conclusion that it sleeps, or at least takes whatever rest it needs, in the air during the months that it spends on the high seas (Ashmole 1962, 1963a). The only other bird for which we have strong evidence of "roosting" on the wing is the Common Swift of Europe (Lack 1956b).

Only exceptionally do birds that breed in open nests use such nests for sleeping. The bird who rests in such a nest is no less exposed to rain than if it roosted amid foliage. Indeed, in wet weather it may stay drier on a perch than within a sodden nest. Accordingly, if a bird roosts on an open nest, it must do so for the sake of protection from below rather than from above. The chief users of open nests for sleeping are the rails, coots, and gallinules, who find it convenient to rest, and to brood their chicks, on a platform above the wet ground or water where many members of the rail family live. In Australian swamps, Magpie-Geese construct somewhat similar platforms for resting, preening, and courting. Their goslings begin to make these platforms when only two or three days old (Davies 1962).

Another bird that constructs an open dormitory inhabits dry rather than wet areas. The Curved-billed Thrasher of the deserts of Mexico and adjacent parts of the United States sleeps on an open nest in a cholla cactus; the sharp thorns protect the thrasher from prowling quadrupeds; the platform of twigs and weed stems protects it from the thorns (Bent 1948). Aside from these, almost the only arboreal birds that I have found, or read about, using open nests as dormitories are certain Southern House Wrens who slept in old, cuplike nests made by other birds, apparently because they could not find suitable nooks or crannies such as they prefer.

The Dormitory Habit

Most birds breed in open nests but they do not sleep in such nests at times when they neither incubate nor brood, except in special circumstances. Of the birds that rear their families in roofed or closed structures, a substantial proportion use them as dormitories. Although it seems obvious that the primary purpose of nests is as receptacles for eggs, might it not be that the refinement of covering over nests was first adopted by birds that needed dry places for sleeping and that later these birds built nests on the improved model for rearing their families? If this were true, we would expect to find in some groups of birds, especially those that had recently acquired the dormitory habit, species that prepare nests for sleeping that are better enclosed than those in which they lay their eggs. No such species is known to me, but a number of birds lay in

covered nests that they do not use as dormitories. Therefore, we must conclude that the first step toward the use of dormitory nests was the provision of a roofed or enclosed receptacle for the eggs and young.

From this point of departure, we may trace successive stages in the development of the dormitory habit:

A. Breeding nest is used as a dormitory while it lasts; none is made only for sleeping.
1. Female sleeps in her nest when it contains neither eggs nor young.
2. Both parents sleep in the nest, sometimes with their helpers.
3. Fledged young return to sleep in the nest, alone or with one or both parents, and sometimes also with helpers.
B. Nest or enclosed space similar to that in which the bird breeds is made or acquired for use as a dormitory throughout the year.
4. Self-supporting individuals sleep singly in the nest at all seasons.
5. Mates sleep together in the nest at all seasons.
6. Dependent young sleep in the nest with one or both parents.
7. Self-supporting young sleep in the nest with their parents.
8. Larger groups occupy a dormitory.

Only in species included under *B* is the dormitory habit fully developed. Birds that use their breeding nests for sleeping but do not, at least occasionally, build or otherwise acquire similar structures for this special purpose will not always have the good fortune to sleep in snug shelters throughout the year. Nests made of vegetable materials in trees and shrubs decay in wet weather; trees with suitable cavities often fall; burrows in banks may collapse; and shelters of all kinds may be taken from the original occupants by some more powerful bird or mammal.

I do not know any family of birds that exemplifies all the stages from 1 to 7 or 8; but numerous examples of each of these stages are available, and some families, such as the woodpeckers and wrens, illustrate segments of this sequence.

PLATE 102. Retort-shaped nest of a Sulphury Flatbill flycatcher hanging from an exposed, thorny twig. The female, who built the nest for eggs and nestlings, may sleep alone in it after their departure, until it becomes dilapidated.

Sleeping in the Breeding Nest While It Lasts

Stage 1. Sometimes we find a bird, usually a female, passing the night in a newly completed nest before the first egg is laid, or even in an unfinished nest; this appears to be more frequent in covered than in open structures. Such behavior has been reported of the Great Tit, Black-capped Chickadee, Violet-green Swallow, Eastern Phoebe, and other birds. A Sulphur-rumped Myiobius slept in her well-enclosed pensile nest four or more nights before she laid her first egg (Skutch 1961a).

In another American flycatcher, the Sulphury Flatbill, the female sleeps in her nest much longer. This small, obscure bird builds a retort-shaped structure that is usually composed of tough, blackish fibers and hangs from the end of a slender, drooping twig or vine, at the woodland's edge or in a shady pasture or plantation (plate 102). The rounded chamber that occupies the body of the retort is reached through the downwardly directed spout, which the flycatcher enters by means of a skillful upward dart. Working without help from her mate, the female takes two or three weeks to construct this remarkable nest, and, as it nears completion, she sleeps in it. By the time she lays her first egg, she has already passed from seven to ten nights in her nest. She continues to sleep in the nest throughout the period of laying, the seventeen or eighteen days of the incubation period, and the twenty-two to twenty-four days of the nestling period. After the fledglings take wing, they do not return to roost in the nest, but their mother continues to do so. I have known a Sulphury Flatbill to lodge in her nest for nearly four months after the departure of the single brood that she reared here in the Valley of El General; but under almost daily rains, most nests become dilapidated sooner than this, or else other small birds claim them to rear their broods.

I am fairly certain that after her old nest becomes uninhabitable the flatbill does not build herself another dormitory, because, until the approach of the following breeding season, no new structures appear among the trees where the old, misshapen ones still hang conspicuously. Female Sulphury Flatbills apparently roost amid the foliage from August or September, if not earlier, until their new breeding nests near completion the following March or April. In the wettest and coolest months, they lack dormitories; the males seem to be without them at all seasons. This American flycatcher, then, exemplifies an early stage in the evolution of the dormitory habit: lodging in the breeding nest as long as it holds its shape, without the construction of special nests for sleeping. As we shall see, the Eye-ringed Flatbill, which builds a similar nest, has advanced a step farther in the use of dormitories.

Stage 2. Sometimes both parents pass the night in the breeding nest while it contains eggs, and they do so more frequently after the nestlings hatch. Since it is improbable that more than one parent is needed for incubating or brooding, the second is evidently using the nest as a comfortable lodging. The Gray-capped Social Weaver of Africa builds a covered nest of coarse, dry grass, which at first has a downwardly directed opening at each end. In this condition, both parents sleep in it before breeding begins. Before the first egg appears, they close one entrance, but both continue to pass the night in the nest during egg laying and incubation. After the young hatch, the second doorway is reopened (Payne 1969b). White-browed Sparrow Weavers also sleep in a covered nest open at both ends (Collias and Collias 1964a). The advantage of this arrangement appears to be that, if an enemy enters by one doorway, the occupants can escape through the other. Moreover, they need not bend their tails while turning around in a confined space. I do not know whether these weavers build special nests for sleeping; possibly they represent a later stage.

In this chapter we shall frequently mention birds that we met in the preceding chapter. This is not a coincidence but a consequence of the fact that one form of close association readily leads to another, so that birds that lodge with others often help at their nests. One of these species is the Black-eared Bushtit (plate 103). From the account given on page 369, it will be recalled that, as the bushtits' downy pouch nears completion, both parents take advantage of the protection that it affords from the frosty night air of the high mountains. Throughout the incubation period, they continue to sleep with the four white eggs. The helpers often join the parents at nightfall. For a while, one pouch sheltered

PLATE 103. The Black-eared Bushtits' exquisite, lichen-covered, downy pouch, with four black-faced nestlings posed upon it. Until the young leave, the two parents and their helpers sleep in the nest with them through chilly mountain nights.

four nestlings, two parents, and two black-faced assistants—a total of eight occupants. After the young take wing, the nests remain deserted. All the bushtits roost amid foliage through the cold and often rainy nights of the wet season and the frosty nights early in the dry season. The snug pouches, which provide warmth during the chilly nights toward the end of the dry season when the bushtits breed, now become sodden and would become unhealthful dormitories. No new ones are built until the following breeding season approaches (Skutch 1935, 1960).

A similar situation is found in the Common Bushtit of the western United States and Long-tailed Tit of Europe. In the former, four adults have been found sleeping with the eggs, but after the breeding season these bushtits roost in trees (Bent 1946). The male and female of the Long-tailed Tit sleep in the oval, feather-lined breeding nest, beginning to use it as a dormitory before it is finished and continuing to lodge in it until the young fly forth (Owen 1945). But in winter these tiny birds, instead of providing snug nests for themselves, huddle together in the open, as already told.

Before an egg is laid, the Common Swift begins to roost in its nest space, a cleft in a precipice, a cranny in a building, or a nest box. Both parents continue to sleep there while they hatch their eggs and rear their nestlings. They also enter the nest to escape rain. After the young swifts fly, the parents may lodge in the nest for a few additional nights, but their young fail to return. Little appears to be known about the way these swifts sleep after they leave their summer home (Lack 1956a).

In Tanzania, White-rumped Swifts often breed in the closed, mud-walled nests that Abyssinian Swallows attach to the sides of buildings in the shelter of the eaves. The swifts line the lower half of the nest's inner wall with feathers and plant down, which they glue in place with their own saliva, applied so thickly that the surface glistens. Both parents sleep in the remodeled nest while the eggs are being laid and incubated and during the five to seven weeks of the nestling period. On quitting the nest, the young promptly fly out of sight and appear never to return to it. The parents, however, continue to sleep in the nest throughout the breeding season, which occupies somewhat more than half the year and in which three broods may be reared. Both parent House Swifts also sleep in their nest, and the fledged young fail to return to it (Moreau 1942*a, b*).

Both parent Brown-headed Nuthatches may roost in the nest cavity, beginning before the eggs are laid and continuing until the young are nearly fledged. After the latter emerge from the hole, they and their parents roost amid vegetation rather than in a cavity. On a cold December night, however, four Brown-headed Nuthatches slept in a nest box. This species foreshadows the more consistent use of dormitories that we find in the Pygmy Nuthatch (R. A. Norris 1958).

Stage 3. Most of our examples of the return of fully fledged young birds to sleep in the nest in which they were hatched and reared are of species that build dormitories, hence they will be considered later. It seems that birds that use only the breeding nest, or the space that held it, as a lodging rarely lead their fledglings to a closed sleeping place. An exception seems to be the Cape Penduline-Tit of South Africa, which builds a felted nest with a side entrance, somewhat like that of the American bushtits. The hanging pouch lacks the lavish encrustation of lichens that embellish the nest of the Black-eared Bushtit, but it has a feature almost unique in bird nests, a door that the tits open as they enter or leave and close behind themselves, as told on page 178.

The Cape Penduline-Tit lays five or six eggs, and both parents sleep in the nest throughout the incubation and nestling periods. Unlike the bushtits, the fledgling penduline-tits return to sleep in the nest with their parents, and they continue to do so while the latter rear a second brood. The fledglings of this second brood also sleep in the nest with their parents and older siblings, and the number of occupants may be swollen by other tits that join them. After the breeding season, up to eighteen tits have been found sleeping together in a single pouch, with the doorway closed above them, even in February, South Africa's hottest month, when the temperature of the outside air at night is 80° to 90°F (27° to 32°C). No nests appear to be constructed specially for dormitories; after mid-May, when the old breeding nests have become dilapidated, the tits have no longer been found sleeping in nests. From May

until the breeding season begins in the following October, all evidently roost in trees or shrubbery (Skead 1959).

Both parent Chimney Swifts sleep on the bracket-like nest glued to the inner wall of a chimney or in some similar situation, and frequently a helper or two rest close by them. After the young have taken their first flights, the whole family returns to sleep in the chimney for about a fortnight. Then, for another week or ten days, only two individuals, evidently the parents, enter the chimney for the night. Soon after this, the swifts gather in large flocks in preparation for their southward flight (Dexter 1952, Sherman 1952). While migrating, hundreds or thousands of Chimney Swifts may pass the night in a disused factory chimney or a hollow tree. In the related Gray-rumped Swift of Brazil, both parents sleep on the nest, which is often in a chimney (Sick 1959*b*). The two parent African Palm-Swifts sleep clinging upright, side by side, on their narrow bracket, which is glued to the surface of a drooping dead palm frond. After their first flights, some young birds return to roost on the nest with their parents for a night or two, but other juveniles do not come back (Moreau 1941).

In numerous swallows, both parents sleep in the breeding nest if it is a closed structure; if it is cup-shaped and placed in a burrow or some sheltered nook in a building or cliff, one parent sleeps on the eggs or young while the other rests on the rim, sometimes with its breast above its mate. Among species that sleep together in one of these ways are the Barn Swallow, Wire-tailed Swallow, Purple Martin, Cliff Swallow, Dusky Sand Martin, House Martin, and Blue-and-white Swallow. The fledglings of many of these swallows return to sleep in their nests for at least a few nights. The young of two earlier broods of the House Martin may sleep with the nestlings of the third brood. The behavior of fledgling Tree Swallows and Violet-green Swallows is variable; some return to sleep in their nests but others do not. Fledgling Bank Swallows are as likely to enter a neighboring burrow as that in which they were reared, and those of two or three families may sleep in the same tunnel. Most of these swallows are migratory; soon after the young can fly, they and their parents gather to roost in trees or marsh vegetation before they leave for warmer regions. Some of those that are permanent residents will receive our attention later (Skutch 1961*a*).

The fledglings of some hole-nesting thrushes may also return to sleep in their nests. A young Red-breasted Chat retired at dusk into the underground hole where it was reared (Macgregor 1950). Juvenile Wheatears may enter burrows or crevices near that in which they hatched (Conder 1956).

Parent and fledgling Gray-headed Waxbills of East Africa return to sleep in the nest for a few nights, although afterward they take shelter in a dense bush or small tree. The same appears to be true of the Crimson-cheeked Blue Waxbill (van Someren 1956).

Parent European Bee-eaters sometimes lead their fledglings back to the nesting colony, where the young birds retire for the night into their nest burrows. The parents, however, go off to sleep in their usual communal roosts in distant trees, and the young often accompany them. In this migratory bird, the dormitory habit is poorly developed. Only one parent, usually the female, passes the night in the burrow with the eggs and nestlings (Swift 1959).

In many other birds, the return of the young to sleep in the nest seems to be related, not to the dormitory habit, but to the gradual way that they sever connection with their nursery. In a number of raptors, herons, ibises, and other large birds that breed in open structures in trees, on cliffs, or on the ground, the young may hop or flit beyond the nest for distances that increase daily as they grow in strength, after each excursion returning to rest or sleep in it, until they are strong enough to fly for considerable distances. Exceptionally, a fledgling pigeon, hummingbird, American flycatcher, jay, or finch will return to pass another night or two in its open nest. It is more likely to do so if another member of its brood is still being fed there.

Sleeping in the Breeding Nest or Similar Structure at All Seasons

Stage 4. The Eye-ringed Flatbill of Central America appears to represent the next step in the development of the dormitory habit. Although it is probably not closely related to the Sulphury Flatbill,

like the latter it attaches a pensile, retort-shaped nest to a slender, drooping shoot. But the Eye-ringed Flatbill's nest is bulkier, for it contains dead leaves and other coarse pieces of vegetation in addition to fibers. At all seasons, I have found these solitary flycatchers sleeping singly in these hanging structures, some of which have scarcely any entrance tube, while the chamber is so shallow that an egg would probably roll out, so that they could not be used for breeding. Unfortunately, the sexes of the Eye-ringed Flatbill are indistinguishable by appearance and voice, and I do not know whether both build and use these dormitories. A female slept for a few nights in a nest that her young had just vacated, but they failed to return to it. The Olivaceous Flatbill of Panama and South America sleeps singly in bulky nests (plate 104) that closely resemble those of the more northern species (Skutch 1960).

One of the best-known representatives of this stage is the Bananaquit, which has been carefully studied in widely separated parts of its vast range that includes nearly all tropical America, continental and insular, at lower altitudes. This tiny, sharp-billed, yellow-breasted honeycreeper builds a small, roughly globular nest of grass blades and similar materials, with a round doorway that faces obliquely downward. Some nests, especially those that will be used for breeding, are built by both sexes, whereas others are made by a single individual. Throughout the year, adults sleep singly in these nests, which they replace with new structures when the old ones become dilapidated (plate 105). No other bird that I know is so frequently seen building. If one member of a pair tries to join its mate in the nest, a struggle ensues and one is ousted. After the departure of the young, the female may continue to sleep in the breeding nest, leaving her fledglings outside. Here they roost until they can find an abandoned nest of another Bananaquit, a wren, or perhaps a flatbill, or until, while still in dull juvenal plumage, each can make a dormitory for itself. If an immature bird builds in the territory of an adult of its kind, the latter may evict the juvenile and occupy the nest itself (Skutch 1954, Biaggi 1955, Gross 1958).

The Troupial, an atypical oriole that is Venezue-

PLATE 104. Dormitory nest in which an Olivaceous Flat-bill flycatcher slept alone in a Panamanian forest. Breeding nests have longer entrance spouts.

PLATE 105. Bananaquit's nest in an orange tree. The entrance, shielded by a visor, is on the right. In nests like this, Bananaquits raise their young and sleep throughout the year, always a single adult male or female in a nest.

la's national bird, has rather similar sleeping habits, with the difference that, instead of building its own nests, it intrudes into the many-chambered nests made of sticks by Rufous-fronted Thornbirds. Instead of using the original doorway of the chamber that it occupies, it tears a wide gap in the side, and it sometimes destroys the thornbirds' eggs or nestlings. A month before she laid an egg, a Troupial

started to sleep in the chamber where her brood was later reared. Repulsed whenever he tried to join his mate in this compartment, the male slept in another chamber of the same nest, and later in the thornbirds' replacement nest hanging nearby. After leaving their nest, the young Troupials did not return to sleep in it, or in any of the neighboring thornbirds' nests. The parent Troupials continued to lodge in these nests, always one in a chamber. Sometimes thornbirds still slept in an adjoining room of the same nest (Skutch 1969a).

Although there is no constant, pronounced difference between nests built by the Bananaquit for breeding and those made solely for use as dormitories, the Chinchirigüí or Plain Wren of Central America has two kinds of nests. The dormitory nests, flimsy horizontal pockets with an opening at one end, are too shallow to hold eggs and are readily distinguished from the deeper, more substantial breeding nests. The adults sleep singly at all seasons. This wren inhabits dense thickets where it is hard to watch; but, as far as I can learn, the parents do not lead their fledglings to a dormitory. Once I found a young wren lodging alone in a Bananaquit's abandoned nest.

The Village Weaver and a number of other weaverbirds resemble the Bananaquit and Plain Wren in that, while the female incubates and broods in one nest, the male sleeps nearby in another. The male weaver's dormitory is one of the several or many covered structures that he builds to attract one or more females; it usually lacks a lining, which is added, often by the female herself, after it has been accepted for rearing a brood. Unlike the Bananaquit and wrens, weaverbirds seem not to build dormitory nests after the termination of the breeding season. Apparently they do not install their fledglings in sleeping nests (van Someren 1956, Collias and Collias 1964a).

We have abundant evidence that many woodpeckers of both the Old World and the New are solitary sleepers in dormitories who fail to provide shelter for their fledglings. Among them are the Red-crowned Woodpecker, Hairy Woodpecker, Lineated Woodpecker, Golden-olive Woodpecker, Ivory-billed Woodpecker, Yellow-shafted Flicker, Black Woodpecker, and apparently many other species. Outside the breeding season, the male carves new holes for himself more frequently than his mate, who is often content to use the cavity that he has abandoned, with the result that her lodging is frequently dilapidated. Since, as a rule, the male has the sounder hole, it is not surprising that the female often lays her eggs in it in preference to her own, although, if necessary, they join forces to exca-

vate a new nest for breeding. With rare exceptions, the male alone stays with the eggs and nestlings through the night; by day, both sexes incubate and brood by turns.

If a fledgling Red-crowned Woodpecker tries to join either of its parents in the adult's dormitory, it is sternly repulsed; if a parent coming to its hole finds a young bird already ensconced within, it evicts the presumptuous juvenile. Accordingly, the fledglings must sleep clinging to the outside of a trunk until each can find an old hole no longer occupied by its parent or some other bird, or until it can carve a lodging for itself. Even in cold highlands, I have watched young Hairy Woodpeckers settle down in the open on stormy evenings, while the parents retired into their shelters. But certain other woodpeckers, which we shall soon consider, carefully lead the fledglings to their dormitories (Skutch 1969b).

The shy brown woodcreepers of tropical America sleep singly at all seasons in a cavity such as they use for nesting, which may be an old woodpecker hole or an obscure cranny in a tree. Often the dormitory is a hollow trunk, open to rain, which suggests that concealment is more important to them than dryness. They enter and leave their sleeping places in such dim light of dusk and dawn that their brown bodies are difficult to distinguish from the dark trunks up which they creep. The Tawny-winged Dendrocincla of this family nests without a mate and is one of the most solitary and truculent of birds; but one evening I watched a female lead her fledglings to two separate hollow trunks, then fly to the dead tree where for several years she slept and nested alone. In the ovenbird family, the Plain Xenops and Red-faced Spinetail seem always to sleep alone, the former in a cavity in a tree, the latter in the bulky, hanging nest in which it breeds or in some similar construction.

Stage 5. The motmots are a lowland family most abundant in hot, arid regions of northern Central America and southern Mexico; but an aberrant member, the Blue-throated Green Motmot, resides in the highlands of Guatemala and adjacent countries up to about 10,000 feet (3,000 meters). This retiring motmot, which lacks the racquetlike central tail feathers of most of its relatives, finds protection from the frigid night air of the high mountains by sleeping throughout the year in long and often very crooked burrows that males and females together dig in roadside banks and other vertical exposures of earth. A tunnel is excavated around the end of June, about a month after the single annual brood has taken wing, but it does not shelter the young from the cold nocturnal rains frequent at this season. The parents retire into the new burrows in the evening twilight; the young appear to roost in the open until they can provide lodgings for themselves.

The mated pair of motmots continue to sleep together in the same burrow through the remainder of the wet season and through the frosty nights of the dry season, which begins in October. They rest cuddled so closely together that they form a single mass of fluffy green plumage. Rarely, in the non-breeding season, one sleeps alone or three sleep together, probably in consequence of the disruption of a pair by death. Early in the following April, the female lays three eggs in the burrow that has already served as a dormitory for about nine months. The pair continue to sleep in it while they incubate their eggs and raise their young. When the latter are about three weeks old, one or both parents sometimes cease to pass the night with them. After the young emerge at the age of about a month, the parents continue, or resume, their old habit of sleeping in the burrow. But the chamber in the earth has been fouled by the nestlings, for motmots give no attention to sanitation. Hence, at their earliest opportunity, the parents dig a new tunnel for themselves (Skutch 1945).

Since the Blue-throated Green Motmot's eggs are laid in a burrow that has long served as a lodging, it appears that the dormitory has become the breeding nest, rather than the reverse. But, at lower altitudes, Blue-diademed Motmots may begin to dig their burrows nearly half a year before they lay an egg in them, although they have not been found sleeping underground before the breeding season, when a single parent accompanies eggs and nestlings by night. Apparently, it was from a family propensity to prepare the nest burrow long before the breeding season that the Blue-throated Green Motmot's peculiar sleeping arrangements evolved, as an adaptation to cool highlands. No lowland motmot is known to use its burrow as a dormitory.

Stage 6. The Southern House Wren well exemplifies this stage in the evolution of the dormitory habit. Adults sleep singly in almost any nook or cranny they can find, including holes in trees, crevices in buildings, niches in banks, birdhouses, and other unexpected spots; in similarly diverse situations they build their nests. Toward the end of the fledglings' first day in the open, the parents lead them to a snug shelter, which may be the nest they have just left but is frequently a nook more accessible to young birds unskilled in flight. The parents direct the young wrens to their sleeping place by flying repeatedly from them to it and going in and out in their presence. After the fledglings are safely ensconced, they may receive a few more meals, and their mother may join them for the night, although sometimes they sleep without a parent. The mother is especially likely to lodge with them if they return to the nest. In this case, they may continue to sleep in the nest space until about the time the next brood hatches, when, as told on page 363, they are evicted by their parents. The last brood of the year may remain with the parents somewhat longer than earlier broods; but, long before the following breeding season, they disperse, and the territory is occupied only by the parents, sleeping in separate nooks (Skutch 1953c).

As far as is known, the sleeping arrangements of the Rufous-browed Wren and Ochraceous Wren of the Central American highlands follow the pattern of their more familiar and widespread congener (Skutch 1960). The Northern House Wren and European or Winter Wren also lead their fledglings to a sheltered nook, at least occasionally (Bent 1948, Sherman 1952, Armstrong 1955). In these northern representatives of a tropical genus, sleeping arrangements are complicated by migration or winter's stress, and these wrens do not properly belong with the year-round users of dormitories.

The Lowland Wood-Wren of tropical American forests builds two kinds of nests, both of which are covered structures with side openings. The deep, substantial breeding nests are so well concealed amid the lowest vegetation that they are rarely found. The thin-walled, pocketlike dormitories, which often have so little sill that an egg would roll out, are built a yard or two up in fairly exposed situations and frequently attract attention. Sometimes these dormitories are occupied by an adult with one or two fledglings, but through most of the year they shelter only a single sleeper, which suggests that families disperse after the young can take care of themselves.

The Riverside Wren builds, often on a branch overhanging a stream, a globular nest with an antechamber that is entered through a downward-facing doorway. It uses such a nest for both breeding and sleeping. As a rule, adults sleep singly, but once I found a dormitory with three occupants, evidently a parent with two young who seemed old enough to feed themselves. The Highland Wood-Wren makes a somewhat similar nest, in which both parents sometimes lodge with one or two fledglings. In this species, however, the parents continue to sleep together after the young wrens separate from them. One who has experienced the penetrating dampness of the forests in the subtropical and temperate altitudinal zones where Highland Wood-Wrens dwell can appreciate the benefit they derive from sleeping with their mates, rather than singly, in the fashion of their lowland relatives (Skutch 1960).

The Blue-and-white Swallow breeds from Patagonia to Costa Rica in such diverse sites as burrows in banks, cavities in trees, crevices in masonry, and beneath the roofs of houses. In the nonmigratory tropical races of this far-ranging species, the nest space also serves the constantly mated pair as a dormitory throughout the year, and they sometimes enter it by day to escape hard rain. After they build their shallowly cupped nest of straws and weed stems and line it with feathers, one member of the pair sleeps in the bowl while its mate rests on the rim, an arrangement usually followed until the nestlings fly. The young swallows return to sleep in the nest space with their parents, and they may continue to do so for some weeks after they can feed themselves. Soon, however, they go elsewhere, leaving the parents to lodge in the cranny where they nested (Skutch 1952*b*).

A pair of Red-winged Starlings of South Africa

often roost throughout the year in their nest site, which may be a hole in a tree, a niche in a cliff, or a shelf or cranny in a building. Here they sleep close together but apparently not in contact. The fledglings of the first brood lodge with their parents until five to seven days before laying is resumed, when they are driven away, usually with much commotion, at roosting time. The young of the second brood are permitted to stay longer with their parents, on the average for five or six weeks, before they are similarly expelled (M. K. Rowan 1955).

Possibly the Verdin, an aberrant member of the titmouse family that inhabits arid parts of the southwestern United States and northern Mexico, should be included here. Adults, apparently at all seasons, sleep singly in closed structures of interlaced twigs with side doorways. Winter lodges of males are usually smaller than breeding nests, with less ample lining of feathers. The young return to sleep in their nursery long after they fledge, but I can find no information as to whether a parent lodges with them (Bent 1946).

Stage 7. The Banded-backed Wren has carried the dormitory habit a step farther than the foregoing members of the family. After the fledglings have been led to sleep in the nest where they grew up, or in a similar structure nearby, they, their parents, and the helpers continue to lodge together throughout the long interval between breeding seasons. The family group, which in one instance numbered eleven, may seek the same nest nightly for months or change domiciles rather frequently, probably as a result of nocturnal disturbance or to avoid vermin. The bulky dormitory nests, sometimes a foot in diameter, are occupied from sea level up to 10,000 feet (3,000 meters) in the mountains and may be the foundation of this wren's wide climatic tolerance (plate 106). In the related Rufous-naped Wren, four individuals slept in a bulky nest in November, when the breeding season had ended. In the forests of Panama, parties of four or five Song Wrens slept in elbow-shaped nests hung over the vertical fork of a sapling in the undergrowth (plate 107). These also appeared to be families whose youngest members had become fully self-supporting (Skutch 1960).

PLATE 106. Dormitory nest of Banded-backed Wrens, in the top of a hawthorn tree in the Guatemalan highlands. As many as eleven wrens have been found sleeping in such nests.

PLATE 107. Dormitory nest in which five grown Song Wrens slept in the undergrowth of heavy rain forest. The entrance is on the right.

The sleeping habits of Black-headed Mannikins in East Africa seem to resemble those of the Banded-backed Wren. They roost in old brood nests, and, when these disintegrate, several birds unite to construct a dormitory nest, in which as many as eleven individuals pass the night (van Someren 1956).

In the large and varied ovenbird family, we find, in addition to such solitary sleepers as the previously mentioned Plain Xenops and Red-faced Spinetail, species with persisting family bonds. An example of the latter is the Rufous-fronted Thornbird, a wren-sized brown bird whose great nests of interlaced sticks, hanging from the ends of slender twigs or dangling vines in open country, are a conspicuous feature of the Venezuelan landscape (plate 108). Each nest consists of two to possibly eight or nine rounded chambers, situated one above the other to a height, rarely, of about seven feet. The chambers do not interconnect, but each has its separate entrance. Although the many-chambered nest suggests an avian apartment house, none of twenty-two that I investigated contained more than one breeding pair. Male and female sleep together in the nest at all times, and they lead back their newly fledged young to sleep with them. Two broods of two or three young may be raised in a year, and thus are built up family groups of six, the largest number of self-supporting birds that I found in any nest.

Although a parent thornbird incubating its second set of eggs once tried to evict its older offspring who lodged in the chamber above, it did not succeed. Many young thornbirds stay with their parents into the following breeding season and help to build a new nest, although I found none feeding a following brood. Sometimes a thornbird that has lost its own nest forces its way at nightfall, against strong opposition, into the nest of another family, where it may pass months as an unwanted guest without becoming integrated with this family (Skutch 1969a). According to Hudson (1920), young of the Firewood-gatherer, another ovenbird, continue for three or four months after fledging to lodge with their parents in the big nest of interlaced sticks.

Four young Pale-headed Jacamars returned to

PLATE 108. A multichambered nest of Rufous-fronted Thornbirds on the Venezuelan *llanos*. It is attached to a slender, dangling vine and is about seven feet long. Such a nest is occupied by a single breeding pair, whose fledged young continue to lodge with them.

sleep with both parents in their natal burrow in a vertical exposure of clay above a Venezuelan ravine. They still lodged with their parents on my last visit, nearly two months after their first flight, when they seemed well able to feed themselves (Skutch 1968).

Although some woodpeckers never tolerate another fledged one in the hole with them, others are more sociable, with more enduring family bonds. In a nest cavity carved by both sexes high in a dead trunk or limb with fairly sound wood, the male and female Golden-naped Woodpeckers sleep together throughout the periods of laying, incubation, and rearing the two or three nestlings, who leave the hole when about five weeks old. In the afternoon of their first day in the open, the young are led by their parents back to the nest cavity, or sometimes they appear to seek it spontaneously. Thereafter the hole, which is kept scrupulously clean, serves as the family dwelling. Some parents call their fledglings to it as a hard shower begins; others go in, leaving their offspring out in the rain, where they come to no harm; or one parent may stay dry in the hole while the other clings to the lower side of a leaning trunk.

If at any time between breeding seasons the Golden-nape family loses its home, such as by the falling of the tree, the woodpeckers promptly carve a new hole and move into it. But, if all goes well, the family continues to sleep in the old nest hole until the following breeding season. A new chamber seems invariably to be made for the eggs; at about the time these are laid, the last of the preceding year's brood departs, leaving the parents alone. Thus, in the Golden-naped Woodpecker, the nest cavity has become the family abode, in which the parents and the grown offspring lodge in comfort and security throughout the year. The closely related Crimson-bellied Black or Yellow-tufted Woodpecker of eastern Ecuador and Peru, of which I found up to five grown birds lodging together, appears to have similar sleeping arrangements.

The family life of piculets of the genus *Picumnus* closely resembles that of the Golden-naped Woodpecker. Among the smallest of birds, these pygmy woodpeckers do not use their tails for support as they climb over trees. With bills that seem too tiny for such work, both sexes carve a small but neatly finished cavity in soft, decaying wood, and both sleep in it while they help each other to incubate and raise the two or three nestlings. They carefully lead the fledglings back to sleep in the nest cavity, and sometimes juvenile Olivaceous Piculets continue to lodge there while their parents rear a second brood (Haverschmidt 1951, Skutch 1969b).

Other woodpeckers in which more than two grown individuals sleep together are the Blood-colored Woodpecker of the Guianas, of which Haverschmidt (1953) found one male and two females occupying the same hole, and the Acorn Woodpecker, of which as many as five may share a dormitory. Further studies of both species are needed to clarify the details of their family life and sleeping arrangements.

Of the few toucans that have been studied, the middle-sized Collared Araçaris and Fiery-billed Araçaris most resemble the Golden-naped Woodpecker and the piculets in their manner of sleeping. A single parent spends the night in the nest hole while it contains eggs; after the eggs hatch, both parents sleep with the nestlings. At a nest of the Collared Araçari, they were joined each evening by three helpers. Fledglings are led back to lodge with their parents and whatever helpers they may have. After the breeding season, five or six araçaris frequently sleep in the same high cavity, sometimes a hole carved by one of the larger woodpeckers, with perhaps three or four other members of the flock in a neighboring hole. As they enter, they fold their long tails over their backs to save space. Araçaris may be unique among toucans in sleeping in holes (Skutch 1958a).

Stage 8. With its curiously swollen, three-pronged bill, the Prong-billed Barbet of highland Costa Rica and western Panama carves a cavity deep in fairly sound dead wood. The male and female share in this task, and when the hole is large enough, both sleep in it. They continue to pass the nights in the nest while they hatch their four or five white eggs and rear their young. After the latter emerge, they are guided back to the nest cavity to sleep with their parents. In all this, these barbets closely resemble the Golden-naped Woodpeckers and the piculets. But, after some months have passed, one begins to find Prong-billed Barbets lodging in groups too large to be single families. Once I discovered sixteen sleeping in a hole that differed in shape from their nest cavities. These birds apparently do not excavate new holes between breeding seasons; if they lose their dormitories, several families may bunch together in any suitable cavity. In this they differ from woodpeckers, most of which preserve their territories and keep their families separate throughout the year. As the breeding season approaches, the barbets again become strongly territorial; the communal dormitories are abandoned; and each pair proceeds to carve a new hole (Skutch 1944).

In Africa, where barbets are numerous, two adult Pied Barbets slept in the hole with the eggs, but later four brought food to the nestlings. Outside the breeding season, up to six lodge together in a hole in a tree (van Someren 1956). Four White-eared Barbets brought food to a hole that contained four nestlings, and five were seen entering a cavity for the night (Moreau and Moreau 1937). The sleeping arrangements of these African barbets seem to resemble those of the American Prong-billed Barbet, with the complication that helpers may join the parents of nestlings, or else two pairs breed together.

As told in the preceding chapter, parent Pygmy Nuthatches continue to lodge in their nest cavity along with their fledged young, and their helper if they have one, sometimes until the following spring. Even after breeding begins, a young male of last year's brood may stay with the parents and share their labors. Sometimes, however, two neighboring families with fledglings join together and sleep in the same hole. This tendency of Pygmy Nuthatches to lodge with members of other families increases during winter, especially at high altitudes where the climate is severe. As many as 150 have been found roosting in an old pine trunk with several holes, at least 100 of them in the same cavity. Some hollow trees where nuthatches slept contained many corpses of birds that may have died of suffocation, although other causes cannot be ruled out (R. A. Norris 1958, Knorr 1957).

In winter, White-breasted Nuthatches usually roost singly, often in an old woodpecker hole. They have the curious habit of emerging at dawn with a dropping that they carry away in the bill, a method of keeping the hole clean rarely employed by adult dormitory users, who as a rule refrain from defecating in their lodgings. Like the more sociable Pygmy Nuthatches, White-breasted Nuthatches occasionally seek bed mates in severe weather, and twenty-nine have been found sleeping together in a hollow pine tree (Kilham 1971b, Bent 1948). Similarly, European or Winter Wrens most of the time sleep singly in old nests of their own or some other bird or even, in mild weather, amid sheltering foliage, but they may bunch together for warmth on bleak nights. As many as forty-six have been found sleeping together in a bird box in England and thirty-one in the northern United States, packed in layers with their heads inward (Armstrong 1955, Bent 1948).

We have considered the barbets and nuthatches after the Banded-backed Wrens, Golden-naped Woodpeckers, and piculets because their larger aggregations of sleeping birds often develop from a situation similar to that which we examined in the preceding section. It would be wrong, however, to conclude that the present stage represents a higher development of the dormitory habit than the preceding stage. On the contrary, in birds like the Banded-backed Wren and Golden-naped Woodpecker, the dormitory habit is more advanced, because they build or excavate a new lodging whenever necessary. The larger aggregations of barbets and nuthatches evidently result from their failure to prepare dormitories except in the breeding season; if they lose their holes, they must share with other families such quarters as they can find. In nuthatches, the desire to cuddle together for warmth in freezing weather doubtless brings many together at nightfall.

Influence of Environment

What are the advantages of sleeping in dormitories, and what are the disadvantages? Are sleeping birds safer in nests or holes or roosting amid foliage? Many studies have shown that nests in holes and crannies are, on average, substantially more successful than open structures in trees, shrubs, or lower vegetation; from this we may conclude that birds who sleep in holes in trees or other cavities are safer than they would be roosting amid foliage.

When the dormitory is a covered nest built in the open, the question is more difficult to answer. The nest is larger and more conspicuous than the sleeping bird, and, since it remains in one spot, nocturnal predators may remember its location. When these predators find their victims by scent rather than by sight, the greater size of the nest probably does not increase the sleeping bird's risk. The very rareness of dormitories probably adds to their safety. Since a small minority of the birds of any region use them, they are rather thinly scattered; if they were more numerous, nocturnal predators might find it profitable to specialize upon plundering them. To minimize the threat of predation, dormitory users often have alternative shelters, so that, if they find one being watched when they are about to retire, they may enter a different one. Another disadvantage of dormitories, the lice or other vermin that may multiply in them, is also reduced by periodic changes of residence and likewise by keeping the nest clean. The fact that many widespread, successful species sleep in dormitory nests built amid foliage points to the conclusion that they do not substantially increase the sleeping bird's risk; if they did, natural selection would eliminate this habit.

The great advantage of the enclosed dormitory is warmth, or the conservation of the bird's vital resources, and the saving is greater the lower the outside temperature. Kendeigh (1961) showed that when a House Sparrow sleeps alone in a wooden box, the inside temperature may be 65.3° F (18.5° C) when the outside temperature is 62.6° F (17° C), and the inside temperature may be 28.8° F (−1.8° C) when the outside temperature is 17.6° F (−8° C). In the first case, the difference is 2.7° F (1.5° C); in the second, 11.2° F (6.2° C). Although, when the outside temperature is 76.5° F (24.7° C) the bird would not save any energy by sleeping in the box, at − 22° F (−30° C) it could preserve its body temperature through the night with a saving of 13.4 percent of its metabolic expenditure. When a number of birds cuddle together in a closed space, the economy is much greater.

This analysis leads us to expect that, the more rigorous the climate, the more widespread the use of dormitories should become. Years ago, I looked into this matter on a Guatemalan mountain, chiefly between 8,000 and 9,000 feet (2,450 and 2,750 meters) above sea level, where, from November to early April, every dawn after a clear, windless night revealed open spaces white with frost. This was caused largely by nocturnal radiation under the starry dry-season sky, for the temperature in the shelter of trees did not fall below the freezing point, as was evident from the persisting greenness of tender foliage. Nevertheless, the thin night air became so penetratingly cold that, even inside the house, a person accustomed to lower altitudes could hardly find enough blankets to keep warm. Despite the nocturnal chill, the majority of the resident birds failed to provide dormitories for themselves: finches, wood warblers, honeycreepers, flycatchers, thrushes, and even tiny hummingbirds all, as far as I could learn, roosted amid foliage, just as most members of these families do in warm lowlands. Nearly all the birds that I found sleeping in dormitories belonged to families, including woodpeckers, woodcreepers, swallows, and wrens, that do so in the warmest regions where they dwell. The drop in temperature with increasing altitude within the tropics (about 3° F per 1,000 feet or 5° C per 1,000 meters) had little effect upon sleeping arrangements, at least up to about 10,000 feet (3,000 meters).

The only exceptions to this that I have discovered, here and in other highlands, are the Blue-throated Green Motmots, who sleep throughout the year in burrows, although their lowland relatives do not, and Highland Wood-Wrens, who double up in their dormitories, whereas their lowland congeners prefer to sleep singly. And perhaps, as already suggested, Banded-backed Wrens are able to reside from sea level up to frosty heights because they cuddle together in well-insulated nests, and Southern House Wrens achieve an almost equally great altitudinal range by seeking sheltered nooks.

The development of the dormitory habit is just the reverse of what we should expect from a consideration of its thermal advantages. Most of our examples of dormitory users are birds, not of the most frigid climates, but of the milder parts of the earth, and this is above all true of those in which the dormitory habit has reached the most advanced stages. A principal reason for this is that birds that breed in seasonally rigorous regions often migrate or wander widely in search of food, whereas those of zones that are more uniformly favorable tend to reside in one spot throughout the year. Only a permanent resident can enjoy a permanent abode. Common Starlings resident in Great Britain sleep in winter in the holes in which they nest, but those that come from the Continent to pass this season take no interest in holes (Bullough 1942). Among the few migratory birds known to sleep in dormitories where they winter are Red-headed Woodpeckers, whose stores of acorns sealed in crannies in trees permit them to remain stationary on defended winter territories, each with its individual sleeping hole, in which it may lodge for as long as six months (Kilham 1958 *a*, *c*).

Another reason why migratory birds, especially those that nest at high latitudes with a short breeding season, rarely sleep in dormitories is that few of them can afford the time to build the elaborate covered nests which, as we have seen, are most often chosen for sleeping. Since migration disrupts families, no bird not permanently resident is known to have attained the more advanced stages, in which the dormitory has become the home of a family that often practices other modes of cooperation, such as the attendance of nestlings and fledglings by their older siblings. To sum up, the use of dormitories is promoted, not by a rigorous climate, but by permanent residence in a continuously favorable climate; by the habit of constructing an elaborate nest for breeding, which is most common in such a climate; and by family ties that persist, as they often do among the permanently resident birds of favored regions.

It would be wrong to leave the impression that birds who must endure frigid nights are insensible of the advantages of sheltered nooks and cuddling. Most of these birds are not, in the strict sense, users of dormitories, which are either nests or spaces used for breeding or similar structures or spaces, but they commonly choose to sleep in the most sheltered places available. We have already noticed how a number of birds burrow into snow to sleep. Rosy Finches, who nest and apparently also roost in crannies among rocks in the almost Arctic climate

of high summits in western North America, descend in winter to lower levels, where in multitudinous companies they sleep in caves, mine shafts, crannies in buildings, or old nests of Cliff Swallows (Shaw 1936, French 1959). In abnormally cold weather in Florida, wintering Tree Swallows clustered in cavities in trees, and Myrtle Warblers, one of the hardiest members of a family not known to use dormitories, sought shelter in outbuildings. Nevertheless, many individuals of both species succumbed to cold and hunger (Christy 1940, Ruff 1940). Although Common Bluebirds nest in boxes and other cavities, in winter they often roost in trees, three or four together in a terminal cluster of dead leaves. In very cold weather, however, two pairs overcame their mutual intolerance to take refuge in the same bird box (Thomas 1946). As we have seen, nuthatches and Winter Wrens, which as a rule sleep singly or in family groups, mass together in such protected places as they can find when the thermometer falls low.

Although on well-wooded tropical mountains, even up to 10,000 feet (3,000 meters), few birds have been found sleeping in dormitories unless this is a widespread habit of their families, in the far more rigorous climate above treeline we find a different situation. In bitterly cold nights high on Mount Kenya in tropical Africa, Johnston's Sunbird seeks protection in deep cavities that the Mountain Chat digs in the dead leaves matted around the trunk of an arboreal groundsel *Senecio keniodendron*. One evening two females and a male entered such a niche. Juvenile sunbirds sought warmth and repose in old covered nests, even by day (J. G. Williams 1951).

Similarly, in the high, frigid puna of the Peruvian Andes, birds find shelter amid the stiff, close-set, swordlike leaves of the tall terrestrial bromeliad *Puya raimondii*. Great numbers of Black-winged Ground-Doves roost on these leaves as close to the central stem as they can cluster, all facing outward for readier escape if attacked. In this same region, Bare-faced Ground-Doves sleep in holes in walls. Five species of finches and an ovenbird, the Bar-winged Cinclodes, huddle in mixed companies of twenty or more beneath projecting rocks. Higher still, above 17,400 feet (5,300 meters) in Bolivia, more than two hundred White-winged Diuca-Finches kept each other warm by crowding together in a crevice in a glacier (Dorst 1957). These roosting finches and doves soil their sleeping places, although most dormitory makers do not. In the high Andes, a wide variety of birds, including hawks, owls, geese, ovenbirds, flycatchers, swallows, and hummingbirds, roost in caves; one of the hummingbirds, the Andean Hillstar, frequently nests in caves and mine shafts. A substantial proportion of the scanty avifauna of these bleak heights seeks covered places for sleeping (O. P. Pearson 1953, French and Hodges 1959).

Finally, we may ask why relatively few birds sleep in dormitories, even among permanent residents that have no lack of time to make them. In tropical lowlands, prolonged rain is often uncomfortably cool, and at altitudes no greater than 5,000 or 6,000 feet (1,500 or 1,800 meters) on tropical mountains, it may chill the unacclimated human to the marrow. The answer seems to be that, in the milder climates most favorable for their development, dormitories are rather redundant. A bird, like a turtle, carries its house with it, and its feathery covering is far more flexible and convenient than the tortoise's rigid carapace. When the bird's plumage is well oiled with the secretion of its preen gland, raindrops roll off, as they do from any unwettable surface. Although I have given instances of woodpeckers, swifts, and swallows sheltering from rain in their dormitories, frequently birds neglect to enter them even in hard downpours. By raising or depressing its feathers, a bird can adjust for changes of temperature. Constantly clothed in a garment that serves as a house with a roof and ventilators that can be opened and shut, with the additional advantage of mobility, most birds seem to find a second domicile of vegetable or mineral materials superfluous. Yet, in times of stress, such a shelter may prevent the disastrous consequences of saturated plumage or depleted vital reserves; and, doubtless, to the more social species the intimacy of a crowded dormitory is pleasant.

CHAPTER 31

Parental Stratagems

The safety of eggs and nestlings depends above all on the protection afforded them by nests that are difficult to find or to reach. Even for powerful birds, the safe placement of the nest is of primary importance because, especially after the nestlings are older and need much food, the parents may not be able to guard it without intermission, and during their absence some hungry prowler may eat their offspring. The active protection of the nest is nearly always secondary to the principles of safety discussed in chapter 7. But the bird who hides its nest, or builds it strongly or in a situation difficult to reach, is courting disaster unless its subsequent behavior supplements what it has already done. It must not betray its well-concealed nest by its own careless movements; it must not confide too much in the impregnability or inaccessibility of its structure, which are only relative and inadequate to deter its more powerful or agile enemies.

If, despite all precautions, a predator discovers and can reach the nest or young that have recently left it, the parent is not always utterly helpless. It may try to divert the enemy's attention from the nest to itself and lure it away with the promise of an easily caught meal; it may try to intimidate the hungry animal by feints of attack; or it may boldly strike the intruder with bill, feet, or wings, and perhaps drive it away. Finally, if parental protection

proves inadequate, or the predator arrives in the parents' absence, the nestling may try to defend itself—a procedure not likely to be effective unless the young bird is endowed with special means of defense. In this chapter we shall survey the stratagems by which parent birds try to divert enemies from their nests and offspring, and in the following chapter we shall consider more direct defense.

Circumspect Approach and Departure

The surest way to prevent the spoliation of a nest is to avoid its detection by a pillager. No matter how well hidden the structure, or how cryptically colored its attendants, they are in danger of betraying its presence by their own approaches and departures, for, as is well known, movement cancels concealing coloration. Many birds are exceedingly reluctant to approach a nest while being watched. In tropical woodland, one may wait for hours, screened by luxuriant vegetation, without seeing a parent come to some unfamiliar nest that he wishes to identify; he must hide himself in a proper blind for success. From such concealment, he may witness a curious phenomenon: The approaching parent, now accustomed to the blind and apparently unaware of the hidden watcher, delays at a distance from its nest, flitting from branch to branch, turning from side to side, displaying the utmost caution, but continuing to repeat loud calls. This behavior, which is especially striking in such birds of the tropical forest as the Red-crowned Ant-Tanager and Blue-black Grosbeak, is absurdly inconsistent, unless the predators chiefly responsible for so much caution hunt by sight rather than by sound. These predators are probably snakes, which in the warmer parts of the earth seem to destroy more nests than any other class of animals.

Some ornithologists have questioned the importance of discreet approaches and departures and of minimal activity of birds at nests with eggs or young. They point out that egg collectors hunt with great success by looking in promising places rather than depending upon the birds to reveal their nests by their own movements. These ornithologists have, apparently, never searched for nests thinly scattered through the luxuriant, epiphyte-burdened vegetation of the humid tropics, where many nests are much smaller than the leaves that screen them and the profuse growths of tree moss that they resemble. Moreover, we have observational evidence that, even in lighter temperate-zone woodlands, some predators are led to nests by the movements of the parent birds. The Common Crow discovers eggs and nestlings of the Scarlet Tanager by waiting and watching or by hunting and frightening the female from her nest (Prescott 1965).

The bird's departure from its nest should be as circumspect as its arrival. Certain small birds, including some euphonias and wood warblers, end a session of incubation or brooding by suddenly dropping like a stone from a nest high in a tree and then, just before they strike the ground, turning sharply to glide over it or to rise into the air. The initial fall easily escapes detection; the bird appears to have begun its flight on or near the ground rather than from a point well above it, and may thereby mislead a watcher. Many birds are as reluctant to leave as to approach a nest while they are aware of being watched. Sometimes they choose the moment when the observer's head is turned aside to steal away unseen. Once, while I watched from a blind, a male Mountain Trogon started to leave his nest hole in a low stub in response to the call of his mate, who had come to replace him. When halfway through the doorway, he noticed a squirrel hopping around

among bushes about twenty feet distant. Cautiously he backed into the cavity and waited there until the rodent vanished. To have flown forth in the animal's presence might have cost his nestlings' lives.

When they see a potential enemy approaching the nest where they sit inconspicuously, birds have two strategically sound procedures: they may leave stealthily the moment the intruder comes into view, thereby probably escaping its notice; or they may continue to sit tightly, relying upon the "invisibility" of the nest and their own dull coloration to avoid detection. Many an obscurely clad bird will permit a human hand to touch it while it sits; yet some very plainly attired birds are so wary that it is most difficult to glimpse them on the nest. Doubtless much depends on the kinds of enemies the bird must elude and its individual experience. Some hole-nesting birds tolerate a good deal of tapping or shaking before they reveal their presence by looking out. They may even refuse to budge while the cavity is cut open.

Flight so prompt and unobtrusive that it fails to catch the approaching eyes and motionless sitting until the last moment consistent with the parent's personal safety are the only sound alternatives when a potential enemy advances toward a nest. To remain sitting until the intruder has come close enough to enjoy an unobstructed view, then to flee while it is still uncertain whether the animal has noticed the nest, is vacillating behavior that will almost certainly betray the eggs or nestlings to destruction. Some birds appear to be instinctively aware that these are the only two sound choices. As species, and even as individuals, they may take one course or the other, scarcely ever showing intermediate behavior; or the same individual may change abruptly from one procedure to the other.

The Slaty Antwren, a tiny, obscure bird that hangs a slight, open cup in the undergrowth of tropical forest, well exemplifies this behavior. At certain nests, the male disappears before you can see him well, while the female sits until you come within arm's length. At other nests, it is just the reverse: the male remains steadfast and his mate flees early. Most revealing was a nest so situated that the sitting antwrens first saw me as I suddenly emerged from a dense tangle of vines sixty feet away. During the first week after the eggs were laid, whichever parent was incubating invariably jumped from the nest and vanished the moment that I came into view. Then, one day, the male watched me advance over the area where the undergrowth was sparse until I was only two or three yards from him, when he jumped from the nest and tried to lure me away. Two days later, the female for the first time sat until I came near. After the nestlings hatched, the parent of either sex continued to brood them until I was almost within reach. Despite the diversity and apparent capriciousness of their behavior, it was quite consistent: They took one or the other of the two courses most compatible with the safety of their eggs and young, shifting abruptly from the first to the second as parental devotion grew stronger. They never left the nest while I was in the midst of the clear space where we could see each other, a course which, if I had not already become familiar with the nest, would have gratuitously betrayed its location.

A diminutive, olive green female Golden-collared Manakin, who had built her slight, cupped nest beside a forest trail in Panama, made a similarly abrupt transition in her response to my approach. At first, she so successfully escaped detection that only the warmth of her eggs assured me that she had just flown from them. But one day she surprised me by continuing calmly to incubate while I approached, bent over her, and almost touched her with my finger tips. I doubt that her character had changed from timorous to bold in a single day; but, with growing devotion to her nest, she had changed her tactics, and, between immediate flight and departure delayed until she was on the point of being caught, there was no procedure consistent with that devotion.

Conspicuous Departure from the Nest

When a bird slips from its nest at the first hint of danger, it leaves as clandestinely as possible, to avoid attracting attention. When, taking the alternative course, it remains until the enemy is almost upon it, the parent often leaves in the exactly opposite way, doing its best to attract attention to itself. Now the only hope of saving its progeny is to divert the animal's attention from the nest, with its immobile eggs or nestlings, to the parent itself, who has wings and may escape. Accordingly, the departing bird makes itself more than ordinarily conspicuous, fluttering its wings and moving with deliberate slowness, trying to lure the animal onward with the prospect of an easily captured meal.

In the most exaggerated forms of such conspicuous departure, the bird acts as though it were crippled or otherwise incapacitated, unable to fly or walk properly. Such behavior of parent birds occurs in the most diverse families in many regions. It has long attracted the attention of those interested in birds, who have called it "feigning injury," "feigning a broken wing," the "broken wing ruse," the "parental ruse," "lure display," "disablement reaction," and so on. Although the first of these designations has perhaps been most widely used by English-speaking people, "injury simulation" is a preferable term for displays that suggest the bird is crippled, for it does not, like "feigning," imply that it is self-consciously acting a part. "Distraction displays" is a good designation for this whole class of reactions, including injury simulation and other displays that have a similar function but hardly suggest disablement. Since they tend to divert the intruder's attention from eggs or young to the parent itself, they may also be called "diversionary displays" (Armstrong 1949).

Concentrating their attention upon the more extreme distraction displays, which often suggest that the distressed parent is in agony or having a fit, early observers offered explanations hardly tenable in the light of present knowledge. They thought that the conflict between fear, impelling the bird to flee, and parental devotion, holding it to eggs or young, resulted in muscular inhibition and actual incapacity to move freely. The displaying bird's behavior was "a compromise between fear and reproductive emotions"; it was "in pain," "deliriously excited," or "hysterical"; its behavior patterns were more or less disorganized (Friedmann 1934, Armstrong 1942, 1947).

At most, the view that the displaying bird is not in full command of its movements would apply to the more extreme forms of distraction displays commonly called "injury feigning." Milder forms hardly suggest any incapacity; to gain a fair view of the

PLATE 109. Common Nighthawk in distraction or "injury-feigning" display (photo by S. A. Grimes).

whole range of stratagems by which parents try to divert an intruder's attention from their progeny, we should consider these first. Often the conspicuous, last-moment departure of a bird who has stuck to its nest, trying to avoid detection, is simply by deliberately slow locomotion, on the ground or in the air, in full view of the intruder that has almost reached it. While I watched a Rufous-sided Towhee from a blind, she left her eggs with a single hop and a flight. But when I walked toward the nest situated on the ground, the female sat quietly until I almost touched her, then jumped out and hopped slowly away over the dead leaves, as though to attract attention. When far enough, she flew up to a low branch and called *tow-he*. She did not wildly beat her wings, fall on her side, or in any other way suggest that she was not in full control of her movements. A Kentucky Warbler, brooding a nestling in

the same woodland, sat until I almost touched her, then leapt abruptly from her nest and walked slowly over the ground with the tips of her wings dragging, chirping excitedly, attracting attention to herself but failing to appear injured. In a scrubby pasture in Colombia, a Common Ground-Dove fluttered from beneath a lantana bush and walked mincingly away from his eggs, much as the warbler had done, but silently.

A more elaborated form of slow withdrawal is the "rodent run" or "rat trick." The parent bird alights on the ground in front of the intruder and runs before him with foreparts depressed in a "hunched" attitude. Its tail is lowered; its trailing, quivering wings simulate the rapidly moving hind-legs of a small quadruped; often, too, its fluffed-out feathers resemble fur; and it may even utter mouse-like squeaks. Looking and acting as much like a small rodent as a bird can, it scurries away in a zig-zag course, from time to time pausing briefly to see if the intruder is pursuing. If not, it may fly back toward him and try again to lure him away. When followed, the bird may continue its rodentlike running for hundreds of feet over the open tundra. This ruse is most common among sandpipers, plovers, sanderlings, and other shore birds that nest far in the north, where rodents, especially lemmings, are a principal food of predators. It may have evolved in response to predation by the Arctic Fox, which presumably is easily tempted to pursue a bird that simulates its preferred prey and so is tolled away from the performer's eggs or chicks (Duffey et al. 1950, R. G. B. Brown 1962, Hobson 1972).

At lower latitudes, variations of the rodent run have been observed in Australian wren-warblers and the Green-tailed Towhee of the western United States. The latter scuttles away from its nest with elevated tail and a fast, even motion; it resembles one of the chipmunks common in the sagebrush where the towhee nests. Apparently, this stratagem is effective in luring coyotes, important mammalian predators of this habitat (Blair et al. 1950, A. H. Miller 1951). Among tropical birds, the Black-striped Sparrow gives a similar performance. Although this finch ordinarily hops over the ground with its feet together, when displaying it runs rapidly with alternately advancing feet, body depressed in a hunched attitude, tail held low, and wings more or less spread. Running in this fashion in front of a fledgling that has incautiously ventured onto open ground, it leads the young bird to cover. This may be the principal function of the display, but I have seen a ground snake and a rat or large mouse pursue a Black-striped Sparrow running in more or less this fashion. That a bird should employ the same display to lead an exposed fledgling and lure an enemy appears strange; but the two situations have one thing in common—the fledglings are in danger. A Gray-striped Brush-Finch also lured a newly emerged fledgling to safety by hopping in front of it with drooping wings and depressed tail.

Conspicuous departure from a higher nest sometimes takes the form of retarded flight rather than slow walking or hopping. A particularly beautiful display of this sort was given repeatedly by a female Tropical Gnatcatcher. Her exquisite, downy, lichen-encrusted nest was situated on a horizontal branch ten feet above a riverside pasture. Whenever I walked beneath while she was incubating, she would drop well below the supporting limb, then rise to the tops of the trees on the river bank by a most peculiar flight. Hanging beneath rapidly vibrating wings and moving with extraordinary slowness, she seemed to float rather than to fly away. The upright position of her long tail evidently furnished the drag to retard her progress. This seemed to be the aerial equivalent of injury simulation on the ground. Although the gnatcatcher appeared to fly with failing strength, to go in this unusual fashion doubtless demanded more skill and muscular coordination than to fly in the habitual manner.

When disturbed, the tiny, greenish female Blue-crowned Manakin often leaves her low, open nest by flying slowly, just above the ground, until lost to view in the undergrowth. Or sometimes she flutters down to a low perch, where she spreads her wings widely and flaps them slowly, often appearing to be precariously balanced and on the point of falling. A soft little trill usually calls attention to this performance, and at times the manakin seems to wait until the watcher's eyes are turned toward her before she begins her display.

Some of the antbirds, another family of tropical American woodlands, have rather similar displays of which the effect is intensified by the disclosure of usually concealed patches of white in the plumage of these generally dull-colored birds. It took much shaking to make a male Tyrannine Antbird abandon the deep pouch in which he sat incubating two eggs above a rivulet at the forest's edge. Finally, however, he dropped from his pensile nest to a fallen log, where, by turning the dark feathers of his back outward, he conspicuously displayed a broad patch of snowy white. As already told, a pair of Slaty Antwrens would either vanish from their nest the instant I came into view, or wait until I was almost within reach, then jump from the nest and display. The most prolonged of these performances was given by the male on the morning his two naked nestlings hatched. Dropping from his nest almost to the ground, he clung to a slender, upright stem, where he spread and slowly waved his blackish

PLATE 110. Ringed Plover reluctantly leaving eggs, about to give a distraction display (photo by E. A. Armstrong).

wings, revealing small white epaulets ordinarily hidden. When I approached, he flitted to a similar upright perch a few feet ahead and repeated this display, but now more briefly. Luring me onward, he flew from sapling to sapling until, when we were at a satisfactory distance from his nest, he vanished into the underwood and called his mate. He did not appear injured, but he seemed to be silently pleading for me to follow him and leave his nest alone.

A very different distraction display is given by a Common Gallinule when driven from its nest. Standing on floating vegetation or in shallow water, it faces the intruder with half-spread wings and stamps its feet alternately, with much noisy splashing to attract attention (Jenni 1961, Fredrickson 1971).

Injury Simulation

Typical injury simulation differs from the foregoing distraction displays in its greater vehemence. The disparity between the often violent movements of the bird's limbs and its usually slow progress creates the impression that it is trying desperately to escape but is too injured to go fast; this encourages the would-be predator to turn his attention from nest or chicks to the agonizing parent, who would make a larger meal. The injury-feigning bird frequently drags itself over the ground or grovels in one spot, beating its widespread wings in apparent helplessness. The display often begins as an incubating or brooding parent leaves the nest, but sometimes it is given by a bird who finds an intruder

PLATE 111. Ringed Plover in full distraction display (photo by E. A. Armstrong).

near its nest when it arrives with food for its young. A parent leading chicks or attending fledglings out of the nest may try to lure away an enemy by simulating injury; it may even leave its young to advance boldly toward an approaching intruder and divert him from its progeny. Although most frequent among birds that nest on or near the ground, injury simulation is practiced by a few birds that build high in trees, including certain wood warblers.

In this family of small, brightly colored birds, injury simulation is frequent. On a scrubby hillside in New York, I found a nest of Chestnut-sided Warblers with two half-grown Brown-headed Cowbirds, who squealed a little as I lifted them, driving the foster parents to frenzied efforts to entice me away. The male warbler fluttered from twig to twig in front of me, vibrating his wings and fanning out his tail, while his mate dropped to the ground and groveled on the fallen leaves with quivering wings and spread tail, moving slowly and lamely. When I squatted down amid the shrubbery to see her better, the male, becoming uncommonly bold and loudly chirping his protests, displayed so close before me that every most delicate marking of his plumage was plainly revealed: the golden green crown, white cheeks, rich chestnut on his sides, every streak of black and white and gray on his wings and back and tail. I might have stretched out my hand and touched him, if he had not moved so rapidly. This display of anguished parenthood should have stirred compassion in a feeling breast, yet it delighted me by its beauty.

Another family in which injury simulation is common is the pigeons, and none performs more earnestly and consistently than the White-fronted Dove of tropical America. Finding a nest of this pigeon above my head in a coffee plantation, I wished to see its contents by lifting up a mirror attached to a long pole. To accomplish this, I urged the sitting bird to leave and disclose what it covered. Shaking the supporting tree did not make the dove budge. When I touched it with the pole, it merely raised its wings straight above its back in a defiant attitude, revealing the beautiful cinnamon of the under wing-coverts. Then it brought down its right wing against the stick with such a resounding whack that I feared the wing was injured.

Thinking that I could glimpse the contents of the nest when the bird rose up in its threat pose, I tied my mirror to the pole and lifted it upward. When the mirror reached the level of the dove, it changed its tactics and dropped to the ground, where, hopping and limping, quivering its wings and vainly beating them, it retreated so slowly that it was easy to believe that it had shattered a bone against the stick and was unable to fly. Nevertheless, the dove easily kept fifteen or twenty feet ahead of me as I followed at a walking pace over the open ground between the coffee shrubs. Proceeding in this fashion to the edge of the plantation, the bird reached a dense growth of bushes and vines that put an end to its demonstration. Now it flew easily over the obstacles, to alight on a log in plain view and resume its antics. Unable to follow through the tangled growth, I counted my paces back to the nest, where my mirror revealed a single half-grown nestling. The dove had led me two hundred feet away from its young. Two features of this performance call for special notice: the attempted defense preceding injury simulation, and the interruption of the display when the dove came to the edge of the open ground.

Shore birds may simulate injury as well as give the rodent run; a single species, such as the Western Sandpiper, may try both types of lure as occasion demands. Like the good strategists they are, they may advance a long distance to meet approaching danger and divert it from their progeny. Taverner

(1936) vividly described the performance of a Killdeer that nested on open prairie in Manitoba: "If a horse or a cow grazed too near the bird lay low until almost the last minute and then flew suddenly into its face with a great outcry. The animal invariably staggered back at the unexpected onslaught and circled the spot in confusion while the bird, its object accomplished, after a few dives and expostulations returned quietly to its nest. However, when a man or dog approached, it flew to meet him from afar, indulged in all the broken-winged helplessness with which we are familiar and led the enemy far from the nest before returning to it. The two schemes of action happened too often that summer to be the result of casual accident. The reaction in respect to the dog was particularly interesting. The dog, slavering and all excitement, pursued the bird closely. Every moment we expected to see its jaws close upon its expected prey but each time that prey escaped by a margin so small that its hopes of success were raised rather than dampened and the chase was hot. As they drew away from the immediate neighborhood the bird took less and less chances until finally the dog was tearing futilely around far from the point of encounter and evidently without thought of returning to it."

F. H. Allen (1936) gave a detailed description of the actions of an injury-feigning Killdeer: "The bird lies on her belly with both wings fluttering and turned front edge down, sometimes rolling over somewhat to one side and fluttering only one wing. When both wings are fluttered, the bird looks at a distance as if it were lying helpless on its back. This is partly because of the angle at which the wings are held and partly because the upper tail-coverts are raised and fluffed out, showing a great deal of whitish about the base of the tail strongly suggestive of the white of the bird's belly. . . . When I first saw this performance to good advantage, I was completely fooled for a time and really thought the bird was on its back. It would obviously be of use to the bird thus to deceive its enemies, whereas actually to turn over on its back would make a getaway difficult."

A few kinds of birds increase the realism of their injury-feigning displays by dropping down as though dead. Van Someren (1956) described the behavior of one of these birds, the White-throated Forest Bulbul of East Africa: "Typically the hen sits deep and close . . . and on occasion I have lifted her up with my finger to examine the eggs. To stroke her as she sits is often possible, but if taken by surprise she slips off, drops to the ground and flutters, then lies on her side or back with wings half open and head lying limply to one side, simulating death. Presently she turns over and 'staggers' a few feet with trailing wings and outspread tail; then she topples over and lies panting; then on another few feet with 'broken-wing'; then a flutter and collapse, and she lies on her side as if dead!" The same author describes a rather similar performance by the Yellow-moustached Olive Bulbul. Another bird that simulates death is the White-eared Honeyeater of Australia (Blair et al. 1950). Unfortunately, I have found no account of the behavior of a death-simulating bird in the presence of an animal intent upon devouring it. Unless the seemingly dead bird is capable of almost instantaneous recovery, its reaction to disturbance at the nest is clearly dysgenic. No instance of simulating death by a New World bird has come to my attention.

Occasionally injury-feigning displays are amazingly prolonged. Ground-nesting goatsuckers are among the most persistent actors; the longest-continued exhibitions that I have watched were given by members of this family. Early one August, I witnessed a surprising demonstration by a female Whip-poor-will in Michigan. As I descended a leaf-strewn slope with scattered low trees and bushes, the goatsucker arose from the ground where she had been resting and alighted on the upturned end of a fallen branch, five or six yards ahead of me. Here, perching crosswise, she puffed out her plumage and uttered a fairly loud, long-continued hiss, punctuated at intervals by a throaty note that suggested the first syllable of the usual *whip-poor-will* call. At the same time, she relaxed her wings, brought them forward until they almost met in front of her breast, and beat them rapidly, making the barred primary feathers strike against the dead branch where she rested. For several minutes she continued this vigorous demonstration, the picture

of distressed parenthood, then flitted to another spot to resume it.

While I continued to stand in the same place, the Whip-poor-will circled around me, always alighting on a branch within a yard of the ground to continue her hissing, moaning, and vigorous wing shaking. How long she might have continued this display, I do not know, for, after she had been performing tirelessly for about fifteen minutes, I left in compassion for her evident distress. Near the point whence she first arose was a single egg, cold and wet, which later proved to be infertile. On every one of our visits to this spot, the female goatsucker behaved in the same way; finally we found a downy chick, about a week old, resting on the ground about 25 feet from the abandoned egg. Twice this female, who was molting, led us for about 150 feet, displaying on successive perches as she flitted ahead of her advancing visitors.

A parent Pauraque guarding feathered young tried just as long to lure me from them. An equally prolonged display was witnessed by the Moreaus (1937) in a tropical African forest. While they examined the nestlings in a frail, low nest of the shrike *Nicator chloris*, the presumed mother arrived and dragged herself about through the ground litter, fluttering as though helpless, and continuing this wonderful demonstration for a quarter of an hour.

The animal before which birds have most often been seen, and recorded, giving distraction displays is, as we would expect, the human observer himself. We should not conclude from this, as some have done, that selection by man has been responsible for the evolution of the more realistic exhibitions of injury simulation. In the first place, man has until recently been very thinly scattered over vast areas of the earth where other predators are far more abundant and more likely to have been the selective agents. Second, given before predatory man, such displays would be disastrous. Certainly, for many thousands of years, men have had enough wit to learn that distraction displays indicate the proximity of eggs or young birds, which the hungry savage would doubtless hunt and find. Moreover, with his missiles, he might kill the parent, who can simulate injury without risking her life only before weaponless animals and sympathetic naturalists. As a selective agent, predatory man would inhibit rather than promote the evolution of distraction displays, for often they would cause the death of parent and young together.

Next to man, the animal before which birds have most frequently been seen giving distraction displays is man's satellite, the dog, who can rarely resist the excitement of chasing a bird who tempts it to pursue. Other mammals before which birds have been known to display are otters, weasels, stoats, foxes, coyotes, agoutis, and deer. Although snakes are a principal enemy of nesting birds, the latter seem more often to attack these inveterate pillagers of their nests than to display before them. Attempts to lure serpents have rarely been recorded. Although I have many times surprised snakes in the act of swallowing eggs or nestlings, I have only once seen a bird display in front of one. This bird was a Black-striped Sparrow with fledged young. In the manner already described, the parent ran over a pasture a few inches ahead of a slender brown ground snake two or three feet long, which pursued the bird for several yards before the latter flew up into a shrub.

Occasionally a bird displays in front of another of a different species. Selous (1927) watched a male Kentish Plover lead away a much bigger European Oystercatcher "scuttling about over the ground just in front of him, with his tail spread and feathers all ruffled out. This conduct quite surprises the big bird, who makes several runs or starts at the little one almost under his feet, then settles himself down quite near the nest, when the latter desists." A female Red-crowned Ant-Tanager (a member of a family that has seldom been seen giving a distraction display) hopped off her nest and fluttered along the supporting branch with wings down and tail spread, in front of a Blue-diademed Motmot that had alighted nearby. Then both parents drove away the larger, occasionally predatory bird (Willis 1961).

We have abundant evidence that injury simulation saves eggs and young from predatory animals, especially mammals. A Killdeer, as has just been told, led a dog far from its nest. One of the most effective acts of injury simulation that I have seen was performed by a Black-striped Sparrow. A pair with two

juveniles from an earlier brood and a fledgling barely able to fly were foraging beneath the palms and colorful shrubbery of a tropical park. Presently a dog came along in company with a nursemaid and child and, espying the birds on the bare ground, rushed at them. I was sure that it would catch the fledgling, who was far from cover; but, quick as a flash, a parent interposed itself between its offspring and the quadruped. Fluttering over the ground just ahead of the eagerly pursuing canine, the bird continued to toll it onward until both passed from view, while the fledgling hopped to safety.

Another successful performance was given by a Pauraque. As I watched from a blind at dawn before a "nest" of this goatsucker where one parent was incubating, the other emerged from the neighboring light woods followed by an opossum. The bird fluttered on the ground in front of the advancing marsupial, waiting until it was about a foot away before rising to flit ahead and grovel again. Ambling onward at a uniform speed, the animal followed the bird in a wide semicircle around the nest, almost bumped into my blind, and finally disappeared into the woods whence it had come. The other parent left the nest to join the display but hardly influenced the opossum's course.

Less successful were the attempts of a male Chestnut-backed Antbird to lure away agoutis that came close to his low nest beneath a forest tree, eating fallen fruits. Largely vegetarian, but perhaps relishing an occasional egg or nestling to vary their fare, these big rodents hardly tried to catch the apparently crippled bird. Nevertheless, his repeated displays seemed to arouse their interest in the nest, which one finally sniffed, and might have upset if I had not intervened (Skutch 1969*b*).

The Incidence of Distraction Displays in Individuals and Species

Scarcely any facet of the reproductive behavior of birds is so individually variable as the performance of distraction displays. Birds of the same species commonly build nests of the same form, incubate their eggs in the same pattern, and feed their young in the same manner. But of two neighboring birds of the same kind, one will display to the human visitor and the other, with similar provocation, will consistently fail to do so. One reason for this variation is the nature of the immediate surroundings of the nest. A bird does not simulate injury on the ground unless it has a clear stage for its performance; to do so amid impeding vegetation might result in its entanglement and capture by the animal to which it displayed. Although a member of an injury-feigning family, the Buff-rumped Warbler rarely displays, probably because its covered nest is usually built on a bank, often facing a stream in which the bird could hardly perform. I first saw a Buff-rumped Warbler give a distraction display at the thirty-fourth nest of the species that I found. This nest was on the side of a mossy rock, with a strip of clean pasture between it and the river bank. On this excellent stage, the female sometimes simulated injury rather unconvincingly.

Yet with similar or identical surroundings, one individual will simulate injury and the other will not. At one nest attended by both parents, the male will display and his mate fail to do so; whereas, at another nest of the same species, it is just the reverse. In the Costa Rican highlands, I found a number of nests of the Slate-throated Redstart, all in niches in a low bank above a path that ran between a pasture and the tangled edge of the forest. The clear pathway below their nests offered equal opportunities for displaying to all these redstarts. Yet, of seven whose nests survived until hatching, only five performed at one time or another; two were never seen to simulate injury. The difference was evidently caused either by the diverse genetic constitution of these female redstarts or by their individual experience.

A bird may give a distraction display at any stage of its reproductive cycle, from the beginning of incubation up to, and often after, the departure of the young from the nest, and rarely even before the eggs have been laid. But the display is most often witnessed, and in its most convincing form, when the eggs are about to hatch or are hatching, while the parents are brooding tender nestlings, and as the young leave the nest. Mouseley (1939) concluded from studies of the Common Snipe and Spotted Sandpiper that the performance "took place only at a time when a cycle of the breeding period was either just beginning, or ending, i.e., the com-

mencement of the incubation period after the laying of the eggs, and the climax of this period, when the young had appeared, or were about to appear." This is not invariably true, even of these species. Miller and Miller (1948), who studied thirty-five nests of Spotted Sandpipers in Michigan, witnessed the full injury-simulation display on every day of the incubation period from that of laying the second egg to that of hatching. Nevertheless, they noticed a marked increase in the number of displays during the interval of two or three days from the hatching of the first egg until the chicks' departure from the nest. Some of these sandpipers, however, were not seen to display at any stage of the whole breeding cycle. Gramza (1967), who studied the distraction displays of brooding female Common Nighthawks by approaching them in a consistent manner, learned that they permitted the closest approach and demonstrated most conspicuously during and shortly after the time of hatching.

Slate-throated Redstarts were somewhat more inclined to simulate injury as they began to incubate, and as the day of hatching approached, than in the middle of the incubation period. One bird displayed soon after she started to incubate, but not thereafter, although she successfully raised her nestlings. A Collared Redstart—a wood warbler so confiding in man that it is sometimes called "the friend of man" by Costa Rican mountaineers—nesting in the same bank, simulated injury from three days before her eggs hatched to three days after. On many visits to numerous nests of the Orange-billed Sparrow of lowland rain forests, I have seen this largely terrestrial bird give a distraction display only once: after brooding newly hatched nestlings until I came very near, their mother gave an excellent demonstration.

Pickwell's study of the Horned Lark threw light on this variability in the performance of distraction displays. At one nest he found that the female would flutter over the ground, as though in distress, if driven off within two minutes after her return. If she had been incubating as long as five minutes, she would leave while her visitor was still a long way off, and as unobtrusively as possible, trying to escape being seen rather than to lure him from her eggs by simulating injury. Such displays were most frequent on very cold days and in the dusk of early morning and evening. Unlike the Slaty Antwrens (p. 402), these larks showed every stage between inconspicuous departure while the intruder was still far from the nest and departure with a distraction display when he had almost reached it (Bent 1942).

Although birds of the most diverse families simulate injury, the habit is by no means uniformly distributed throughout the whole avian class. Injury simulation and other distraction displays occur with some frequency among ostriches (Struthionidae), loons (Gaviidae), ducks (Anatidae), grouse (Tetraonidae), quails and partridges (Phasianidae), cranes (Gruidae), rails (Rallidae), jaçanas (Jacanidae), oystercatchers (Haematopodidae), plovers (Charadriidae), sandpipers (Scolopacidae), avocets and stilts (Recurvirostridae), phalaropes (Phalaropodidae), thick-knees (Burhinidae), seedsnipes (Thinocoridae), coursers and pratincoles (Glareolidae), jaegers (Stercorariidae), skimmers (Rhynchopidae), sandgrouse (Pteroclidae), pigeons and doves (Columbidae), owls (Strigidae), goatsuckers or nightjars (Caprimulgidae), antbirds (Formicariidae), larks (Alaudidae), titmice (Paridae), babblers (Timaliidae), bulbuls (Pycnonotidae), thrushes (Turdidae), Old World warblers (Sylviidae), wren-warblers (Maluridae), Old World flycatchers (Muscicapidae), wagtails and pipits (Motacillidae), honeyeaters (Meliphagidae), wood warblers (Parulidae), and finches (Fringillidae). In many other families distraction displays have been less frequently recorded; in still others, displays that are more than rudimentary seem to be absent (correspondence in *Auk*, 1935 and 1936; Frederick V. Hebard, MSS).

Of the factors that determine the occurrence of distraction displays, none is more influential than the character and situation of the nest. Ostriches, loons, ducks, grouse, quails and partridges, cranes, rails, oystercatchers, plovers, sandpipers, avocets and stilts, phalaropes, thick-knees, coursers and pratincoles, jaegers, skimmers, goatsuckers, and larks commonly nest on or near the ground, in an open nest or none at all. Pigeons nest from ground level to high in trees. Low, open nests are frequent in the passerine families listed above. Many kinds of wood warblers build on or near the ground;

others nest high in trees. It is of great interest to find some of the latter, such as the Black-throated Green Warbler, Pine Warbler, and Yellow-throated Warbler, falling from their high nests to creep with drooping wings over the ground, then sometimes returning to flutter along horizontal branches near their nests. Among finches, injury simulation occurs chiefly in species that breed on or near the ground. In the largely arboreal oriole family (Icteridae), the habit has been noticed principally in the ground-nesting Bobolink and meadowlarks. Even the ground-nesting Ostrich grovels and beats its wings like some little songbird. Equally surprising, the ruse is sometimes practiced by raptors, particularly the owls, including not only the Short-eared and the Snowy that nest on the ground but also the Long-eared and Great Horned, which breed in trees, often high. Diurnal raptors less often give distraction displays, but they have been reported in the European Sparrow Hawk, Hobby, and Merlin.

Birds that nest in burrows, including petrels, kingfishers, motmots, bee-eaters, rollers, and many swallows and ovenbirds, seem never to simulate injury, and they rarely give any other form of demonstration, such as darting at the person who disturbs their nests. Nevertheless, many of these underground nesters are most devoted parents, sometimes clinging so steadfastly to their eggs or nestlings that they may be lifted from them, while after being driven from the burrow they evince concern for their progeny by loud cries. Their lack of all but vocal demonstrations is evidently related to the situation of their nests. An animal digging down to the nest chamber from above would not be likely to notice the parent flying from the burrow's mouth some feet away; whereas, if the predator crept into the burrow through its entrance, the parent would be trapped. Accordingly, distraction displays by burrow-nesting birds would be either ineffectual or impossible.

Likewise, nesters in holes in trees, termitaries, and similar cavities, including trogons, swifts, hornbills, toucans, barbets, woodpeckers, woodcreepers, and many cotingas and swallows, almost invariably fail to do anything more effective than emit cries of distress when their nests are threatened. Exceptional among the hole nesters are the titmice and chickadees, which occasionally simulate injury while clinging to a branch, especially when their young cry in pain or fright; and the White-breasted Nuthatch, which with wings widely spread sways slowly from side to side while perching upright near its nest cavity (Kilham 1968*b*).

The virtual absence of the broken-wing stratagem in families of birds whose open nests are seldom built on the ground, including hummingbirds, American flycatchers, crows and jays, vireos, honeycreepers, and tanagers, suggests that among pigeons and wood warblers the habit may have arisen among ancestors that built on or near the ground and persisted while they gradually extended their nesting higher into the trees. Flycatchers, including some of the smallest and weakest, often dart at an intruder and threaten with angry, castanetlike snappings of their bills; while hummingbirds, honeycreepers, and tanagers sometimes attack a snake or other small assailant of their nests but usually fail to demonstrate in the presence of man.

The widespread occurrence of distraction displays among pigeons is of special significance. The whiteness of the one or two eggs, laid on open and often frail nests, appears a glaring exception to the rule that eggs laid in exposed situations are pigmented to make them less conspicuous. However, from the day the first is laid, the two parents keep them almost constantly covered, sitting steadfastly and often, when threatened, leaving only in time to save their lives. With them, an early and inconspicuous departure would be highly imprudent, for it would leave eye-catching eggs exposed to the predator's gaze. After the intruder has come near, the white eggs are less likely to hold its attention than the departing pigeon, who makes a final attempt to draw the enemy from its nest by its usually proficient imitation of a wounded bird. Such displays persist even in the Galápagos Dove, which is quite fearless of man, yet even on islands devoid of indigenous enemies it simulates injury like other doves when approached at its nest by a human (Swarth 1935).

Attempts to lure predators from a nest are rare or absent among birds that breed in crowded colonies on the ground, rock ledges, or even in trees, including penguins, pelicans, cormorants, gannets, boobies, flamingos, auks, murres, most gulls and terns, and, among arboreal passerines, oropéndolas in America and colonial weavers in Africa. In a dense colony, a predator could be enticed from one nest only at the price of passing over others, causing confusion and havoc. Among the few colonial ground nesters that give distraction displays are the Black Skimmer, Parasitic Jaeger or Arctic Skua, and American Avocet, all three of which breed in loose rather than compact communities.

A number of these loosely colonial birds may simulate injury simultaneously. Williamson (1949) frequently saw groups of two to five Parasitic Jaegers "grovelling and whimpering in various places at a radius of from thirty to fifty yards from the observer." Gibson (1971) watched as many as thirty-five American Avocets simultaneously engaged in distraction displays; he saw a deer and a coyote effectively distracted by these demonstrations. Even less gregarious birds sometimes help the parents of threatened young to distract an intruder. Neighboring Semipalmated Plovers join the parents in distraction displays, and the same sometimes occurs among Hooded Warblers (Swarth 1935, Grimes 1936). Five McCown's Longspurs lined up almost in a row, leading away a man who had caught and then released one of the nestlings (Bent et al. 1968). In Australia, when a nesting pair of certain chats and honeyeaters flutter over the ground uttering distressful cries, they are sometimes reinforced by other individuals of the same species, who may or may not be nesting, and all simulate injury (Chisholm 1936). To this meager list of cooperative injury simulators may be added a New Zealand stilt and an Indian pratincole (Armstrong 1947).

Such sympathetic injury simulation is certainly not common. The nearest approach to it that I know among the inland birds of tropical America is the distraction display of the Creamy-breasted Canastero or Black-winged Spinetail. This member of the ovenbird family builds a massive covered nest amid the sword-leaves of the *Puya* bromeliad on the bleak Peruvian puna. When a man approaches a nest, the parents skip about on the leaves, rapidly and with great agility. Raising and lowering their closed tails, they utter a loud, harsh, metallic *brreee*, not heard in other circumstances, which draws neighboring canasteros, alone or in pairs, until as many as fifteen may be hopping and calling among the leaves. They continue until the intruder departs (Dorst 1957). This collective display is the more surprising because it occurs in a family of birds that generally live in pairs or alone, and in which distraction displays of any sort have rarely been recorded.

The Origin of Distraction Displays

We can only speculate about the origin of distraction displays as, naturally, the fossil record preserves no trace of this behavior. Their occurrence through the whole length of the phylogenetic tree, from great, primitive, flightless Ostriches to tiny, advanced, arboreal warblers and finches, tempts one to postulate that the displays are inherited from the original progenitors of birds, possibly even from a reptilian ancestor. However, displays of this sort apparently are not known in contemporary reptiles, and it is doubtful whether they could be profitably employed by flightless animals much less powerful than Ostriches. Their success depends not only upon the parent's diverting the would-be predator from the young but also upon preserving its own life, for it is needed to attend its dependent offspring and continue to propagate its kind. By timely flight, the bird can almost invariably escape the pursuing terrestrial predator, but flightless parents lack this safeguard.

Another difficulty of the theory that all distraction displays are derived from a common ancestral display is their great diversity. The rodent run of a sandpiper or a towhee, for example, involves movements so different from those of a goatsucker groveling and beating the ground with its wings that they do not appear to be homologous or derived from the same source. Moreover, if certain phylogenetic views are correct—for example, that the wood warblers, which often simulate injury, are derived from the vireos, which do not—then we must conclude that the habit has been lost and regained in the evolutionary history of the former. On any phylogenetic view, it is highly probable that as environments and nesting habits changed, making injury simulation now profitable and now futile, it has been practiced and neglected in a long lineage, perhaps repeatedly.

A more promising approach to the problem is to seek a foundation widespread, if not universal, in the avian class on which distraction displays could be built rather rapidly, as evolutionists count time, when they become profitable. This foundation appears to consist of emotional attachment to nest and young and a modicum of insight or intelligence, by which we understand the individual's capacity to adjust its behavior to shifting circumstances. That the parent bird's attachment to its offspring is not just a mechanical reaction but charged with emotion is hard to doubt, when we witness its reluctance to flee even when self-preservation prompts this course and hear the cries of distress that many utter, even when they can do nothing effective to protect their young.

As to intelligence or judgment, scarcely anything a bird does requires more of it than injury simulation. To prevent disastrous entanglement amid vegetation, the bird must avoid performing unless it has a clear stage. It must time its movements with precision to avoid discouraging the pursuing predator, on the one hand, or being caught by it, on the other. In scarcely any other situation in its life does it exercise such agility and alertness. Where intelligence enters, we expect individually variable conduct, and, as we have seen, distraction displays are far more variable than other aspects of reproductive behavior. Compared with their distraction displays, direct attacks by birds upon intruders at their nests are often stupid; when they treat us that way, we might often catch or kill them with our hands, if so inclined. But what man or dog can catch the injury-feigning bird? Far from suffering from delirium or muscular inhibition, birds are never in fuller command of all their movements than when they act as though crippled or helpless.

Injury simulation is much less likely to be the expression of a "blind instinct" than are, for example, incubation or migration. A bird incubating, especially for the first time, can hardly foresee the outcome of its patient sitting; what is happening inside the opaque shell is hidden from it. It would be surprising if a young bird setting out on its first long migratory journey has any idea of its destination. But when a bird gives a distraction display, all the pertinent elements—nest, predator, luring parent—are patent, permitting the bird to be cognizant of exactly what it is doing and to adjust its movements accordingly.

Strong emotion and thwarted impulse generally take overt expression, such as cries or movements of the limbs. A bird abandoning its nest with reluctant slowness and wings aquiver with excitement may be hopefully pursued by a hungry predator, even without any special display. If the nest is thereby saved, the bird may remember this experience, and it may repeat the performance, perhaps with elaboration, when its nest is next threatened. This bird's progeny should be equally able to learn the favorable consequences of a conspicuously slow retreat from a threatened nest, with the result that its lineage increases more rapidly than that of less perceptive individuals of its species.

The similarity of distraction displays within a species and even within a family of birds is proof that the patterns are determined genetically. We have no evidence that a behavior trait, individually acquired by a parent, can be inherited by its offspring. But if an action, frequently repeated by individuals of similar genetic constitution who react similarly in similar circumstances, promotes survival, it may in the course of generations be supported by mutations that fix it firmly in the genotype—a process known as "organic selection," which simulates Lamarckian evolution but does not require the inheritance of individually acquired characters. Armstrong (1949) pointed out how the effectiveness of the distraction display is increased when a bird retreating from a threatened nest incorporates into its movements elements taken from other contexts. Thus, when luring enemies from their nests, some birds simulate chicks begging for food; others perform courtship movements or those used to threaten rivals. Probably a relatively slight genetic alteration would suffice to transfer such innate patterns from their original context to that of slow retreat when a nest is imperiled.

Whatever the sources of some of the movements that enter into distraction displays, and however much the conflict of tendencies to attack and to escape, or to brood and to flee, have influenced their genesis, today the more specialized distraction displays are integrated patterns of behavior, like courtship or nest building, although they are used more

sporadically, since the circumstances that evoke them are more variable. To conclude, as we do here, that mind has influenced the evolution of distraction displays helps us to understand the evolution of mind itself. If intelligence never affected the course of evolution, evolution could never promote intelligence (Skutch 1954–1955).

Other Parental Stratagems

The parental stratagems of birds comprise all those expedients, those lures, wiles, and maneuvers, by which they try to save their eggs and young from predatory animals, except threats and actual attacks. In addition to the diversionary displays that we have just surveyed, they include covering or hiding the eggs during the parents' absences and shielding flight, which we earlier considered (pp. 176, 301). A few other stratagems worthy of notice remain.

We have seen that two strategically sound procedures are available to a brooding bird who spies a potential enemy: it should either flee immediately, while the intruder is far away and perhaps unaware of its presence, or it should wait until the approaching animal has obviously discovered its nest. Birds of tropical forests, where predators abound, generally choose one of these alternatives. But there is one special contingency to which neither is quite applicable. The parent may be surprised while standing on the nest's rim, feeding or guarding its babies. It is then much more conspicuous than while sitting in a nest, but the movement of jumping down into the cup, like the movement of flight, may betray the nest to a nearby predator. The bird is in a peculiarly compromising situation.

The only bird that I have seen rise in a special way to meet this special contingency is the Cedar Waxwing. Revisiting a nest ten feet up in a pine tree, I found one parent brooding the two full-grown nestlings, while the other rested on the rim. The latter stood quite erect, with bill inclined upward and crest laid back, and he preserved this immobile, statuesque pose as long as I watched. On a later visit, I again found a parent resting on the rim in the same sentinellike attitude. Evidently it had just arrived to feed the nestlings, for its throat was distended with food for them. For twenty minutes it remained stretched up rigidly, immobile except for a slight swallowing movement. Only toward the end of this interval did its head begin to sag sideward, doubtless from fatigue. Then I shook the small tree, making it vibrate to its topmost shoot, but the waxwing resolutely maintained its statuesque pose. Finally, to further test its constancy, I climbed the tree. The bird remained steadfast until my head was almost level with it, then flew silently away.

The waxwing was evidently trying to escape detection by taking a statuesque posture, as bitterns, herons, and owls sometimes do. Its chance of avoiding notice, already good because of immobility alone, was increased by its stiff, erect, unbirdlike attitude. Herrick (1905) described the same behavior of the Cedar Waxwing and suggested that "the olive-gray, rod-like body of this bird might . . . when surrounded by foliage, have been readily mistaken for a short stub or a truncated branch of a tree." When surprised in a leafless tree by Mountfort (1957), a Hawfinch instantly assumed a slim, upright stance and stood immobile against the trunk, permitting a close approach in an effort to escape detection, much in the manner of a waxwing at its nest.

A similar stratagem is habitually employed in a somewhat different context by the Common Potoo and related species. These large, goatsuckerlike birds of tropical America lay a single big egg on the broken end of a stub or in a knot hole on a leaning branch, often in a depression so shallow that one wonders how it remains there during a month or more of incubation. When incubating or brooding at ease, the potoo sits upright with its body contracted, feathers fluffed out, and head horizontal. The moment it notices a potential enemy, whether a man, quadruped, or big bird flying overhead, it elongates and compresses its body until it appears about one-half longer and only half as thick. At the same time, it brings its neck and head into line with its vertical body and tail. Its big, yellow eyes are almost closed. In its cryptic plumage of blended grays, browns, and black, it is easily mistaken for part of the stub on which it rests (plate 112). The transformation is effected about as rapidly as it can be made without perceptible movement, which

PLATE 112. Common Potoo incubating a single egg that is precariously lodged in a knot-hole on the side of a nearly vertical branch. In the elongated cryptic posture, the grayish bird resembles a stub.

would attract attention. When the alarm has passed, the potoo becomes short and stout much more gradually than it became tall and thin. Even when perching away from a nest, the potoo assumes the cryptic pose if disturbed (Skutch 1970).

One of the most extraordinary parental stratagems was practiced by some big Striped-backed Wrens while Cherrie (1916) collected their nest in the Orinoco valley. It was one of half a dozen shapeless masses of fine rootlets, feathers, and seed down of the silk-cotton tree. This seemingly dilapidated structure was softly lined and held four fresh eggs, while the neighboring masses were empty. Instead of vainly crying or scolding the collector, the wrens immediately set to work carrying mouthfuls of feathers and seed down from one nest mass to another, without going near the real nest. But when they realized that their enemy was not to be deceived by this artifice, they turned their attention to the nest with the eggs and worked so efficiently, packing its doorway with the same soft materials even while he was cutting down the supporting branch, that by the time the nest was on the ground no entrance was visible.

The same author told how, in the same region, the Bicolored Wren selects for its own use the abandoned nest of some other bird, usually the oven-shaped structure of a flycatcher. A single isolated tree on the savannas may support from three to eight nests of this type. The birds line one of the older, more dilapidated of these structures with soft, dry grasses and seed down, and lay from three to five eggs in it. If their nest tree is attentively watched, the parent wrens become active taking material into one of the newer nests, but do not approach the old one that shelters their family. Their object seems to be to divert attention from the old, apparently abandoned structure—their true nest—by making a show of building a new one.

Do these wrens resort to deliberate deception to save their eggs and young? Apparently their behavior is an example of the "false nest building" discussed by Armstrong (1947). When a bird's impulses are thwarted, or it has been intensely excited, it often performs acts that are obviously not purposive and frequently most incongruous—the so-

called displacement or substitute activities. Thus, after escape from danger, or in the intervals between fights, or while its nest is being looted, it may sing or perform the movements of preening, eating, or drinking. A female Yellow Bunting, after having driven an intruder from her territory, collected grass only to drop it. Possibly the wrens' curious stratagem developed from substitute activities of this sort. We do not know whether it is employed in the presence of potential enemies other than man, or whether it ever averts destruction from eggs or nestlings. The interpretation of such behavior is fraught with such great difficulty that we cannot be sure whether the wrens act with deliberate intention, or merely discharge, in a manner that appears intelligent to the human watcher, nervous energy diverted by external influences from its normal channels.

Wrens have been credited with another stratagem. Many kinds of these dynamic birds build more nests than they can use for reproduction, and some naturalists have supposed that the extra structures serve as "dummies" to divert predators from eggs and nestlings, which are often in a nest that is better hidden. The nests that never contain eggs are built at all seasons. Some are used as dormitories, as told in chapter 30, while others are built by polygamous males, especially of the Long-billed Marsh Wren, to entice prospective mates with a choice of sites. Dormitory nests are rarely so near the breeding nest that the artifice described by Cherrie would be noticed by a predator intent upon the true nest.

Allied in motivation to false nest building is false brooding. Often a Lapwing, Little Ringed Plover, or some other shore bird whose eggs or young seem to be in danger, squats down at a distance from them and remains sitting quietly as though covering its chicks. Such false brooding might well deceive a predator and cause it to search in the wrong place. Sometimes a parent bird, kept from its eggs by a human visitor, covers the eggs or nestlings of another bird, perhaps of a different species, as when a female Ross' Gull sat on one set of tern's eggs after another, and a Dunlin brooded Redshank chicks (Williamson 1952a). At other times, shore birds peck at the ground, as though eating, while a potential enemy is watching them, but they do not actually pick up food. This false feeding is another ruse that might mislead a predator searching for their young.

Sociable helmet-shrikes practice a quite different ruse. One coming to replace its mate on the nest is often accompanied by a whole flock. The incubating or brooding bird slips from the nest to join the party, while the mate goes on. Then the party continues onward, as though it had passed through the nest tree while foraging (Ames 1963).

A closed nest with a single narrow opening, whether it be a cavity in a tree, a construction of sticks, or a tunnel in the ground, provides better protection for its occupants than an open nest can give, but it exposes them to a peculiar peril. The enemy that takes possession of the exit has them trapped and may proceed to devour parents and offspring at its leisure. One would expect these hole nesters to develop some special measures for overcoming this danger, yet, surprisingly, in very few of them have such measures been discovered. The female hornbill, with her big beak in the narrow slit in the plug that seals her in the hole where her mate brings her food, seems to be in an advantageous situation for repulsing any monkey, squirrel, or snake that covets her eggs or young. Likewise, the larger woodpeckers are well able to defend their nests with their sharp, chisellike beaks when a toucan or other marauder comes to molest them; and doubtless the powerful crushing bills of large parrots keep at least their smaller enemies from their nest holes.

These stronger hole-nesting birds protect their offspring not by stratagems but by force. Some of the smaller ones, however, have resorted to deception. When a titmouse or chickadee is molested in its nest hole, it emits a loud puff or hiss that resembles the hiss of a snake. At the same time, it may open its mouth widely, spread its wings as far as the cavity permits, and sway slowly from side to side. At intervals it may suddenly jump upward, emitting another hiss from its gaping mouth. A display consisting of some or all of these features has been noticed in a number of species of *Parus* in both the Eastern and Western hemispheres and perhaps occurs throughout the genus. Even nestlings may hiss when disturbed. Observers have testified

to the alarming character of the snakelike sound, which persists even after one knows its true source. A small animal that heard this hiss issuing from a dim nest hole might well be deceived, and, if it looked inside, the vaguely seen, gaping mouth swaying from side to side, or occasionally lunging forward, would strengthen the illusion that a serpent lurked within.

Another hole-nesting bird that displays in similar fashion is the Wryneck, an aberrant woodpecker. Not only does it hiss when disturbed in its nest, but it also darts out its tongue at the intruder, thereby heightening its resemblance to a snake. If taken in hand, it may fall into deathlike immobility, a reaction rare among birds. Nestling Wood Warblers of the Old World also hiss explosively when disturbed in their covered nest with a side entrance. Sibley (1955) regarded such snakelike behavior as a variety of Batesian mimicry. Just as a harmless insect secures protection from predators by resembling a stinging one, or an edible butterfly escapes predation by acquiring the colors and patterns of an unpalatable species, so, apparently, do these birds that breed in closed spaces hold enemies aloof by making snakelike sounds and movements.

Chemical repellents, now increasingly used by technological man, seem only rarely to be employed by birds to hold aloof their enemies, large or small. A possible example, already mentioned (p. 108), is the use of the hydrocyanic acid released in minute amounts by withering leaves to reduce insect pests in nests. Quite different is the "bill sweeping" of the White-breasted Nuthatch. Swinging its whole body from side to side in a wide arc, the nuthatch of either sex sweeps its bill over the inside of its nest cavity or the outside of the tree that contains it, sometimes for considerable distances from its doorway. This activity, which may continue for many minutes, is most frequent during nest building and the last half hour before the nuthatch retires to cover its eggs or nestlings through the night. The bird sometimes sweeps with empty bill, or while holding a tuft of fur or fragments of a plant. Most often, however, it holds an insect that it rubs against the wood or bark until little of the body remains. Sometimes the insect so used is a large blister beetle, whose oily secretions blister the human skin. The article employed for sweeping may have been previously stored by the nuthatch in a crevice. Since the approach of a tree squirrel or chipmunk incites active bouts of bill sweeping, Kilham (1968*b*, 1971*c*), who was the first to study this curious behavior, believed that it serves to spread substances deterrent to these rodents, who compete with nuthatches for holes in trees.

A few hole-nesting birds surround their doorways with resin, apparently to hold certain predators aloof. Red-cockaded Woodpeckers regularly carve the cavities in which they sleep and nest in living pines whose heartwood has been softened by a fungal infection. Around their doorways, often for a distance of several feet, they peck through the bark, causing resin to exude freely. Usually they do this just before they retire into their holes in the evening. So important to them is the ring of resin around the entrance that, when it ceases to flow in response to their pecking, the woodpeckers desert the hole and make a new one. Although it does not prevent ants and flying squirrels from invading the cavity, the resin deters snakes, as was proved experimentally by Jackson (1974).

Red-breasted Nuthatches smear the resinous exudation of coniferous trees around the entrance to their nest cavity. During the incubation and nestling periods they continue to add drops of resin to this sticky barrier, until it becomes so thick that the gum oozes down the trunk. The parent nuthatches themselves avoid getting stuck by deftly shooting through the orifice, often without touching its edges. Nevertheless, they occasionally gum their plumage, which becomes increasingly disheveled as nesting progresses. And, just as the wasps upon which some birds depend to protect their nests occasionally prove fatal to them, so a Red-breasted Nuthatch sometimes becomes so firmly stuck to the pitch that it hangs there until it dies—evidence that the resinous barrier can prevent small creatures from reaching the doorway. When the young are about to emerge, the parents cover the floor of the entrance way with litter, which may save the fledglings from getting stuck as they pass through (Kilham 1972).

CHAPTER 32

Direct Defense of Nests and Young

To safeguard their progeny, birds have adopted the most diverse nesting habits. Some hang strongly woven pouches high in trees; some lay their eggs deep in the ground; some build great fortresses of closely interwoven sticks; some make inconspicuous nests barely large enough to hold a single egg on a tree limb; some breed on forbidding seaside cliffs and others on remote oceanic islands. Yet none has achieved absolute safety; in every case, some of the innumerable hungry mouths of the animal kingdom have found them out and exploited this source of food. To divert or confuse these enemies, birds have developed all the stratagems that we considered in the preceding chapter; but none is invariably successful. When the predator finds and is about to devour the eggs or young, they can be saved only by direct attack upon the enemy, or by threat so intimidating that it retreats. Only the larger and more powerful birds are likely to save their progeny by defying the intruder, but not for this reason do many small and weak birds fail to try.

Attack Combined with Distraction Display

Armstrong (1956) pointed out that birds who give distraction displays tend to be bold. They permit a closer approach by a potential enemy than do birds that discreetly disappear at the first hint of danger; they may attack or try to intimidate the intruder before they attempt to lure him away, or, if this ruse fails, they may return and bravely assail him. It will be recalled that the White-fronted Dove who led me so far through a coffee plantation tried first to defend its nest with vigorous blows of its wings. When I wished to see the newly hatched nestlings of a Yellow-thighed Finch, their mother faced me with one wing raised in a defiant attitude. Since I was not deterred by this threat, she dropped to the ground and hopped deliberately away with wings elevated above her back, trying to lure me onward. Neither her first nor her second act was as well done as those of the dove, but she was a much smaller, weaker bird.

A Ringed Plover that gave the most vigorous and persistent distraction display that Armstrong had ever seen was the only individual of this species bold enough to attack him. A Common Eider duck will bite the hand that touches her while she incubates, then give a distraction display if forced from her eggs. A Swainson's Warbler, who would not leave her eggs until pushed off, fluttered over the ground like a crippled bird. Within a few minutes she was back on her nest, accepting deerflies from her visitor's fingers (Grimes 1936). Such large, aggressive birds as the Great Horned Owl and Parasitic Jaeger may, according to circumstances, either strike a man or simulate injury (Bent 1938, Williamson 1949).

Sometimes, instead of giving a distraction display after a futile attempt to hold the intruder aloof by threat or attack, the parent attacks after an unsuccessful distraction display. Walking along a forest trail, I met a Bicolored Antbird holding a large insect in its bill. Its scolding *churr* suggested that it had a nest nearby; when I turned to search, the bird darted past me, so close that its wing brushed my leg. While I hunted through the undergrowth, the antbird repeatedly dropped to the ground and beat its half-spread wings against the fallen leaves, in various spots a yard or two from me. When finally I found two nestlings that had fallen from their collapsed nest, I held them on my hand in front of

their parent, who lunged forward and bit a finger three or four times. Then, spreading a handkerchief on the ground, I laid the nestlings upon it while I pondered how to accommodate them in some substitute for their ruined nest. For at least ten minutes, the parent sat on the ground nearby, guarding them. The whole was an amazing display of boldness by a small bird that had followed me closely through the woodland, catching insects stirred up by my passage (Skutch 1969*b*).

Very rarely, one parent tries to lure the intruder away, while the other resorts to direct attack. The only birds that I have known to practice this "division of labor" were a pair of Slaty Antshrikes in a Panamanian forest. When I visited nestlings of this antbird, their mother, who had been guarding them, dropped to the ground and fluttered slowly and ap-

PLATE 113. Brown Pelican defending its chicks in a threatening pose (photo by S. A. Grimes).

PLATE 114. American Bittern protecting its eggs. The widely spread wings increase the bird's apparent size and make it appear more formidable (photo by S. A. Grimes).

parently painfully away. After enticing me onward for about twenty feet, she "recovered" and called her mate. While I inspected the nestlings, he perched nearby in a belligerent attitude, spread wings fluttering, tail expanded into a fan bordered with white spots, black crown feathers erect and bristling, and the slaty feathers of his back turned outward to reveal a broad central patch of pure white, like that of the Tyrannine Antbird. Looking as big and impressive as such a small bird could, he lunged forward and nipped my fingers thrice, doubtless as hard as he could, but so gently that I hardly felt his bites.

Direct Attack by Parents

Just as most recorded distraction displays have been given before the human observer, so he has been the object of most of the recorded attacks and feints of attack by parent birds. When man is the object of the parents' anger, the feints are perhaps more common than attacks that are carried home. Even in species of which the two parents take almost equal shares in all the operations of the nest, sometimes one, sometimes the other, is much bolder in its defense. At one of three nests of the Golden-fronted Woodpecker in Guatemala, neither parent was much excited by my visits to their nestlings. At the second nest, the male became greatly agitated, although he never flew at me, while his mate appeared apathetic. At the third nest, the male remained aloof, but the female did her best to drive me away. Flying in long loops between trees on opposite sides of the trunk that contained the nest hole, she came so straight toward me that for a moment I feared that her sharp bill would pierce a cheek or transfix an eye, but always she veered aside just in time to avoid collision. On one of her swoops, however, her wing brushed the hand that I had raised to the doorway of her nest. At several nests of the Golden-naped Woodpecker, the female, but never her prudent mate, darted similarly toward me, sometimes lightly striking my head, shoulder, or arm. Since she attacked only when I was not looking at her, I could never learn with what part of her body she touched me.

Groove-billed Anis, especially at nests attended by several pairs, vigorously buffet the head of a person who views their nestlings, at times almost knocking off his hat. In Maryland, a pair of Gray Catbirds cooperated in nest defense in an interesting manner. While the male repeatedly struck the back of my head, his mate pecked with her bill and struck with her feet the hand placed over her nestlings. She appeared to understand how the human hand works. When I held it palm upward, she always attacked it from the side, as though she feared being caught by the sudden closure of the fingers. But when the back of my hand was upward, she stood upon it while, with her slightly deformed bill, she showered blows upon it, once until she drew a little blood.

The Gray Catbird's larger relative, the Brown Thrasher, is a much more formidable assailant when its nest is disturbed. Sometimes it permits itself to be touched while it broods, then jumps off, scolds and squeals as though in intense pain, and dashes at the intruder's head or hands with such vehemence that blood flows. At times the enraged bird appears to strike at the eyes. Snakes and dogs are attacked with equal vigor (Bent 1948). When Grimes (1940) touched a nest of the Scrub Jay in Florida, two of the attendants pecked his hand, drawing blood, while the third punished his ear. When he removed

the nestlings, a parent attacked the empty nest with such energy that he feared it would fall to pieces. Jays and other birds that fear to attack the intruder at their nests not infrequently direct their rage upon a surrogate. When I climbed to a nest of Brown Jays, they screamed and darted close around me. Hesitating to attack, some of them pecked branches of the supporting willow tree or ripped up nearby banana leaves. In similar circumstances, Ravens hammer on a dead limb, pull up grass by the roots, or settle on the back of an unoffending sheep and tear out its wool (Gilbert and Brooks 1924).

As one intrudes into a populous colony of such terns as the Common, Arctic, and Least, the graceful birds rise on all sides from their terrestrial nests and circle overhead in a dizzying cloud, from which they dive at the interloper with such directness and speed that he is bewildered. Sometimes their sharp bills actually strike him. Jaegers, skuas, and some of the bolder gulls also fly threateningly toward an intruder and may hit him with wing or feet. While wild geese incubate, the gander stands guard nearby and is well able to defend the nest with powerful blows of his wings. A male Canada Goose twice struck Audubon's (1840) right arm so hard that it felt as though it had been broken.

Strong-footed cassowaries, who can disembowel a man with sharp toenails, are doubtless well able to defend their nests. But it is above all the more powerful raptors who punish severely those who dare to meddle with their progeny. A female African Crowned Eagle became increasingly aggressive toward L. H. Brown (1966) as, year after year, he climbed into a neighboring tree to inspect her nest. At first, like my Golden-naped Woodpecker, she could be diverted in full attack by turning to face her. Finally, she swooped down from a distance of three hundred yards and struck the climber before he could descend to cover, tearing his shirt and furrowing his back with three deep talon wounds that had a spread of eight inches. Since a stroke on the head by those fearsome talons could have serious consequences, a foremost student of diurnal raptors prudently desisted from exposing himself to further attacks. Parent Red-tailed Hawks can be hardly less formidable (Fitch et al. 1946). Climbers to nests of the Great Horned Owl, including Bent

(1938) himself, have been severely mauled, sometimes by both parents together, who struck stunning blows, ripped clothing to shreds, and tore long, deep, profusely bleeding gashes into head, torso, and arms. Yet, at other times, these owls simulate injury like a little, harmless songbird! Snowy Owls often hang in the air precisely between the bright sun and the intruder before they dash down to strike him (G. M. Sutton in litt.).

Certainly no squirrel, monkey, or other climbing mammal less able to defend itself than one of the big cats could attack with impunity the nest of one of these powerful raptors, unless they happened to be absent. As a Coyote trotted with lowered head toward a terrestrial nest of a pair of Ferruginous Hawks, the female struck it nearly full in the face, turning the animal from its course. A moment later, the male hawk raked its right flank with his talons. With loud screams, the raptors rose to dive at the quadruped again and again, until it was far from their eyrie (Angell 1969). But when a Tayra, a big, black weasel of tropical American forests, climbed to the cavity high in a great tree where a downy young Laughing Falcon dwelt, its parent failed to save it. Guarding on a bough in front of the eyrie, she watched the animal ascend the long trunk, then with a low cry darted toward it at the last moment. As she came near, the Tayra bared its sharp teeth and she dropped away, permitting it to kill her nestling with a stroke of a forepaw (Skutch 1971).

Considering the relative sizes of the protagonists, a pair of Yellow-bellied Sapsuckers were more valiant than the snake-eating falcon. Thrice they knocked to the ground a weasel that persistently tried to climb to their nestlings in a hole thirty feet above the ground (R. A. Johnson 1947). A female Scarlet Tanager attacked with equal energy a Chipmunk that came near a fledgling who had just fluttered to the ground. Uttering loud screeches, she struck the small rodent again and again with her wings (Prescott 1965).

Common Mockingbirds, like the closely related thrashers, spiritedly drive intruders from the vicinity of their nests; they are perhaps as remarkable for their pugnacity as for their vocal mimicry. Gosse (1847) wrote an amusing account of the troubles of parent mockingbirds who attacked pigs that came to eat oranges fallen beneath the tree where they nested: "The Mocking-bird feeling nettled at the intrusion, flies down and begins to peck the hog with all his might:—Piggy, not understanding the matter, but pleased with the titillation, gently lies down and turns up his broad side to enjoy it; the poor bird gets into an agony of distress, pecks and pecks again; but only increases the enjoyment of the luxurious intruder, and is at last compelled to give up the effort in despair."

Although sometimes parent birds try to toll snakes from their progeny with a distraction display, perhaps more often they directly attack these most voracious despoilers of their nests. The disparity in size of attacker and attacked often wins admiration. When a six-foot *tigra* snake climbed into a tree whose branches interlocked with those of the tree where a nest of Rufous-fronted Thornbirds hung, the sparrow-sized parents bit or pecked its hindparts while it stretched out; when it coiled up, they hopped around a few inches from it but were not seen to touch it. Repeating sharp *chip*'s of alarm, they continued for half an hour to watch and worry the reptile, never approaching their nestlings in this interval. Despite their close watch, the serpent showed clearly that it would have plundered their nest if I had not driven it away.

The slightly larger Scarlet-rumped Black Tanagers lose so many nests to snakes that they have become the most pertinacious enemies of these reptiles. Many a time, their complaining cries have warned us that a snake was creeping through the grass or lurking amid the shrubbery around our house. With harsh notes, the tanagers follow the serpent as it glides through trees and bushes, from time to time advancing to peck or nip it, never ceasing until it has disappeared or we have killed it. Once, worrying a lance-headed tree snake in a hedge, a female tanager ventured so close that this somewhat venomous serpent struck at, and apparently hit, her. At their nests, females boldly attack these pillagers. The only instance of a snake catching an adult bird that I have seen occurred when a green tree snake seized a tanager who had assailed it when it was on the point of devouring her egg. Fluttering wildly with loud cries, the bird dangled by one wing from the serpent's jaws until I rescued her. Although this

PLATE 115. Male Red-winged Blackbird attacking a stuffed crow near the nests of his mates (photo by S. A. Grimes).

snake was nearly four feet long, it was only finger-thick, and I doubted that it could have swallowed the bird. Despite their courageous attacks, the tanagers seem rarely to save their eggs or young from devouring serpents, which when intent on prey are not easily deterred, even by bullet wounds from which they will soon die, as I learned when I shot a Mica that was beyond my reach.

My experience with these tanagers raises the question of which birds should be called shy and which bold. Although for thirty years they have nested freely around our house, enjoying protec-

tion, they remain as shy of us as they were at first, fleeing from their nests the moment they notice anybody approaching and never giving a distraction display. Yet in their dealings with snakes they are bold enough; their attacks upon yard-long snakes seem no less daring than Charles Waterton's bare-handed seizure and struggle with a large boa, as told in his *Wanderings in South America.*

Somewhat stronger birds assail their reptilian enemies more effectively. Ionides (1954) watched a pair of Black-headed Bush-Shrikes attack something in a large mango tree, from which presently fell a four-foot Bird Snake, so badly wounded on the head and an eye that it soon died. When a Pilot Blacksnake entered a hole in which Pileated Woodpeckers were apparently nesting, the male clung in front and calmly, methodically delivered hard blows into the cavity. After about twenty minutes and fifty or sixty pecks with the bird's strong bill, the intruding serpent was evidently dead (Nolan 1959). For over half an hour, a Common Mockingbird attacked a Blacksnake two and a half feet long, delivering vicious blows on its head and neck without, apparently, injuring it much (Hicks 1955). As is well known, snakes, often of considerable size, are standard prey of certain marsh birds, raptors, and other large birds. The strife between birds and snakes, which prey mutually on each other, is one of the fiercest and most unrelenting in a world of conflict.

Birds must protect their nests and young from avian as well as mammalian and reptilian predators (plate 115). One of the largest families of birds, the Tyrannidae or American flycatchers, is so called because of the zeal with which one of its members, the Eastern Kingbird, *Tyrannus tyrannus*, repulses all creatures that it regards as dangerous. With far-seeing vigilance, the kingbird perches on some high lookout near his nest, and the moment he spies a large and potentially injurious bird—hawk, vulture, crow, or sometimes a less formidable feathered creature—he flies afar to meet and harass it, sometimes striking its back or even riding upon it and pulling out feathers. The stout-hearted bird has been known to attack a low-flying airplane and repeatedly strike a bird photographer (Bent 1942).

The Tropical Kingbird is an equally zealous

guardian of his nest. Although he tolerates small, innocuous avian visitors to the nest tree, he is tireless in pursuit of predatory birds, his special enemy being the Swallow-tailed Kite, which tears the young from nests in exposed sites this kingbird often chooses. A larger relative, the Boat-billed Flycatcher, advances to assail its particular foe, the toucan, as soon as this great-beaked pillager comes within view of its nest. Like the kingbird, the Boat-bill dares to strike the toucan only in flight, when it is unable to defend its back; the moment the toucan alights and can bring its huge, brightly colored bill into action, the assailant maintains a respectful distance, even if the intruder is in the act of plundering its nest. Some of the smaller of the "tyrant flycatchers," such as the Tropical Pewee, display admirable spirit in the defense of their nests, darting with clacking bills close by the heads of intruding birds as large as the Great Kiskadee, one of the biggest of the flycatchers, and menacingly swooping over an inquisitive ornithologist.

More gregarious birds unite in attacking their flying enemies. A great cloud of enraged Common Terns pursued the Short-eared Owls that flew over their colony to prey on them. Noddies assail, and sometimes rout, intruding frigatebirds (Bent 1921). Magpies join forces to pursue hawks. One who recalls how often small, weak birds fly furiously at the much bigger despoilers of their nests, reads with amazement that Great White Pelicans watch with passive indifference while Egyptian Vultures smash and eat their eggs a few feet away (L. H. Brown and Urban 1969). Although the pelican's huge, pouched beak is doubtless more efficient for fishing than for repelling its enemies, one might suppose that so great a bulk could drive away the smaller predator.

The Limits of Parental Devotion

Although birds of many kinds exhibit great distress when a predator approaches their nest, and make frenzied efforts to drive it away, they frequently fail to save their progeny. Commonly, the parents survive, while helpless nestlings fill hungry maws. One might expect more heroic conduct by parents who display so much concern. If they felt as intensely as they appear to do, would they not fight unto death to save their cherished offspring? Are they, perhaps, merely acting a part, with no intention to sacrifice one precious feather in defense of their family?

To understand the behavior of the parents, one must consider the strategy of predatory animals. Some appear to live most hazardously, preying upon large ungulates with sharp horns and formidable hoofs, on snakes including venomous species, or on the young brood of fiercely stinging wasps. Yet in every case, the predator enjoys a clear superiority over its victim; rarely does it engage with its habitual prey in a struggle of doubtful outcome. Except when driven by extreme hunger, in the absence of its usual food, the predatory animal seldom risks its life to procure a meal. The reason for this prudence of predators is obvious, especially in the case of raptorial birds. To nourish itself and rear its young, each must make hundreds of captures in the course of a year; if as much as 1 percent of its attacks on intended victims cost its own life, these birds, with their usually slow rate of reproduction, would become extinct. Predators must become expert in overcoming their prey; the bunglers soon vanish from the earth.

The plunderers of bird nests are of two kinds: the sneak thieves, which carry off eggs or nestlings only when parents are out of sight, and the habitual predators, which pillage nests even in the presence of their owners. The former will not risk encounters with the parents; the latter are too strong for the parents, who are restrained by innate caution from coming to grips with them. Since without parental care the young are unlikely to survive, nothing is to be gained, and all might be lost, by sacrificing parental lives in defense of progeny. Not the race in which parents die nobly protecting their offspring, but that in which they preserve their own lives, so that they may breed again even if they lose their present young, will prosper and multiply over the years. Accordingly, natural selection inevitably sets limits to parental devotion, fomenting its growth up to the highest point consistent with the parents' survival, but forbidding it to pass beyond this point in any species whose nests are frequently pillaged by predators able to kill the adults. When we notice with what zeal many birds defend their young from more powerful enemies, we suspect that avian species have from time to time become extinct because parents took too-great risks in behalf of their offspring.

Not only among birds does innate prudence set a limit to parental devotion. In social insects and other animals of which the young are dependent upon adult care, we witness the same restraint in the presence of habitual predators. Salmon, eels, and other animals that never care for their progeny may lay their eggs and die; but wherever parental care is indispensable, parental sacrifice is dysgenic (Skutch 1971).

Self-defense by Nestlings

The nestlings of passerines and other small birds are nearly always too weak and harmless to hold aloof the animals that crave their flesh. While they are very young and sightless, the shaking of their nest, or almost any movement close above it, stimulates them to lift up their gaping mouths for food, just as they do when a parent arrives to feed them. Doubtless they greet in this trustful manner the snake or squirrel that comes to devour them. After their eyes open and they can distinguish their parents, the approach of anything strange causes the nestlings to crouch down in the nest and make themselves as inconspicuous as possible. They may clutch the lining of the nest so tightly that a person who removes them for examination or photography must be careful not to cut their toes on the horsehairs or other fine strands from which he separates them. If feathered young have disappeared from a nest between the watcher's visits and he is uncertain whether they left spontaneously or were forcibly removed, the examination of the lining will often provide the answer. If it is pulled up, the young have probably been torn away by a predator; if it lies flat, they have probably left spontaneously. But if visits were widely spaced, a building bird may have disarranged the lining by extracting pieces for its nest.

Often, if one touches or even looks at nestlings that have crouched down at his approach, they burst suddenly from the nest. At times a whole nestful of young birds leaves suddenly in this fashion, shooting off so swiftly in various directions that they seem to explode. Even if they can scarcely fly, they flutter to the ground and hop rapidly away. If they are caught and replaced in the nest, they often refuse to stay, but perversely jump out again as soon as they are released. Sometimes they can be persuaded to remain by holding a hand over them until they become quiet, then gently withdrawing it.

In contrast to these small birds that shrink and flee, many larger nestlings stand their ground and fight. On a visit to an islet in the Caribbean Sea where Brown Boobies and Red-billed Tropicbirds nested, I was impressed with the ability of the larger young of both species to defend themselves. As told in chapter 23 (p. 291), when I approached a booby chick clad in thick, soft, white down or an older one with dark gray feathers covering much of its body but still flightless, it vigorously bit whatever I placed within its reach. The tropicbirds nested on ledges and in niches on the steeper parts of the islet. Some of the young were already well feathered. When picked up, they uttered deafening screams, at the same time trying to bite and struggling so violently that I was glad to put them down.

The only other nestlings that I have known to struggle as wildly and make as much noise as the Red-billed Tropicbirds were those of the Ringed Kingfisher, the largest member of its family in the Western Hemisphere (plate 116). While I examined a burrow of kingfishers with a single feathered nestling, it fled down the entrance tunnel and jumped into the river, where it spread its wings, turned upstream, and flapped its way slowly against the current. When it encountered obstacles, such as stranded branches, it hooked its bill over them and scrambled across. Although I pursued through the muddy shallows of the river, the flightless young bird managed to keep ahead of me until the great leaves of a fallen banana plant halted its progress. When captured, it screamed loudly and tried fiercely to bite. For ten agonized minutes, it resisted with all its might, but at last it became reconciled to me and perched on my hand until its feathers dried and I replaced it in the burrow.

Older nestlings of night-herons defend themselves with their bills, which they direct not only at birds of other species but also against adults of their own kind who come too near their arboreal nests of

PLATE 116. Ringed Kingfisher, four weeks old, almost ready to fly from its burrow in the bank of a Guatemalan river. Few nestlings defend themselves so vigorously.

sticks. To avoid being mistaken for a stranger and treated accordingly, a parent arriving with food greets its young with an appeasement ceremony and recognition cry, which causes the nestlings to shift from aggressive behavior to begging.

Nestlings of some of the larger pigeons have a number of reactions that seem to serve for their defense. Although Scaled Pigeons spend most of their time high in the treetops of tropical American forests, they often build their thin, open platform nests in tangled second-growth thickets, a yard or two above the ground. Here they usually lay a single white egg, at least in Costa Rica. When a solitary young Scaled Pigeon is disturbed, it rises in the nest, stretches up its neck, puffs out its breast, and lifts its wings, all of which make it look much bigger. In this attitude it sways upward and backward, downward and forward, with each forward and downward movement making a loud clicking or clacking sound with its bill. As long as it feels itself menaced, it continues to perform rhythmically in this fashion.

The young pigeon's clack is produced by its mandibles in a peculiar manner. The lower mandible is pushed slightly forward, until its apex rests against the downwardly bent tip of the upper mandible. The bill is then slightly open. Apparently the two mandibles are pressed together until the lower one suddenly slips back into its normal position; the two, striking together along their whole length, emit the sharp sound. The nestling also darts forward to peck an intruding hand, and, after its feathers begin to expand, it strikes with its wings. If taken in hand, it struggles vigorously without ceasing to clack its bill, and at the same time it hisses slightly. Doubtless all this bellicose display intimidates small mammals and possibly even snakes, yet some nestlings are taken by predators.

Older nestling Red-billed Pigeons make a similar clacking sound with their bills, but not so loud as that of the Scaled Pigeon. They also strike and bite intruding fingers, but not hard enough to be painful. Downy owl nestlings hiss and clack their bills when disturbed, and sometimes they fall on their backs to present their talons to the intruder. Nestling chickadees lunge forward and hiss, much as the incubating or brooding female does (Brewer 1961).

Very different from the biting and striking of the foregoing nestlings is the young Fulmar's defense. Although many members of the petrel family, to which Fulmars belong, nest underground in burrows, chiefly in the Southern Hemisphere, Fulmars lay their single eggs on cliffs above cool northern waters. On their seaside ledges, the nestlings are exposed to attack by aerial predators, such as Great Black-backed Gulls, Great Skuas, and Ravens. When approached by a flying bird or even a human visitor, the Fulmar chick regurgitates an oily fluid that it shoots toward the intruder with considerable force and often great accuracy. The spray from a fasting chick consists of oil alone; but, if the nestling has just had a meal regurgitated by a parent, both oil and partly digested food may be ejected.

The anatomical researches of L. H. Matthews (1949) indicated that the ejected oil is secreted by glandular cells in the mucous membrane of the forestomach. That the oil is a true secretion rather than recently ingested food is evident from the observation of a bird bander who picked up a Fulmar's egg on the point of hatching. The enclosed chick made spasmodic movements, chirped softly, and ejected a small amount of clear yellow oil from its bill, which protruded from a narrow hole that it had just made in the shell. Obviously, it had not yet been fed. Such precocious ejection of oil is, however, exceptional (Lees 1950, Williamson 1952*b*).

To defend itself against a bird that would snatch it up while in flight, the young Fulmar must act promptly. Hence it shoots out its protective fluid without delaying to ascertain whether the approaching object is bird or man, friend or foe. As Duffey (1951) saw, even its own parent arriving incautiously may receive the oily spray. The chick defends all the space above and around itself within a radius of two or three feet from the point where it rests; to avoid a hostile reception, the parent alights at the edge of, or beyond, this defended territory. The parent cackles, then approaches cautiously, stopping if the young bird shows signs of alarm, but advancing with more cackling when it calms down. Finally, it reaches the chick's side and nibbles the down on its forehead, face, and neck to stimulate it to take food, for which at first it does not seem eager. The nibbling is succeeded by bill fencing between parent and chick, which resembles two people rubbing noses; then the parent opens its mouth widely and regurgitates to the chick. In the third week of its life, the young Fulmar recognizes almost immediately its approaching parent, who need now exercise less caution to avoid an oily reception.

The same method of defense is used by nestlings of certain other petrels, but apparently not by albatrosses, in which the stomach oil has a nutritive function (Rice and Kenyon 1962). Adult Fulmars and other petrels either do not spit oil or they shoot it less forcefully and accurately than their chicks. Since they can escape by flying, this method of defense is less necessary for them. Probably the habit of spitting at an intruder evolved from the reaction, widespread among sea birds, of regurgitating food before taking flight when alarmed. The Fulmar chick's ability to defend itself has contributed greatly to the success and wide distribution of its species of southern origin, for it permits Fulmars to nest in open sites that are more abundant than cavities; moreover, it frees both parents to forage simultaneously, for the better nourishment of their young (Armstrong 1951).

Hoopoes and wood hoopoes, all of which are confined to the Eastern Hemisphere, have a different way of discouraging predators. When disturbed in their hole in a tree, stone wall, or bank, nestlings first hiss and huff. If further annoyed, they turn around, scuttle across the floor of their nest, and press their breasts against the rear wall with their tails spread out over their backs. Then they discharge a foul-smelling fluid that consists of their excrement mixed with a stinking secretion of their preen gland. The female parent has similar methods of defending herself when cornered in the nest. She huffs violently and discharges a dark, musky liquid at her persecutor. When molested, nestling Hackle-necked Coucals also hiss like snakes and excrete a black, nauseating fluid (van Someren 1956).

CHAPTER 33

The Rate of Reproduction

We have now followed in some detail the long, arduous efforts that birds make to perpetuate their kind. We have studied the formation of pairs, acquisition of territories, building of nests, laying and incubation of eggs, feeding and brooding of nestlings, education of young birds, and the varied means that parents employ to save their progeny from many enemies. We have still to examine the outcome of all this strenuous endeavor. How many living young do a pair of birds produce in a year? This depends, in the first place, not only upon the number of eggs in a set, which we considered in chapter 11, but also upon the number of broods in a season. No less important is the success of these broods: from what proportion of the eggs laid do young hatch and survive until they become self-supporting? When we have answered, to the best of our ability, these difficult questions, for which we need immense amounts of painstakingly gathered data, there remains a problem of great theoretical interest, which has lately given rise to much speculation and discussion among ornithologists: What determines the number of eggs that a bird lays in a single set or in a season? Do birds rear as many offspring as they can, regardless of what happens to the excess, or as many as they need to maintain their population?

Number of Broods

Birds differ greatly not only in the number of eggs that they incubate at one time but also in the number of times they lay in a season. The number of broods that parents can rear depends on the time required to raise a single brood and the length of the breeding season. As we have seen (p. 54), a few large birds, including certain albatrosses, frigatebirds, and eagles, take so many months to rear a single young that they can produce only one brood in two years. Much more usual is a single annual brood, such as is undertaken by the majority of fairly large birds of slow development, including albatrosses, petrels, shearwaters, penguins, pelicans, many waterfowl, most raptors, the larger toucans, and many others. Where the season favorable for nesting is short, as in the Arctic, practically all the birds, including the smallest, are single brooded. Likewise on high tropical mountains, few birds rear second broods. In a year of bird study in the Guatemalan highlands, above 8,000 feet (2,500 meters), I found no evidence that any birds, except hummingbirds, tried to raise two broods. Among small birds, the wood warblers are mostly single brooded, and the same is true of a number of small tropical flycatchers with long incubation and nestling periods.

To avoid confusion, we must distinguish between second (or later) broods and repeated attempts to rear a first (or second or third) brood. The best practice is to restrict the designation "second brood" to a set of eggs or nestlings following the successful fledging of a first brood in the same breeding season. When a bird, having failed to rear its first brood, tries again, its eggs may be designated a replacement set, or a re-laying. We earlier noticed that certain marine birds who fail to rear a chick may lay again in less than a year—in about six months, in certain populations of the Sooty Tern. Is this new egg a replacement, and should not these birds be considered to make two nesting attempts in a single breeding season? Since they molt in the long interval between their first and second attempts, they evidently begin a new breeding season rather than lay twice in the same season. Similarly, the marine birds with a less-than-annual breeding season should be regarded as single brooded (p. 68). Apparently few of them even lay a replacement set if their first nesting attempt fails. Manx Shearwaters occasionally lay a second egg if the first is lost immediately after being laid, but such behavior appears to be most exceptional among the tube-nosed swimmers (Harris 1966).

Except where the favorable season is very short, probably the majority of small birds not only replace lost first broods but also try to raise one or more additional broods after a successful nesting. This is true of many passerines, hummingbirds, swifts, small woodpeckers, small kingfishers, small toucans, trogons, and pigeons. I have known Southern House Wrens and Smooth-billed Anis to lay six sets in a season, but such a large number was caused by repeated failures. Among birds that have laid five sets of eggs and raised at least four broods are the Palm Tanager in Trinidad, Mourning Dove in Iowa, and Crimson Fire Finch in Senegal (Snow and Snow 1964, McClure 1942, Morel 1967). A Ruddy Ground-Dove in Trinidad laid five sets of eggs in the same nest, but how many of these broods were successful is not recorded. Among birds that have laid four sets of eggs are Song Sparrows in

Ohio, White-crowned Sparrows in California, Blackbirds in England, and Pale-breasted Thrushes in Surinam; at least the last of these has raised all four broods (Nice 1937, Blanchard 1941, D. W. Snow 1958, Haverschmidt 1959). But the most prolific birds of which I have found a record were a pair of Least Grebes nesting on a pond in a botanic garden in Cuba. From July to the following April, they laid eight sets, a total of thirty-five eggs. Six of these broods were successful, and, of the twenty-seven chicks that hatched, twenty-four were reared (A. O. Gross 1949a).

The same, or closely related, species may raise quite different numbers of broods in different regions. At 3,000 feet (900 meters) above sea level in Costa Rica, Southern House Wrens sometimes successfully rear four broods from January to September; but at 8,500 feet (2,600 meters) in Guatemala, they raised only one, in April and May. In Ohio, the closely related, perhaps conspecific, Northern House Wren raises only two broods. In eastern North America, the Common Bluebird raises a single brood in Canada but often three, and occasionally it attempts four, in the central part of its range in the United States. Rather surprisingly, in Florida and the Gulf states its breeding season is hardly longer than in Canada and New England. In the central part of its range, where its nesting season is longest and it raises most broods, its sets average slightly larger than elsewhere and its population is densest (Peakall 1970).

The interval between consecutive broods in the same season is most variable. It tends to be shorter between the loss of eggs or nestlings and laying the first egg of the replacement set than between the successful departure from the nest of one brood and the start of laying for the following brood. Perhaps most commonly, in passerines and other small land birds, from one to five weeks elapse from the departure of a successful brood to the resumption of laying, but the interval may be shorter or longer. When it is much over a month, it is difficult to rule out the possibility that the pair has tried to rear a brood in some undiscovered spot in the period between its known nestings. In conformity with the general acceleration of the breeding schedule at higher latitudes, temperate-zone birds tend to space their broods more closely than do tropical birds. According to a compilation made by Ricklefs (1969*b*), among temperate-zone passerines the mean interval between a nest failure and the start of a new brood was 7.8 days, while that between successful fledging and the resumption of laying was 8.2 days. The corresponding figures for tropical passerines were 13.3 and 25.8 days, but the overlap of the two sets of values was great.

Occasionally parent birds begin to prepare for the next brood even before their current brood has left the nest, and they are more likely to do this if few nestlings survive. Among birds that have been discovered building a new nest while still attending nestlings are the Acadian Flycatcher, Rusty-margined Flycatcher, Northern House Wren, Common Bluebird, and American Goldfinch (Mumford 1964, Haverschmidt 1971, Kendeigh 1941, Thomas 1946, Stokes 1950). These birds did not lay in the new nest until the earlier brood had left the first nest, but sometimes the overlap of broods is greater than this. Wood Pigeons sometimes lay in a neighboring nest before the departure of their first brood, and Rock Doves deposit a second set of eggs in the nest cavity where the young of the first brood still remain (Murton and Isaacson 1962, Petersen et al. 1949). A Chestnut-collared Swift laid the first egg of her second brood about four days before the departure of the single nestling of her first brood (D. W. Snow 1962*b*). A pair of Cedar Waxwings, still attending their first brood, commonly built their second nest, in which the female began to lay in the interval extending from the day before the first brood fledged until three days after this brood left the nest. Meanwhile, the male waxwing fed the young of this brood (Putnam 1949). A female Great Tit laid the first egg of her second brood a full day before her first brood flew from the nest, and three other tits resumed laying on the day their first broods left their nest boxes (Gibb 1950). But the greatest overlap of broods that has come to my attention was achieved by an Oilbird, who laid the first egg of a new brood thirteen days before the departure of the last young of the preceding brood (D. W. Snow 1962*c*).

All the foregoing examples of overlapping broods are of species in which both parents attend the young. In such birds, the male is often chiefly or wholly responsible for feeding the fledglings while his mate incubates the following brood. More surprising is the diligence of certain hummingbirds who, unaided by a mate, undertake two broods simultaneously. A female Ruby-throat built a second nest while attending her first and incubated two eggs while she fed the single nestling of the earlier brood (Nickell 1948). The related Black-chinned Hummingbird has repeatedly been found undertaking this double task (Cogswell 1949). Although I have never known a hummingbird to attend two nests simultaneously, I once watched a White-eared Hummingbird feed a well-grown juvenile of an earlier brood in the intervals of warming her second set of eggs.

Many tropical birds have longer breeding seasons and can lay more sets of eggs than related temperate-zone birds, but they do not for this reason have a higher reproductive potential, for their more numerous broods are smaller. I have no information of any tropical American finch that undertakes as many as four broods in a season, as Song Sparrows and White-crowned Sparrows do. The usual size is two eggs; sets of three are much rarer. Even on the liberal allowance of four broods of two, or three broods of three, a female would lay only eight or nine eggs in a season. A Song Sparrow in Ohio may lay four sets of four; a White-crowned Sparrow in California laid four sets totaling ten eggs in one year and four sets totaling thirteen in the following year. Tropical wood warblers typically raise a single brood of two, three, or rarely four; northern members of the family raise a single brood of three to five, rarely six. The previously cited record of five sets laid in a single nest by a Ruddy Ground-Dove in Trinidad was matched by a similar record made by a Mourning Dove in Iowa. Rather exceptional in having a slightly greater reproductive potential than its temperate-zone counterpart is the Southern House Wren, which I have known to lay four sets of four eggs, whereas the two larger sets of the Northern House Wren rarely total more than fourteen. The Southern House Wren is a widespread, adaptable species, capable of expanding its range to keep pace with the rapid replacement of tropical forest, which it strictly avoids, by farms on which it flourishes.

The Success of Nests

When we know how many eggs are laid in a set and how many broods a bird is capable of rearing in a season, we can readily calculate how many young each pair might produce, if all went well. But rarely does all go well with nesting birds, and to learn the actual number of progeny reared to independence, we must take into account the losses of eggs, nestlings, and fledglings or precocial chicks. This subject requires an approach quite different from that followed in the preceding chapters. When we watch from beginning to end a single nesting of a species not hitherto studied, it is highly probable that the behavior we observe is typical and that other nestings of this species differ only in minor details. We have learned facts worthy of publication as an addition to the sum of ornithological knowledge. But to know whether a single nest was a success or failure tells us little about the rate of recruitment of the species. We must deal with numbers large enough to have value as statistics. Even a sample of a hundred nests may give an erroneous picture of the nesting success of the species as a whole. The nest losses of the same species may differ widely in different parts of its range, or in the same locality in successive years or even months. A study of thousands of nest record cards for the Common Bluebird showed that in New Jersey, Delaware, and Maryland 78.3 percent of the nests were at least partly successful, but in the central United States only 52.1 percent yielded at least one living young (Peakall 1970). In England, only 8 percent of 167 Wood Pigeons' eggs laid in May produced fledged young, while in August, when food was more abundant, 44 percent of 463 eggs did so (Murton 1965).

The study of nesting success is fraught with difficulties and uncertainties. We wish to know what proportion of nests are successful without interference by the observer; but often it is scarcely possible to visit a nest at intervals and examine its contents without in some measure altering its surroundings—by trampling a path through dense herbage, cutting our way through a thicket, or pushing aside concealing foliage. We do not know whether the trail we make, or perhaps lingering traces of human scent, lead a sharp-eyed or keen-nosed predator to a nest that it might otherwise have missed. So many occupied nests that I found in tropical lowland forest were empty when I revisited them a day or two later, that I could hardly resist the suggestion that I was followed by an unseen animal, ready to devour the eggs or young as soon as my back was turned.

The degree to which nest losses are increased by our visits, however circumspectly made, necessarily remains a matter of conjecture; but we can hardly doubt that nests so well hidden that we miss them are just the ones most likely to escape predators, so that our figure for nesting success may be too low. On the other hand, when we study nests in our gardens where birds are protected, or in boxes that we put up for hole-nesting kinds, we accumulate records more rapidly than is often possible in natural situations, but they may show abnormally high success. Ideally, to compute this success, we should use only nests that we find not later than the start of laying, for those with full sets, and even more those that already have nestlings, have survived perils to which neighboring nests have succumbed. They are already a selected group and will yield too high an estimate of success. Yet nests of some kinds and in some habitats are so hard to find that, unless we include these more favored ones, our sample will be too small to have much significance. In view of all these difficulties and uncertainties, we must regard most statements of nesting success as, at best, rough approximations of what happens under natural conditions. But, if we accumulate enough of these approximations, they point to certain important conclusions.

One of the first broad generalizations to be drawn from a comparison of records of nesting success was that nests in holes in trees, birdhouses, and other enclosed spaces are much safer than those built in the open, amid the foliage of trees and shrubs or on the ground. The figures in table 14 showing the relative success of open nests and those in holes and boxes in the North Temperate Zone (46 and 66 percent) are from a large compilation by Nice (1957). Lack (1954), using a somewhat different set of studies,

TABLE 14

Variations of Nesting Success of Altricial Birds with Region, Habitat, and Type of Nest

Region	Habitats	Nest	No. of Species or Studies	No. of Nests	Success of Nests[a] %	Eggs Laid	Success of Eggs[b] %
North Temperate	all	open	24 st.	7,788	49.3	—	—
North Temperate	all	open	29 st.	—	—	21,951	45.9
Wet Tropics–Costa Rica	all	open[c]	49 spp.	885	34.8	—	—
Wet Tropics–Costa Rica	all	open[c]	42 spp.	483	33.3	978	29.4
Dry Tropics–S.W. Ecuador	all	open[c]	14 spp.	1,538	50.1	3,618	45.3
North Temperate	all	hole[d]	13 spp.	—	—	94,400	66.0
Wet Tropics–Costa Rica	all	hole[d]	16 spp.	145	60.6	—	—
Wet Tropics–Costa Rica	rain forest	open[c]	28 spp.	129	24.0	—	—
Wet Tropics–Costa Rica	nearby garden and pasture	open[c]	37 spp.	208	40.9	—	—

[a] A nest is considered successful if at least one young survives to the usual time for departure.

[b] Eggs that produced young who survived to the time of nest leaving.

[c] Includes roofed and pensile nests amid vegetation.

[d] Includes nests in cavities in trees, birdhouses, termitaries, burrows, etc.

SOURCES: North Temperate Zone, Nice 1957; southwestern Ecuador, Marchant 1969; Costa Rica, Skutch 1966a (recalculated).

calculated that on average about 45 percent of eggs of altricial birds in open nests give rise to flying young, but about 67 percent of those laid in holes and nest boxes do so. The success of eggs in these closed spaces is nearly 50 percent greater than that of eggs in open nests. Smaller samples from Costa Rica show an even greater difference, nests in holes in trees, termitaries, burrows, nest boxes, and the like being nearly twice as successful as open or roofed nests built in trees and shrubbery. Because so many of the nests of woodpeckers and other birds that I studied carefully were in inaccessibly high holes in dead trees, I could not always learn how many eggs they held.

The reasons for the greater safety of nests in holes and crannies are fairly obvious. Less conspicuous than nests in the open, they are often also less accessible, and the narrow entrance excludes some of the larger predators. Yet snakes can climb a smooth, branchless trunk to plunder a woodpecker's hole, and powerful arboreal mammals may tear away the wood around the doorway to reach the eggs or young.

Why do so relatively few birds take advantage of the greater safety that holes and crannies give? Suitable cavities are often in short supply. Woodpeckers and some barbets can carve holes for themselves in more or less sound wood, and a few other birds, including some titmice and ovenbirds, can do so in wood softened by decay; but the majority of birds are incapable of digging into dead trees or termitaries, even if enough such sites were available. A flourishing forest contains innumerable leafy boughs on which nests can be built but relatively few dead trees and limbs suitable for carving holes, and in many tropical woodlands termitaries are rather widely spaced. Birds that nest in holes but cannot make them often try to wrest them from their original occupants, or they contend for their possession with other pairs of the same or different species in a similar predicament. Their contests are often not as mild as that of the Masked Tityras described on page 49, and sometimes one is injured or even killed.

Dead trees or branches in which woodpeckers or other hole nesters are rearing their young often fall, but perhaps the greatest disadvantage of hole nesting is the danger of being evicted by some other bird that covets the hole. In Michigan, where Northern House Wrens competed with Prothonotary Warblers for birdhouses and natural crannies, only 23 percent of 121 nests of the latter were successful; but in Tennessee, where the warbler was not molested by the wren, it reared young in 63 percent of 30 nests (Walkinshaw 1941). Among wood warblers, the Prothonotary is almost unique in nesting in holes, and correlated with this are a degree of pugnacity unusual in the family and exceptionally large sets of eggs. Hole nesters are not infrequently prevented from nesting by lack of a suitable cavity; but those that succeed in establishing themselves in these safer sites rear broods larger than those of similar open-nesting birds, thereby compensating for the failure of other members of their species to obtain receptacles for their eggs.

As more studies are made elsewhere than in the temperate zones where most ornithologists live and work, certain regional trends in nesting success are becoming apparent. Marchant's (1960) extensive records from arid southwestern Ecuador suggest that in the dry tropics nesting is about as successful as in temperate regions. In the rainy tropics, nest losses tend to be much higher, as shown by my records for numerous species in the Valley of El General in southern Costa Rica. Here nests within the rain forest were even less successful than those in neighboring gardens, pastures, plantations, second-growth thickets, and other deforested areas (table 14).

The astoundingly high rate of failure of nests in tropical forest is borne out by studies in other regions. Of 35 nests that I followed actually within the forest on Barro Colorado Island in 1935, only 5 (14 percent) yielded at least one living fledgling. In Belize, or British Honduras, Willis (1961) studied 53 nests of Red-crowned and Red-throated anttanagers, of which only 8 (15 percent) escaped premature destruction. From at least 147 eggs laid in these nests, only sixteen young were fledged, an egg success of 11 percent. In Trinidad, only 19 percent of 227 nests of the Black-and-white Manakin, all found before the last egg was laid, survived until one or two young fledged. In five years, the nest success of this diminutive bird, largely a forest dweller, ranged from 10 to 41 percent (D. W. Snow

1962a). On the same island, a detached part of the South American continent, nests of White-necked Thrush and Cocoa Thrush in forest were, respectively, 20 and 23 percent successful, whereas the success of nests of the Cocoa Thrush and Bare-eyed Thrush in plantations was 40 and 33 percent (Snow and Snow 1963).

These authors show that thrush nests in English woodlands were even less successful than those in the forest of Trinidad, although those in English gardens were substantially more successful than those in plantations on Trinidad, as were those of American Robins in parkland in the United States. They suggested that the real contrast may lie, not between tropical and temperate-zone birds, but between nests in relatively undisturbed habitats, where predators abound, and those in areas where man's activities have reduced the number of natural predators. To substantiate this, more studies are needed. Yet the birds themselves seem somehow aware of the difference, for birds of tropical forest frequently enter neighboring clearings to nest, whereas open-country birds far more rarely seek the forest for breeding.

Just as the nesting success of small altricial land birds increases with latitude, so it appears to increase with altitude within the tropics. Since most studies of tropical birds have been made at no great altitude, the data to show this are by no means as abundant as we could wish, but the point is so interesting that in table 15 I give what is available. The localities included in this table are not all strictly comparable in features other than altitude, because the nests on Barro Colorado were in rain forest and a narrow clearing in its midst, whereas all of those in the Motagua Valley were on a cattle and banana farm at a distance from forest. The areas where I worked in El General in 1939, at Vara Blanca in 1937 and 1938, and on the Sierra de Tecpán in 1933, all had extensive tracts of woodland adjoining cultivation and pastures; the nests were in both the woods and the clearings, so that comparison of the data from these localities probably gives a fair picture of what happens as one ascends into tropical highlands, both as to the number of nests that one is likely to find and study in a season and as to their success. At high altitudes predators, es-

TABLE 15
Nesting Success in Six Central American Localities[a]

Locality	Altitude in Meters	Period	Nests Found[b]	Species Repre- sented	Nests of Known Outcome	Success of Nests %
Barro Colorado Island, Canal Zone	25–150	Feb.–June, 1935	83	38	62	21
Motagua Valley, Guatemala	60–240	Feb.–June, 1932	96	41	68	43
Valley of El General, Costa Rica	610–700	Jan.–June, 1939	136	61	85	33
Vara Blanca, Costa Rica	1,525–1,830	July, 1937–Aug., 1938	123	47	80	53
Los Cartagos, Costa Rica	1,980–2,285	Feb.–July, 1963	81	27	41	44
Sierra de Tecpán, Guatemala	2,440–3,050	Jan.–Dec., 1933	82	28	67	55

[a] From Skutch 1966a.
[b] Excludes inaccessible nests of colonial icterids.

TABLE 16
Nesting Success of Precocial and Subprecocial Birds

Species	No. of Nests	Success of Nests %	No. of Eggs	Eggs Hatched %	Reference and Locality
Pied-billed Grebe	138	70.4	—	—	Glover 1953, Iowa
Pied-billed Grebe	107	89.6	—	—	Chabreck 1963, Louisiana
Magpie-Goose	41	—	304	28.0	Frith & Davies 1961, Australia
Canada Goose	104	46	—	—	Klopman 1958, Manitoba
Wood Duck	63	—	868	80.0	F. Leopold 1951, Iowa
Blue-winged Teal	223	59.6	—	—	Bennett 1938, Iowa
Redhead Duck	160	56.2	1,516	45.0	Low 1945, Iowa
Ruddy Duck	71	73.2	546	69.4	Low 1941, Iowa
Sage Grouse	238	35.0	—	—	Keller et al. 1941, Colorado
Bobwhite	602	36.0	—	—	Stoddard 1946, Georgia & Florida
Ring-necked Pheasant	445	23.0	—	—	Hamerstrom 1936, Iowa
Sora Rail	36	61.1	266	66.5	Walkinshaw 1940, Michigan
Clapper Rail	56	89.3	—	—	Kozicky & Schmidt 1949, New Jersey
American Coot	16	75.0	119	48.0	Gullion 1954, California
Oystercatcher	140	—	412	61.0	Harris 1967, Wales
Piping Plover	174	—	668	91.7	Wilcox 1959, New York
Spotted Sandpiper	35	71.5	119	64.7	Miller & Miller 1948, Michigan
Dunlin	42	81.0	—	—	Holmes 1966, Alaska
Namaqua Sandgrouse	23	69.5	69	68.1	Maclean 1968, South Africa

NOTE: A few of these studies also give the fledging success, i.e., the percentage of young that survived until they could fly. Oystercatcher: 47% of 186 chicks in 75 broods survived to fly. Redhead: about 70% of the ducklings lived to independence. American Coot: 26% of 119 eggs laid, or 54% of those hatched, produced independent young.

pecially snakes, are much fewer than in the low-lands; but this advantage is in some degree offset by the danger that nestlings will die from exposure while their parents hunt food in the hard, cold rains on high mountains. The greater nesting success at high altitudes compensates for the shorter breeding season and reduced number of broods. The size of broods does not increase as we ascend into tropical highlands, as it does when we travel poleward.

Table 16 gives the nesting success of some precocial and subprecocial birds. Pertinent studies of such birds are less numerous than those of altricial passerines, and generalizations seem premature, but available information suggests that birds of ponds and marshes, including grebes, ducks, rails, and coots, and likewise shorebirds, nest more successfully than gallinaceous birds (Ricklefs 1969a). Probably the aquatic habitat supports fewer predators than do woods and fields. Although snakes abound in some marshes, the larger rails may be able to protect their eggs and young from at least the smaller of these reptiles. When comparing precocial and subprecocial birds with altricials, we should remember that their relatively larger, more slowly developing eggs are exposed to predation and other perils for a longer interval, but their young leave the nest and disperse much sooner. Because of its mobility, a newly hatched precocial or subprecocial chick has a better chance of surviving than a newly hatched altricial nestling.

Among sea birds (table 17) some of those that breed on cliffs or stacks, including the Gannet, Shag, Kittiwake, and Thick-billed Murre, like the burrow-nesting Manx Shearwater, have nesting success comparable to that of hole-nesting altricial birds, whereas the success of other species is more like that of passerines with open nests. The low success of the two boobies appears to have been a consequence of unusual scarcity of food in the waters surrounding Ascension Island in the years when they were studied.

What counts for the perpetuation of any species or population is the number of young that reach reproductive maturity. Most of our studies record only the number of fledglings or chicks that leave the nest, because this is relatively easy to determine. But what we need to know in order to evaluate the true breeding success of any species is the number that survive until they are no longer dependent on parental care and, even more, the proportion of the eggs laid that produce individuals who live to reproduce in turn. This is much harder to learn, because after the young leave the nest they wander and are difficult to find. If we assume that the population of any species remains constant from year to year, and know the mortality rate of adults, we can estimate the number of individuals from each annual generation who survive to replace the lost members. To learn by direct observation how well the young survive after leaving the nest seems possible only in especially favorable conditions, such as in parks or light, open woodland—certainly not in dense tropical forests. Only a few pertinent studies are available.

As we would expect, the young are most vulnerable in the first days after they leave the nest and their rate of survival improves with increasing age, strength, and experience. In the Oxford Botanic Garden, 81 percent of 140 Blackbirds that left the nest survived for at least five days, and 66 percent to independence at fifteen to twenty days after fledging. Old males raised 70 percent of their prog-

TABLE 17
Nesting Success of Sea Birds

Species	No. of Nests	Success of Nests %	No. of Eggs	Success of Eggs %	Reference and Locality
Adelie Penguin	122	—	225	50.2	R. H. Taylor 1962, Antarctica
Manx Shearwater	131	75.0	131	75.0	Harris 1966, Wales
Banded-rumped Storm-Petrel	154	33.1	154	33.1	Allan 1962, Ascension Island
Red-billed Tropicbird	328	51.5	328	51.5	Stonehouse 1962, Ascension Island
Yellow-billed Tropicbird	821	30.3	821	30.3	Stonehouse 1962, Ascension Island
Shag	294	—	893	61.8	B. K. Snow 1960, England
Gannet	500	74.0	500	74.0	Nelson 1966, Scotland
Brown Booby	185	9.7	370	4.9	Dorward 1962, Ascension Island
White Booby	699	9.7	1,398	4.8	Dorward 1962, Ascension Island
South Polar Skua	67	—	128	24.2	E. C. Young 1963, Antarctica
California Gull	100	—	293	60.5	Behle & Goates 1957, Utah
Kittiwake	—	—	348	60.6	E. Cullen 1957, England
Fairy Tern	178	29.4	178	29.4	Dorward 1963, Ascension Island
Thick-billed Murre	1,200	59.9	1,200	59.9	Tuck 1960, Canadian Arctic

eny to independence, but yearling males only 56 percent (D. W. Snow 1958). In a tree-shaded residential area and nearby woods in British Columbia, S. M. Smith (1967) watched 95 fledglings of the Black-capped Chickadee. Twelve disappeared during their first week out of the nest, but the remaining 83 survived until the families dispersed three or four weeks after nest leaving. Eleven of the 12 casualties vanished during an exceptionally severe storm during their first week in the open. Here, as in the Oxford Botanic Garden, the young birds lived in an unusually favorable locality. Of 70 Ovenbirds fledged in Michigan, only 39 lived to leave, when thirty to forty days old, the woods where they were reared (Hann 1937).

As a rule, birds that have recently become self-supporting survive less well than more experienced and dominant adults, who by territorial behavior may relegate the former to marginal situations where they are exposed to greater perils and who take precedence at concentrated sources of food. According to Lack (1954), among temperate-zone songbirds, as likewise the Lapwing, the young survive about as well as older individuals after they are about six months old, but in certain larger birds the difference in survival rate persists for a year or more. In the Yellow-eyed Penguin, yearlings suffer an annual loss of 58 percent, but for individuals in their second and third years the mortality rate falls to 17 or 18 percent, and for experienced adults it is only 10 percent.

The Causes of Nesting Failure

Eggs fail to yield fledged young from intrinsic inadequacies but more often from external causes. Among the former are infertility, death of the embryo due to genetic defects or improper incubation, failure of the completely formed young bird to break its way out of the shell, and death of the nestling from developmental flaws or inadequate parental care, such as insufficient brooding or inequitable partition of food. Of all losses, those of the first three classes are easiest to attribute to their true source because nearly all birds, except woodpeckers, leave unhatched eggs in the nest, so that they can be opened to learn whether they contain a dead embryo or none at all. These are, as a rule, minor

causes of failure. Of 5,131 eggs of six passerine species, only 260, or 5.1 percent, failed to hatch from intrinsic causes (Ricklefs 1969a). Of 1,516 eggs of the Redhead Duck, 4.7 percent remained undeveloped and were probably infertile (Low 1945). However, when Tricolored Blackbirds tried to breed in autumn, not enough males were in breeding condition to inseminate the females able to lay eggs, with the consequence that, in a sample of 110, 89 were apparently infertile (Payne 1969a).

Repeated disturbances or a single bad fright may cause a bird to desert its nest. Difficulty in finding food, such as occurs when a late snowstorm overtakes early nesters in the north or when the drought is renewed after a shower stimulates nesting in arid regions, may cause birds to neglect their nests until eggs chill or nestlings starve. When we find an abandoned nest with spoiled eggs or dead young, it is often impossible to tell whether the parent has died or has deserted in consequence of disturbances or scarcity of food. Parental attachment increases after the young hatch, and much more provocation is needed to make the parents abandon their young than to cause them to desert their eggs. In the above-mentioned sample of six species of passerines, 2.8 percent of the eggs were deserted, but of the 2,966 young that hatched, only 0.7 percent were abandoned.

Sometimes colonial birds desert their nests in a body. Black-faced Weavers may abandon their huge colonies during nest building, while they are incubating, and even after the young hatch; why they do so has not been explained (Lack 1966). Mass desertion has been repeatedly observed in the Tricolored Blackbird, not only in their exceptional autumn colonies, but also in spring, when in one instance all but fifteen of over one thousand nests with eggs were suddenly abandoned, apparently because of food shortage. A week later, only a few devoted females remained to feed their young (Orians 1961, Payne 1969a). Lesser Flamingos nest in immense colonies of many thousands of birds. After the majority of their neighbors had finished nesting and departed, a belated group of about six thousand pairs became very nervous and deserted en masse, leaving hatching eggs and newborn young. Large groups may abandon their nests if disturbed by hyenas (L. H. Brown and Root 1971). Greater Flamingos may desert as many as 20 percent of late nests (L. H. Brown 1958). When a general exodus of young begins in a Common Murre colony in the far north, belated nesters often join the movement, leaving eggs and unfledged chicks to perish (J. Greenwood 1964). Likewise, Sooty Shearwaters abandon chicks that hatched late, when the urge to migrate overtakes the whole colony (Richdale 1954a).

When a colony of Caspian Terns was visited by men, the excited birds plunged down from the air and broke many of their own eggs (Bent 1921). The premature onset of the molt causes some birds to neglect their young. Unable to feed their chicks, molting Black-footed Penguins leave them to starve (Kearton 1930). Similarly, male Pied Flycatchers cease to feed their nestlings when they begin to molt (Creutz 1955). Before abandoning their eggs, both White Boobies and Brown Boobies remained incubating them for exceptionally long spells, waiting to be relieved by mates whose absence at sea was greatly prolonged by difficulty in finding enough to eat. About this time, many chicks starved.

When parents desert their nestlings, the whole brood succumbs to starvation and exposure. When they continue under adverse conditions to attend their family, they may save part of the brood. In this case, it is usually the younger nestlings who are undernourished, drop behind their nest mates in weight and development, and eventually die. Colonial-nesting marsh birds, such as Yellow-headed Blackbirds and Tricolored Blackbirds, seem not infrequently to suffer from food shortage, which results in death of part of the brood even when not sufficiently acute to cause wholesale desertion. In prolonged spells of cold, wet weather in England, Common Swifts are unable to pursue their aerial flycatching and many of their young die in the nest (Lack 1956a). Among the birds of humid tropical lowlands, which have small broods, starvation seems rarely to occur, except among woodpeckers, which appear to have difficulty distributing meals equitably among all the nestlings in their dim holes. Parents of large precocial families sometimes lead away their brood while belated chicks are still hatching, leaving the unfortunate ones to die. This is less likely to happen when both parents attend the nest, as in grebes and rails, than when a single parent is in charge, as in ducks.

Mechanical accidents and storms cause the failure of many nests. Badly situated structures fall from their supports. Those attached wholly or in part to monocotyledonous plants with basal growth may be tilted sideways or overturned by the unequal elongation of their supports, spilling out the eggs, as happens to Red-winged Blackbirds nesting amid growing cattails and Great-tailed Grackles that tie their nests to immature fronds of coconut palms (A. A. Allen 1914, Skutch 1954). Occasionally a falling branch knocks down a nest. Burrows may cave in or, if in river banks, be washed out by rising water. An unusually high tide, especially when augmented by shoreward-blowing winds, may bring widespread disaster to nests on beaches and coastal marshes (Pettingill 1961). Eggs and small chicks of cliff-nesting birds may be lost by falling into the sea, onto the rocks below, or into crevices from which the parents cannot retrieve them; this is the greatest cause of mortality in colonies of murres (Tuck 1960). Occasionally a small land bird, jumping from its nest, brushes out a very young nestling, or less often an egg, which it cannot replace.

In certain regions, parasitic birds cause heavy losses to their hosts. European Cuckoos, Brown-headed Cowbirds, and other species commonly remove an egg from the nest into which they drop their own. Soon after they hatch, some kinds of cuckoos heave all their foster siblings, along with any unhatched eggs, from the nest; some species of honey-guides murder their nest mates with the sharp hooks on the tips of their mandibles (p. 291). Some kinds of cuckoos and, as far as is known, all species of African widow-birds and American parasitic cowbirds may grow up along with the nestlings of their host species. Nevertheless, for each Brown-headed Cowbird that they raise, the foster parents lose on average one of their own young (Nice 1937, R. T. Norris 1947). A source of loss not greatly different from that inflicted on nest mates by European Cuckoos and Lesser Honey-guides is the cannibalism or murder of siblings by the nestlings of certain raptors (p. 287).

Parasitic insects and other arthropods appear to cause fewer losses of nestlings than parasitic birds where the latter are abundant, but occasionally these small creatures are troublesome. Maggots of blowflies of the genus *Protocalliphora* or *Apaulina* suck much blood from the toes, legs, and bare areas of the bodies of nestlings, and often they creep into the ears and nasal cavities of the young birds. They have been found on nestlings ranging in size from warblers and goldfinches to hawks and eagles, in open nests as well as in holes. At times they cause heavy mortality among nestling Mourning Doves by so persecuting them that they squirm, kick, and roll about until they fall from their shallow nests (Neff 1945*b*, Hill and Work 1947). After four nestling Purple Martins succumbed to the attacks of these maggots, 294 large ones, weighing 21.5 grams, were found in the nest (R. W. Allen and Nice 1952). In tropical America, parasitic flies of another kind (*Philornis* spp.) deposit on naked nestlings eggs or living larvae that burrow beneath the skin and cause conspicuous swellings. I have known nestlings of a variety of small birds, including a young Vermilion-crowned Flycatcher with eleven of these larvae, to recover from heavy infestations; but N. G. Smith (1968) found that nestlings of the large oropéndolas usually succumb when more than seven parasitize them. The wasps and, even better, stingless bees whose nests are often in the tree that supports the oropéndola colony hold the noisy parasitic flies aloof, so that nestlings near these social insects are rarely infested.

Tiny, blood-sucking black flies are troublesome not only in northern woods but also in tropical regions where they breed in rapidly flowing streams. These insects beset downy young Red-tailed Hawks in such multitudes that, despite the nestlings' frantic efforts to defend themselves by snapping with their bills and flapping their short wings, their eyelids were swollen shut, their bodies became covered with scabs and flakes of dried blood, and finally seven succumbed to the attacks of *Eusimulium clarum* (Fitch et al. 1946).

Ticks often attach themselves to birds that forage on or near the ground, and in colonies of ground nesters they may swarm in tremendous numbers. After almost a thousand ticks had dropped from

each of four young Sooty Terns in five days, Worth (1967) calculated that in aggregate these parasites sucked one-fifth of a cubic centimeter of blood, or about 4 percent of that in the host's body, every twenty-four hours. This would correspond to the daily loss of one pint of blood from a man or woman. Ticks may have been the principal, or at least a contributing, cause of the death of many young terns in the colony. Occasionally nests, especially those in closed spaces, are alive with lice or fleas. As a rule, however, arthropod parasites do not kill nestlings unless they are undernourished, overheated, or otherwise in poor condition. The construction of a new nest for each brood helps to reduce the numbers of these pests.

In a few species of birds, cannibalism reduces the number of young. White Storks sometimes throw one of their young from the nest, or even kill and eat it, a practice that has been called "chronism," after the Titan Chronos who, in Greek mythology, devoured all but one of his own children. Lack (1966) suggested that the storks kill primarily those nestlings that fail to make appropriate responses because they are already weak from starvation and would have died in any case. The reduction of the number of nestlings helps to adjust the size of the brood to the food supply. I have never myself seen a parent bird eat its living offspring, but once I watched a Laughing Falcon devour her chick that had just been killed by a Tayra (Skutch 1971).

Herring Gulls sometimes become rogues and prey heavily on chicks of their own kind. In the colony of these gulls on the Isle of May, Scotland, where no other natural predators occur, four of these cannibals ate 164 chicks of their own species, or 12 percent of those available (Parsons 1969). A minority of breeding South Polar Skuas eat their neighbors' chicks when prevented from foraging in the sea by bad weather, and parent skuas occasionally devour their own chicks (E. C. Young 1963, Spellerberg 1971). Small chicks of frigatebirds on Ascension Island were invariably swallowed by neighbors or interlopers if exposed for even a few seconds (Stonehouse and Stonehouse 1963). Male Boat-tailed Grackles have also been accused of devouring tender nestlings of their own kind (McIlhenny 1937). Although colonial sea birds, other than gulls, skuas, and frigatebirds, seem not to be guilty of cannibalism, some do, as we earlier noticed (p. 377), mercilessly persecute chicks that wander from their nests, sometimes causing their death. One advantage of nesting on a narrow ledge, in the manner of the Kittiwake, Black Noddy, and Fairy Tern, is that the chicks are unable to stray from their natal spots, thereby exposing themselves to attack by neighbors.

On remote islands, where marine birds find freedom from flightless predators (other than those introduced by man), their young fall prey to frigatebirds in warm seas and to skuas, gulls, and Giant Petrels in colder waters. Disturbance by man increases mortality by driving parents from their nests, leaving their eggs and young exposed to aerial marauders or causing the chicks to wander away from their nests and run the gantlet of hostile beaks. But it is in continental areas with their varied fauna that predators take the greatest toll of eggs and young. Many studies of temperate-zone birds attribute 50 or 60 percent of all losses to this source, and in tropical forests it can be consistently higher. Since we rarely surprise the plunderers in the act, their identification is usually difficult; but, when whole sets of eggs, or whole broods of unfledged young, disappear from a nest, we are fairly safe in concluding that a predator has looted it. Doubt may arise when a brood almost ready to leave has vanished from a nest between our visits. If the lining lies flat, the young birds have probably left spontaneously; if it is pulled up, they have probably been torn from the nest while trying in vain to save themselves by clinging to it.

The eaters of eggs and nestlings are innumerable. From remote times, men have regularly taken a rich harvest of eggs from sea bird colonies and the incubation mounds of megapodes. Doubtless savages wandering through forests or over steppes did not hesitate to devour such eggs and nestlings as they found, including those that our well-fed contemporaries would spare. Carnivorous quadrupeds, from weasels and stoats to foxes, coyotes, mongooses, and perhaps even larger kinds, prey heavily on bird nests. Even largely vegetarian mammals, such as squirrels and certain monkeys, do not disdain eggs and nestlings when they find them. Monkeys doubtless knock many eggs from their nests as

they pass with tremendous leaps through the treetops, violently shaking the boughs. Certain large bats are known to prey on birds; I suspect that these nocturnal mammals are responsible for the loss of young from hummingbirds' nests attached beneath slippery palm or banana leaves, where no snake or quadruped seems able to reach them. Mice sometimes evict small birds from their nests, especially covered ones, in which they ensconce themselves, closing the doorway with shredded materials but often leaving the eggs unharmed beneath their bed. The diminutive marsupials known as marmosas likewise make themselves comfortable in roofed nests captured from birds.

Avian predators on bird nests include not only hawks and owls but also storks, jaegers, ravens, crows, jays, magpies, and many others. Swallowtailed Kites pluck nestlings from arboreal nests so exposed that they can be reached while hovering. Although toucans are largely frugivorous, they plunder many bird nests, intimidating the outraged parents with their huge, brightly colored bills. But by far the worst enemies of nesting birds, especially in tropical forests, are snakes, whose slender, sinuous bodies can reach nests far out on thin twigs and creep into cavities with narrow doorways. These are the destroyers of bird nests that I have most often caught in the act. In a Panamanian forest, two nestlings were taken successively from a nest of Crimson-backed Tanagers by two snakes of different species in less than a day. Another snake reached a colony of Yellow-rumped Caciques and, apparently hiding in plundered pouches by day, emptied nest after nest until I finally shot it in the night (Skutch 1954). In arid no less than in humid tropical regions, snakes are a major source of nest losses (Marchant 1960, Lloyd 1960).

Among predators upon nestlings we must include ants, of which fire ants (*Solenopsis*) are, in my experience, the most destructive. Absent from closed woodland, they at certain seasons abound in open places, such as lawns, where they frequently invade low nests, including those of woodpeckers in trees and kingfishers in banks, stinging the nestlings to death and stripping all flesh from their bones. Although they seem unable to penetrate an intact egg, they sometimes make the incubating parent fidget until it cracks the shell, or they wait until the hatching chick pierces it, then pour in to devour the contents. On the other hand, the hordes of army ants, which seem such a peril to low nests in tropical American woodlands, appear seldom to harm them, as they prey principally upon insects, spiders, and other arthropods. I have watched both *Eciton* and *Labidus* swarm over the mouths of puffbirds' burrows in the forest floor without harming the nestlings; a tiny seedeater continued to cover her eggs while army ants crawled over her nest's rim.

Beset by perils, woodland birds rarely rear fledged young from half of the eggs that they lay, and in lowland tropical forest only one egg in five or six may produce a flying juvenile. Predation, above all by snakes, not only thwarts most of the ornithologist's attempts to study the nesting of birds whose breeding habits are unknown but also, if he be sensitive, saddens him with its tremendous waste of life and effort. Yet, when he reflects, he may draw a consoling conclusion from this situation. If it is so difficult for these birds to reproduce, yet their numbers remain more or less constant from year to year,

the adults must have fairly long lives. For at least one tropical bird, this deduction from observed nest losses has been confirmed by D. W. Snow (1962*a*) in Trinidad. In a season, each female Black-and-white Manakin starts from two to four nests, each with two eggs, but she rears on average only one young. The adults' annual survival rate, at least of the males, who are easier to band and recapture, is 89 percent, which indicates a longevity substantially greater than that of any small temperate-zone bird whose mortality rate is known. This seems a better life than that of many small birds of northern lands that lay more eggs, nest more successfully, but lose about half the adult population every year.

A similar situation prevails in certain tropical forests of the Orient. Of 167 nests of a variety of species (excluding hole nesters) in a dipterocarp forest and neighboring areas in Sarawak, 144, or 86 percent, were lost by predation. But of 232 banded adult birds, 200 (again 86 percent!) were alive one year later, from which Fogden (1972) estimated an annual mortality of no more than 10 percent for adults of the species he studied.

CHAPTER 34

Regulation of the Rate of Reproduction

The species that consistently fails to produce enough progeny to replace its annual losses disappears from the earth. This has been the immediate cause of the extinction of every animal that no longer exists, whatever may have been the factors underlying this failure. On the other hand, the animal that reproduces too profusely creates difficulties for itself. Too many mouths make excessive demands upon available resources, leading to hunger, starvation, and diseases that spread most freely among undernourished animals. A too-dense population may destroy its sources of food and ruin its habitat, perhaps thereby bringing about its own extinction by excessive rather than by deficient reproduction. These are the Scylla and Charybdis between which all animals, including man, steer a precarious course. How do they manage to avoid these disastrous extremes? The question is difficult to answer but of the greatest importance; it can hardly fail to be of interest to a humanity that is ruining its environment by its own excessive fecundity.

Theories of the Control of Rate of Reproduction

The two theories of the regulation of reproduction in birds that are now widely discussed by ornithologists may be briefly designated as the theory of maximum reproduction and the theory of adjusted reproduction. The reasoning underlying the former

is brief and clear. If, in a certain species or population of animals, one strain or genotype consistently produces more young than its neighbors, without detriment to the progeny's fitness to survive or to the parents' capacity for continued reproduction, then this more prolific genotype must, by sheer force of numbers, eventually supplant its less fertile neighbors and become the prevailing type. Like every other character, the fertility of animals is subject to mutation, so that we may expect that, in every species, individuals that produce a greater number of young will from time to time appear. Accordingly, we must conclude that every kind of animal, whatever its size or taxonomic position or habitat, is actually reproducing just as rapidly as it can, regardless of what happens to its perhaps excessive progeny.

This theory appeals to those who concentrate upon the simple mathematics of the situation. Other things being equal, the genotype that consistently produces three young at a time will, in the course of a few generations, so greatly outnumber the genotype that produces only two that the latter will tend to disappear. Although mathematically impeccable, the theory of maximum reproduction often fails to give sufficient attention to some of the harmful consequences of excessive breeding.

These evil consequences weigh heavily in the

thinking of the proponents of the theory of adjusted reproduction. They recognize that a successful species must, in an average year, produce enough young to balance its annual mortality, with perhaps a small surplus to allow for contingencies, such as a succession of unfavorable years or the opportunity to expand its range. Yet, even with this margin of safety, the species may be reproducing at a rate far below its maximum potential. If it produced all the young that it could, it might cause severe intraspecific competition for food and territory, deplete the resources of its habitat, and increase mortality from predation, disease, and starvation—all without strengthening its position in the realm of life. To avoid all this useless waste of life and effort, many birds, particularly those in stable environments, appear to limit their reproductive rate; and this limitation must be regarded as an adaptation, like any other that gives an organism more perfect adjustment to its environment (Skutch 1967*b*).

Even when we admit the reality of this limitation —for which evidence will presently be given—it is difficult to understand just how it is effected. Evidently the factors involved are subtle and complex, less obvious than the simple mathematical proposition that continued multiplication by a larger number soon yields a product much greater than that which will result from multiplication by a smaller number.

A fundamental difference between these two points of view becomes clear when we consider them in relation to the mortality rate. According to the theory of adjusted reproduction, the reproductive rate is primarily determined by the mortality rate; the animal produces enough young to replace inevitable losses, with a small margin for safety. But it follows from the theory of maximum reproduction that the mortality is determined by the reproductive rate, whenever this exceeds the adjusted rate. When a population rises above the carrying capacity of its habitat, the excess will necessarily be removed by death in one of its many forms, so that the greater the number of superfluous individuals that are born, the greater the number that will each year die. The optimum rate of reproduction would be that which avoids this gratuitous mortality. It

might be defined as the rate beyond which any increase would be balanced, in an average year, by increased losses. By reproducing above the optimum rate, an animal effects only a transient rise in its population density; before the next breeding season, this will be reduced to a level determined by the carrying capacity of the habitat. When it maintains the optimum rate, the animal holds its population at a more constant level, avoiding violent fluctuations in its numbers and making the best sustained use of its resources.

Among all the complex problems that confront the evolutionist, that of the regulation of the reproductive rate is unique—and uniquely difficult. It is widely accepted that any mutation that increases an animal's efficiency or improves its adaptation to its environment will spread through the population. The favorable mutation may take the form of better protection from enemies, an improved mode of foraging, greater resistance to cold or heat, or any of countless other possible modifications. But a mutation that confers excessive fertility benefits neither parent nor offspring. Only if the offspring will eventually aid the parent, as by cooperating with it in defense or helping it in its old age, does the parent derive substantial benefit from its progeny; such assistance is rare among birds and other animals. The parent may, indeed, derive pleasure or satisfaction from attending its young, but this is a subjective aspect that we cannot assess and which is of doubtful value in the harsh struggle to survive. In any case, a parent's enjoyment of its offspring is probably greatest when they are few and it does not have to work so hard for them—as with ourselves.

Apart from these problematic benefits, animals individually derive no advantage from reproducing. Quite the contrary, they prepare for their own replacement by the next generation; each "signs its own death-warrant, makes its will, and institutes its heir," as George Santayana picturesquely said. Moreover, parents may weaken themselves by squandering their energy in the care of many young and diminish their chances of escaping enemies while devoting their attention to their offspring rather than vigilantly guarding their own safety. As to the progeny, when produced in excessive numbers they enter the world with the prospect of a hard life prematurely cut off by starvation, predation, or some other misfortune. And what fails to benefit individuals brings no advantage to the species, which may weaken its position by overexploiting its food resources and preparing a fertile bed for the spread of disease.

In organisms of all kinds, most mutations turn out to be harmful and that for excessive fertility must be included among them, for it benefits neither individuals nor the species as a whole. Although other harmful mutations tend to be eliminated by natural selection, this one has a perverse capacity to increase, simply because the gene (or genes) responsible for it can multiply itself faster than competing genes—its alleles. Moreover, it does not confine its deleterious effects to its bearers, as most undesirable genes do, but it extends them to the whole population by depleting the food supply on which all depend. Unable to multiply as rapidly as the more prolific mutants, the less fertile members of the population are equally exposed to death from starvation or some other consequence of overpopulation. The gene for excessive fertility tends to spread like some pestilent disease, like some pathogenic virus, which has some of the characteristics of genes.

What can stop the mutation for greater fertility from suppressing its rivals and invading the whole population? I believe that only in a stable environment, where populations remain essentially constant from year to year, can it be stopped. Such environments are found chiefly in the rainier parts of the tropics and perhaps in certain maritime climates beyond the tropics, and to them we shall presently return. Wherever the climate exhibits violent extremes, as in vast areas of the North Temperate Zone, and populations of birds are subject to drastic reductions at irregular intervals, as by winter killing or disasters on migration, maximum reproduction is likely to prevail. The reason for this is that, after each catastrophic reduction, the species enjoys an interval of free expansion, when competition between individuals is reduced and the most rapidly reproducing genotypes contribute most to the recovery of the population. With repetitions of this situation, the reproductive rate should be brought up to the maximum compatible with the rearing of sturdy offspring, even if this will result in great and wasteful overproduction after a succession of favorable years. But in a stable population of essentially sedentary individuals, the more prolific strain may never gain this ascendancy. Here the advantage of economy, of husbanding rather than squandering resources, is likely to prevail over that of the ability to recover quickly from a disastrous crash. Here the birds may increase their survival by devoting to self-maintenance, or to providing better care for each of their young, some of the energy that they save by refraining from producing an unnecessarily large number of them (Cody 1966).

What Limits the Size of Broods?

Although we have sound theoretical reasons for believing that in certain circumstances the reproductive rate of some species will rise to a maximum, they do not produce infinite offspring, so we must ask what limits the number. At what stage of the reproductive cycle—nest building, egg laying, or rearing the young—does the bottleneck lie?

The size of the nest may be the limiting factor. The female hummingbird fits snugly into her downy cup, the better to keep her two tiny eggs warm. Before they fly, at the age of three weeks or more, the two nestlings grow nearly as large as their mother. They are so crowded in the space made for one that they tend to flatten out the nest and sometimes burst it asunder. The nest could hardly accommodate a third nestling. The Rufous Piha of Central American forests and some other cotingas build amid foliage a tiny platform barely wide enough to support the single big egg and the nestling that hatches from it, and crested swifts glue to the side of a lofty branch a cup just big enough to accommodate their solitary eggs. Doubtless, if it were imperative for them to raise more young in order to balance the mortality, the hummingbirds, pihas, and other birds in similar circumstances could build larger nests to hold more eggs and nestlings; but this might bring certain disadvantages, such as less efficient warming by the hummingbirds and greater conspicuousness of the piha's nest.

More frequently, broods seem to be limited by the difficulty of mobilizing enough material to form eggs, which when many, or large in relation to the bird who lays them, may approach or even surpass her in total weight. Among waterfowl and gallinaceous birds, there is a tendency for the number of eggs in the set to vary inversely with their size, suggesting that the female has the alternative of laying a few big eggs or a larger number of small ones (Lack 1968). Wagner (1957) believed that deficient nutrition limits the number of eggs laid by tropical birds, which do not store up reserves of fat like migratory northern birds and in certain cases do not continue to forage through the hot middle of the day. They might compensate for a low daily food intake by laying their eggs at longer intervals, but this would increase the time that the nest is exposed to predation and also the risk that the earliest eggs in the set would spoil. By feeding his mate, the male of some species helps her to form her eggs, and by more generous courtship feeding she might be enabled to produce larger sets.

It appears that scarcity of materials limits the formation of eggs by certain Arctic birds. Geese that nest on the tundra, such as Ross' Goose, arrive on their breeding ground while much ice and snow remain and food is still scarce. To form their eggs, and to incubate without long interruptions for foraging, these geese are largely dependent upon reserves that they have stored in their bodies before starting on the long migration of thousands of miles. If they became too heavy by eating abundantly before beginning their journey, they could not accomplish their long flights. If they reduced the size of their eggs in order to lay more, the goslings that hatch from them might lack sufficient reserves of fat to see them through their first days, before they can forage well (see p. 232). If the geese depleted their body reserves by forming too many large eggs, they would have to reduce their incubation time in order to forage—as it is, they may lose up to 44 percent of their body weight during egg laying and incubation. The goose seems to compromise between these limitations, laying three or four eggs large enough to contain reserves of food for the goslings and retaining enough nourishment in her body to sit rather constantly through the cool Arctic days (Ryder 1970).

Although in certain cases physiological exhaus-

tion may restrict the number of eggs that a bird can lay, this does not appear to be the factor that most often limits the size of the set. We can be sure that it is not in those birds that have been demonstrated to be indeterminate layers (p. 140), for we have only to remove eggs from their nests to make them continue laying, sometimes almost indefinitely. It would be most surprising if this turned out to be the primary reason for the small sets of two or three eggs laid by so many tropical birds. In a region and at a time when migratory birds are storing nutritive reserves for a long journey, these resident birds might accumulate materials for the prompt formation of more eggs, if there were any advantage in doing so.

When we view the nestful of eggs of a duck or partridge, or even of a northern titmouse or wren, they spread over so much space that we wonder how the incubating parent can apply heat to all of them. It would seem that, if she laid more eggs, some of them must certainly fail to hatch because of inadequate warming. But the English Partridge, which most commonly lays twelve to seventeen eggs in a set, can hatch more than twenty with no loss of efficiency (Lack 1954). When Wood Ducks in Iowa incubated eighteen or twenty eggs, they hatched an even higher percentage than when they covered their usual set of eleven to fourteen (Leopold 1951). Moreover, the view that the size of sets is limited by the bird's ability to incubate the eggs does not help to explain one of the puzzling facts of ornithology, why a bird in the warm tropics lays and hatches fewer eggs than a northern bird of the same size. Only in certain special cases does the number of eggs appear to be limited by the parent's ability to incubate them. Among these are the Gannet, which clutches its egg with its webbed feet and can hatch at most two; the Emperor Penguin, which holds its egg above the ice on its feet; and albatrosses. The incubation patch of the Laysan Albatross can accommodate only a single egg. When given a second, the bird seemed uncomfortable, usually pushed an egg from the nest, or, incubating the two by turns, hatched neither (Rice and Kenyon 1962).

Although the dimensions of the nest, the female's capacity to form eggs, or the parents' ability to hatch them restricts the size of the set in certain species, the upper limit is usually set, especially among altricial birds, by the parents' ability to nourish the young. Lack (1947–1948, 1954, 1966) and his co-workers have spent many years and countless hours trying to prove, especially among European birds, that the most common size of the set of any species has been adjusted by natural selection to correspond to the greatest number of young that the parents can, in a normal year, raise to healthy independence. If the parents try to rear too large a brood, the young are undernourished, and, if they do not die in the nest, they fledge while underweight and their chances of survival are reduced; therefore, individuals who lay unusually large sets may leave fewer, rather than more, progeny to join the breeding population in a subsequent year than do individuals of the same species who lay the usual number of eggs. By contributing the greatest number of individuals to the population, the genotype that lays the "correct" number of eggs tends to supplant those that lay too few or too many. Underlying this proposition is the assumption that clutch size is hereditary. Doubtless it has a hereditary foundation, but, for many birds that lay sets greater than two, it is subject to considerable modification by the environment, the birds laying more when conditions are favorable than when they are adverse.

Some birds fit this formulation admirably, others fail to conform. Among the former are certain European swifts, which are well suited for studies of survival because their nests in holes in trees or crannies in buildings are difficult for predators to reach, so that almost all losses of nestlings are by starvation. In Switzerland, the Alpine Swift usually lays three eggs, often two, rarely one, and exceptionally four. In a colony that was studied for sixteen years, 97 percent of the young that hatched in broods of one lived to fly. When the brood consisted of two, 87 percent of the young fledged; from broods of three, 79 percent fledged. From the few broods of four, however, only 60 percent of the nestlings lived to fly. Because of the higher mortality in the larger families, the number of swifts raised per nest was the same in broods of three and broods of four; this suggests that three, which is by far the most frequent number of eggs, is also the largest number of young that the parents can usually rear.

In England, where the summer weather is often less favorable for insect-catching swifts than it is in Switzerland, the Common Swift lays two eggs more often than three, and sets of four are rare. In a six-year study of four colonies in southern England, 84 percent of the young flew from broods of two but only 58 percent from broods of three. In each case, the average number of young fledged per brood was 1.7. The most common brood size, two, was also the most efficient, because by laying one more egg the swifts did not, on the average, increase the number of young that they reared. However, in the finer summers, when the amount of sunshine was average or better, the swifts raised 95 percent of their young from broods of two and 75 percent from broods of three. Now the average number of young fledged per brood was 2.3 for the larger as against 1.9 for the smaller broods. The advantage of the broods of three in the sunnier summers seems to explain why a substantial minority of the female swifts persist in laying three eggs (Lack 1954).

In small passerine birds, unlike swifts, a significantly greater proportion of the nestlings do not die in large than in small broods. But, except in years of exceptionally favorable food supply, the young of large broods may, in some species, leave the nest weighing, on the average, less than the young of small broods, a consequence of the more adequate nourishment that the parents were able to provide for the smaller families. The latter would appear to have a better chance of surviving in the following months than the undernourished fledglings from large broods. The investigation of this point is difficult, for, after leaving the nest, the young birds scatter and cannot often be found. A small proportion, however, are recovered, either in the bird bander's traps or by being picked up dead. If they were banded before they left the nest, the bander's records will reveal the size of the brood from which they came.

A study of Common Starlings banded in Switzerland showed that, three or more months after leaving the nest, the proportion of survivors was the same for broods of three, four, and five; but for larger broods it was lower, and the decrease almost balanced the advantage of starting with a larger brood. As a result, the average number of survivors per brood was about the same, whether the brood originally consisted of five, six, seven, or eight nestlings. Hence five appears to be the most efficient, as it is also the most frequent, brood size among the starlings of this region. Similar results were obtained at other seasons and in other places. It appears that this correspondence between the most frequent brood size and the most efficient size is an adaptation effected by natural selection (Lack 1954).

Difficulties of the Theory of Maximum Reproduction

However, this subject is most complex, and its investigation is beset by the greatest difficulties. In the first place, alternative explanations merit consideration. Possibly the mortality during the first weeks after leaving the nest is greater among the young of large than of small broods, not because they leave the nest underweight, but because the former are harder to keep together and busy parents temporarily or permanently lose contact with some of them. Second, the available evidence is not consistent. After Lack published his painstaking statistical study of survival in young Common Starlings, Dunnet (1955) showed that in Scotland the weights of nestling starlings did not differ significantly in broods of different sizes or in relation to the amount of food available. At Oxford, England, D. W. Snow (1958) discovered that nestling Blackbirds of large broods averaged heavier rather than lighter than those of small broods. He attributed this to the fact that the large broods belong chiefly to older parents, who attend their nestlings more efficiently than do pairs breeding for the first time. In this thrush, the youngest and lightest nestlings often die in the nest before they are eight days old, and less often after eight days. But the handicap of being the youngest and lightest member of the family does not last long. Continued observation of banded families revealed that, until they dispersed at the age of about twenty days, young Blackbirds that as nestlings had different weights survived about equally well.

Another way to answer the question of whether birds are rearing as many young as they can adequately nourish is by experimentally increasing the size of their broods. When each of forty-two pairs of Manx Shearwaters was given a second newly hatched young as soon as its own chick hatched,

thirty-three of these pairs reared both young, and nine other pairs each raised a single chick. Accordingly, pairs with twins raised 1.79 young per pair, while forty-five control pairs raised only 42 young, or 0.93 young per pair. The twins left the nesting burrows weighing somewhat less than lone nestlings, yet their post-fledging survival rate of 23 percent was not significantly lower than the survival rate of 29 percent of solitary young banded at the same time in the same area. This leads one to ask why Manx Shearwaters do not regularly lay two eggs, thereby greatly increasing their reproductive rate. The answer appears to be that at the normal time of laying the female may not be able to obtain enough food to produce two large eggs, and to delay laying would not yield more surviving young, because they would fledge too late in the season to have a good chance for survival (Perrins et al. 1973).

In other tube-nosed swimmers, the outcome of twinning experiments has been different. Norman and Gottsch (1969) found that Short-tailed Shearwaters on islands in Bass Strait, Australia, were unable to rear two young. In similar tests with the Laysan Albatross, Rice and Kenyon (1962) demonstrated that the two parents were unable to rear two chicks. Part of the difficulty was the mutual antagonism of the chicks; but the same conclusion can be drawn from the usual inability of a single parent to provide enough food for a single chick.

Another bird that lays a single egg, the Gannet, is, however, frequently able to attend two young so adequately that they leave the nest only slightly lighter than single chicks, with good chances for survival, as Nelson (1964) showed by repeated tests on Bass Rock, Scotland. Curiously enough, closely related boobies lay two eggs but only with great rarity, in exceptionally favorable seasons, rear two young (Simmons 1965). As we earlier noticed (p. 290), one young booby often evicts its sibling from the nest; but doubtless this habit would have been suppressed by natural selection if these boobies were often able to provide for a brood of two. In the genus *Sula*, at least, there is no close correlation between the number of eggs and the parents' ability to rear the young that hatch from them.

Likewise, we fail to find correspondence between

the number of nestlings and the number of adults who attend them. If the parents are working as hard as they can, two should be able to rear twice as many as a single one. This point can be tested by comparing the brood size of closely related species, in which the male does or does not help the female to feed the nestlings, as has been done for several European and North American families by Haartman (1955). In the genus *Emberiza*, both sexes feed the nestlings in the Yellow Bunting, Reed Bunting, and Ortolan, but mostly only the female in the Corn Bunting and Cirl Bunting; yet there is little difference in the number of eggs laid by these two groups. Similarly, in the genus *Phylloscopus*, both sexes of the Wood Warbler and Willow Warbler nourish the young, but chiefly or only the female in the Chiffchaff and Bonelli's Warbler, and again the broods of these two groups are of about the same size. It certainly appears that, in both the buntings and the warblers, the species in which both sexes regularly attend the nest are not rearing as many young as they could adequately feed. Among icterids that breed in the United States and Canada, the young are regularly fed by both parents in the Common Meadowlark and Baltimore Oriole, but male Red-winged Blackbirds, Yellow-headed Blackbirds, and Boat-tailed Grackles are at best undependable providers; yet the brood size of these five species is much the same. We shall presently notice a similar situation among tropical birds.

The large broods of hole-nesting birds are likewise difficult to reconcile with the theory of maximum reproduction. According to figures published by Nice (1957), eighteen species of European hole nesters lay an average of 6.9 eggs in a set, whereas fifty-four species of European birds with open nests lay only 5.1 eggs. In North America, the corresponding figures are 5.4 and 4.0. Since hole nesters enjoy no special advantages for feeding their young not available to open nesters, if the latter rear as many offspring as they can nourish, how does it happen that hole nesters can bring up larger families of sturdy fledglings? The nestling periods of hole-nesting birds are substantially longer than those of open nesters, the average age of departure of the young of these two groups being, respectively, 17.3 and 13.2 days in Europe and 18.8 and 11.0 days in North America. Defenders of the maximum reproduction theory try to escape the difficulty that the large broods of hole nesters present by appealing to their longer nestling periods. They say that the hole nesters can feed more young than open nesters because their nestlings grow more slowly and need less nourishment each day. They rear more young in more time rather than fewer young in less time.

This way of trying to account for the larger families of hole nesters overlooks important facts. Hole-nesting passerines gain weight as rapidly, or almost as rapidly, as nestlings of about the same size in open nests. In small birds of both groups, the period of most rapid increase in weight covers the first ten or twelve days of life. The difference is that the young in open nests leave at an earlier stage of development, finding safety in scattering, whereas those that grow up in holes have a better chance of surviving if they stay in their safer nests until their plumage is well developed and they can fly well. During their final days in the cavity they may lose weight, after having become heavier than their parents. Barn Swallows, for example, when ten days old weigh about 19.5 grams, as much as their parents. Their maximum weight of about 21 grams is attained at the age of twelve days, but this declines to about 17.5 grams while the young swallows delay in the nest for nearly a week longer (Stoner 1935). The loss in weight is caused chiefly by the drying and hardening of their plumage (Ricklefs 1968). Tree Swallows, Pied Flycatchers, titmice, and other hole nesters show similar growth curves (Paynter 1954, Haartman 1957, Gibb 1950). Hole-nesting Black-capped Chickadees do not exert themselves to their utmost to feed large broods of seven young (Kluyver 1961).

If the size of broods is determined simply by the parents' ability to feed the young, it is hard to understand why it should be larger in hole-nesting birds than in open nesters. If, on the other hand, it depends upon the species' need for recruitment, explanations are available. Haartman (1957) pointed out that in northern Europe hole-nesting passerines include a greater proportion of permanent residents than is found among open-nesting passerines, which are more largely migratory. The winter mortality of the resident birds is much heavier than that of species that migrate to warmer climates. Accordingly, the hole nesters need larger broods to recuperate from their winter losses. Moreover, as suggested on page 438, they need to raise more young to compensate for delays in establishing themselves in a suitable cavity and for the members of their species that are excluded from breeding by failure to acquire a hole. The greater safety of the enclosed nest makes it advantageous for the young to remain within until they can fly well, so that after emerging they should suffer fewer losses than do open nesters that often leave when they can barely flutter. This double gain makes hole nesting profitable, so long as it is practiced by only a minority of the avifauna.

Despite the theoretical reasons for believing that in unstable climates, where at irregular intervals the population is drastically reduced, the reproductive rate should reach a maximum and some observational evidence that it does so, there are so many exceptions and puzzling cases, even among the birds of northern lands where we should expect maximum reproduction to prevail, that the theory is far from firmly established. Its defenders can often find reasons for discounting contrary evidence: the observations were not made in a normal year, they were not sufficiently extended to be significant, the eggs in excess of the number of young that can be reared are insurance against the loss of some during incubation, and so forth. One reason why a species has not brought its rate of reproduction up to the maximum may be that the disasters that decimate its population and permit a subsequent period of free expansion have not been frequent enough. In any event, we should not expect a single-factor explanation to account for a phenomenon so complex as the reproductive rate of birds. Undoubtedly, many influences are at work here, some retarding and some accelerating the rate, and they have different effects on different species.

Evolution is a crude, brutal process, an aeonian game of chance, depending upon mutations that are as unpredictable as a throw of dice and mostly harmful, along with the ruthless elimination of all the resulting misfits. It does, to be sure, frequently hit the mark with a fine adaptation, but it seems to make a high score only because no record is kept of its far more numerous misses. It would be surprising if such a mindless process should often finely adjust a bird's reproductive rate to its food supply, which fluctuates erratically with the vagaries of the weather.

Moreover, a high rate of reproduction can be self-defeating. During a cold winter, many birds depend upon supplies of food, such as resting seeds, fruits preserved by cold, and hibernating insects, that will not be renewed until warm weather returns. They are in the situation of mariners shipwrecked upon a desert island with no provisions except the very inadequate stores that they have managed to salvage from their sunken vessel. The fewer the shipwrecked men, the more will survive until they are rescued, because those that die have already consumed irreplaceable food that might have kept others alive. The very high mortality, often exceeding 50 percent and falling chiefly on the youngest individuals, that birds resident in northern lands frequently suffer during the winter months, might be considerably less if they reared fewer young in the preceding summer and entered the inclement period with a smaller population. By laying fewer eggs and raising fewer young, they might hold their population at a consistently higher level, without such violent annual fluctuations. But, because the more prolific genotypes tend to predominate in the population, they seem to be caught in a trap from which they cannot escape. Misspent effort in raising too many young, stress, strife with other members of their species, premature death from starvation, predation, and disease are the high price that these birds pay for evolution's failure to adjust their reproductive rate, not to their ability to rear many young by their utmost exertion, but to their actual need for recruitment.

Why Do Tropical Birds Lay Few Eggs?

It is above all in mild, equable climates, where the birds reside throughout the year, enjoying sufficient food even in the leanest season and maintaining a stable population that is rarely, if ever, decimated by a widespread catastrophe, that we should expect the reproductive rate to be adjusted to a low and fairly constant annual mortality. Such climates are found chiefly in tropical regions generously supplied with rain. According to Wagner (1957), the German explorers in South America, Prince Wied (1830) and Schomburgk (1848), first called attention to the fact that tropical songbirds lay smaller sets of eggs than their European counterparts. This has been confirmed by many subsequent observers, not only in America but also in the tropics of the Old World, who have shown that the difference in clutch size with latitude is by no means confined to songbirds, although it does not appear in every family. It is also independent of the origin of the birds. When birds of tropical origin, such as wrens, wood warblers, American orioles, and tanagers, extend their range into temperate latitudes, they lay more eggs; when birds of northern origin, such as juncos and towhees, invade the tropics, they lay fewer. Often the difference is evident even in a single species of wide latitudinal range. On the other hand, altitude appears not to affect the number of eggs that birds lay; the birds of high tropical mountains lay sets no larger than those of their relatives in neighboring lowlands.

Even within the tropics, the size of sets tends to increase with distance from the equator, the difference becoming noticeable at about 12° or 15°, in both Africa (Moreau 1944) and America. This is at first sight puzzling, as with similar rainfall the tropics exhibit a broad uniformity of ecological conditions from end to end. Doubtless the principle of the continuity of nature—*natura non fecit saltum*—provides the explanation. What would surprise us would be an abrupt change in the size of egg sets as one passed a certain degree of latitude, say the twenty-third.

Hesse (1922) appears to have been the first to suggest that birds at low latitudes lay fewer eggs than those at higher latitudes because days are shorter. Other things being equal, parents with a working day of about twelve hours, as occurs near the equator, could hardly supply their young with as much food daily as those that remain active through sixteen or eighteen hours of daylight, as is the case at high latitudes in summer. Even where daylight is continuous, as it is above the Arctic Circle in June, many birds are not continuously active but rest for a few hours when the sun is lowest in the sky.

Although it seems impossible that, by working their hardest, parents with an active day of twelve or thirteen hours could adequately feed as many young as those that enjoy one-half more daylight, this hardly accounts for the whole difference in the size of their broods. The tropical birds might compensate for the smaller daily ration, and consequent slower growth, by prolonging their nestling periods. To a certain extent they do this; but, unless the nest is unusually inaccessible to predators, the prolongation of the nestling period increases losses. This would soon set a limit to the degree that it would be profitable to retard the nestlings' development in order to rear a larger brood.

Even if we assume that northern birds rear all the young that they can feed, we have good reasons for believing that many tropical birds, with smaller families, do not. Consider first the time available to the parents. We cannot proceed so simply as by saying that, if two parents with a working day of 18 hours can supply six nestlings with food, those with a 12-hour day should be able to feed four (4:6::12:18). This loses sight of the fact that the parents must devote some time to self-maintenance; in the first case, the whole family contains eight mouths; in the second, six. We do not know whether the maximum daily requirement of food of a nestling is greater or less than that of the parent who attends it. The nestling needs nourishment for growth, but the parent burns up more energy seeking food and bringing it to the nest. We shall probably not err greatly if we assume that the needs of a parent and an older nestling are about equal. With a 17-hour working day, as in the high forties or fifties of latitude in midsummer, two parents feeding six nestlings (not an unusually large brood in these regions) have a total of 34 bird-hours for meeting the needs of eight mouths, or 4.25 hours for each. With the same allowance, a pair of birds at 10°, where the active day in the main breeding season hardly exceeds 12.75 hours, should be able to feed four nestlings ($6 \times 4.25 = 2 \times 12.75$). Yet many of the finches, tanagers, wrens, and other birds at this latitude rear no more than two.

Calculations like those just given, which might be refined for different latitudes, months, and bird families, are at best suggestive rather than convincing. Perhaps food, especially insects for nestlings, is more readily gathered in temperate regions than in the tropics. Although the teeming insect life of the humid tropics is more varied, a single species might be present in the simpler plant communities of a temperate zone in much greater numbers. By concentrating upon one or a few kinds of insects, temperate-zone birds might gather them more rapidly than do tropical birds that feed their young a greater variety, many of which are cryptically colored and hard to detect.

However, we have stronger evidence that tropical birds do not, as a rule, rear as many young as they could adequately feed. Birds of the tropical forest, such as trogons and antbirds, commonly bring infrequent but large meals to their nestlings, thereby reducing the number of visits that might guide hostile eyes to the nest. Often they bring items so large that the nestlings can hardly gulp them down. Yet they are capable of supplying food at a greatly accelerated rate, as happened with a pair of Tyrannine Antbirds who, apparently detained by some excitement off in the woods, returned after an absence of nearly two hours to find their neglected nestlings very hungry. In the next sixty-nine minutes, they brought food twenty-three times, which was about four and a half times their average rate of 4.5 meals per hour. In the quarter-hour when they were most active, the antbirds delivered 9 meals, which was eight times the average rate. Now many of the insects were smaller than they customarily delivered, but others were large, so that the total amount of food given to the nestlings in an hour was substantially increased. Ordinarily, these antbirds, like certain other parent birds, appeared promptly to eat the smaller articles that they found, reserving only the larger ones for delivery at the nest. After the nestling antbirds showed by their somnolence that they were no longer hungry, their parents fell back to their usual leisurely rate of bringing food.

When I increased the brood in a nest of Scarlet-rumped Black Tanagers from two (the almost invariable number) to three, the parents maintained for five hours a rate of feeding 41 percent greater than before. After the disappearance of the male parent of three nestling Wire-tailed Swallows in Africa, their mother continued until they fledged to feed this brood of the most frequent size at a rate slightly

above the average for both parents together (Moreau 1947). In Mexico, Wagner (1957) doubled the broods in eight nests of House Finches and Yellow-eyed Juncos. Two of these nests were prematurely destroyed, and from another a nestling inexplicably vanished. In the remaining five nests, the parent juncos and finches with a double load, four or five young instead of two or three, reared all of them. They appeared in no way to suffer from insufficient attention.

In addition to these indications that parent birds of a number of tropical species are not working as hard as they might to feed their young, we have even more convincing evidence in the lack of correlation between the number of young in a nest and the number of attendants. Altricial species in which only the female cares for the nest and nestlings are more abundant in the New World tropics than in the North Temperate Zone. These solitary mothers include the great majority of hummingbirds that have been adequately studied, all manakins as far as known, several cotingas, many flycatchers, a few woodcreepers and ovenbirds, and certain icterids, including the large oropéndolas and caciques. Most of these birds lay two eggs, which is the number most frequent among their neighbors of which both parents attend the young. A few of the cotingas, including the Rufous Piha, Calfbird, Bearded Bellbird, and Amazonian Umbrellabird, lay only one. It might be supposed that females working alone can rear as many young as do neighboring pairs because they select food that is especially nourishing or unusually abundant and easily collected. But the diets and modes of foraging of these solitary parents are hardly less diverse than those of the more numerous species that breed in pairs. Thus it is most unlikely that special dietary advantages permit them to rear, alone, as many nestlings as pairs of their neighbors raise (Skutch 1949).

The evidence is most convincing when the solitary parents and the paired parents belong to the same family. Among the woodcreepers, nests of the Tawny-winged Dendrocincla are attended by one parent but those of the Streaked-headed Woodcreeper by both parents, yet in each case the set consists of two white eggs. These two kinds of brown birds sometimes breed in neighboring trees. Among the

TABLE 18
Number of Eggs Laid by Some Smaller Flycatchers in the Humid Tropical Zone of Costa Rica and Panama

For each species, the number of nests examined and the average number of eggs are given.
Nestlings Fed by:

Female Alone	Male and Female
Oleaginous Pipromorpha	Yellow-bellied Elaenia
23 nests 2.7 eggs	45 nests 2.0 eggs
Sulphur-rumped Myiobius	Torrent Flycatcher
11 nests 2.0 eggs	21 nests 2.0 eggs
Northern Royal Flycatcher	Paltry Tyranniscus
7 nests 1.9 eggs	18 nests 1.9 eggs
Black-tailed Myiobius	Golden-crowned Spadebill
5 nests 2.0 eggs	17 nests 2.0 eggs
Eye-ringed Flatbill	Black-fronted Tody-Flycatcher
5 nests 1.8 eggs	16 nests 2.6 eggs
Northern Bentbill	Sulphury Flatbill
3 nests 1.3 eggs	12 nests 2.4 eggs
Ruddy-tailed Flycatcher	Slate-headed Tody-Flycatcher
2 nests 2.0 eggs	7 nests 2.0 eggs
Southern Bentbill	Yellow Flycatcher
1 nest 2.0 eggs	5 nests 2.0 eggs

American flycatchers that build hanging nests are paired as well as solitary species. The solitary nesters include the Ruddy-tailed Flycatcher, Northern Bentbill, Eye-ringed Flatbill, Sulphur-rumped Myiobius, and Oleaginous Pipromorpha. We may add to this list the Northern Royal Flycatcher, of which the female, although attended by a mate, receives no help in feeding and brooding the nestlings. The paired breeders, in which the male feeds the young, include the Black-fronted Tody-Flycatcher, Slate-headed Tody-Flycatcher, Sulphury Flatbill, and Paltry Tyranniscus. In both groups, the set usually consists of two eggs, except in the pipromorpha and the Black-fronted Tody-Flycatcher, which lay three eggs more often than two. As table 18 shows, there is little difference in the number of young reared by these two groups of related birds, despite the fact that in one the nests have twice as many attendants as in the other.

Nocturnal birds have a longer active period in the tropics than at higher latitudes in the milder half of the year when nights are short, so that we might expect them to show the "latitude effect" in reverse, rearing larger families in the tropics than in the temperate zones. Nevertheless, in tropical Africa owls lay sets only slightly more than one-third as large as those of European owls (Lack 1947–1948). Apparently, the size of their broods is determined by something other than foraging time.

Precocial birds, whose young pick up their own food under parental guidance, should not be affected by the length of the parents' working day in the same manner as altricial birds, who carry food to the nest. It is not evident why a quail or pheasant cannot scratch for as large a brood in tropical forest as in northern woodland or steppe. Nevertheless, in no category of birds does the size of the brood change more strikingly with latitude. Compare the family of four or five chicks of a Marbled Wood-Quail in tropical America with the ten to eighteen chicks of the Bobwhite in the United States or the nine to twenty of the Partridge in England. In the Eastern Hemisphere, broods in the pheasant family are also much smaller in the tropics than at middle latitudes, where the number of eggs in a set averages more than twice as great. An almost equally great difference is found in the rail family, in which the average set of four species in equatorial Africa is 3.8 eggs, while that of seven mid-European species is 8.6 eggs (Lack 1947–1948). However, since parent rails and gallinules pick up food for the downy young who follow them, the time factor may be more influential here than with gallinaceous birds.

Evidence from the most diverse families, with quite different diets and methods of rearing their young, points to the conclusion that the number of offspring of tropical birds is not correlated with the number of parents who feed them or closely controlled by their ability to supply food. If it were, we should expect altricial broods attended by both parents to average about twice as large as those attended by only one, while there is no obvious reason why precocial broods should not be as large in the tropics where days are short as at higher latitudes where they are long. To what, then, is the reproductive rate of tropical birds adjusted? Evidently it is adjusted to the annual mortality, which in birds of similar size and habitat should hardly be affected by whether one or two parents attend the nest. Those flycatchers, for example, that live in pairs seem no more vulnerable to predation than the members of this family that are always solitary, hence they have no need to produce larger broods, although, with twice the labor force, they should be able to raise families twice as large.

Birds constantly resident in mild climates with a fairly dependable food supply throughout the year have evolved a strategy for racial survival that contrasts with that of birds that periodically face widespread starvation or succumb in large numbers to climatic extremes and disasters on migration, as in many avian populations at middle and high latitudes. The latter, however, often nest more successfully during the favorable season, and they raise larger broods than do birds at low latitudes. They seldom take prolonged care of the offspring that with relative ease they produce in large numbers. Cast at an early age on their own resources, the young birds often suffer high mortality. Tropical birds lay smaller sets of eggs and lose a higher proportion of their nests; they must often try repeatedly before they succeed in rearing one of their small broods. Accordingly, each fledged young of the tropical birds represents a much larger investment of time and energy than does a fledged young of birds at higher latitudes. Like a prudent investor of hard-earned money, the tropical birds carefully guard their investment, often keeping their young with them for months or years and thereby increas-

ing the young birds' chances of survival. These contrasting strategies provide the key to understanding the different breeding systems of tropical and temperate-zone birds.

Behavioral Patterns
That Limit the Reproductive Rate

We have abundant evidence from many sources, only some of which could be given in the preceding section, that birds in equable climates, especially those in the rainier parts of the tropics, do not rear as many young as they could adequately nourish. Nevertheless, a problem remains. Since the number of eggs a bird lays is subject to mutation, and the more rapidly reproducing genotypes should prevail over the less prolific members of their species, how can a reproductive rate below the maximum be maintained? Should not tropical birds, as has been alleged of northern birds and indeed of the whole animal kingdom, breed just as rapidly as they can, regardless of their real need of recruitment?

Although we can still give no universally valid answer to this question, and certainly none that will satisfy all biologists, we can point to a number of cases in which the behavioral patterns of the birds have, so to speak, locked the reproductive rate below the point that it might have attained. Although the number of eggs that a bird lays is subject to mutation, so are all its other characters, both structural and behavioral. Indeed, some of these characters appear to be more labile or subject to change than the clutch size, which, as we earlier noticed (p. 138), is remarkably constant in whole families and orders. An alteration in behavior that would limit the bird's reproductive capacity might arise before the genetically determined number of eggs increased, making ineffective any subsequent mutation for greater fertility. If the species in which this occurred could barely replace its losses, this alteration of behavior would doubtless be promptly eliminated by natural selection; but, if the species were well able to maintain its population, it might become firmly established.

An evolutionary change of this sort is that which releases the male from attendance at the nest. The ancestors of all those species, previously mentioned, in which the female attends the nest alone were almost certainly conventionally paired, with both parents feeding the young. But somehow the males were relieved of this task, and, since the females were quite able, without their help, to rear all the young needed, the habit persisted and became firmly established in whole genera and families of birds. Indeed, many of these solitary females breed so successfully that their progeny are among the most abundant birds in tropical America, while the males, after emancipation from the nest, developed all the curious and spectacular courtship antics that we surveyed in chapter 3. It is obvious that, without a complete and highly improbable reversal of the whole course of evolution in families like the manakins and hummingbirds and in many genera of cotingas and birds of paradise, no mutation that conferred greater fertility upon the female could raise the rate of reproduction to the level that it could attain if a male helped her to feed the young.

Another behavioral development that may limit the reproductive rate is the construction of large and elaborate nests, some of which were described in chapter 7. These nests take much longer to complete than simple open cups; their construction consumes energy and time that might be devoted to forming eggs and rearing young. If, by decreasing predation, these elaborate nests increase nesting success, their construction might be compatible with maximum reproduction. Unfortunately, it is most difficult to learn how successful these nests are, because some commonly hang high out of reach, while others are so well enclosed that their contents cannot be ascertained without alterations that seem to make them more susceptible to predation. My impression is that covered nests dangling from slender twigs, such as some flycatchers build, are more successful than open cups in the same locality, but that the great castles erected by certain ovenbirds are raided as frequently as open nests. Some of the elaborate pensile nests are built by unaided female flycatchers, so that in these cases two behavioral traits—time-consuming nest construction and aloofness of the male—may be contributing to a reduced reproductive rate.

A widespread means of reducing the reproductive rate is by raising the age when breeding begins. This retardation has both a physiological and a be-

havioral aspect, as some birds, especially males, refrain from nesting even after their reproductive organs are well developed. Although delayed breeding occurs in a number of small tropical land birds, it has been most thoroughly investigated in sea birds, where it is most prolonged. Scarcely any aspect of bird life is more surprising, and more in need of explanation, than this. Why should the maturation of a bird's reproductive behavior be delayed for years after it has ceased to grow in size; why this pause in total development, when in mammals sexual ripeness often precedes the termination of growth?

Ashmole (1963b) suggested that certain birds, especially among tropical marine species, cannot breed sooner because they need years to develop enough skill in foraging, in warm seas where food may be scarce, to feed their young in addition to themselves. Yet Sandwich Terns about seven to nine months old, practicing the difficult art of plunge fishing in their winter home on the coast of Africa, caught fish on 13.4 percent of their dives, while birds in adult plumage, doubtless including many that had already nested, were successful on 16.6 percent of their plunges. The fish caught by the two age groups did not differ noticeably in size. Already the foraging efficiency of the young terns was only 3.2 percent less than that of members of their species who had enjoyed one or more additional years of practice (Dunn 1972).

Is it probable that any bird who, at the age of a year or less, has become adept enough to feed itself should require two or more additional years to improve sufficiently to provide, with a mate's help, for the single, slow-growing nestling that many of these tardily maturing seafowl rear? And, if the usual brood consists of more than one, beginners might, if their contribution to the population were needed, start their reproductive life by rearing a single nestling. We earlier noticed (p. 278) that in a number of species older parents have been demonstrated to be more efficient and to rear larger broods than those nesting for the first time. This increased efficiency is doubtless gained by actual participation in parental activities and cannot otherwise be acquired. But practice in foraging, whether for self or offspring, begins long before the bird starts to breed and should increase steadily from month to month.

The long delay in reaching reproductive maturity is hardly understandable otherwise than as an adaptation to avoid excessive multiplication and the depletion of resources that it causes (Skutch 1953*b*).

In a number of species of land birds, yearlings who do not breed help mated pairs at their nests, thereby gaining valuable experience. But when, as often happens, the number of attendants exceeds the number of young, these birds are evidently not rearing as many as they could feed. The practice of helping at the nests of older birds seems to increase the safety and longevity of individuals but to decrease the birth rate (see chapter 29).

Another pattern of behavior that limits reproduction, especially among marine birds, is the restriction of nesting to certain traditional sites. From time immemorial, some of these birds have returned, year after year, to raise their young on the same small island or seaside cliff, which may be crowded to capacity with nesting birds. At the same time, other apparently suitable sites, at no great distance, remain occupied. Thus on Nightingale Island, one of the Tristan da Cunha group in the South Atlantic, Greater Shearwaters occupied every burrow in their immense, crowded colony. Hundreds of thousands of birds unable to obtain burrows laid their eggs on the surface of the ground, where they did not hatch. An estimated 10 percent of the whole colony failed in this manner to reproduce. Nevertheless, breeding grounds on neighboring islands were more sparsely occupied, and other apparently appropriate areas were quite neglected by the shearwaters (M. K. Rowan 1965). Like delayed reproductive maturity, stubborn adherence to traditional nesting sites appears to be an adaptation for controlling the increase of long-lived seabirds, which might otherwise bear too heavily upon their food resources in the wide oceans, with disastrous consequences to themselves.

Since the above was written, a similar restriction of nesting to a limited part of the apparently suitable sites has been reported of a colonial terrestrial bird. In Colombia, a very dense colony of Cattle Egrets occupied only a fraction of an extensive stand of tall timber bamboos. Instead of building the colony outward, as appeared feasible, nestless adults destroyed eggs, and even more frequently nestlings of all ages, in order to capture occupied nests in its midst. In these struggles for sites, many nestlings, possibly thousands, were pecked to death (Borrero 1972).

Among the behavior patterns that restrain reproduction is territoriality, but since this was considered in chapter 4, it seems unnecessary to do more than call attention to it here. In a large and stimulating book, Wynne-Edwards (1962) examined a number of other social interactions by which birds might become aware of their own abundance and regulate their reproductive effort accordingly, making it the more intense the greater the need for recruitment. He applied the term "epideictic" to all those gatherings and displays whereby the members of a species might gauge their own numbers. Epideictic manifestations among birds include the simultaneous singing of all the males of a district at special times of day, as of thrushes at dawn and in the evening; communal roosting by birds that forage dispersedly; and courtship assemblies of manakins, hummingbirds, and many other birds. Wynne-Edwards recognized a very important secondary division of labor among the sexes, quite distinct from the primary one of producing sperms or eggs. To the male, who contributes so little of his own substance to posterity, has accrued the primary burden of population control in its widest sense, while the female supplies the major part of the material necessary for reproduction. The testing of many of these ideas awaits intensive work in the field.

Although we can point to certain patterns of behavior, such as the male bird's aloofness from the nest, the time-consuming construction of elaborate nests, and delayed sexual maturity, that limit the rate of increase and even lock it at a level well below that which it might otherwise attain, these fail to account for the persistence of small broods in many species of tropical land birds that seem able to rear larger families. These birds build simple open nests in which both parents feed only two or rarely three young. In these cases, it is more difficult to understand why the more fertile genotypes, which must from time to time arise by mutation, do not supplant the current slowly reproducing population. It appears that in these species the reproductive rate is held below the maximum by a number of subtle influences less obvious than the developments that we have just considered. A population well adjusted to the productivity of its habitat is more firmly established than one that exceeds it. Individuals who restrict their reproductive effort to needed recruitment are stronger than those who squander their energy producing superfluous progeny destined soon to perish. In a stable environment, the conservative birds well established on continuously held territories should be able to hold their ground against encroachments by spendthrift stocks that might from time to time arise among them, although this may not be possible where wandering or migratory birds are exposed to climatic extremes that sometimes decimate them. This, I believe, is the direction in which we must seek the explanation of the low reproductive rates of birds constantly resident in our planet's more benign regions.

A restrained reproductive rate is, to be sure, not the only means by which populations of birds, or other organisms, are adjusted to their habitats. When they become too dense, nature has other, harsher methods to redress the balance. Since these methods of removing excess individuals act the more strongly the more concentrated the population becomes, they are said to be density-dependent. The principal forms of density-dependent mortality are starvation, predation, and death from disease. Probably no animal that by excessive breeding overcrowds its environment, whether this be a small island, a narrow valley, or a planet, can by strength or cunning or technological competence permanently escape the misery of density-dependent mortality. If the animal be wise, it will neglect no opportunity to learn from other creatures, however humble, how populations may be adjusted to their environment by means less mercilessly severe than density-dependent mortality.

BIBLIOGRAPHY

Addicott, A. B. 1938. Behavior of the Bush-tit in the breeding season. *Condor* 40:49–63.

Alcock, J. 1969. Observational learning in three species of birds. *Ibis* 111:308–321.

Alexander, W. B. 1946. The Woodcock in the British Isles. *Ibis* 88:1–24.

Allan, R. G. 1962. The Madeiran Storm Petrel *Oceanodroma castro*. *Ibis* 103b:274–295.

Allen, A. A. 1914. The Red-winged Blackbird: A study in the ecology of a cat-tail marsh. *Abs. Proc. Linn. Soc. N.Y.*, pp. 43–128.

Allen, C. S. 1893. The nesting of the Black Duck on Plum Island. *Auk* 10:53–59.

Allen, F. H. 1936. Correspondence [Injury-feigning]. *Auk* 53:125–127.

Allen, R. P., and Mangels, F. P. 1940. Studies of the nesting behavior of the Black-crowned Night Heron. *Proc. Linn. Soc. N.Y.*, No. 50–51:1–28.

Allen, R. W., and Nice, M. M. 1952. A study of the breeding biology of the Purple Martin (*Progne subis*). *Amer. Midl. Nat.* 47:606–665.

Allen, T. T. 1961. Notes on the breeding behavior of the Anhinga. *Wilson Bull.* 73:115–125.

Alley, R., and Boyd, H. 1950. Parent-young recognition in the Coot *Fulica atra*. *Ibis* 92:46–51.

Altmann, S. A. 1956. Avian mobbing behavior and predator recognition. *Condor* 58:241–253.

Alvarez del Toro, M. 1970. Notas para la biología del Pájaro Cantil (*Heliornis fulica*). *Icach* 2:7–13.

———. 1971. On the biology of the American Finfoot in southern Mexico. *Living Bird* 10:79–88.

Amadon, D., and Eckelberry, D. R. 1955. Observations on Mexican birds. *Condor* 57:65–80.

Ames, P. L. 1963. The helmet shrike. *Animal Kingdom* 66:13–14.

Anderson, A. H., and Anderson, A. 1962. Life history of the Cactus Wren. Part V: Fledging to independence. *Condor* 64:199–212.

———. 1963. Life history of the Cactus Wren. Part VI: Competition and survival. *Condor* 65:29–43.

Andrew, R. J. 1961. The displays given by passerines in courtship and reproductive fighting: A review. *Ibis* 103a:315–348, 549–579.

Andrews, M. I., and Naik, R. M. 1965. Some observations on flocks of the Jungle Babbler, *Turdoides striatus* (Dumont) during winter. *Pavo* 3:47–54.

Angell, T. 1969. A study of the Ferruginous Hawk: Adult and brood behavior. *Living Bird* 8:225–241.

Antevs, A. 1947. Towhee helps Cardinals feed their fledglings. *Condor* 49:209.

———. 1948. Behavior of the Gila Woodpecker, Ruby-crowned Kinglet, and Broad-tailed Hummingbird. *Condor* 50:91–92.

Armstrong, E. A. 1942. *Bird display: An introduction to the study of bird psychology*. Cambridge: At the University Press.

———. 1947. *Bird display and behaviour*. London: Lindsay Drummond.

———. 1949. Diversionary display. *Ibis* 91:88–97, 179–188.

———. 1951. Discharge of oily fluid by young fulmars. *Ibis* 93:245–251.

————. 1955. *The Wren.* London: Collins.

————. 1956. Distraction display and the human predator. *Ibis* 98:641–654.

————. 1963. *A study of bird song.* London: Oxford University Press.

Ashmole, N. P. 1962. The Black Noddy *Anous tenuirostris* on Ascension Island. Part 1: General biology. *Ibis* 103*b*:235–273.

————. 1963*a*. The biology of the Wideawake or Sooty Tern *Sterna fuscata* on Ascension Island. *Ibis* 103*b*: 297–364.

————. 1963*b*. The regulation of numbers of tropical oceanic birds. *Ibis* 103*b*:458–473.

Ashmole, N. P., and Tovar, S. H. 1968. Prolonged parental care in Royal Terns and other birds. *Auk* 85: 90–100.

Atwood, H. 1929. A study of the time factor in egg production. *Bull. Agric. Expt. Station, College of Agric., West Virginia Univ.* 223:1–11.

Audubon, J. J. 1940. *The birds of America.* Vol. 1.

Bailey, A. M.; Niedrach, R. J.; and Baily, A. L. 1953. The Red Crossbills of Colorado. *Denver Mus. Nat. Hist., Mus. Pictorial* No. 9:1–63.

Bailey, R. E. 1952. The incubation patch of passerine birds. *Condor* 54:121–136.

————. 1955. The incubation patch in tinamous. *Condor* 57:301–303.

Baird, S. F., and Stansbury, H. 1852. Exploration and survey of the Valley of the Great Salt Lake of Utah. Philadelphia.

Balda, R. P., and Bateman, G. C. 1971. Flocking and annual cycle of the Piñon Jay, *Gymnorhinus cyanocephalus. Condor* 73:287–302.

Banks, R. C. 1957. Birds mobbing snake skin. *Condor* 59:213.

Barth, E. K. 1955. Egg-laying, incubation and hatching of the Common Gull (*Larus canus*). *Ibis* 97:222–239.

Bartholomew, G. A. 1966. The role of behavior in the temperature regulation of the Masked Booby. *Condor* 68:523–535.

Bartholomew, G. A., Jr., and Dawson, W. R. 1952. Body temperature in nestling Western Gulls. *Condor* 54: 58–60.

Bartonek, J. C., and Hickey, J. J. 1969. Selective feeding by juvenile diving ducks in summer. *Auk* 86:443–457.

Beck, J. R. 1969. Food, moult, and age of first breeding in the Cape Pigeon, *Daption capensis. Bull. British Antarc. Surv.* 21:33–44. Abstract in *Ibis* (1970) 112: 407–408.

Beck, J. R., and Brown, D. W. 1971. The breeding biology of the Black-bellied Storm Petrel *Fregetta tropica. Ibis* 113:73–90.

Beebe, W. 1925. The Variegated Tinamou, *Crypturus variegatus. Zoologica* (N.Y. Zool. Soc.) 6:195–227.

Beer, J. R.; Frenzel, L. D.; and Hansen, N. 1956. Minimum space requirements of some nesting passerine birds. *Wilson Bull.* 68:200–209.

Behle, W. H., and Goates, W. A. 1957. Breeding biology of the California Gull. *Condor* 59:235–246.

Bell, B. D., and Hornby, R. J. 1969. Polygamy and nest-sharing in the Reed Bunting. *Ibis* 111:402–405.

Bell, J.; Bruning, D.; and Winnegar, A. 1970. Black-necked Screamer seen feeding a chick. *Auk* 87:805.

Bellrose, F., Jr. 1943. Two Wood Ducks incubating in the same nesting box. *Auk* 60:446–447.

Belt, T. 1888. *The naturalist in Nicaragua.* 2d ed. London: Edward Bumpus.

Bené, F. 1946. The feeding and related behavior of hummingbirds, with special reference to the Black-chin *Archilochus alexandri* (Bourcier and Mulsant). *Mem. Boston Soc. Nat. Hist.* 9:397–478.

Bennett, L. J. 1938. *The Blue-winged Teal: Its ecology and management.* Ames, Iowa: Collegiate Press.

Bent, A. C. 1919–1958. *Life histories of North American birds.* U.S. Natl. Mus. Bulls. Washington, D.C.

Bent, A. C., and collaborators. 1968. Life histories of North American cardinals, grosbeaks, buntings, towhees, finches, sparrows, and allies. *U.S. Natl. Mus. Bull.* 237:1–1889.

Berger, A. J. 1968. Clutch size, incubation period, and nestling period of the American Goldfinch. *Auk* 85: 494–498.

Bergtold, W. H. 1917. *A study of the incubation period of birds.* Denver: Kendrick-Bellamy Co.

Betts, F. N. 1952. The breeding seasons of birds in the hills of South India. *Ibis* 94:621–628.

Biaggi, V., Jr. 1955. Life history of the Puerto Rican Honeycreeper, *Coereba flaveola portoricensis* (Bryant). Univ. Puerto Rico Agric. Expt. Station.

Blackford, J. L. 1950. Wildwood adventure. *Nature Mag.* 43:233–236.

Blair, H. M. S.; Oliver, H. R.; and Chisholm, A. H. 1950. "Rodent-run" distraction-behaviour in birds. *Ibis* 92:476–477.

Blanchard, B. D. 1941. The White-crowned Sparrows (*Zonotrichia leucophrys*) of the Pacific seaboard: Environment and annual cycle. *Univ. Calif. Publ. Zool.* 46:1–178.

Blume, D. 1961. Über die Lebensweise einiger Spechtarten (*Dendrocopos major, Picus viridis, Dryocopus martius*). *Jour. f. Ornith.* 102:1–115.

Bock, C. E. 1970. The ecology and behavior of the Lewis Woodpecker (*Asyndesmus lewis*). *Univ. Calif. Publ. Zool.* 92:1–91.

Borrero H., J. I. 1972. Historia natural de la Garza del Ganado, *Bubulcus ibis*, en Colombia. *Cespedesia* (Cali) 1:387–479.

Bourlière, F. 1954. *The natural history of mammals.* New York: Alfred A. Knopf.

Bourne, W. R. P. 1955. The birds of the Cape Verde Islands. *Ibis* 97:508–556.

Boyd, H. J., and Alley, R. 1948. The function of the head-coloration of the nestling Coot and other nestling Rallidae. *Ibis* 90:582–593.

Brackbill, H. 1944. Juvenile Cardinal helping at a nest. *Wilson Bull.* 56:50.

———. 1952. A joint nesting of Cardinals and Song Sparrows. *Auk* 69:302–307.

Brewer, R. 1961. Comparative notes on the life history of the Carolina Chickadee. *Wilson Bull.* 73:348–373.

Brooks, W. S., and Garrett, S. E. 1970. The mechanism of pipping in birds. *Auk* 87:458–466.

Brosset, A., and Darchen, R. 1967. Une curieuse succession d'hôtes parasites des nids de *Nasutitermes.* *Biologica Gabonica* 3:153–168. Abstract in *Ibis* (1969) 111:264.

Brown, J. 1970. Environmental setting, Barrow, Alaska. *Proc. Conference on Productivity and Conservation in Northern Circumpolar Lands*, pp. 50–64.

Brown, J. L. 1963. Aggressiveness, dominance and social organization in the Steller Jay. *Condor* 65:460–484.

———. 1969. Territorial behavior and population regulation in birds. *Wilson Bull.* 81:293–329.

———. 1970. Cooperative breeding and altruistic behaviour in the Mexican Jay, *Aphelocoma ultramarina.* *Animal Behaviour* 18:366–378.

Brown, L. H. 1958. The breeding biology of the Greater Flamingo *Phoenicopterus ruber* at Lake Elmenteita, Kenya Colony. *Ibis* 100:388–420.

———. 1966. Observations on some Kenya eagles. *Ibis* 108:531–572.

Brown, L. H., and Root, A. 1971. The breeding behaviour of the Lesser Flamingo *Phoeniconaias minor.* *Ibis* 113:147–172.

Brown, L. H., and Urban, E. K. 1969. The breeding biology of the Great White Pelican *Pelecanus onocrotalus roseus* at Lake Shala, Ethiopia. *Ibis* 111:199–237.

Brown, R. G. B. 1962. The aggressive and distraction behaviour of the Western Sandpiper *Ereunetes mauri.* *Ibis* 104:1–12.

Buckley, F. G. 1968. Behaviour of the Blue-crowned Hanging Parrot, *Loriculus galgulus* with comparative notes on the Vernal Hanging Parrot, *L. vernalis. Ibis* 110:145–164.

Buckley, F. G., and Buckley, P. A. 1972. The breeding ecology of Royal Terns *Sterna* (*Thalasseus*) *maxima maxima. Ibis* 114:344–359.

Bullough, W. S. 1942. The reproductive cycles of the British and Continental races of the Starling. *Philos.*

Trans. Roy. Soc. London Ser. B, no. 580, vol. 231: 165–246. Abstract in *Auk* (1943) 60:295–296.

Burger, J. W. 1949. A review of experimental investigations on seasonal reproduction in birds. *Wilson Bull.* 61:211–230.

Burke, V. E. M., and Brown, L. H. 1970. Observations on the breeding of the Pink-backed Pelican *Pelecanus rufescens. Ibis* 112:499–512.

Buxton, J. 1950. *The Redstart.* London: Collins.

Cade, T. J., and Maclean, G. L. 1967. Transport of water by adult sandgrouse to their young. *Condor* 69:323–343.

Calder, W. A. 1971. Temperature relationships and nesting of the Calliope Hummingbird. *Condor* 73:314–321.

Carr, T., and Goin, C. J., Jr. 1965. Bluebirds feeding Mockingbird nestlings. *Wilson Bull.* 77:405–407.

Cayley, N. W. 1949. *The fairy wrens of Australia: Blue birds of happiness.* Sydney and London: Angus and Robertson.

Chabreck, R. H. 1963. Breeding habits of the Pied-billed Grebe in an impounded coastal marsh in Louisiana. *Auk* 80:447–452.

Chapin, J. P. 1954. The calendar of Wideawake fair. *Auk* 71:1–15.

Chapman, F. M. 1928. The nesting habits of Wagler's Oropéndola (*Zarhynchus wagleri*) on Barro Colorado Island. *Bull. Amer. Mus. Nat. Hist.* 58:123–166.

———. 1929. *My tropical air castle.* New York and London: D. Appleton and Co.

———. 1935. The courtship of Gould's Manakin (*Manacus vitellinus vitellinus*) on Barro Colorado Island, Canal Zone. *Bull. Amer. Mus. Nat. Hist.* 68:471–525.

Cherrie, G. K. 1916. A contribution to the ornithology of the Orinoco region. *Mus. Brooklyn Inst. Arts and Sci., Sci. Bull.* 2:133a–374.

Chisholm, A. H. 1936. Injury feigning in birds. *Auk* 53:251–253.

———. 1952a. Bird-insect nesting associations in Australia. *Ibis* 94:395–405.

———. 1952b. Strange relations of birds and insects. *Nature Mag.* 45:526–528, 550.

———. 1954. The use by birds of "tools" or "instruments." *Ibis* 96:380–383.

Christy, B. H. 1940. Mortality among Tree Swallows. *Auk* 57:404–405.

Clark, G. A., Jr. 1960. Notes on the embryology and evolution of the megapodes (Aves: Galliformes). *Postilla, Yale Peabody Mus. Nat. Hist.* No. 45:1–7.

———. 1961. Occurrence and timing of egg teeth in birds. *Wilson Bull.* 73:268–278.

———. 1964. Ontogeny and evolution in the megapodes (Aves: Galliformes). *Postilla, Yale Peabody Mus. Nat. Hist.* No. 78:1–37.

———. 1969. Oral flanges of juvenile birds. *Wilson Bull.* 81:270–279.

Cody, M. L. 1966. A general theory of clutch size. *Evolution* 20:174–184.

Cogswell, H. L. 1949. Alternate care of two nests in the Black-chinned Hummingbird. *Condor* 51:176–178.

Collias, N. E. 1952. The development of social behavior in birds. *Auk* 69:127–159.

Collias, N. E., and Collias, E. C. 1962. An experimental study of the mechanisms of nest building in a weaverbird. *Auk* 79:568–595.

———. 1964a. Evolution of nest-building in the weaverbirds (Ploceidae). *Univ. Calif. Publ. Zool.* 73:1–162.

———. 1964b. The development of nest-building behavior in a weaverbird. *Auk* 81:42–52.

Collins, C. T. 1963. The "downy" nestling plumage of swifts of the genus *Cypseloides*. *Condor* 65:324–328.

Collins, V. B., and de Vos, A. 1966. A nesting study of the Starling near Guelph, Ontario. *Auk* 83:623–636.

Conder, P. J. 1956. The territory of the Wheatear *Oenanthe oenanthe*. *Ibis* 98:453–459.

Contino, F. 1968. Observations on the nesting of *Sporophila obscura* in association with wasps. *Auk* 85:137–138.

Conway, W. G., and Bell, J. 1968. Observations on the behavior of Kittlitz's Sandplovers at the New York Zoological Park. *Living Bird* 7:57–70.

Coombs, C. J. F. 1960. Observations on the Rook *Corvus frugilegus* in southwest Cornwall. *Ibis* 102:394–419.

Coppinger, R. P. 1970. The effect of experience and novelty on avian feeding behavior with reference to the evolution of warning coloration in butterflies. *Amer. Nat.* 104:323–335. Abstract in *Auk* (1971) 88:465.

Cott, H. B. 1954. The palatability of the eggs of birds: Mainly based upon observations of an egg panel. *Proc. Zool. Soc. London* 124:335–463.

Cottrille, W. P., and Cottrille, B. D. 1958. Great Blue Heron: Behavior at the nest. *Univ. Mich. Mus. Zool., Misc. Publ.* No. 102:1–15.

Coulson, J. C. 1965. The functions of the pair bond in the Kittiwake. *Ibis* 107:427.

Coulson, J. C., and White, E. 1958. The effect of age on the breeding biology of the Kittiwake *Rissa tridactyla*. *Ibis* 100:40–51.

Coulter, J. M., and Chamberlain, C. J. 1910. *Morphology of Gymnosperms*. Chicago: University of Chicago Press.

Coward, T. A. 1928. *The birds of the British Isles and their eggs*. 3d ed. London: Frederick Warne.

Cramp, S. 1947. Notes on territory in the Coot. *British Birds* 40:194–198.

Creutz, G. 1955. Der Trauerschnäpper (*Muscicapa hy-*

poleuca [Pallas]). Eine Populationsstudie. *Jour. f. Ornith.* 96:241–326.

Crook, J. H. 1960. Nest form and construction in certain West African weaverbirds. *Ibis* 102:1–25.

———. 1963. A comparative analysis of nest structure in the weaverbirds (Ploceinae). *Ibis* 105:238–262.

Crossin, R. S. 1967. The breeding biology of the Tufted Jay. *Proc. Western Found. Vert. Zool.* 1:265–299.

Crouch, J. E. 1936. Nesting habits of the Cedar Waxwing (*Bombycilla cedrorum*). *Auk* 53:1–8.

Cullen, E. 1957. Adaptations in the Kittiwake to cliff-nesting. *Ibis* 99:275–302.

Cullen, J. M. 1957. Plumage, age and mortality in the Arctic Tern. *Bird Study* 4:197–207. Abstract in *Ibis* (1960) 102:336.

———. 1962. The pecking response of young Wideawake Terns *Sterna fuscata*. *Ibis* 103b:162–173.

Cullen, J. M., and Ashmole, N. P. 1963. The Black Noddy *Anous tenuirostris* on Ascension Island. Part 2. Behaviour. *Ibis* 103b:423–446.

Cushing, J. E., and Ramsay, A. O. 1949. The non-heritable aspects of family unity in birds. *Condor* 51:82–87.

Cushing, J. E., Jr. 1944. The relation of non-heritable food habits to evolution. *Condor* 46:265–271.

Cuthbert, N. L. 1954. A nesting study of the Black Tern in Michigan. *Auk* 71:36–63.

Danforth, R. E. 1930. The Sparrow-hawk's first flight. *Sci. Mon.* 30:81–84.

Darwin, C. 1871. *The descent of man and selection in relation to sex.* (Reprinted in The Modern Library, New York.)

Davies, S. J. J. F. 1962. The nest-building behaviour of the Magpie Goose *Anseranas semipalmata*. *Ibis* 104:147–157.

Davis, D. E. 1940a. Social nesting habits of the Smooth-billed Ani. *Auk* 57:179–218.

———. 1940b. Social nesting habits of *Guira guira*. *Auk* 57:472–484.

———. 1942. The phylogeny of social nesting habits in the Crotophaginae. *Quarterly Rev. Biol.* 17:115–134.

———. 1945. The occurrence of the incubation-patch in some Brazilian birds. *Wilson Bull.* 57:188–190.

———. 1955. Determinate laying in Barn Swallows and Black-billed Magpies. *Condor* 57:81–87.

Davis, J. 1960. Nesting behavior of the Rufous-sided Towhee in coastal California. *Condor* 62:434–456.

Davis, M. 1945. A change in breeding season by Australian gulls. *Auk* 62:137.

———. 1952. Captive Raven carries food to non-captive Black Vulture. *Auk* 69:201.

Davis, T. A. W. 1953. An outline of the ecology and breeding seasons of birds of the lowland forest region of British Guiana. *Ibis* 95:450–467.

Dawson, W. R., and Evans, F. C. 1960. Relation of

growth and development to temperature regulation in nestling Vesper Sparrows. *Condor* 62:329–340.

Deck, R. S. 1945. The neighbors' children. *Nature Mag.* 38:241–242, 272.

De Garis, C. F. 1936. Notes on six nests of the Kentucky Warbler (*Oporornis formosus*). *Auk* 53:418–428.

Delacour, J. 1951–1952. *The pheasants of the world.* 2 vols. London: Country Life.

Delacour, J., and Amadon, D. 1973. *Curassows and related birds.* New York: American Museum of Natural History.

Delacour, J., and Mayr, E. 1945. The family Anatidae. *Wilson Bull.* 57:3–55.

Delius, J. D. 1965. A population study of Skylarks *Alauda arvensis. Ibis* 107:466–492.

Deusing, M. 1939. Nesting habits of the Pied-billed Grebe. *Auk* 56:367–373.

Dexter, R. W. 1952. Extra-parental cooperation in the nesting of Chimney Swifts. *Wilson Bull.* 64:133–139.

Diamond, J. M. 1972. Further examples of dual singing by southwest Pacific birds. *Auk* 89:180–183.

Diamond, J. M., and Terborgh, J. W. 1968. Dual singing by New Guinea birds. *Auk* 85:62–82.

Diesselhorst, G. 1950. Erkennen des Geschlechts und Paarbildung bei der Goldammer (*Emberiza c. citrinella* L.). *Ornith. Berichte* 3:69–112.

Dilger, W. C. 1956. Nest-building movements performed by a juvenile Olive-backed Thrush. *Wilson Bull.* 68:157–158.

Dixon, K. L. 1949. Behavior of the Plain Titmouse. *Condor* 51:110–136.

———. 1954. Some ecological relations of chickadees and titmice in central California. *Condor* 56:113–124.

Donohoe, R. W.; McKibben, C. E.; and Lowry, C. B. 1968. Turkey nesting behavior. *Wilson Bull.* 80:103–104.

Dorst, J. 1956. Etude biologique des Trochilidés des hauts plateaux Péruviens. *L'Oiseau et R. F. O.* 26:165–193.

———. 1957. The puya stands of the Peruvian high plateaux as a bird habitat. *Ibis* 99:594–599.

———. 1962. Nouvelles recherches biologiques sur les Trochilidés des hautes Andes Péruviennes (*Oreotrochilus estella*). *L'Oiseau et R. F. O.* 32:95–126.

Dorward, D. F. 1962. Comparative biology of the White Booby and the Brown Booby *Sula* spp. at Ascension. *Ibis* 103b:174–220.

———. 1963. The Fairy Tern *Gygis alba* on Ascension Island. *Ibis* 103b:365–378.

Dow, D. D. 1970. Communal behaviour of nesting Noisy Miners. *Emu* 70:131–134. Abstract in *Ibis* (1971) 113:385.

Drinkwater, H. 1963. An observation of parental behavior of a Rough-winged Swallow. *Wilson Bull.* 75:277.

Driver, P. M. 1967. Notes on the clicking of avian egg-young, with comments on its mechanism and function. *Ibis* 109:434–437.

Duffey, E. 1951. Field studies on the Fulmar *Fulmarus glacialis. Ibis* 93:237–245.

Duffey, E.; Creasey, N.; and Williamson, K. 1950. The rodent-run distraction-behaviour of certain waders. *Ibis* 92:27–33.

Dunn, E. K. 1972. Effect of age on the fishing ability of Sandwich Terns *Sterna sandvicensis. Ibis* 114:360–366.

Dunnet, G. M. 1955. The breeding of the Starling *Sturnus vulgaris* in relation to its food supply. *Ibis* 97:619–662.

Durango, S. 1949. The nesting associations of birds with social insects and with birds of different species. *Ibis* 91:140–143.

Eddinger, C. R. 1967. Feeding helpers among immature white-eyes. *Condor* 69:530–531.

———. 1970. The white-eye as an interspecific feeding helper. *Condor* 72:240.

Eisner, E. 1960. The biology of the Bengalese Finch. *Auk* 77:271–287.

Elder, W. H., and Elder, N. L. 1949. Role of the family in the formation of goose flocks. *Wilson Bull.* 61:133–140.

Emlen, J. T., Jr. 1941. An experimental analysis of the breeding cycle of the Tricolored Red-wing. *Condor* 43:209–219.

———. 1942. Notes on a nesting colony of Western Crows. *Bird-Banding* 13:143–154.

Epple, A.; Orians, G. H.; Farner, D. S.; and Lewis, R. A. 1972. The photoperiodic testicular response of a tropical finch, *Zonotrichia capensis costaricensis. Condor* 74:1–4.

Erickson, M. M. 1938. Territory, annual cycle, and numbers in a population of Wren-Tits (*Chamaea fasciata*). *Univ. Calif. Publ. Zool.* 42:247–334.

Euler, C. 1867. Beiträge zur Naturgeschichte der Vögel Brasiliens. *Jour. f. Ornith.* 15:177–198, 217–233, 399–420.

Evans, R. M. 1970. Parental recognition and the "mew call" in Black-billed Gulls (*Larus bulleri*). *Auk* 87:503–513.

Farkas, T. 1969. Notes on the biology and ethology of the Natal Robin *Cossypha natalensis. Ibis* 111:281–292.

Farner, D. S. 1958. Incubation and body temperatures in the Yellow-eyed Penguin. *Auk* 75:249–262.

Farner, D. S., and Serventy, D. L. 1959. Body temperature and the ontology of thermoregulation in the Slender-billed Shearwater. *Condor* 61:426–433.

Ficken, M. S. 1965. Mouth color of nestling passerines and its use in taxonomy. *Wilson Bull.* 77:71–75.

Finley, W. L. 1907. *American birds*. New York: C. Scribner's Sons.

Fisher, H. I. 1958. The "hatching muscle" in the chick. *Auk* 75:391–399.

———. 1961. The hatching muscle in North American grebes. *Condor* 63:227–233.

———. 1962. The hatching muscle in Franklin's Gull. *Wilson Bull.* 74:166–172.

———. 1969. Eggs and egg-laying in the Laysan Albatross. *Condor* 71:102–112.

———. 1971. The Laysan Albatross: Its incubation, hatching, and associated behaviors. *Living Bird* 10:19–78.

———. 1972. Sympatry of Laysan and Black-footed Albatrosses. *Auk* 89:381–402.

Fisher, H. I., and Dater, E. E. 1961. Esophagal diverticula in the Redpoll, *Acanthis flammea*. *Auk* 78:528–531.

Fitch, H. S. 1949. Sparrow adopts kingbirds. *Auk* 66:368–369.

Fitch, H. S.; Swenson, F.; and Tillotson, D. F. 1946. Behavior and food habits of the Red-tailed Hawk. *Condor* 48:205–237.

Fleay, D. 1961. Gouras of New Guinea. *Animal Kingdom* 64:106–110.

Fogden, M. P. L. 1972. The seasonality and population dynamics of equatorial forest birds in Sarawak. *Ibis* 114:307–342.

Forbush, E. H. 1929. *Birds of Massachusetts and other New England states*. Vol. 3. Boston: Massachusetts Department of Agriculture.

Forsythe, D. M. 1971. Clicking in the egg-young of the Long-billed Curlew. *Wilson Bull.* 83:441–442.

Foster, W. L., and Tate, J., Jr. 1966. The activities and coactions of animals at sapsucker trees. *Living Bird* 5:87–113.

Fournier O., L. A., and Salas D., S. 1966. Algunas observaciones sobre la dinámica de la floración en el bosque tropical húmedo de Villa Colon. *Rev. Biol. Trop.* 14:75–85.

Fowler, J. M., and Cope, J. B. 1964. Notes on the Harpy Eagle in British Guiana. *Auk* 81:257–273.

Fox, W. 1952. Behavioral and evolutionary significance of the abnormal growth of beaks of birds. *Condor* 54:160–162.

Fraga, R. M. 1972. Cooperative breeding and a case of successive polyandry in the Bay-winged Cowbird. *Auk* 89:447–449.

Franks, E. C. 1967. The responses of incubating Ringed Turtle Doves (*Streptopelia risoria*) to manipulated egg temperatures. *Condor* 69:268–276.

Fredrickson, L. H. 1971. Common Gallinule breeding biology and development. *Auk* 88:914–919.

French, N. R. 1959. Life history of the Black Rosy Finch. *Auk* 76:159–180.

French, N. R., and Hodges, R. W. 1959. Torpidity in cave-roosting hummingbirds. *Condor* 61:223.

Friedmann, H. 1929. *The cowbirds: A study in the biology of social parasitism*. Springfield, Ill.: Charles C. Thomas.

———. 1934. The instinctive emotional life of birds. *Psychoanal. Rev.* 21(3 and 4).

———. 1955. The honey-guides. *U.S. Natl. Mus. Bull.* 208.

———. 1960. The parasitic weaverbirds. *U.S. Natl. Mus. Bull.* 223.

———. 1963. Host relations of the parasitic cowbirds. *U.S. Natl. Mus. Bull.* 233.

———. 1968. The evolutionary history of the avian genus *Chrysococcyx*. *U.S. Natl. Mus. Bull.* 265.

Friedmann, H., and Smith, F. D., Jr. 1950. A contribution to the ornithology of northeastern Venezuela. *Proc. U.S. Natl. Mus.* 100:411–538.

———. 1955. A further contribution to the ornithology of northeastern Venezuela. *Proc. U.S. Natl. Mus.* 104:463–524.

Frisch, O. Von. 1966. Wiesenweihe *Circus pygargus* trägt Junge ein. *Z. Tierpsychol.* 23:581–583. Abstract in *Ibis* (1968) 110:387.

Frith, H. J. 1956. Breeding habits in the family Megapodiidae. *Ibis* 98:620–640.

———. 1962. *The Mallee-Fowl: The bird that builds an incubator*. Sydney: Angus and Robertson.

Frith, H. J., and Davies, S. J. J. F. 1961. Ecology of the Magpie Goose, *Anseranas semipalmata* Latham (Anatidae). *C.S.I.R.O. Wildlife Research* (Australia) 6:91–141.

Fry, C. H. 1967. Studies of bee-eaters. *Nigerian Field* 32:4–17. Abstract in *Auk* (1967) 84:454–455.

———. 1972. The social organisation of bee-eaters (Meropidae) and co-operative breeding in hot-climate birds. *Ibis* 114:1–14.

Ganier, A. F. 1964. The alleged transportation of its eggs or young by the Chuck-will's-widow. *Wilson Bull.* 76:19–27.

Gatter, W. 1971. Wassertransport beim Flussregenpfeifer (*Charadrius dubius*). *Vogelwelt* 92:100–103. Abstract in *Ibis* (1972) 114:415.

Genelly, R. E. 1955. Annual cycle in a population of California Quail. *Condor* 57:263–285.

Gibb, J. 1950. The breeding biology of the Great and Blue Titmice. *Ibis* 92:507–539.

———. 1954. Feeding ecology of tits, with notes on Treecreeper and Goldcrest. *Ibis* 96:513–543.

———. 1956. Territory in the genus *Parus*. *Ibis* 98:420–429.

Gibson, F. 1971. The breeding biology of the American Avocet (*Recurvirostra americana*) in central Oregon. *Condor* 73:444–454.

Gilbert, H. A., and Brook, A. 1924. *Secrets of bird life.* London: Arrowsmith.

Gilliard, E. T. 1956. *Exotic birds of the South Pacific.* Garden City, N.Y.: Nelson Doubleday.

———. 1958. *Living birds of the world.* Garden City, N.Y.: Doubleday and Co.

———. 1959a. Notes on some birds of northern Venezuela. *Amer. Mus. Novit.* No. 1927:1–33.

———. 1959b. The courtship behavior of Sanford's Bowerbird (*Archboldia sanfordi*). *Amer. Mus. Novit.* No. 1935:1–18.

———. 1959c. Notes on the courtship behavior of the Blue-backed Manakin (*Chiroxiphia pareola*). *Amer. Mus. Novit.* No. 1942:1–19.

———. 1969. *Birds of paradise and bower birds.* London: Weidenfeld and Nicolson.

Glover, F. A. 1953. Nesting ecology of the Pied-billed Grebe in northwestern Iowa. *Wilson Bull.* 65:32–39.

Goertz, J. W. 1962. An opossum-titmouse incident. *Wilson Bull.* 74:189–190.

Goodwin, D. 1947. Breeding behaviour in domestic pigeons four weeks old. *Ibis* 89:656–658.

———. 1948a. Incubation habits of the Golden Pheasant. *Ibis* 90:280–284.

———. 1948b. Tameness in birds. *Ibis* 90:316–318.

———. 1953. Observations on the voice and behaviour of the Red-legged Partridge *Alectoris rufa*. *Ibis* 95:581–614.

———. 1965. A comparative study of captive blue waxbills (Estrildidae). *Ibis* 107:285–315.

Gosse, P. H. 1847. *The birds of Jamaica.* London: John Van Voorst.

Graber, R. R. 1955. Artificial incubation of some non-Galliform eggs. *Wilson Bull.* 67:100–109.

Gramza, A. F. 1967. Responses of brooding nighthawks to a disturbance stimulus. *Auk* 84:72–86.

Graul, W. D. 1971. Observations at a Long-billed Curlew nest. *Auk* 88:182–184.

Greenwood, J. 1964. The fledging of the Guillemot *Uria aalge* with notes on the Razorbill *Alca torda*. *Ibis* 106:469–481.

Greenwood, R. J. 1969. Mallard hatching from an egg cracked by freezing. *Auk* 86:752–754.

Grey of Fallodon, Viscount. 1927. *The charm of birds.* New York: Frederick A. Stokes.

Grier, J. W. 1971. Pre-attack posture of the Red-tailed Hawk. *Wilson Bull.* 83:115–123.

Grimes, S. A. 1936. "Injury-feigning" by birds. *Auk* 53:478–480.

———. 1940. Scrub Jay reminiscences. *Bird-Lore* 42:431–436.

Grimmer, J. L. 1962. Strange little world of the Hoatzin. *Natl. Geographic Mag.* 122:391–400.

Gross, A. O. 1949a. The Antillean Grebe at Central Soledad, Cuba. *Auk* 66:42–52.

———. 1949b. Nesting of the Mexican Jay in the Santa Rita Mountains, Arizona. *Condor* 51:241–249.

———. 1958. Life history of the Bananaquit of Tobago Island. *Wilson Bull.* 70:257–279.

Gross, W. A. O. 1935. The life history of Leach's Petrel (*Oceanodroma leucorhoa leucorhoa*) on the outer sea islands of the Bay of Fundy. *Auk* 52:382–399.

Grubb, T. C., Jr. 1970. Burrow digging techniques of Leach's Petrel. *Auk* 87:587–588.

Guhl, A. M., and Ortman, L. L. 1953. Visual patterns in recognition of individuals among chickens. *Condor* 55:287–298.

Gullion, G. W. 1954. The reproductive cycle of American Coots in California. *Auk* 71:366–412.

Gurney, J. H. 1913. *The Gannet, a bird with a history.* London: Witherby.

Haartman, L. Von. 1953. Was reizt den Trauerfliegenschnäpper (*Muscicapa hypoleuca*) zu füttern? *Vogelwarte* 16:157–164.

———. 1955. Clutch size in polygamous species. *Acta XI Int. Ornith. Congr.* (Basel, 1954), pp. 450–453.

———. 1956. Territory in the Pied Flycatcher *Muscicapa hypoleuca*. *Ibis* 98:460–475.

———. 1957. Adaptation in hole-nesting birds. *Evolution* 11:339–347.

———. 1958. The incubation rhythm of the female Pied Flycatcher (*Ficedula hypoleuca*) in the presence and absence of the male. *Ornis Fennica* 35:71–76.

———. 1969a. Nest-site and evolution of polygamy in European passerine birds. *Ornis Fennica* 46:1–12.

———. 1969b. The nesting habits of Finnish birds. *Commentationes Biologicae Societas Scientiarum Fennica* 32:1–187.

Hagar, J. A. 1966. Nesting of the Hudsonian Godwit at Churchill, Manitoba. *Living Bird* 5:5–43.

Hailman, J. P. 1962. Pecking of Laughing Gull chicks at models of the parental head. *Auk* 79:89–98.

Hales, H. 1896. Peculiar traits of some Scarlet Tanagers. *Auk* 13:261–263.

Hall, K. R. L. 1960. Egg-covering by the White-fronted Sandplover *Charadrius marginatus*. *Ibis* 102:545–553.

Hamerstrom, F. 1957. The influence of a hawk's appetite on mobbing. *Condor* 59:192–194.

Hamerstrom, F. N., Jr. 1936. A study of the nesting habits of the Ring-necked Pheasant in northwest Iowa. *Iowa State College Jour. Sci.* 10:173–203.

Hamilton, G. D. 1952. English Sparrow feeding young Eastern Kingbirds. *Condor* 54:316.

Hamilton, W. J., Jr. 1943. Nesting of the Eastern Bluebird. *Auk* 60:91–94.

Hamilton, W. J., III. 1965. Sun-oriented display of the Anna's Hummingbird. *Wilson Bull.* 77:38–44.

Hammer, D. A. 1970. Trumpeter Swan carrying young. *Wilson Bull.* 82:324–325.

Haneda, K., and Shinoda, T. 1969. A study of breeding

biology of the Japanese Wagtail. *Misc. Repts. Yamashina Inst. Ornith.* 5:602–622. Abstract in *Auk* (1971) 88:952.

Hann, H. W. 1937. Life history of the Oven-bird in southern Michigan. *Wilson Bull.* 49:145–237.

Hardy, J. W. 1957. The Least Tern in the Mississippi Valley. *Mich. State Univ. Mus. Biol. Ser.* 1:1–60.

———. 1961. Studies in behavior and phylogeny of certain New World jays (Garrulinae). *Univ. Kansas Sci. Bull.* 42:13–149.

Harris, M. P. 1966. Breeding biology of the Manx Shearwater *Puffinus puffinus*. *Ibis* 108:17–33.

———. 1967. The biology of Oystercatchers *Haematopus ostralegus* on Skokholm Island, S. Wales. *Ibis* 109:180–193.

———. 1969. The biology of storm petrels in the Galápagos Islands. *Proc. Calif. Acad. Sci.* 37:95–165. Abstract in *Ibis* (1970) 112:409–410.

———. 1970. Breeding ecology of the Swallow-tailed Gull, *Creagrus furcatus*. *Auk* 87:215–243.

———. 1973. Biology of the Waved Albatross *Diomedea irrorata* of Hood Island, Galápagos. *Ibis* 115:483–510.

Harrison, C. J. O. 1967. Sideways-throwing and sideways-building in birds. *Ibis* 109:539–551.

Hartley, P. H. T. 1949. The biology of the Mourning Chat in winter quarters. *Ibis* 91:393–413.

———. 1950. An experimental analysis of interspecific recognition. *Symposia Soc. Expt. Biol.* 4:313–336.

Hartshorne, C. 1958. The relation of bird song to music. *Ibis* 100:421–445.

Hatch, J. J. 1966. Collective territories in Galápagos mockingbirds, with notes on other behavior. *Wilson Bull.* 78:198–206.

Haverschmidt, F. 1951. Notes on the life history of *Picumnus minutissimus* in Surinam. *Ibis* 93:196–200.

———. 1952. Nesting behavior of the Southern House Wren in Surinam. *Condor* 54:292–295.

———. 1953. Notes on the life history of the Blood-colored Woodpecker in Surinam. *Auk* 70:21–25.

———. 1958. Notes on the breeding habits of *Panyptila cayennensis*. *Auk* 75:121–130.

———. 1959. Notes on the nesting of *Turdus leucomelas* in Surinam. *Wilson Bull.* 71:175–177.

———. 1971. Notes on the life history of the Rusty-margined Flycatcher in Surinam. *Wilson Bull.* 83:124–128.

Hawksley, O., and McCormack, A. P. 1951. Doubly-occupied nests of the Eastern Cardinal, *Richmondena cardinalis*. *Auk* 68:515–516.

Hebb, D. O. 1953. Heredity and environment in mammalian behaviour. *British Jour. Animal Behaviour* 1:43–47.

Heinroth, O., and Heinroth, K. 1959. *The birds.* London: Faber and Faber.

Herbert, K. G. S. 1971. Starling feeds young Robins. *Wilson Bull.* 83:316–317.

Herrick, F. H. 1905. *The home life of wild birds.* New York: Knickerbocker Press.

Hesse, R. 1922. Die Bedeutung der Tagesdauer für die Vögel. *Sitzber. Naturhist. Ver. Bonn für 1922–23*, pp. A13–17.

Hickling, R. A. O. 1959. The burrow-excavation phase in the breeding cycle of the Sand Martin, *Riparia riparia*. *Ibis* 101:497–500.

Hicks, T. W. 1955. Mockingbird attacking Blacksnake. *Auk* 72:296–297.

Hill, H. M., and Work, T. H. 1947. Protocalliphora larvae infesting nestling birds of prey. *Condor* 49:74–75.

Hinde, R. A. 1956. The biological significance of the territories of birds. *Ibis* 98:340–369.

Hindwood, K. A. 1955. Bird/wasp nesting associations. *Emu* 55:263–274.

Hobson, W. 1972. The breeding biology of the Knot (*Calidris c. canutus*) with special reference to Arctic non-breeding. *Proc. Western Found. Vert. Zool.* 2:5–26.

Hochbaum, H. A. 1960. The brood season. *Nat. Hist.* 69:54–61.

Hoffmann, A. 1949. Über die Brutpflege des polyandrischen Wasserfasans, *Hydrophasianus chirurgus* (Scop.). *Zool. Jahrb.* 78:367–403.

Höglund, N. H. 1955. [Body temperature, activity and reproduction of the Capercaillie.] *Viltrevy* 1:1–87. Abstract in *Auk* (1961) 78:459.

Holmes, R. T. 1966. Breeding ecology and annual cycle adaptations of the Red-backed Sandpiper (*Calidris alpina*) in northern Alaska. *Condor* 68:3–46.

Holyoak, D. 1969. The function of the pale egg colour of the Jackdaw. *Bull. British Ornith. Club* 89:159. Abstract in *Auk* (1970) 87:852.

Hopcraft, J. B. D. 1968. Some notes on the chick-carrying behavior in the African Jaçana. *Living Bird* 7:85–88.

Hori, J. 1964. The breeding biology of the Shelduck *Tadorna tadorna*. *Ibis* 106:333–360.

Hornocker, M. G. 1969. Goslings descend from aerial nest, attacked by Bald Eagle. *Auk* 86:764–765.

Horobin, J. M. 1969. The breeding biology of an aged population of Arctic Terns. *Ibis* 111:443.

Howard, H. E. 1920. *Territory in bird life.* London: John Murray.

Howard, L. 1952. *Birds as individuals.* London: Collins.

Howell, T. R. 1959. A field study of temperature regulation in young Least Terns and Common Nighthawks. *Wilson Bull.* 71:19–32.

Howell, T. R., and Bartholomew, G. A. 1961. Temperature regulation in Laysan and Black-footed Albatrosses. *Condor* 63:185–197.

———. 1962a. Temperature regulation in the Red-tailed Tropic Bird and the Red-footed Booby. *Condor* 64:6–18.

———. 1962b. Temperature regulation in the Sooty Tern *Sterna fuscata. Ibis* 104:98–105.

Howell, T. R., and Dawson, W. R. 1954. Nest temperatures and attentiveness in the Anna Hummingbird. *Condor* 56:93–97.

Hoyt, J. S. Y. 1948. Observations on nesting associates. *Auk* 65:188–196.

Hudson, W. H. 1920. *Birds of La Plata.* 2 vols. London: J. M. Dent and Sons.

Hunt, J. H. 1971. A field study of the Wrenthrush, *Zeledonia coronata. Auk* 88:1–20.

Hutson, H. P. W. 1947. Observations on nesting of some birds around Delhi. *Ibis* 89:569–576.

Huxley, J. S. 1914. The courtship habits of the Great Crested Grebe (*podiceps cristatus*); with an addition to the theory of sexual selection. *Proc. Zool. Soc. London,* pp. 491–562.

———. 1938a. Darwin's theory of sexual selection and the data subsumed by it, in the light of recent research. *Amer. Nat.* 72:416–433.

———. 1938b. "The present standing of the theory of sexual selection." In *Evolution,* edited by G. R. de Beer. Oxford: Clarendon Press.

———. 1941. On the habit of brooding on the perch in birds. *Proc. Zool. Soc. London* Ser. A, 111:37–39.

———. 1947. Notes on the problem of geographical differences in tameness in birds. *Ibis* 89:539–552.

Huxley, J. S.; Buxton, A.; Ingram, C.; and Goodwin, D. 1948. Tameness in birds. *Ibis* 90:312–318.

Huxley, T. H., and Huxley, J. 1947. *Evolution and ethics 1893–1943.* London: Pilot Press.

Immelmann, K. 1962. Beiträge zu einer vergleichenden Biologie australischer Prachtfinken (Spermestidae). *Zool. Jahrb. Syst.* 90:1–196.

———. 1963. Tierische Jahresperiodik in ökologischen Sicht. *Zool. Jahrb. Syst.* 91:91–200.

———. 1966. Beobachtungen an Schwalbenstaren. *Jour. f. Ornith.* 107:37–69.

———. 1967. Periodische Vorgänge in der Fortpflanzung tierischer Organismen. *Studium Generale* 20:15–33.

Ingram, C. 1959. The importance of juvenile cannibalism in the breeding biology of certain birds of prey. *Auk* 76:218–226.

———. 1962. Cannibalism by nestling Short-eared Owls. *Auk* 79:715.

———. 1966. *In search of birds.* London: H. F. and G. Witherby.

Ionides, C. J. P. 1954. Passerines attacking snakes. *Ibis* 96:310–311.

Irving, L. 1960. Birds of Anaktuvuk Pass, Kobuk, and Old Crow: A study in Arctic adaptation. *U.S. Natl. Mus. Bull.* 217.

Ivor, H. R. 1944a. Birds' fear of man. *Auk* 61:203–211.

———. 1944b. Aye, she was Bonnie. *Nature Mag.* 37:473–476.

———. 1952. Hatching of eggs by hand-reared Wood Thrushes. *Auk* 69:284–288.

Jackson, J. A. 1974. Gray Rat Snake versus Red-cockaded Woodpeckers: Predator-prey adaptations. *Auk* 91:342–347.

Janzen, D. H. 1969. Birds and the ant × acacia interaction in Central America, with notes on birds and other myrmecophytes. *Condor* 71:240–256.

Jehl, J. R., Jr. 1968a. The breeding biology of Smith's Longspur. *Wilson Bull.* 80:123–149.

———. 1968b. The egg tooth of some Charadriiform birds. *Wilson Bull.* 80:328–330.

Jenni, D. A. 1961. Distraction display of the Common Gallinule. *Wilson Bull.* 73:387–388.

Johnsgard, P. A., and Kear, J. 1968. A review of parental carrying of young by waterfowl. *Living Bird* 7:89–102.

Johnson, N. K. 1963. Biosystematics of sibling species of flycatchers in the *Empidonax hammondii–oberholseri–wrightii* complex. *Univ. Calif. Publ. Zool.* 66:79–238.

Johnson, R. A. 1947. Rôle of male Yellow-bellied Sapsucker in the care of the young. *Auk* 64:621–623.

———. 1969. Hatching behavior of the Bobwhite. *Wilson Bull.* 81:79–86.

Johnston, R. F. 1960. Behavior of the Inca Dove. *Condor* 62:7–24.

Johnston, R. F., and Hardy, J. W. 1962. Behavior of the Purple Martin. *Wilson Bull.* 74:243–262.

Kale, H. W., II. 1962. A captive Marsh Wren helper. *Oriole* (June).

Kear, J. 1963. Parental feeding in the Magpie Goose. *Ibis* 105:428.

Kearton, C. 1930. *The island of penguins.* New York: Longmans, Green.

Keller, R. J.; Shephard, H. R.; and Randall, R. N. 1941. Report of the Sage Grouse survey: Pittman-Robertson Project, Colorado 4-R, season 1941, with comparative data of previous seasons. *Sage Grouse Survey Colorado* 3:31.

Kelso, L. 1940. Antipathy in the Screech Owl. *Auk* 57:252–253.

Kendeigh, S. C. 1941. Territorial and mating behavior of the House Wren. *Illinois Biol. Monogr.* 18:1–120.

———. 1945. Nesting behavior of wood warblers. *Wilson Bull.* 57:145–164.

———. 1952. Parental care and its evolution in birds. *Illinois Biol. Monogr.* No. 22.

———. 1961. Energy of birds conserved by roosting in cavities. *Wilson Bull.* 73:140–147.

———. 1963. New ways of measuring the incubation period of birds. *Auk* 80:453–461.

Kendeigh, S. C.; Kramer, T. C.; and Hamerstrom, F. 1956. Variations in egg characteristics of the House Wren. *Auk* 73:42–65.

Kessler, F. 1962. Measurement of nest attentiveness in the Ring-necked Pheasant. *Auk* 79:702–705.

Kilham, L. 1957. Egg-carrying by the Whip-poor-will. *Wilson Bull.* 69:113.

———. 1958*a*. Sealed-in winter stores of Red-headed Woodpeckers. *Wilson Bull.* 70:107–113.

———. 1958*b*. Pair-formation, mutual tapping and nest hole selection of Red-bellied Woodpeckers. *Auk* 75:318–329.

———. 1958*c*. Territorial behavior of wintering Red-headed Woodpeckers. *Wilson Bull.* 70:347–358.

———. 1959. Early reproductive behavior of flickers. *Wilson Bull.* 71:323–336.

———. 1960. Courtship and territorial behavior of Hairy Woodpeckers. *Auk* 77:259–270.

———. 1961. Reproductive behavior of Red-bellied Woodpeckers. *Wilson Bull.* 73:237–254.

———. 1968*a*. Reproductive behavior of Hairy Woodpeckers. II: Nesting and habitat. *Wilson Bull.* 80:286–305.

———. 1968*b*. Reproductive behavior of White-breasted Nuthatches. I: Distraction display, bill-sweeping, and nest hole defense. *Auk* 85:477–492.

———. 1971*a*. Reproductive behavior of Yellow-bellied Sapsuckers. I: Preference for nesting in *Fomes*-infected aspens and nest hole interrelations with Flying Squirrels, Raccoons, and other animals. *Wilson Bull.* 83:159–171.

———. 1971*b*. Roosting habits of White-breasted Nuthatches. *Condor* 73:113–114.

———. 1971*c*. Use of blister beetle in bill-sweeping by White-breasted Nuthatch. *Auk* 88:175–176.

———. 1972. Death of a Red-breasted Nuthatch from pitch around nest hole. *Auk* 89:451–452.

King, J. R. 1972. Postnuptial and postjuvenal molt in Rufous-collared Sparrows in northwestern Argentina. *Condor* 74:5–16.

Klopman, R. B. 1958. The nesting of the Canada Goose at Dog Lake, Manitoba. *Wilson Bull.* 70:168–183.

Kluijver [Kluyver], H. N. 1950. Daily routines of the Great Tit, *Parus m. major* L. *Ardea* 38:99–135.

———. 1951. The population ecology of the Great Tit, *Parus m. major* L. *Ardea* 39:1–135.

———. 1955. Das Verhalten des Drosselrohrsangers, *Acrocephalus arundinaceus* (L.), am Brutplatz mit besonderer Berücksichtigung der Nestbautechnik und der Revierbehauptung. *Ardea* 43:1–50.

———. 1961. Food consumption in relation to habitat in breeding chickadees. *Auk* 78:532–550.

Knorr, O. A. 1957. Communal roosting of the Pygmy Nuthatch. *Condor* 59:398.

Koskimies, J., and Lahti, L. 1964. Cold-hardiness of the newly hatched young in relation to ecology and distribution in ten species of European ducks. *Auk* 81:281–307.

Kozicky, E. L., and Schmidt, F. V. 1949. Nesting habits of the Clapper Rail in New Jersey. *Auk* 66:355–364.

Kramer, G. 1950. Der Nestbau beim Neuntöter (*Lanius collurio* L.). *Ornith. Berichte* 3:1–14.

Kuerzi, R. G. 1941. Life history studies of the Tree Swallow. *Proc. Linn. Soc. N.Y.* 52–53:1–52.

Lack, D. 1940. Courtship feeding in birds. *Auk* 57:169–178.

———. 1947. *Darwin's finches*. Cambridge: At the University Press.

———. 1947–1948. The significance of clutch-size. *Ibis* 89:302–352; 90:25–45.

———. 1948. Further notes on clutch and brood size in the Robin. *British Birds* 41:98–104, 130–137.

———. 1950. The breeding seasons of European birds. *Ibis* 92:288–316.

———. 1953. *The life of the Robin*. London: Penguin Books.

———. 1954. *The natural regulation of animal numbers*. Oxford: Clarendon Press.

———. 1956*a*. *Swifts in a tower*. London: Methuen and Co.

———. 1956*b*. Seaward flights of Swifts at dusk. *Bird Study* 3:37–42.

———. 1958. The significance of the colour of turdine eggs. *Ibis* 100:145–166.

———. 1966. *Population studies of birds*. Oxford: Clarendon Press.

———. 1968. *Ecological adaptations for breeding in birds*. London: Methuen and Co.

Lack, D., and Lack, E. 1952. The breeding behaviour of the Swift. *British Birds* 45:186–215.

———. 1958. The nesting of the Long-tailed Tit. *Bird Study* 5:1–19.

Lamm, D. W. 1948. Notes on the birds of the States of Pernambuco and Paraiba, Brazil. *Auk* 65:261–283.

Lancaster, D. A. 1964*a*. Biology of the Brushland Tinamou, *Nothoprocta cinerascens*. *Bull. Amer. Mus. Nat. Hist.* 127:269–314.

———. 1964*b*. Life history of the Boucard Tinamou in British Honduras. Part II: Breeding biology. *Condor* 66:253–276.

———. 1970. Breeding behavior of the Cattle Egret in Colombia. *Living Bird* 9:167–194.

Lanyon, W. E. 1956. Territory in the meadowlarks, genus *Sturnella*. *Ibis* 98:485–489.

———. 1960. "The ontology of vocalizations in birds." In *Animal sounds and communication*. Amer. Inst. Biol. Sci.

Lapham, H. 1970. A study of the nesting behavior of the Rufous-vented Chachalaca (*Ortalis r. ruficauda*) in Venezuela. *Bol. Soc. Venezolana Cien. Nat.* 28:291–329.

Laskey, A. R. 1936. Fall and winter behavior of Mockingbirds. *Wilson Bull.* 48:241–255.

———. 1939. A study of nesting Eastern Bluebirds. *Bird-Banding* 10:23–32.

———. 1940. The 1939 nesting season of Bluebirds at Nashville, Tennessee. *Wilson Bull.* 52:183–190.

———. 1944. A Mockingbird acquires his song repertory. *Auk* 61:211–218.

———. 1948. Some nesting data on the Carolina Wren at Nashville, Tennessee. *Bird-Banding* 19:101–121.

———. 1958. Blue Jays at Nashville, Tennessee: Movements, nesting, age. *Bird-Banding* 29:211–218.

Lawrence, L. de K. 1966. A comparative life-history study of four species of woodpeckers. *Amer. Ornith. Union, Ornith. Monogr.* No. 5.

———. 1968. Notes on hoarding nesting material, display, and flycatching in the Gray Jay (*Perisoreus canadensis*). *Auk* 85:139.

Leach, F. A. 1925. Communism in the California Woodpecker. *Condor* 27:12–19.

Leck, C. F. 1972. Seasonal changes in feeding pressures of fruit- and nectar-eating birds in Panama. *Condor* 74:54–60.

Lees, J. 1950. Stomach oil in Fulmars. *Ibis* 92:152–153.

Lehrman, D. S., and Wortis, R. P. 1967. Breeding experience and breeding efficiency in the Ring Dove. *Animal Behaviour* 15:223–228.

Lemmons, P. 1956. Cardinal feeds fishes. *Nature Mag.* 49:536.

Leopold, A. S. 1944. The nature of heritable wildness in Turkeys. *Condor* 46:133–197.

Leopold, F. 1951. A study of nesting Wood Ducks in Iowa. *Condor* 53:209–220.

Levick, G. M. 1914. *Antarctic penguins.* London: William Heinemann.

Lewis, H. F. 1961. "The eider and her helpers." In *Discovery*, edited by J. K. Terres. Philadelphia: J. B. Lippincott Co.

Lewis, R. A., and Orcutt, F. S., Jr. 1971. Social behavior and avian sexual cycles. *Scientia* (Milan) 65:447–472.

Ligon, J. D. 1970. Behavior and breeding biology of the Red-cockaded Woodpecker. *Auk* 87:255–278.

———. 1971. Late summer-autumnal breeding of the Piñon Jay in New Mexico. *Condor* 73:147–153.

Lind, E. A. 1960. Zur Ethologie und Ökologie der Mehlschwalbe, *Delichon u. urbica* (L.). *Ann. Zool. Soc. 'Vanamo'* 21(2):1–123.

———. 1964. Nistzeitliche Geselligkeit der Mehlschwalbe *Delichon u. urbica* (L.). *Ann. Zool. Fennici* 1:7–43.

Lloyd, M. 1960. Statistical analysis of Marchant's data on breeding success and clutch-size. *Ibis* 102:600–611.

Lockley, R. M. 1942. *Shearwaters.* London: J. M. Dent and Sons.

———. 1952. Notes on the birds of the islands of the Berlengas (Portugal), the Desertas and Baixo (Madeira) and the Salvages. *Ibis* 94:144–157.

———. 1953. *Puffins.* London: J. M. Dent and Sons.

———. 1954. *The seals and the curragh.* London: J. M. Dent and Sons.

Lofts, B. 1962. Photoperiod and the refractory period of reproduction in an equatorial bird, *Quelea quelea. Ibis* 104:407–414.

Logan, S. 1951. Cardinal, *Richmondena cardinalis*, assists in feeding Robins. *Auk* 68:516–517.

Lorenz, K. Z. 1952. *King Solomon's ring: New light on animal ways.* London: Methuen and Co.

Low, J. B. 1941. Nesting of the Ruddy Duck in Iowa. *Auk* 58:506–517.

———. 1945. Ecology and management of the Redhead (*Nyroca americana*) in Iowa. *Ecol. Monogr.* 15:35–69.

Lustick, S. 1970. Energy requirements of molt in cowbirds. *Auk* 87:742–746.

McClure, H. E. 1942. Mourning Dove production in southwestern Iowa. *Auk* 59:64–75.

———. 1945. Reaction of the Mourning Dove to colored eggs. *Auk* 62:270–272.

MacDonald, S. D. 1970. The breeding behavior of the Rock Ptarmigan. *Living Bird* 9:195–238.

Macgregor, D. E. 1950. Notes on the breeding of the Red-breasted Chat *Oenanthe heuglini. Ibis* 92:380–383.

McIlhenny, E. A. 1937. Life history of the Boat-tailed Grackle in Louisiana. *Auk* 54:274–295.

Maciula, S. J. 1960. Worm-eating Warbler "adopts" Ovenbird nestlings. *Auk* 77:220.

Maclaren, P. I. R. 1950. Bird-ant nesting associations. *Ibis* 92:564–566.

Maclean, G. L. 1967. The breeding biology and behaviour of the Double-banded Courser *Rhinoptilus africanus* (Temminck). *Ibis* 109:556–569.

———. 1968. Field studies of the sandgrouse of the Kalahari Desert. *Living Bird* 7:209–235.

———. 1969. A study of seedsnipe in southern South America. *Living Bird* 8:33–80.

McNab, B. K. 1966. An analysis of the body temperatures of birds. *Condor* 68:47–55.

Maher, W. J. 1962. Breeding biology of the Snow Petrel near Cape Hallett, Antarctica. *Condor* 64:488–499.

Marchant, S. 1959. The breeding season in S.W. Ecuador. *Ibis* 101:137–152.

———. 1960. The breeding of some S.W. Ecuadorian birds. *Ibis* 102:349–382, 584–599.

————. 1963. The breeding of some Iraqi birds. *Ibis* 105:516–557.

Marshall, A. J. 1952. Non-breeding among Arctic birds. *Ibis* 94:310–333.

————. 1954. *Bower-birds: Their displays and breeding cycles.* Oxford: Clarendon Press.

————. 1959. Internal and environmental control of breeding. *Ibis* 101:456–478.

Marshall, N. 1943. Factors in the incubation behavior of the Common Tern. *Auk* 60:574–588.

Mathews, G. M. 1924. *The birds of Australia.* Vol. 11. London: Witherby and Co.

Matthews, G. V. T. 1954. Some aspects of incubation in the Manx Shearwater *Procellaria puffinus*, with particular reference to chilling resistance in the embryo. *Ibis* 96:432–440.

Matthews, L. H. 1949. The origin of stomach oil in the petrels, with comparative observations on the avian proventriculus. *Ibis* 91:373–392.

Mayr, E. 1935. Bernard Altum and the territory theory. *Proc. Linn. Soc. N.Y.* (1933–1934) 45–46.

————. 1939. The sex ratio in wild birds. *Amer. Nat.* 73:156–179.

————. 1942. *Systematics and the origin of species.* New York: Columbia University Press.

————. 1966. Hummingbird caught by Sparrow Hawk. *Auk* 83:664.

Meanley, B. 1953. Nesting of the King Rail in the Arkansas rice fields. *Auk* 70:261–269.

————. 1956. Food habits of the King Rail in the Arkansas rice fields. *Auk* 73:252–258.

Medway, Lord. 1962. The swiftlets (*Collocalia*) of Niah Cave, Sarawak. *Ibis* 104:45–66, 228–245.

Meinertzhagen, R. 1954. The education of young Ospreys. *Ibis* 96:153–155.

Merritt, J. H. 1951. Little orphan ani. *Audubon Mag.* 53:225–231.

Mertens, J. A. L. 1969. The influence of brood size on energy metabolism and water loss in nestling Great Tits *Parus major major*. *Ibis* 111:11–16.

Mertens, R. 1960. *The world of amphibians and reptiles.* New York: McGraw-Hill Book Co.

Mewaldt, L. R. 1956. Nesting behavior of the Clark Nutcracker. *Condor* 58:3–23.

Meyerriecks, A. J. 1960. Success story of a pioneering bird. *Nat. Hist.* 69:46–57.

Mickey, F. W. 1943. Breeding habits of McCown's Longspur. *Auk* 60:181–209.

Mihelsons, H. A., ed. 1968. *Ecology of waterfowl of Latvia.* Ornith. Study 5. Inst. Biol., Acad. Sci. Latvian S.S.R. Riga: Zinātne Publ. House. Abstract in *Ibis* (1969) 111:407–408.

Miller, A. H. 1951. The "rodent-run" of the Green-tailed Towhee. *Ibis* 93:307–308.

————. 1959a. Reproductive cycles in an equatorial sparrow. *Proc. Natl. Acad. Sci.* 45:1095–1100.

————. 1959b. Response to experimental light increments by Andean Sparrows from an equatorial area. *Condor* 61:344–347.

————. 1960. Adaptation of breeding schedule to latitude. *Proc. XII Int. Ornith. Congr.* (Helsinki, 1958), pp. 513–522.

————. 1962. Bimodal occurrence of breeding in an equatorial sparrow. *Proc. Natl. Acad. Sci.* 48:396–400.

————. 1963. Seasonal activity and ecology of an American equatorial cloud forest. *Univ. Calif. Publ. Zool.* 66:1–78.

————. 1965. Capacity for photoperiodic response and endogenous factors in the reproductive cycles of an equatorial sparrow. *Proc. Natl. Acad. Sci.* 54:97–101.

Miller, J. R., and Miller, J. T. 1948. Nesting of the Spotted Sandpiper at Detroit, Michigan. *Auk* 65:558–567.

Miller, L. 1952. Auditory recognition of predators. *Condor* 54:89–92.

Mills, E. A. 1931. *Bird memories of the Rockies.* Boston: Houghton Mifflin Co.

Milne, H. 1969. Eider biology. *Ibis* 111:278.

Modha, M. L., and Coe, M. J. 1969. Notes on the breeding of the African Skimmer, *Rynchops flavirostris*, on Central Island, Lake Rudolf. *Ibis* 111:593–598.

Moffett, G. M., Jr. 1970. A study of nesting Torrent Ducks in the Andes. *Living Bird* 9:5–27.

Moreau, R. E. 1936a. The breeding biology of certain East African hornbills (*Bucerotidae*). *Jour. E. Africa and Uganda Nat. Hist. Soc.* 13:1–27.

————. 1936b. Breeding seasons of birds in East African evergreen forest. *Proc. Zool. Soc. London* Pt. 3: 631–653.

————. 1937. Migrant birds in Tanganyika Territory. *Tanganyika Notes and Records* No. 4:1–34.

————. 1939. Numerical data on African birds' behaviour at the nest: *Hirundo s. smithii* Leach, the Wiretailed Swallow. *Proc. Zool. Soc. London* Ser. A, 109: 109–125.

————. 1940. Numerical data on African birds' behaviour at the nest. II: *Psalidoprocne holomelaena massaica* Neum., the Rough-wing Bank-Martin. *Ibis* Ser. 14, 4:234–248.

————. 1941. A contribution to the breeding biology of the Palm-Swift, *Cypselus parvus*. *Jour. E. Africa and Uganda Nat. Hist. Soc.* 15:154–170.

————. 1942a. *Colletoptera affinis* at the nest. *Ostrich* 13:137–147.

————. 1942b. The breeding biology of *Micropus caffer streubelii* Hartlaub, the White-rumped Swift. *Ibis* Ser. 14, 6:27–49.

————. 1942c. The nesting of African birds in association with other living things. *Ibis* Ser. 14, 6:240–263.

————. 1944. Clutch size: A comparative study, with special reference to African birds. *Ibis* 86:286–347.

————. 1947. Relations between number in brood, feeding-rate and nestling period in nine species of birds in Tanganyika Territory. *Jour. Animal Ecol.* 16: 205–209.

————. 1949. The African Mountain Wagtail *Motacilla clara* at the nest. *Ornith. als biologische Wissenschaft*, pp. 183–191.

————. 1950. The breeding seasons of African birds. 1: Land birds. *Ibis* 92:223–267.

Moreau, R. E., and Moreau, W. M. 1937. Biological and other notes on some East African birds. *Ibis* Ser. 14, 1: 321–345.

————. 1940. Hornbill studies. *Ibis* Ser. 14, 4:639–656.

————. 1941. Breeding biology of the Silvery-cheeked Hornbill. *Auk* 58:13–27.

Morehouse, E. L., and Brewer, R. 1968. Feeding of nestling and fledgling Eastern Kingbirds. *Auk* 85:44–54.

Morel, G.; Morel, M-Y.; and Bourlière, F. 1957. The Blackfaced Weaver Bird or Dioch in West Africa: An ecological study. *Jour. Bombay Nat. Hist. Soc.* 54: 811–825.

Morel, M-Y. 1967. Les oiseaux tropicaux elevent-ils autant de jeunes qu'ils peuvent en nourrir? Le cas de *Lagonosticta senegala*. *Terre et Vie* No. 1:77–82.

Morley, A. 1950. The formation and persistence of pairs in the Marsh-Tit. *British Birds* 43:387–393. Abstract in *Ibis* (1951) 93:476–477.

Mountfort, G. 1957. *The Hawfinch.* London: Collins.

Mouseley, H. 1939. Nesting behavior of Wilson's Snipe and Spotted Sandpiper. *Auk* 56:129–133.

Mumford, R. E. 1964. The breeding biology of the Acadian Flycatcher. *Univ. Mich. Mus. Zool., Misc. Publ.* No. 125:1–50.

Murphy, R. C. 1936. *Oceanic birds of South America.* New York: American Museum of Natural History.

Murton, R. K. 1965. *The Wood-Pigeon.* London: Collins.

Murton, R. K., and Isaacson, A. J. 1962. The functional basis of some behaviour in the Woodpigeon *Columba palumbus*. *Ibis* 104:503–521.

Neff, J. A. 1944. A protracted incubation period in the Mourning Dove. *Condor* 46:243.

————. 1945a. Foster parentage of a Mourning Dove in the wild. *Condor* 47:39–40.

————. 1945b. Maggot infestation of nestling Mourning Doves. *Condor* 47:73–76.

Nelson, J. B. 1964. Factors influencing clutch-size and chick growth in the North Atlantic Gannet *Sula bassana*. *Ibis* 106:63–77.

————. 1966. The breeding biology of the Gannet *Sula bassana* on the Bass Rock, Scotland. *Ibis* 108:584–626.

————. 1971. The biology of Abbott's Booby *Sula abbotti*. *Ibis* 113:429–467.

Nero, R. W. 1956. A behavior study of the Red-winged Blackbird. *Wilson Bull.* 68:5–37, 129–150.

Newbigin, M. I. 1898. *Colour in nature.* London: John Murray.

Newton, I. 1967. The adaptive radiation and feeding ecology of some British finches. *Ibis* 109:33–98.

Nice, M. M. 1922–1923. A study of the nesting of Mourning Doves. *Auk* 39:457–474; 40:37–58.

————. 1937. Studies in the life history of the Song Sparrow I. *Trans. Linn. Soc. N.Y.* 4:vi. + 247 pp.

————. 1941. The role of territory in bird life. *Amer. Midl. Nat.* 26:441–487.

————. 1943. Studies in the life history of the Song Sparrow II. *Trans. Linn. Soc. N.Y.* 6:viii + 328 pp.

————. 1953. Some experiences in imprinting ducklings. *Condor* 55:33–37.

————. 1954. Incubation periods throughout the ages. *Centaurus* (Copenhagen) 3:311–359.

————. 1957. Nesting success in altricial birds. *Auk* 74:305–321.

————. 1962. Development of behavior in precocial birds. *Trans. Linn. Soc. N.Y.* 8:xii + 211 pp.

Nice, M. M., and Nice, L. B. 1932. A study of two nests of the Black-throated Green Warbler. *Bird-Banding* 3:95–105, 157–172.

Nice, M. M., and Ter Pelkwyk, J. 1941. Enemy recognition by the Song Sparrow. *Auk* 58:195–214.

Nice, M. M., and Thomas, R. H. 1948. A nesting of the Carolina Wren. *Wilson Bull.* 60:139–158.

Nicholson, E. M. 1930. Field notes on Greenland birds. *Ibis* Ser. 12, 6:280–313, 395–428.

Nickell, W. P. 1948. Alternate care of two nests by a Ruby-throated Hummingbird. *Wilson Bull.* 60:242–243.

Noble, G. K. 1936. Courtship and sexual selection of the Flicker (*Colaptes auratus luteus*). *Auk* 53:269–282.

————. 1938. Sexual selection among fishes. *Biol. Rev. Cambridge Phil. Soc.* 13:133–158.

Noble, G. K., and Curtis, B. 1939. The social behavior of the Jewel Fish, *Hemichromis maculatus* Gill. *Bull. Amer. Mus. Nat. Hist.* 76:1–46.

Noble, G. K., and Lehrman, D. S. 1940. Egg recognition by the Laughing Gull. *Auk* 57:22–43.

Noble, G. K., and Vogt, W. 1935. An experimental study of sex recognition in birds. *Auk* 52:278–286.

Noble, G. K., and Wurm, M. 1942. Further analysis of the social behavior of the Black-crowned Night Heron. *Auk* 59:205–224.

Nolan, V., Jr. 1958. Anticipatory food-bringing in the Prairie Warbler. *Auk* 75:263–278.

————. 1959. Pileated Woodpecker attacks Pilot Black Snake at tree cavity. *Wilson Bull.* 71:381–382.

Nolan, V., Jr., and Schneider, R. 1962. A Catbird helper at a House Wren nest. *Wilson Bull.* 74:183–184.

Norman, F. I., and Gottsch, M. D. 1969. Artificial twinning in the Short-tailed Shearwater *Puffinus tenuirostris. Ibis* 111:391–393.

Norris, R. A. 1958. Comparative biosystematics and life history of the nuthatches *Sitta pygmaea* and *Sitta pusilla. Univ. Calif. Publ. Zool.* 56:119–300.

Norris, R. T. 1947. The cowbirds of Preston Frith. *Wilson Bull.* 59:83–103.

Norton, D. W. 1972. Incubation schedules of four species of Calidridine sandpipers at Barrow, Alaska. *Condor* 74:164–176.

Oberholzer, A., and Tschanz, B. 1969. Zum Jagen der Trottellumme (*Uria aalge aalge*) nach Fisch. *Jour. f. Ornith* 110:465–470. Abstract in *Ibis* (1970) 112:413.

Odum, E. P. 1941–1942. Annual cycle of the Black-capped Chickadee. *Auk* 58:314–333, 518–535; 59:499–531.

Odum, E. P., and Kuenzler, E. J. 1955. Measurement of territory and home range size in birds. *Auk* 72:128–137.

Olrog, C. C. 1965. Diferencias en el ciclo sexual de algunas aves. *Hornero* 10:269–272.

Orians, G. H. 1960. Autumnal breeding in the Tricolored Blackbird. *Auk* 77:379–398.

————. 1961. The ecology of blackbird (*Agelaius*) social systems. *Ecol. Monogr.* 31:285–312.

Orr, Y. 1970. Temperature measurements at the nest of the Desert Lark (*Ammomanes deserti deserti*). *Condor* 72:476–478.

Owen, J. H. 1945. The nesting of the Long-tailed Tit. *British Birds* 38:271–273. Abstract in *Bird-Banding* (1946) 17:48.

Parkes, K. C. 1953. The incubation patch in males of the suborder Tyranni. *Condor* 55:218–219.

Parkes, K. C., and Clark, G. A., Jr. 1964. Additional records of avian egg teeth. *Wilson Bull.* 76:147–154.

Parmelee, D. F. 1970. Breeding behavior of the Sanderling in the Canadian high Arctic. *Living Bird* 9:97–146.

Parmelee, D. F.; Greiner, D. W.; and Graul, W. D. 1968. Summer schedule and breeding biology of the White-rumped Sandpiper in the central Canadian Arctic. *Wilson Bull.* 80:5–29.

Parsons, J. 1969. Chick mortality in Herring Gulls. *Ibis* 111:443–444.

Payne, R. B. 1969a. Breeding seasons and reproductive physiology of Tricolored Blackbirds and Redwinged Blackbirds. *Univ. Calif. Publ. Zool.* 90:1–115.

————. 1969b. Nest parasitism and display of Chest-nut Sparrows in a colony of Grey-capped Social Weavers. *Ibis* 111:300–307.

————. 1969c. Giant Cowbird solicits preening from man. *Auk* 86:751–752.

Paynter, R. A., Jr. 1954. Interrelations between clutch-size, brood-size, prefledging survival, and weight in Kent Island Tree Swallows. *Bird-Banding* 25:35–58, 102–110, 136–148.

Peakall, D. B. 1970. The Eastern Bluebird: Its breeding season, clutch size, and nesting success. *Living Bird* 9:239–255.

Pearson, A. K., and Pearson, O. P. 1955. Natural history and breeding behavior of the tinamou *Nothoprocta ornata. Auk* 72:113–127.

Pearson, O. P. 1953. Use of caves by hummingbirds and other species at high altitudes in Peru. *Condor* 55:17–20.

Peek, F. W.; Franks, E.; and Case, D. 1972. Recognition of nest, eggs, nest site, and young in female Red-winged Blackbirds. *Wilson Bull.* 84:243–249.

Pennycuick, C. J. 1956. Observations on a colony of Brünnich's Guillemot *Uria lomvia* in Spitzbergen. *Ibis* 98:80–99.

Perrins, C. M. 1970. The timing of birds' breeding seasons. *Ibis* 112:242–255.

Perrins, C. M.; Harris, M. P.; and Britton, C. K. 1973. Survival of Manx Shearwaters *Puffinus puffinus. Ibis* 115:535–548.

Perry, R. 1946. *Lundy, isle of Puffins.* London: Lindsay Drummond.

Petersen, N. F.; á Botni; and Williamson, K. 1949. Polymorphism and breeding of the Rock Dove in the Faeroe Islands. *Ibis* 91:17–23.

Pettingill, O. S., Jr. 1960. Crèche behavior and individual recognition in a colony of Rockhopper Penguins. *Wilson Bull.* 72:213–221.

————. 1961. "Honeymoon on Cobb Island." In *Discovery*, edited by J. K. Terres. Philadelphia: J. B. Lippincott Co.

Pitelka, F. A. 1951. Breeding seasons of hummingbirds near Santa Barbara, California. *Condor* 53:198–201.

————. 1958. Timing of molt in Steller Jays of the Queen Charlotte Islands, British Columbia. *Condor* 60:38–49.

Pokrovskaya, I. V. 1968. Observations on nest site selection in some passerines. *Ibis* 110:571–573.

Powell, H. 1946. Nuthatch feeds nestling Starlings. *British Birds* 39:316. Abstract in *Ibis* (1947) 89:152.

Power, D. M. 1966. Antiphonal dueting and evidence for auditory reaction time in the Orange-chinned Parakeet. *Auk* 83:314–319.

Prescott, K. W. 1965. The Scarlet Tanager *Piranga olivacea. New Jersey State Mus. Investigations* No. 2.

————. 1967. Unusual activities of a House Sparrow

and a Blue Jay at a Tufted Titmouse nest. *Wilson Bull.* 79:346–347.

———. 1971. Unusual activity of Starlings at Yellow-shafted Flicker nest. *Wilson Bull.* 83:195–196.

Prévost, J. 1955. Observations écologiques sur le Manchot Empereur (*Aptenodytes forsteri*). *Acta XI Int. Ornith. Congr.* (Basel, 1954), pp. 248–251.

Putnam, L. S. 1949. The life history of the Cedar Waxwing. *Wilson Bull.* 61:141–182.

Quaintance, C. W. 1941. Voice in the Brown Towhee. *Condor* 43: 152–155.

Ralph, C. J., and Pearson, C. A. 1971. Correlation of age, size of territory, plumage, and breeding success in White-crowned Sparrows. *Condor* 73:77–80.

Ramsay, A. O. 1950. Conditioned responses in Crows. *Auk* 67:456–459.

———. 1951. Familial recognition in domestic birds. *Auk* 68:1–16.

Ramsay, A. O., and Hess, E. H. 1954. A laboratory approach to the study of imprinting. *Wilson Bull.* 66: 196–206.

Rand, A. L. 1940. Breeding habits of the birds of paradise: *Macgregoria* and *Diphyllodes*. Results of the Archbold expeditions. No. 26. *Amer. Mus. Novit.* No. 1073:1–14.

———. 1941. Development and enemy recognition of the Curve-billed Thrasher *Toxostoma curvirostre*. Results of the Archbold expeditions. No. 34. *Bull. Amer. Mus. Nat. Hist.* 78:213–242.

Raney, E. C. 1939. Robin and Mourning Dove use same nest. *Auk* 56:337–338.

Rea, G. 1945. Black and White Warbler feeding young of Worm-eating Warbler. *Wilson Bull.* 57:262.

Rice, D. W., and Kenyon, K. W. 1962. Breeding cycles and behavior of Laysan and Black-footed Albatrosses. *Auk* 79:517–567.

Rice, O. O. 1969. Record of female Cardinals sharing nest. *Wilson Bull.* 81:216.

Richardson, F. 1965. Breeding and feeding habits of the Black Wheatear *Oenanthe leucura* in southern Spain. *Ibis* 107:1–16.

Richdale, L. E. 1943. The Kuaka or Diving Petrel, *Pelecanoides urinatrix* (Gmelin). *Emu* 43:24–48, 97–107.

———. 1943–1944. The White-faced Storm Petrel or Takahi-Karemoana. *Trans. Roy. Soc. New Zealand* 73:97–115, 217–232, 335–350.

———. 1944a. The Titi Wainui or Fairy Prion *Pachyptila turtur* (Kuhl.). *Trans. Roy. Soc. New Zealand* 74:32–48, 165–181.

———. 1944b. The Sooty Shearwater in New Zealand. *Condor* 46:93–107.

———. 1945. The nestling of the Sooty Shearwater. *Condor* 47:45–62.

———. 1947. The pair bond in penguins and petrels: A banding study. *Bird-Banding* 18:107–117.

———. 1951. *Sexual behavior in penguins.* Lawrence: University of Kansas Press.

———. 1952. Post-egg period in albatrosses. *Biol. Monogr.* No. 4. Dunedin: Otago Daily Times and Witness Newspapers.

———. 1954a. Duration of parental attentiveness in the Sooty Shearwater. *Ibis* 96:586–600.

———. 1954b. The starvation theory in albatrosses. *Auk* 71:239–252.

Richmond, S. M. 1953. The attraction of Purple Martins to an urban location in western Oregon. *Condor* 55:225–249.

Ricklefs, R. E. 1968. Weight recession in nestling birds. *Auk* 85:30–35.

———. 1969a. An analysis of nesting mortality in birds. *Smithsonian Contrib. Zool.* No. 9:1–48.

———. 1969b. The nesting cycle of songbirds in tropical and temperate regions. *Living Bird* 8:165–175.

Ridpath, M. G. 1964. The Tasmanian Native Hen. *Australian Nat. Hist.* 14:346–350.

Riney, T. 1951. Relationships between birds and deer. *Condor* 53:178–185.

Ripley, S. D. 1957a. Notes on the Horned Coot, *Fulica cornuta* Bonaparte. *Postilla, Yale Peabody Mus. Nat. Hist.* No. 30:1–8.

———. 1957b. Additional notes on the Horned Coot, *Fulica cornuta* Bonaparte. *Postilla, Yale Peabody Mus. Nat. Hist.* No. 32:1–2.

———. 1961. Aggressive neglect as a factor in interspecific competition in birds. *Auk* 78:366–371.

Robin, P. 1969. L'Engoulevent du Sahara (*Caprimulgus aegyptius saharae*) dans le sud marocain. *Oiseau, Paris* 39:1–7. Abstract in *Ibis* (1970) 112:132.

Rollin, N. 1957. Incubation by drake Wood Duck in eclipse plumage. *Condor* 59:263–265.

Rollo, M., and Domm, L. V. 1943. Light requirements of the weaver finch. 1: Light period and intensity. *Auk* 60:357–367.

Root, R. B. 1969. The behavior and reproductive success of the Blue-gray Gnatcatcher. *Condor* 71:16–31.

Rothschild, M., and Lane, C. 1960. Warning and alarm signals by birds seizing aposematic insects. *Ibis* 102: 328–330.

Rowan, M. K. 1952. The Greater Shearwater *Puffinus gravis* at its breeding grounds. *Ibis* 94:97–121.

———. 1955. The breeding biology and behaviour of the Redwinged Starling *Onychognathus morio*. *Ibis* 97: 663–705.

———. 1965. Regulation of sea-bird numbers. *Ibis* 107:54–59.

———. 1969. A study of the Cape Robin in southern Africa. *Living Bird* 8:5–32.

Rowan, W. 1925. Relation of light to bird migration and developmental changes. *Nature* 115:494–495.

Rowe, E. G. 1947. The breeding biology of *Aquila verreauxi* Lesson. *Ibis* 89:387–410, 576–606.

Rowley, I. 1957. Co-operative feeding of young by Superb Blue Wrens. *Emu* 57:356–357.

———. 1965. The life history of the Superb Blue Wren *Malurus cyaneus*. *Emu* 64:251–297.

Royall, W. C., Jr., and Pillmore, R. E. 1968. House Wren feeds Red-shafted Flicker nestlings. *Murrelet* 49:4–6.

Royama, T. 1966. Factors governing feeding rate, food requirement and brood size of nestling Great Tits *Parus major*. *Ibis* 108:313–347.

Rudebeck, G. 1955. Some observations at a roost of European Swallows and other birds in the south-eastern Transvaal. *Ibis* 97:572–580.

Ruff, F. J. 1940. Mortality among Myrtle Warblers near Ocala, Florida. *Auk* 57:405–406.

Ruschi, A. 1949. [Observations on the Trochilidae.] (Portuguese). *Bull. Mus. Biol. Prof. Mello Leitao* (Santa Teresa, Brazil) No. 7.

Russell, W. C. 1947. Mountain Chickadees feeding young Williamson Sapsuckers. *Condor* 49:83.

Rust, H. J. 1947. Migration and nesting of nighthawks in northern Idaho. *Condor* 49:177–188.

Ruttledge, R. F. 1946. Roosting habits of the Irish Coal-Tit, with some observations on other habits. *British Birds* 39:326–333. Abstract in *Ibis* (1947) 89:380.

Ryder, J. P. 1970. A possible factor in the evolution of clutch size in Ross' Goose. *Wilson Bull.* 82:5–13.

Ryser, F. A., and Morrison, P. R. 1954. Cold resistance in the young Ring-necked Pheasant. *Auk* 71:253–266.

Sabine, W. S. 1959. The winter society of the Oregon Junco: Intolerance, dominance, and the pecking order. *Condor* 61:110–135.

Sargent, T. D. 1965. The role of experience in the nest building of the Zebra Finch. *Auk* 82:48–61.

Sauer, E. G. F., and Sauer, E. M. 1966. The behavior and ecology of the South African Ostrich. *Living Bird* 5:45–75.

Schäfer [Schaefer], E. 1953. Contribution to the life history of the Swallow-Tanager. *Auk* 70:403–460.

———. 1954a. Sobre la biologia de *Colibri coruscans*. *Bol. Soc. Venezolana Cien. Nat.* 15:153–162.

———. 1954b. Der Vogel mit dem Stein auf dem Kopf. *Kosmos* 50:9–13, 118–124.

———. 1954c. Zur Biologie des Steisshuhnes *Nothocercus bonapartei*. *Jour. f. Ornith.* 95:219–232.

Schaller, G. B., and Emlen, J. T., Jr. 1961. The development of visual discrimination patterns in the crouching reactions of nestling grackles. *Auk* 78:125–137.

Schantz, W. E. 1944. All-day record of an incubating Robin. *Wilson Bull.* 56:118.

Schifter, H. 19–. "Familie Bartvögel." In *Grzimeks*

Tierleben 9:63–75. München: Kindler Verlag. English translation, *Grzimek's Animal Life Encyclopedia*. New York: Van Nostrand Reinhold, 1972–.

Schomburgk, M. R. 1847–1848. *Reisen in British-Guiana in den Jahren 1840–1844*. Leipzig: J. J. Weber.

Schrantz, F. G. 1943. Nest life of the Eastern Yellow Warbler. *Auk* 60:367–387.

Schreiber, R. W., and Ashmole, N. P. 1970. Sea-bird breeding seasons on Christmas Island, Pacific Ocean. *Ibis* 112:363–394.

Scott, P. 1970. The wild swans of Slimbridge. *Publ. Wildfowl Trust*. 12 pp. Abstract in *Auk* (1971) 88: 698.

Sealy, S. G. 1970. Egg teeth and hatching methods in some alcids. *Wilson Bull.* 82:289–293.

Seel, D. C. 1969. Food, feeding rates and body temperature in the nestling House Sparrow *Passer domesticus* at Oxford. *Ibis* 111:36–47.

Selander, R. K. 1959. Polymorphism in Mexican Brown Jays. *Auk* 76:385–417.

———. 1964a. Speciation in wrens of the genus *Campylorhynchus*. *Univ. Calif. Publ. Zool.* 74:1–259.

———. 1964b. Behavior of captive South American cowbirds. *Auk* 81:394–402.

Selander, R. K., and Giller, D. R. 1959. Interspecific relations of woodpeckers in Texas. *Wilson Bull.* 71: 107–124.

———. 1961. Analysis of sympatry in Great-tailed and Boat-tailed Grackles. *Condor* 63:29–86.

Selander, R. K., and La Rue, C. J., Jr. 1961. Interspecific preening invitation display of parasitic cowbirds. *Auk* 78:473–504.

Selous, E. 1927. *Realities of bird life*. London: Constable and Co.

Semenov-Tian-Shansky, O. 1959. [On ecology of tetraonids.] (Russian, English summary). Moscow.

Serventy, D. L. 1956. Age at first breeding of the Short-tailed Shearwater *Puffinus tenuirostris*. *Ibis* 98:532–533.

Seton-Thompson, E. 1898. *Wild animals I have known*. New York: Charles Scribner's Sons.

Shaw, W. T. 1936. Winter life and nesting studies of Hepburn's Rosy Finch in Washington State. *Auk* 53: 9–16, 133–149.

Shchepanek, M. J. 1968. Flying duck transports young on its back. *Canadian Field-Naturalist* 82:223. Abstract in *Auk* (1969) 86:793.

Sherman, A. R. 1910. At the sign of the Northern Flicker. *Wilson Bull.* 22:135–171.

———. 1952. *Birds of an Iowa dooryard*. Boston: Christopher Publishing House. (Contains reprint of Sherman 1910.)

Short, L. L., Jr. 1964. Extra helpers feeding young of Blue-winged and Golden-winged Warblers. *Auk* 81: 428–430.

Sibley, C. G. 1955. Behavioral mimicry in the titmice (Paridae) and certain other birds. *Wilson Bull.* 67: 128–132.

———. 1957. The evolutionary and taxonomic significance of sexual dimorphism and hybridization in birds. *Condor* 59:166–191.

Sick, H. 1959*a*. Die Balz der Schmuckvögel (Pipridae). *Jour. f. Ornith.* 100:269–302.

———. 1959*b*. Notes on the biology of two Brazilian swifts, *Chaetura andrei* and *Chaetura cinereiventris*. *Auk* 76:471–477.

———. 1960. Vergleichende Beobachtungen über die Balz brasilianischer Pipriden. *Proc. XII Int. Ornith. Congr.*, pp. 672–680.

———. 1967. Courtship behavior in the manakins (Pipridae): A review. *Living Bird* 6:5–22.

Sielmann, H. 1958. *Das Jahr mit den Spechten*. Berlin: Verlag Ullstein.

Simmons, K. E. L. 1951. Interspecific territorialism. *Ibis* 93:407–413.

———. 1965. Breeding periodicity of the Brown Booby at Ascension. *Ibis* 107:429.

———. 1967. Ecological adaptations in the life history of the Brown Booby at Ascension Island. *Living Bird* 6:187–212.

Skead, C. J. 1954. A study of the Cape Wagtail *Motacilla capensis*. *Ibis* 96:91–103.

———. 1959. A study of the Cape Penduline Tit *Anthoscopus minutus minutus* (Shaw and Nodder). *Proc. First Pan-African Ornith. Congr.*, pp. 274–288.

Skead, C. J., and Ranger, G. A. 1958. A contribution to the biology of the Cape Province white-eyes (*Zosterops*). *Ibis* 100:319–333.

Skutch, A. F. 1931. The life history of Rieffer's Hummingbird (*Amazilia tzacatl tzacatl*) in Panama and Honduras. *Auk* 48:481–500.

———. 1935. Helpers at the nest. *Auk* 52:257–273.

———. 1944. Life history of the Prong-billed Barbet. *Auk* 61:61–88.

———. 1945. Life history of the Blue-throated Green Motmot. *Auk* 62:489–517.

———. 1947. Life history of the Marbled Wood-Quail. *Condor* 49:217–232.

———. 1949. Do tropical birds rear as many young as they can nourish? *Ibis* 91:430–455.

———. 1950. The nesting seasons of Central American birds in relation to climate and food supply. *Ibis* 92: 185–222.

———. 1951. Congeneric species of birds nesting together in Central America. *Condor* 53:3–15.

———. 1952*a*. On the hour of laying and hatching of birds' eggs. *Ibis* 94:49–61.

———. 1952*b*. Life history of the Blue and White Swallow. *Auk* 69:392–406.

———. 1953*a*. The White-throated Magpie-Jay. *Wilson Bull.* 65:68–74.

———. 1953*b*. Delayed reproductive maturity in birds. *Ibis* 95:153–154.

———. 1953*c*. Life history of the Southern House Wren. *Condor* 55:121–149.

———. 1953*d*. How the male bird discovers the nestlings. *Ibis* 95:1–37, 505–542.

———. 1954. Life histories of Central American birds. *Pacific Coast Avif.* No. 31. Berkeley: Cooper Ornithological Society.

———. 1954–1955. The parental stratagems of birds. *Ibis* 96:544–564; 97:118–142.

———. 1958*a*. Roosting and nesting of araçari toucans. *Condor* 60:201–219.

———. 1958*b*. Life history of the White-whiskered Softwing *Malacoptila panamensis*. *Ibis* 100:209–231.

———. 1959*a*. Life history of the Blue Ground Dove. *Condor* 61:65–74.

———. 1959*b*. Life history of the Groove-billed Ani. *Auk* 76:281–317.

———. 1960. Life histories of Central American birds. II. *Pacific Coast Avif.* No. 34. Berkeley: Cooper Ornithological Society.

———. 1961*a*. The nest as a dormitory. *Ibis* 103a:50–70.

———. 1961*b*. Helpers among birds. *Condor* 63:198–226.

———. 1962*a*. The constancy of incubation. *Wilson Bull.* 74:115–152.

———. 1962*b*. Life history of the White-tailed Trogon *Trogon viridis*. *Ibis* 104:301–313.

———. 1963*a*. Life history of the Little Tinamou. *Condor* 65:224–231.

———. 1963*b*. Habits of the Chestnut-winged Chachalaca. *Wilson Bull.* 75:262–269.

———. 1964*a*. Life histories of Central American pigeons. *Wilson Bull.* 76:211–247.

———. 1964*b*. Life history of the Blue-diademed Motmot *Momotus momota*. *Ibis* 106:321–332.

———. 1966*a*. A breeding bird census and nesting success in Central America. *Ibis* 108:1–16.

———. 1966*b*. Life histories of three tropical American cuckoos. *Wilson Bull.* 78:139–165.

———. 1967*a*. Life histories of Central American highland birds. *Publ. Nuttall Ornith. Club.* No. 7.

———. 1967*b*. Adaptive limitation of the reproductive rate of birds. *Ibis* 109:579–599.

———. 1968. The nesting of some Venezuelan birds. *Condor* 70:66–82.

———. 1969*a*. A study of the Rufous-fronted Thornbird and associated birds. *Wilson Bull.* 81:5–43, 123–139.

———. 1969*b*. Life histories of Central American birds.

III. *Pacific Coast Avif.* No. 35. Berkeley: Cooper Ornithological Society.

⸻. 1970. Life history of the Common Potoo. *Living Bird* 9:265–280.

⸻. 1971. *A naturalist in Costa Rica.* Gainesville: University of Florida Press.

⸻. 1972. Studies of tropical American birds. *Publ. Nuttall Ornith. Club.* No. 10.

⸻. 1973. *The life of the hummingbird.* New York: Crown Publishers.

Sladen, W. J. L. 1953. The Adelie Penguin. *Nature* 171: 952–961.

⸻. 1955. Some aspects of the behaviour of Adelie and Chinstrap Penguins. *Acta XI Int. Ornith. Congr.* (Basel, 1954), pp. 241–247.

Slessers, M. 1970. Bathing behavior of land birds. *Auk* 87:91–99.

Smith, F. V. 1969. *Attachment of the young: Imprinting and other developments.* Edinburgh: Oliver and Boyd.

Smith, K. D. 1955. The winter breeding season of landbirds in eastern Eritrea. *Ibis* 97:480–507.

Smith, N. G. 1968. The advantages of being parasitized. *Nature* 219:690–694.

Smith, S. 1950. *The Yellow Wagtail.* London: Collins.

Smith, S. M. 1967. Seasonal changes in the survival of the Black-capped Chickadee. *Condor* 69:344–359.

Snow, B. K. 1960. The breeding biology of the Shag *Phalacrocorax aristotelis* on the Island of Lundy, Bristol Channel. *Ibis* 102:554–575.

⸻. 1961. Notes on the behavior of three Cotingidae. *Auk* 78:150–161.

⸻. 1970. A field study of the Bearded Bellbird in Trinidad. *Ibis* 112:299–329.

⸻. 1972. A field study of the Calfbird *Perissocephalus tricolor. Ibis* 114:139–162.

Snow, D. W. 1956. Territory in the Blackbird *Turdus merula. Ibis* 98:438–447.

⸻. 1958. The breeding of the Blackbird *Turdus merula* at Oxford. *Ibis* 100:1–30.

⸻. 1961. The natural history of the Oilbird, *Steatornis caripensis,* in Trinidad, W. I. Part 1: General behavior and breeding habits. *Zoologica* (N.Y. Zool. Soc.) 46:27–48.

⸻. 1962a. A field study of the Black and White Manakin, *Manacus manacus,* in Trinidad. *Zoologica* (N.Y. Zool. Soc.) 47:65–104.

⸻. 1962b. Notes on the biology of some Trinidad swifts. *Zoologica* (N.Y. Zool. Soc.) 47:129–139.

⸻. 1962c. The natural history of the Oilbird, *Steatornis caripensis,* in Trinidad, W. I. Part 2: Population, breeding ecology and food. *Zoologica* (N.Y. Zool. Soc.) 47:199–221.

⸻. 1963. The evolution of manakin displays. *Proc. XIII Int. Ornith. Congr.,* pp. 553–561.

⸻. 1965a. The breeding of the Red-billed Tropic Bird in the Galápagos Islands. *Condor* 67:210–214.

⸻. 1965b. The breeding of Audubon's Shearwater (*Puffinus lherminieri*) in the Galápagos. *Auk* 82:591–597.

⸻. 1971. Observations on the Purple-throated Fruit-crow in Guyana. *Living Bird* 10:5–17.

Snow, D. W., and Collins, C. T. 1962. Social breeding behavior of the Mexican Tanager. *Condor* 64:161.

Snow, D. W., and Snow, B. K. 1963. Breeding and the annual cycle in three Trinidad thrushes. *Wilson Bull.* 75:27–41.

⸻. 1964. Breeding seasons and annual cycles of Trinidad land birds. *Zoologica* (N.Y. Zool. Soc.) 49: 1–39.

⸻. 1966. The breeding season of the Madeiran Storm Petrel *Oceanodroma castro* in the Galápagos. *Ibis* 108:283–284.

⸻. 1967. The breeding cycle of the Swallow-tailed Gull *Creagrus furcatus. Ibis* 109:14–24.

Soper, J. D. 1946. Ornithological results of the Baffin Island expeditions of 1928–1929 and 1930–1931, together with more recent records. *Auk* 63:1–24.

Southern, W. E. 1968. Further observations on fosterfeeding by Purple Martins. *Wilson Bull.* 80:234–235.

Spalding, D. 1873. Instinct, with original observations on young animals. (Reprinted in *British Jour. Animal Behaviour* [1954] 2:2–11.)

Spellerberg, I. F. 1971. Aspects of McCormick Skua breeding biology. *Ibis* 113:357–363.

Spingarn, E. D. W. 1934. Some observations on the Semipalmated Plover (*Charadrius semipalmatus*) at St. Mary's Islands, Province of Quebec, Canada. *Auk* 51:27–36.

Stefanski, R. A. 1967. Utilization of the breeding territory in the Black-capped Chickadee. *Condor* 69: 259–267.

Stenger, J., and Falls, J. B. 1959. The utilized territory of the Ovenbird. *Wilson Bull.* 71:125–140.

Steven, D. M. 1955. Transference of "imprinting" in a wild gosling. *British Jour. Animal Behaviour* 3:14–16.

Stewart, P. A. 1959. The "romance" of the Wood Duck. *Audubon Mag.* 61:63–65.

Stoddard, H. L. 1946. *The Bob-white Quail.* New York: Charles Scribner's Sons.

Stokes, A. W. 1950. Breeding behavior of the Goldfinch. *Wilson Bull.* 62:107–127.

⸻. 1960. Nest-site selection and courtship behaviour of the Blue Tit *Parus caeruleus. Ibis* 102: 507–519.

⸻. 1971. Parental and courtship feeding in the Red Jungle Fowl. *Auk* 88:21–29.

Stokes, A. W., and Williams, H. W. 1968. Antiphonal calling in quail. *Auk* 85:83–89.

Stonehouse, B. 1953. The Emperor Penguin *Aptenodytes*

forsteri Gray. I: Breeding behaviour and development. *Falkland Islands Dependencies Survey Sci. Rep.* 6: 1–33.

————. 1960. The King Penguin, *Aptenodytes patagonica*, of South Georgia. I: Breeding behaviour and development. *Falkland Islands Dependencies Survey Sci. Rep.* 23:1–81.

————. 1962. Tropic birds (genus *Phaethon*) of Ascension Island. *Ibis* 103b:124–161.

Stonehouse, B., and Stonehouse, S. 1963. The Frigate Bird *Fregata aquila* of Ascension Island. *Ibis* 103b: 409–422.

Stoner, Dayton. 1935. Temperature and growth studies on the Barn Swallow. *Auk* 52:400–407.

————. 1945. Temperature and growth studies of the Northern Cliff Swallow. *Auk* 62:207–216.

Storer, R. W. 1952. A comparison of variation, behavior and evolution in the sea bird genera *Uria* and *Cepphus*. *Univ. Calif. Publ. Zool.* 52:121–222.

————. 1969. The behavior of the Horned Grebe in spring. *Condor* 71:180–205.

Summers-Smith, D. 1958. Nest-site selection, pair formation and territory in the House-Sparrow *Passer domesticus*. *Ibis* 100:190–203.

Sutton, G. M., and Parmelee, D. F. 1956. Breeding of the Snowy Owl in southeastern Baffin Island. *Condor* 58:273–282.

Swainson, G. W. 1970. Co-operative rearing in the Bell Miner. *Emu* 70:183–188. Abstract in *Ibis* (1971) 113: 386.

Swanberg, P. O. 1956. Territory in the Thick-billed Nutcracker *Nucifraga caryocatactes*. *Ibis* 98:412–419.

Swarth, H. S. 1935. Injury-feigning in nesting birds. *Auk* 52:352–354.

Swift, J. J. 1959. Le Guêpier d'Europe *Merops apiaster* L. en Camargue. *Alauda* 27:97–143.

Taverner, P. A. 1936. Injury feigning by birds. *Auk* 53:366.

Taylor, P. M. 1972. Hovering behavior by House Finches. *Condor* 74:219–221.

Taylor, R. H. 1962. The Adelie Penguin *Pygoscelis adeliae* at Cape Royds. *Ibis* 104:176–204.

Thomas, R. H. 1946. A study of Eastern Bluebirds in Arkansas. *Wilson Bull.* 58:143–183.

Thomson, A. L. 1950. Factors determining the breeding seasons of birds: An introductory review. *Ibis* 92: 173–184.

————, ed. 1964. *A new dictionary of birds*. London: Nelson.

Thoresen, A. C. 1964. The breeding behavior of the Cassin Auklet. *Condor* 66:456–476.

Thorpe, W. H. 1956. *Learning and instinct in animals*. London: Methuen and Co.

————. 1958. The learning of song patterns by birds, with especial reference to the song of the Chaffinch *Fringilla coelebs*. *Ibis* 100:535–570.

Tinbergen, N. 1939. The behavior of the Snow Bunting in spring. *Trans. Linn. Soc. N.Y.* 5:1–94.

————. 1951. *The study of instinct*. Oxford: Clarendon Press.

————. 1953. *The Herring Gull's world*. London: Collins.

————. 1958. *Curious naturalists*. London: Country Life.

Tompa, F. S. 1962: Territorial behavior: The main controlling factor of a local Song Sparrow population. *Auk* 79:687–697.

————. 1971. Catastrophic mortality and its population consequences. *Auk* 88:753–759.

Truslow, F. K. 1967. Egg-carrying by the Pileated Woodpecker. *Living Bird* 6:227–236.

Tschanz, B. 1968. Trottellummen. *Z. Tierpsychol.* Suppl. 4, 103 pp. Abstract in *Ibis* (1969) 111:127–128.

Tuck, L. M. 1960. The murres: Their distribution, populations and biology. *Canadian Wildlife Service Monogr. Ser.* No. 1.

————. 1972. The snipes: A study of the genus *Capella*. *Canadian Wildlife Service Monogr. Ser.* No. 5.

Turner, D. A., and Gerhart, J. 1971. "Foot-wetting" by incubating African Skimmers *Rynchops flavirostris*. *Ibis* 113:244.

Turner, E. L. 1924. *Broadland birds*. London.

Turpin, G. M. 1918. The nesting habits of the hen. *Bull. Agric. Expt. Station, Iowa State College Agric.* 178: 211–232.

Van Someren, V. G. L. 1947. Onset of sexual activity. *Ibis* 89:51–56.

————. 1956. Days with birds: Studies of habits of some East African species. *Fieldiana: Zoology.* Vol. 38.

Van Tyne, Josselyn. 1929. The life history of the toucan *Ramphastos brevicarinatus*. *Univ. Mich. Mus. Zool., Misc. Publ.* No. 19:1–43.

Verner, J. 1961. Nesting activities of the Red-footed Booby in British Honduras. *Auk* 78:573–594.

Verner, J., and Willson, M. F. 1969. Mating systems, sexual dimorphism, and the role of male North American passerine birds in the nesting cycle. *Amer. Ornith. Union, Ornith. Monogr.* No. 9.

Vince, M. A. 1969. How quail embryos communicate. *Ibis* 111:441.

Vogt, W. 1957. "Will and Kate." In *The bird watcher's anthology*, edited by R. T. Peterson. New York: Harcourt, Brace and Co.

Voous, K. H. 1950. The breeding seasons of birds in Indonesia. *Ibis* 92:279–287.

Wagner, H. O. 1945. Notes on the life history of the Mexican Violet-ear. *Wilson Bull.* 57:165–187.

————. 1954. Versuch einer Analyse der Kolibribalz. *Z. Tierpsychol.* 11:182–212.

————. 1957. Variation in clutch size at different latitudes. *Auk* 74:243–250.

————. 1959. Beitrag zum Verhalten des Weissohrkolibris (*Hylocharis. leucotis* Vieill.). *Zool. Jahrb. Syst.* 86:253–302.

————. 1966. *Meine Freunde, die Kolibris: Streifzüge durch Mexico.* Berlin and Hamburg: Verlag Paul Parey.

Wagner, H. O., and Stresemann, E. 1950. Über die Beziehungen zwischen Brutzeit und Ökologie mexikanischer Vögel. *Zool. Jahrb. Syst.* 79:273–308.

Walkinshaw, L. H. 1940. Summer life of the Sora Rail. *Auk* 57:153–168.

————. 1941. The Prothonotary Warbler: A comparison of nesting conditions in Tennessee and Michigan. *Wilson Bull.* 53:3–21.

————. 1944. The Eastern Chipping Sparrow in Michigan. *Wilson Bull.* 56:193–205.

————. 1947. Some nesting records of the Sarus Crane in North American zoological parks. *Auk* 64:602–615.

————. 1965. Attentiveness of cranes at their nests. *Auk* 82:465–476.

Wallace, A. R. 1871. *Contributions to the theory of natural selection.* 2d ed. New York: Macmillan and Co.

Walters, J. 1959. [Observations on two broods of Kentish Plovers, *Charadrius alexandrinus*, on Texel.] (Dutch, German summary.) *Ardea* 47:48–67. Abstract in *Ibis* (1960) 102:338.

Warham, J. 1958a. The nesting of the shearwater *Puffinus carneipes*. *Auk* 75:1–14.

————. 1958b. The nesting of the Little Penguin *Eudyptula minor*. *Ibis* 100:605–616.

————. 1960. Some aspects of breeding behaviour in the Short-tailed Shearwater. *Emu* 60:75–87.

————. 1963. The Rockhopper Penguin, *Eudyptes chrysocome*, at Macquarie Island. *Auk* 80:229–256.

————. 1971. Body temperatures of petrels. *Condor* 73:214–219.

————. 1972. Breeding seasons and sexual dimorphism in Rockhopper Penguins. *Auk* 89:86–105.

Watson, J. B., and Lashley, K. S. 1915. An historical and experimental study of homing. *Papers Dept. Marine Biol., Carnegie Inst. Washington* 7:7–60.

Weeden, J. S. 1966. Diurnal rhythm of attentiveness of incubating female Tree Sparrows (*Spizella arborea*) at a northern latitude. *Auk* 83:368–388.

Weller, M. W. 1958. Observations on the incubation behavior of a Common Nighthawk. *Auk* 75:48–59.

————. 1961. Breeding biology of the Least Bittern. *Wilson Bull.* 73:11–35.

————. 1968. The breeding biology of the parasitic Black-headed Duck. *Living Bird* 7:169–207.

Weston, H. G., Jr. 1947. Breeding behavior of the Black-headed Grosbeak. *Condor* 49:54–73.

W[estwood], R. W. 1946. Contents noted. *Nature Mag.* 39:399.

Wetherbee, D. K. 1962. Egg teeth and shell rupture of the American Woodcock. *Auk* 79:117.

Wied, Maximilian Prinz Zu. 1825–1833. *Beiträge zur Naturgeschichte von Brasilien.* Weimar: Landes-Industrie-Comptoirs.

Wilbur, H. M. 1969. The breeding biology of Leach's Petrel, *Oceanodroma leucorhoa*. *Auk* 86:433–442.

Wilcox, L. 1959. A twenty year banding study of the Piping Plover. *Auk* 76:129–152.

Williams, G. R. 1960. The Takahe (*Notornis mantelli* Owen, 1848): A general survey. *Trans. Roy. Soc. New Zealand* 88:235–258.

Williams, J. G. 1951. *Nectarinia johnstoni*: A revision of the species, together with data on plumages, moults and habits. *Ibis* 93:579–595.

Williams, L. 1942. Interrelations in a nesting group of four species of birds. *Wilson Bull.* 54:238–249.

————. 1952. Breeding behavior of the Brewer Blackbird. *Condor* 54:3–47.

Williamson, K. 1948. Field-notes on nidification and distraction-display in the Golden Plover. *Ibis* 90:90–98.

————. 1949. The distraction behaviour of the Arctic Skua. *Ibis* 91:307–313.

————. 1952a. Regional variation in the distraction displays of the Oyster-catcher. *Ibis* 94:85–96.

————. 1952b. The spitting reaction of nestling Fulmars. *Ibis* 94:165–166.

Willis, E. O. 1961. A study of nesting ant-tanagers in British Honduras. *Condor* 63:479–503.

————. 1967. The behavior of Bicolored Antbirds. *Univ. Calif. Publ. Zool.* 79:1–127.

————. 1972. The behavior of Spotted Antbirds. *Amer. Ornith. Union, Ornith. Monogr.* No. 10.

Winn, H. E. 1950. The Black Guillemots of Kent Island, Bay of Fundy. *Auk* 67:477–485.

Witherby, H. F.; Jourdain, F. C. R.; Ticehurst, N. F.; and Tucker, B. W. 1938. *The handbook of British birds.* Vols. 1 and 2. London: H. F. and G. Witherby.

Wood-Gush, D. G. M. 1955. The behaviour of the domestic chicken: A review of the literature. *British Jour. Animal Behaviour* 3:81–110.

Worth, C. B. 1940. Egg volumes and incubation periods. *Auk* 57:44–60.

————. 1967. *A naturalist in Trinidad.* Philadelphia: J. B. Lippincott Co.

Wynne-Edwards, V. C. 1952. Zoology of the Baird Expedition (1950). I: The birds observed in central and south-east Baffin Island. *Auk* 69:353–391.

————. 1962. *Animal dispersion in relation to social behaviour.* Edinburgh: Oliver and Boyd.

Yarrow, R. M. 1970. Changes in Redstart breeding territory. *Auk* 87:359–361.

Young, E. C. 1963. The breeding behaviour of the South Polar Skua *Catharacta maccormicki*. *Ibis* 105:203–233.

Young, H. 1956. Territorial activities of the American Robin *Turdus migratorius*. *Ibis* 98:448–452.

Zahavi, A. 1971. The social behaviour of the White Wagtail *Motacilla alba alba* wintering in Israel. *Ibis* 113:203–211.

Zonov, G. B. 1967. [On the winter roosting of Paridae in Cisbaikal.] (Russian.) *Ornitologiya* 8:351–354. Abstract in *Ibis* (1968) 110:391.

INDEX

We are still far from achieving the ideal of a uniform and stable ornithological nomenclature. The names, both Latin and English, used in the writings consulted in the preparation of this book and listed in the bibliography are, in many cases, different from those that appear in some of the more recent publications. To avoid confusion, I have, as a rule, included these alternative names in this Index.

Illustrations are indicated by numbers in italics.